Y0-BCK-967

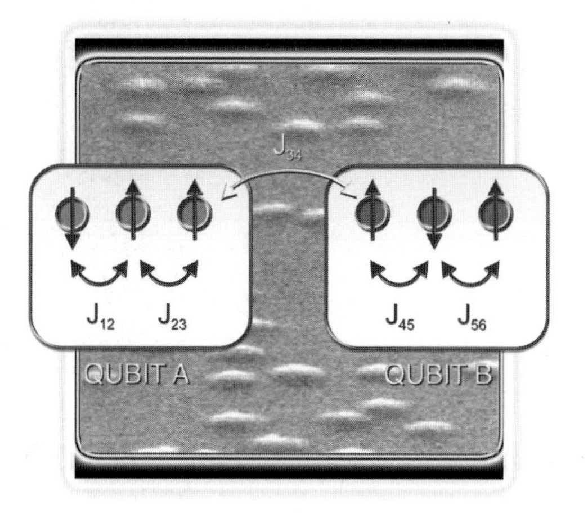

SEMICONDUCTOR QUANTUM BITS

SEMICONDUCTOR QUANTUM BITS

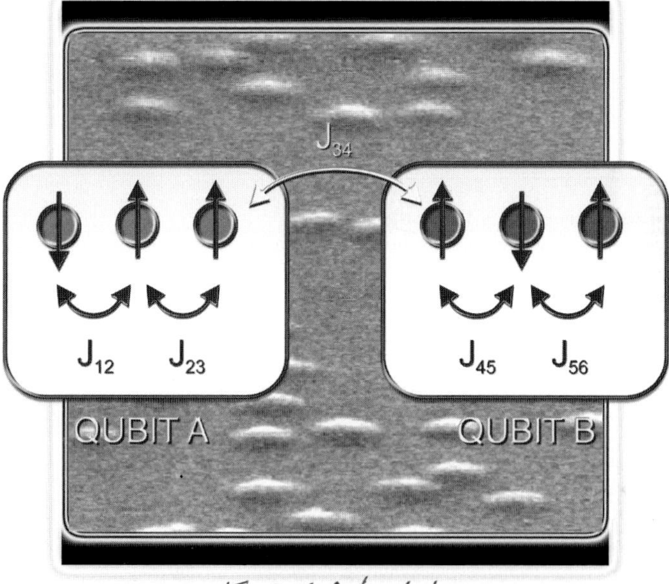

[edited by]
Fritz Henneberger
Oliver Benson
Humboldt-Universität zu Berlin, Germany

Published by

Pan Stanford Publishing Pte. Ltd.
5 Toh Tuck Link
Singapore 596224

Distributed by

World Scientific Publishing Co. Pte. Ltd.
5 Toh Tuck Link, Singapore 596224
USA office: 27 Warren Street, Suite 401-402, Hackensack, NJ 07601
UK office: 57 Shelton Street, Covent Garden, London WC2H 9HE

British Library Cataloguing-in-Publication Data
A catalogue record for this book is available from the British Library.

SEMICONDUCTOR QUANTUM BITS

ISBN-13 978-981-4241-05-2
ISBN-10 981-4241-05-9

Printed in Singapore by B & Jo Enterprise Pte Ltd.,
E-mail: sankaran@rpsonline.com.sg

Preface

Quantum information processing and computation (QIPC) is one of today's most active research fields. Many researchers worldwide are motivated by the fascinating prospects of controlling complex quantum systems and by the potentially huge impact of QIPC applications on our information society.

David DiVincenzo formulated five prerequisites for successful QIPC technology:

(1) A scalable physical system with well characterized quantum bits or *qubits*;
(2) The ability to initialize the state of each qubit to a simple initial state;
(3) Long relevant decoherence times, much longer than gate operations;
(4) A universal set of quantum gates (single and two-qubit gates); and
(5) A qubit-specific measurement capability.

According to these requirements there have been numerous attempts to implement and control qubits in different systems utilizing: nuclear spin in molecules, electronic states in single atoms or ions, polarization and spatial modes of photons, charge or flux quanta in superconductors, as well as charge and spin in semiconductor quantum dots. The latter two examples are of particular interest as they represent implementations in the solid-state where very rapid decoherence has long been considered a severe obstacle. Indeed, the progress in the fabrication of artificial nanostructures during the last years has enabled researchers to overcome previous limitations. Individual charge and spin carriers can be strongly decoupled from their environment so that the processes destroying the quantum coherence in the bulk are largely suppressed. In this way, the road towards miniaturized and integrated quantum logical circuits compatible with the existing semiconductor technology is opened.

In this book, we have concentrated on semiconductors as they offer spin as well as charge to implement qubits. Also, the radiative recombination of electrons and holes in semiconductors allows to use optical methods for qubit read-out and manipulation. Moreover, an efficient interface between flying and stationary qubits can be envisioned, e.g. with the help of quantum electrodynamical effects.

Our book aims to provide an overview of recent exciting results, both in experiment as well as in the theory. Leading experts in the field have provided chapters covering the many aspects of QIPC in semiconductors. We have organized the contributions as follows: The first part addresses explicit implementations and properties of charge and spin based qubits. Their manipulation, read-out, and control is the main subject of the second part. A third part is devoted to decoherence, still a central problem in solid-state systems. The generation of flying qubits and routes towards interfacing with stationary qubits are discussed in the fourth part. The last part introduces concepts, building blocks, and first demonstrations of QIPC using semiconductor based single photon sources and microcavities.

Finally, we would like to express our thanks to all authors who have provided excellent contributions to this book.

Oliver Benson and Fritz Henneberger

Contents

PART 1

Spin and Charge Qubits

Chapter 1

Coded Qubit Based on Electron Spin

Marek Korkusinski and Pawel Hawrylak

Quantum Theory Group, Institute for Microstructural Sciences, National Research Council of Canada, Ottawa, Canada

We review here the progress toward the realisation of a coded qubit based on electron spin in lateral gated devices. We show that by encoding the qubit in the states of several coupled electron spins one combines the coherence of spin with the advantages of using voltage for their coherent manipulation. The theoretical and experimental methods used to design and manipulate the systems of up to three coupled electron spins in lateral gated devices are reviewed in detail.

Contents

1.1. Introduction

The central issue in applications of quantum information processing [1–3] is the realisation of scalable quantum bits (qubits) and gates. The spins of individual electrons localized in semiconductor nanostructures are considered as viable candidates for qubits due to their quantum two-level nature and measured long spin coherence times [4–6]. In this design, the states $|0\rangle$ and $|1\rangle$ of the qubit are identified as the spin down and up states of the electron spin, respectively, and an array of N localized interacting spins forms the quantum register [7, 8]. Quantum computation with such a quantum register involves a coherent time evolution of its state, driven by a sequence of quantum operations. Since the single- and two-qubit operations form a minimal set sufficient to implement any quantum algorithm [3], the quantum computer can be modeled by the Heisenberg Hamiltonian:

$$\hat{H} = \frac{\mu_B}{\hbar} \sum_{i=1}^{N} g_i \vec{B}_i \vec{S}_i + \frac{1}{2} \sum_{i \neq j}^{N} \vec{S}_i J_{ij} \vec{S}_j. \tag{1.1}$$

Here μ_B is the Bohr magneton, g_i and $\vec{B}_i$ are, respectively, the local Landé factor and the local magnetic field in the vicinity of the spin $\vec{S}_i$, and J_{ij} is the pairwise exchange coupling between the ith and jth spins. As described by the first term in the above Hamiltonian, the single-qubit rotations are performed by coupling each spin to an external magnetic field. In this approach the individual qubits are addressed either by using local magnetic fields, or by tuning the Landé factor g_i of each spin separately. On the other hand, the two-qubit operations are implemented by controlling the exchange interactions J_{ij} between pairs of spins [8, 9].

The key technological steps required to implement this design include (i) the ability to localize individual electrons and perform coherent single-qubit rotations on their spins, (ii) coherent control of the exchange coupling between two spin qubits constituting a quantum gate, and (iii) scalability of the design to a larger number of qubits.

In the solid-state environment the localization of individual electrons is achieved by confining them in quantum dots (QDs). In early attempts the carrier confinement was created by the conduction band offset of semiconductor QDs embedded in a wider-bandgap material [10], while the number of electrons populating the QD was controlled by an external gate. Using the Single Electron Capacitance Spectroscopy, Ashoori [11, 12] has demonstrated a controlled localization of a single electron in such sample. Tarucha and co-workers [13] modified the sample design by depositing metalic contacts above and below the QD, which allowed to probe the properties of the system with a tunneling current in the Coulomb Blockade regime. The shell-filling effects and signatures of Hunds rules [14] detected in these transport spectra demonstrated the artificial atomic character of the QDs. A step towards a two-qubit gate, involving the creation and controlled charging of a vertically coupled double-dot molecule, was further reported by Pi and co-workers [15]. However, in the vertical geometry the control of coupling between dots, necessary for the two-qubit operations, is limited.

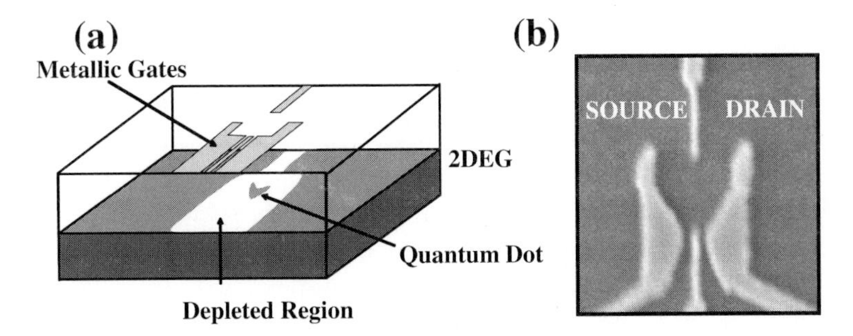

Fig. 1.1. (a) A schematic diagram of a lateral gated quantum dot device. (b) Surface gate layout allowing to empty the dot completely and fill it with exactly one electron [17].

An improved control of the confinement and interdot coupling is achieved in the lateral technology. Here the precise control of electronic position is accomplished by confining the carriers in an electrostatic potential created by an arrangement of gates in the split-gate geometry [16]. A schematic diagram of such a single-dot lateral gated device is shown in Fig. 1.1(a), and the gate layout used by Ciorga and co-workers to confine precisely one electron [17] is presented in Fig. 1.1(b). The sample is composed of a heterojunction of AlGaAs/GaAs semiconductor materials created at a distance of $D \approx 90$ nm below the surface. The barrier material between the heterojunction and the surface is doped with donor atoms, of which some become ionized. The excess electrons are confined in the triangular quantum well created by the band offset of the materials and the Coulomb potential of the donors. The density of the two-dimensional electron gas (2DEG) created in this geometry is further locally modulated by the electrostatic field from the surface gates. With a sufficiently large negative gate voltage the area directly underneath is depleted of electrons. As the voltage is tuned towards zero, the surface area between the gates (see Fig. 1.1(b)) translates into a shallow local potential minimum, which can be populated by electrons tunneling from the edges of the 2DEG.

The energy scale of such a lateral gated quantum dot is defined predominantly by the Coulomb interaction between electrons, and renormalized by the single-particle energy quantization of the confined carriers. As a result, as the gate voltage is tuned, only one electron at a time can tunnel into the dot. Next electron is admitted only when the modification of the dot potential overcomes the excess charging energy, so that the chemical potential of the dot becomes equal to that of the 2DEG. This Coulomb blockade (CB) effect translates into the appearance of discrete conduction peaks in the current tunneling through the device, measured as a function of the gate voltage [18–21], and appearing as subsequent electrons enter the dot. The CB effect together with a careful design of the gate pattern makes it possible to empty the dot completely of electrons, and then readmit exactly one electron [17].

Access to the spin of the confined electron was achieved by applying an external magnetic field perpendicular to the 2DEG plane [22, 23]. As a result, the edges of the 2DEG become spin-polarized: the edge of the dominant spin domain (say, spin down) is closer to the dot than the other domain edge (spin up). This means that most of the electrons tunneling into the dot will have spin down, because they have to tunnel across a narrower potential barrier. With this additional constraint, the tunneling current may be suppressed not only by the CB effect, but also by the Pauli exclusion principle, which will come into effect when the first unoccupied QD orbital can only hold an electron spin up. This spin blockade (SB) effect is manifested as an additional modulation of the tunneling current amplitude in the transport measurement, and can be used to probe the spin of the electronic droplet confined in the dot with the number of electrons ranging from several tens down to precisely one carrier [24–26].

In order to use the spin of the localized electron as a qubit we have to be able to manipulate its states in a coherent manner. In the Hamiltonian (1.1) the single-qubit rotations are carried out by addressing each spin individually with a magnetic field. The selection of a specific qubit for the quantum operation is achieved in two ways: by using local magnetic fields or by g-factor engineering. Unfortunately, at present the implementation of each of these approaches is prohibitively difficult.

Indeed, the local magnetic fields necessary to manipulate each qubit individually would have to be constrained at most to the area of a single dot (in the case of the design presented in Fig. 1.1, about 200×200 nm). Although solutions to this problem, e.g., involving microscopic wires coiled in the vicinity of the dot have been proposed [27], generation of tightly focused and switchable fields of order of 1 T remains a challenge.

In the second case the same magnetic field $\vec{B}$ is applied to the entire register, but each qubit couples to it differently due to the locally engineered Landé factor g_i [28, 29]. This idea is implemented by preparing the heterojunction in Fig. 1.1(a) with a composition gradient, which translates into a gradient of the g factor along the vertical direction. The coupling of the electronic spin to the field $\vec{B}$ can be then modulated by shifting the position of the electron vertically using additional surface gates. Although this proposal eliminates the problem of local magnetic fields, the added complication of the sample geometry makes it equally challenging technologically.

These difficulties question the first term in the Hamiltonian (1.1) as an effective means of performing quantum operations on our quantum register. We are left with the Heisenberg term, accounting for the exchange interaction between spins. On one hand, as we shall show later on, the exchange constants J_{ij} can be tuned selectively by adjusting voltage on the appropriate gates defining the structure. On the other hand, the exchange term involves pairs of spins, and therefore it cannot be used to perform rotations of individual spins without perturbing the rest of the register. In this situation we abandon the idea of identifying single spins as qubits. Instead, we define a quantum register of coded qubits. Here the states $|0_L\rangle$ and $|1_L\rangle$ of each coded qubit are encoded in terms of the selected states of several spins. In the remainder of this chapter we shall present a discussion of the design of the voltage-controlled coded qubit and review its implementations in the lateral gated technology.

1.2. Encoding and Manipulation of Three-Spin Coded Qubit States

In constructing the coded qubit we first determine the minimal number of spins necessary to encode the logical states. We can establish this by analyzing the symmetries of the system of spins interacting via exchange. The Heisenberg Hamiltonian describing this system commutes with the total spin operator $\hat{S}^2$ and its projection S_z on the z axis. Therefore, both total spin and total spin projection are conserved quantities, and the states with different sets of spin quantum numbers are not mixed in the time evolution defined by the exchange Hamiltonian. As a result, we have to map the logical qubit states $|0_L\rangle$ and $|1_L\rangle$ onto the eigenstates of the few-spin system characterized by the same pair (S, S_z).

Using these criteria we find that the system composed of two spins is not sufficient to realize the coded qubit. Indeed, the two spins can be arranged in four configurations. We have two spin-polarized triplets, $|T_-\rangle = |\downarrow\downarrow\rangle$ with the quantum numbers $(S, S_z) = (1, -1)$, and $|T_+\rangle = |\uparrow\uparrow\rangle$ with $(S, S_z) = (1, +1)$. The two remaining unpolarized configurations can be arranged in the eigenstates of the total spin as $|S\rangle = \frac{1}{\sqrt{2}}(|\downarrow\uparrow\rangle - |\uparrow\downarrow\rangle)$, a spin singlet with $(S, S_z) = (0, 0)$, and $|T_0\rangle = \frac{1}{\sqrt{2}}(|\downarrow\uparrow\rangle + |\uparrow\downarrow\rangle)$, a spin triplet with $(S, S_z) = (1, 0)$. As we can see, each of the four states is characterized

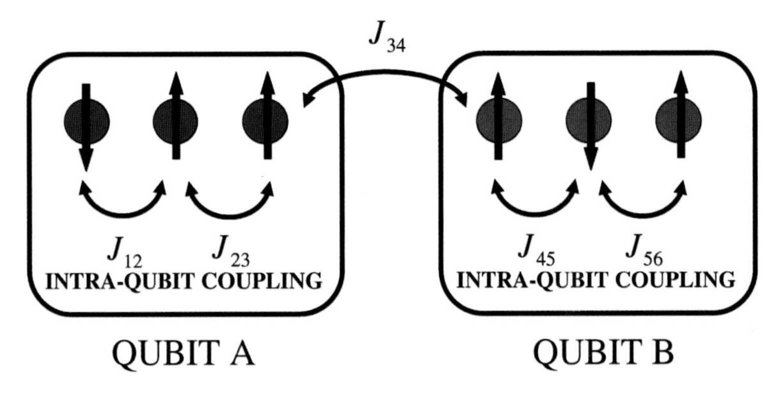

Fig. 1.2. A model three-spin coded qubit A coupled to a neighboring coded qubit B in a linear layout [9].

by a unique set of quantum numbers, and therefore the two-spin states are not suitable to encode the logical states of our exchange-based coded qubit.

As demonstrated by DiVincenzo and co-workers [9], the minimal system suitable for our needs is that of three spins. One of the possible layouts of such system is shown schematically in Fig. 1.2. The general form of the Hamiltonian of such a system is:

$$\hat{H}_{CQ} = \frac{g\mu_B}{\hbar} B \sum_{i=1}^{3} S_{z,i} + J_{12}\vec{S}_1\vec{S}_2 + J_{23}\vec{S}_2\vec{S}_3 + J_{31}\vec{S}_3\vec{S}_1. \tag{1.2}$$

Apart from the pairwise exchange terms, here we have also accounted for the presence of an external, uniform magnetic field $\vec{B} = [0, 0, B]$ coupling to each spin in the same manner.

Let us begin our analysis of the three-spin system by enumerating all possible configurations [30]. Their total number is $2^3 = 8$. Among them we have two spin-polarized states: one, $|\downarrow\downarrow\downarrow\rangle$, with $(S, S_z) = (3/2, -3/2)$, and the other one, $|\uparrow\uparrow\uparrow\rangle$, with $(S, S_z) = (3/2, +3/2)$. These configurations are not suitable for encoding of the logical qubit states.

The other six configurations are characterized by a lower spin polarization. We have three states with $S_z = -1/2$, and three with $S_z = +1/2$. Either of these subspaces can be considerd for encoding. Note that the exchange Hamiltonian alone does not distinguish between the two spin polarizations, leading to the appearance of spin degeneracies in the spectra. This is detrimental to the performance of our coded qubit, because it allows to construct the eigenstates of the system in the form of linear combinations of the states with $S_z = -1/2$ and $S_z = 1/2$. The application of the magnetic field $\vec{B}$ as described in the first term of the Hamiltonian (1.2) removes this degeneracy and allows us to focus on one spin subspace only.

In the following we consider the subspace with $S_z = -1/2$. In Ref. [31] we have shown that in the basis $\{|\downarrow\downarrow\uparrow\rangle, |\downarrow\uparrow\downarrow\rangle, |\uparrow\downarrow\downarrow\rangle\}$ the Hamiltonian (1.2) takes the form:

$$H_{CQ} = \frac{\hbar^2}{4}\begin{bmatrix} J_{12} - J_{23} - J_{31} & 2J_{23} & 2J_{31} \\ 2J_{23} & -J_{12} - J_{23} + J_{31} & 2J_{12} \\ 2J_{31} & 2J_{12} & -J_{12} + J_{23} - J_{31} \end{bmatrix}. \quad (1.3)$$

Here we have neglected the global magnetic field, which modifies the energy of each configuration in the same manner, in this case lowering it by the Zeeman energy $\Delta E_Z = g\mu_B B/2$.

The central point of this analysis is the rotation of the three-spin basis states into the Jacobi basis: $|A\rangle = \frac{1}{\sqrt{2}}(|\downarrow\downarrow\uparrow\rangle - |\downarrow\uparrow\downarrow\rangle)$, $|B\rangle = \frac{1}{\sqrt{6}}(|\downarrow\downarrow\uparrow\rangle + |\downarrow\uparrow\downarrow\rangle - 2|\uparrow\downarrow\downarrow\rangle)$, and $|C\rangle = \frac{1}{\sqrt{3}}(|\downarrow\downarrow\uparrow\rangle + |\downarrow\uparrow\downarrow\rangle + |\uparrow\downarrow\downarrow\rangle)$. In this new basis the Heisenberg Hamiltonian takes the form:

$$H_{CQ} = \frac{3\hbar^2}{4}\begin{bmatrix} -J_{av} - (J_{23} - J_{av}) & \frac{1}{\sqrt{3}}(J_{12} - J_{31}) & 0 \\ \frac{1}{\sqrt{3}}(J_{12} - J_{31}) & -J_{av} + (J_{23} - J_{av}) & 0 \\ 0 & 0 & J_{av} \end{bmatrix}, \quad (1.4)$$

where $J_{av} = (J_{12} + J_{23} + J_{13})/3$ is the average exchange constant. We see that the state $|C\rangle$ is decoupled from the other two basis states. This occurs because this state is an eigenstate of the total spin described by the quantum numbers $(S, S_z) = (3/2, -1/2)$, while the other two states have $(S, S_z) = (1/2, -1/2)$. Therefore we encode the coded qubit state $|0_L\rangle$ as the state $|A\rangle$, and the coded qubit state $|1_L\rangle$ as the state $|B\rangle$.

The form of the Hamiltonian (1.4) allows to determine the steps that need to be taken to perform any rotation of the coded qubit. Specifically, if all exchange constants are equal, the two states $|A\rangle$ and $|B\rangle$ are degenerate and both assume the energy $-3\hbar^2 J_{av}/4$. Detuning the exchange constant J_{23} from J_{av} leads to the splitting of the two energies, however without mixing of the two states (the offdiagonal elements remain zero). This situation is analogous to that appearing for a single spin in an external magnetic field $\vec{B} = [0, 0, B]$, and therefore by manipulating the exchange constant J_{23} we perform the σ_z operation on our coded qubit.

On the other hand, by setting $J_{12} \neq J_{31}$ but choosing $J_{23} = J_{av}$ we introduce the offdiagonal matrix elements mixing the configurations $|0_L\rangle$ and $|1_L\rangle$. This case is analogous to that of a single spin in an external magnetic field $\vec{B} = [B, 0, 0]$, and therefore leads to the σ_x rotation of our coded qubit. Note that the above two quantum operations are controlled by *differences* of the pairwise exchange constants rather than their absolute values. Therefore, to perform them it is sufficient to detune our three-spin system from the symmetric arrangement $J_{12} = J_{23} = J_{31}$, while still maintaining finite values of the exchange parameters.

Let us now demonstrate that using these two operations we can prepare the state of our coded qubit in the form of any linear combination $|\Psi\rangle = \alpha|0_L\rangle + \beta|1_L\rangle$ of its basis states. We start with the σ_x operation and denote the 2×2 Hamiltonian describing it by $\hat{H}_{SX}$. This Hamiltonian can be easily diagonalized in the basis of our coded qubit configurations, giving the ground state $|-\rangle = \frac{1}{\sqrt{2}}(|0_L\rangle - |1_L\rangle)$ with the energy $E_- = -3\hbar^2 J_{av}/4 - \Delta E_{SX}$, and the excited state $|+\rangle = \frac{1}{\sqrt{2}}(|0_L\rangle + |1_L\rangle)$ with the energy $E_- = -3\hbar^2 J_{av}/4 + \Delta E_{SX}$. Here we assumed that $\Delta E_{SX} = \sqrt{3}\hbar^2(J_{12} - J_{31})/4 > 0$, i.e., $J_{12} > J_{31}$. The time evolution of the qubit state driven by the $\hat{H}_{SX}$ operator can then be written as:

$$|\Psi(t)\rangle = \exp\left(-\frac{i}{\hbar}\hat{H}_{SX}t\right)|0_L\rangle = \cos(\Omega_x t)|0_L\rangle - i\sin(\Omega_x t)|1_L\rangle, \qquad (1.5)$$

with $\Omega_x = \Delta E_{SX}/\hbar$. In the above equation we assumed that the initial state of the coded qubit $|\Psi(t=0)\rangle = |0_L\rangle = \frac{1}{\sqrt{2}}(|-\rangle + |+\rangle)$.

From the above time evolution we see that we can indeed perform a rotation of our qubit on a time scale defined by the parameter Ω_x. However, the σ_x rotation imposes a phase constraint between the coefficients α and β: if one is real, the other one must be imaginary. To remove this constraint we need to apply the σ_z operation. To demonstrate this let us assume that we have carried out the time evolution as defined above for the time t_0 such that $\Omega_x t_0 = \pi/4$. This leads to the state of the qubit in the form $|\Psi(t_0)\rangle = \frac{1}{\sqrt{2}}(|0_L\rangle - i|1\rangle)$. If we now turn the σ_x rotation off and apply the σ_z rotation, using $-i = e^{i3\pi/2}$ we get:

$$|\Psi(t_0 + t)\rangle = \frac{1}{\sqrt{2}}\left[|0_L\rangle + \exp(i3\pi/2)\exp\left(-i2\Omega_z t\right)|1\rangle\right], \qquad (1.6)$$

with $\Omega_z = 3\hbar(J_{23} - J_{av})/4$. Thus, the σ_z operation can be seen as a means of tuning the phase of the coefficient β independently of that of the coefficient α. As a result, we can prepare our qubit in any state $|\Psi\rangle = \alpha|0_L\rangle + \beta|1_L\rangle$ by an appropriate superposition of the two rotation operators.

DiVincenzo and co-workers [9] also demonstrate the two-qubit operations implemented with logical qubits. Coupling between the logical qubits

A and B can be induced simply by coupling the individual spins of the two physical qubits found at the boundary, as shown in Fig. 1.2, however more complicated topologies are possible. In particular, the Authors present an implementation of the CNOT gate, whose application influences the state of the second logical qubit depending on the state of the first qubit: if the first qubit is in the state $|0_L\rangle$, the second qubit is not altered, while if the first qubit is in the state $|1_L\rangle$, the second qubit is flipped. The gate is implemented as a series of 19 operations involving only neighboring pairs of spins.

As we have shown, the three exchange-coupled spins is the simplest system that can be treated as the voltage-controlled coded qubit. However, it is also possible to encode the logical qubit states in systems composed of more spins. Bacon and co-workers [32] have shown that such an encoding can be carried out on the states from the $S_z = 0$ subspace of four spins. Apart from the possibility of exchange-driven single- and two-qubit operations, such system also provides an improved immunity to the decoherence effects caused by the environment.

1.3. Two-spin Coded Qubit

1.3.1. *Principles of encoding and manipulating the two-spin qubit*

In the previous section we have demonstrated that it is possible to construct a coded qubit with a complete set of quantum operations accomplished by controlled exchange interactions. However, this flexibility and universality of the qubit comes at a price of increased structural complexity. In the following we shall describe a less robust, but simpler coded qubit design, which involves a gated lateral double quantum dot device.

The states of this coded qubit are encoded as configurations of two coupled spins. This design, put forward by Levy [33], is defined by the following Hamiltonian:

$$\hat{H}_{2S} = \frac{\mu_B}{\hbar} B(g_1 S_{z(1)} + g_2 S_{z(2)}) + J_{12} \vec{S}_1 \vec{S}_2. \tag{1.7}$$

Each of the spins is localized in a different local environment, characterized by Landé factors $g_1 \neq g_2$. In contrast to the qubit design based on the single spin defined by the Hamiltonian (1.1), these factors are static, i.e., not tunable. The system is placed in a uniform, global magnetic field $\vec{B} = [0, 0, B]$.

The qubit states are encoded in two-spin configurations as $|0_L\rangle = |\downarrow\uparrow\rangle$ and $|1_L\rangle = |\uparrow\downarrow\rangle$. In the basis of these two states the Hamiltonian (1.7) can be written in the following matrix form:

$$\hat{H}_{2S} = \begin{bmatrix} -\frac{1}{2}\mu_B B(g_1 - g_2) - \frac{1}{4}\hbar^2 J_{12} & \frac{1}{2}\hbar^2 J_{12} \\ \frac{1}{2}\hbar^2 J_{12} & +\frac{1}{2}\mu_B B(g_1 - g_2) - \frac{1}{4}\hbar^2 J_{12} \end{bmatrix}. \quad (1.8)$$

The general form of the above Hamiltonian is similar to that of the three-spin coded qubit presented in Eq. (1.4). However, its matrix elements are controlled in a different manner. The σ_z rotation is carried out by the magnetic field-dependent diagonal term rather than by the exchange interactions. The characteristic time scale of the time evolution associated with this operation is set by the magnitude of the field and the asymmetry of the Landé factors of the two spins. The exchange interaction term appearing as the offdiagonal element is responsible for the σ_x rotation. However, the time scale of this rotation is defined by the *magnitude* of the parameter J_{12} rather than the detuning of exchange parameters, as was the case for the three-spin qubit. This means that in order to turn the σ_x rotation off we have to bring the exchange interaction parameter to zero.

The two-qubit operations involving the coded qubits are carried out by exchange operators connecting pairs of spins, each of which belongs to a different qubit. Such operations, however, drive the two-qubit system out of the four-dimensional computational space spanned by the coded qubit states. To see this, let us prepare our two-qubit state as $|0_L\rangle_1|0_L\rangle_2 = |\downarrow\uparrow\rangle_1 |\downarrow\uparrow\rangle_2$. If we apply the exchange operator $\hat{H}_{23} = J_{23}\vec{S}_2\vec{S}_3$, connecting the second and third spin, that is the neighboring edge spins of the two qubits, the state will eventually evolve to a configuration $|\downarrow\downarrow\rangle_1|\uparrow\uparrow\rangle_2$. As we can see, although the system did not acquire a net spin, each coded qubit became spin-polarized. However, Levy [33] demonstrated that by carefully designing the two-qubit operations one can still coherently drive the two-qubit system, and, for example, create entanglement between the coded qubits.

1.3.2. *Implementation of the two-spin coded qubit*

The two electronic spins necessary to construct the coded qubit can be localized and coupled in the double dot device, forming a diatomic artificial molecule. Such system has been demonstrated by several groups [34–39]. Here we will briefly describe the lateral gated device fabricated at the Institute for Microstructural Sciences, NRC Canada [40]. In this system, the

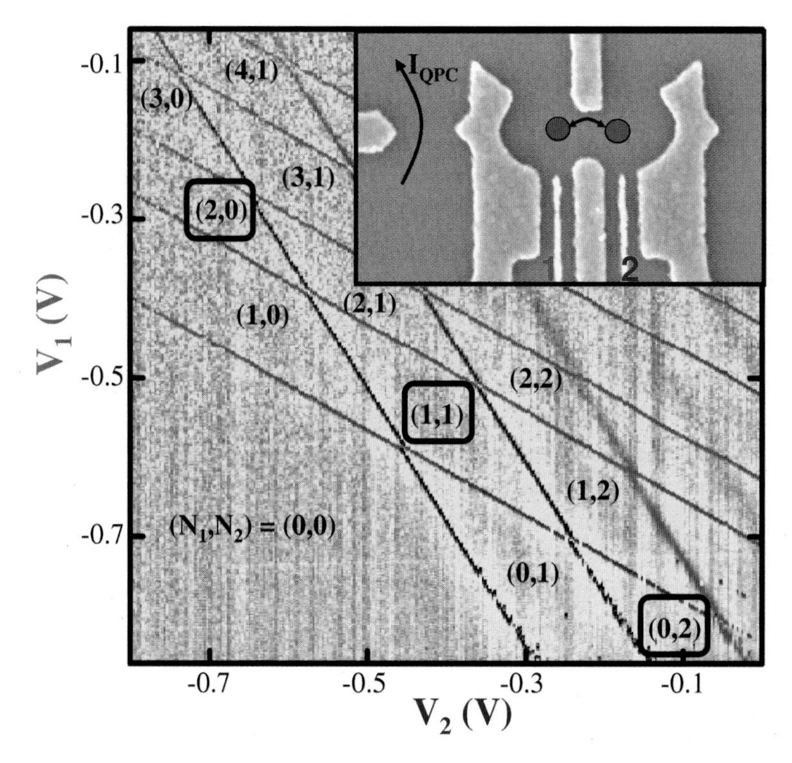

Fig. 1.3. Charging diagram of the double dot device as a function of the voltage V_1 and V_2 on the gates 1 and 2 measured using the charge detection spectroscopy. Inset shows an SEM image of the layout of gates defining the device [40].

double-dot electrostatic confinement was created by the arrangement of metalic gates shown in the inset to Fig. 1.3. The number and distribution of confined electrons is monitored using the charge-detection spectroscopy, as introduced in single-dot measurements by Field and co-workers [41]. This technique takes advantage of the nonlinear current-voltage characteristics of the quantum point contact (QPC) built on the left-hand side of the device. Small changes of the QPC voltage are translated into large modulation of the QPC current I_{QPC}. Such changes occur whenever an extra electron enters the device, or is transferred from one dot to another. Therefore, by measuring the current I_{QPC} as a function of the gate voltages V_1 and V_2 we can map out the stability diagram of the double-dot device. The addition spectrum of the device measured as a function of V_1 and V_2 is shown in Fig. 1.3 [40, 42]. Here each charge addition or redistribution is rendered as a line. In this characteristic honeycomb pattern the regions between the addition lines correspond to constant numbers (N_L, N_R) of electrons on each dot. The double-spin coded qubit is realized with two confined electrons in the configuration $(N_L, N_R) = (1, 1)$. However, the addition diagram also contains the two-electron configurations with both electrons on the same dot. These configurations mix with the singly-occupied state, leading to distortions in the operation of the coded qubit. They have to be included in the detailed analysis of the properties of the double-dot molecule [43–45].

Coherent manipulations of the system of two electrons confined in the double dot were carried out by Petta and co-workers [39]. The device used in this experiment, similar to that shown in Fig. 1.3, did not implement the g-factor anizotropy. Coherent manipulation of the system was carried out by adjusting the voltage difference $V_2 - V_1$. As we shall demonstrate, control of this voltage difference translated into electrostatic control of the effective exchange interaction between the two spins.

A good qualitative understanding of the system can be obtained within the Hubbard approach. We assume that each of the two dots is a Hubbard site with one spin-degenerate electronic orbital with energy E_i $(i = 1, 2)$. The two orbitals form an orthogonal single-particle basis. Let us denote the creation (annihilation) operator of an electron with spin σ (up or down) on site i by $c^+_{i\sigma}$ $(c_{i\sigma})$. Then the Hubbard Hamiltonian for our double dot attains the form:

$$\hat{H}_{DD} = \sum_{i=1}^{2}\sum_{\sigma} E_i n_{i\sigma} + \frac{E_Z}{2}\sum_{i=1}^{2}(n_{i\uparrow} - n_{i\downarrow}) - \sum_{\sigma} t_{12}(c^+_{1\sigma}c_{2\sigma} + c^+_{2\sigma}c_{1\sigma}) + V_{12}\varrho_1\varrho_2 + \sum_{i=1}^{2} U_i n_{i\uparrow}n_{i\downarrow}. \tag{1.9}$$

Here $n_{i\sigma} = c^+_{i\sigma}c_{i\sigma}$ is the number operator for the site i and spin σ, and $\varrho_i = n_{i\uparrow} + n_{i\downarrow}$ measures the total electron number on the site i. Moreover, $E_Z = g\mu_B B$ is the Zeeman energy and $t_{12} > 0$ is the tunneling matrix element responsible for the hybridization of single-dot orbitals. The parameter U_i describes the Coulomb repulsion between the electrons on the same site i, while V_{12} measures the Coulomb repulsion between electrons on different sites. Note that with a single spin-degenerate orbital per site we can analyze the double-dot system charged with up to four electrons.

The analysis of the system defined by the Hamiltonian (1.9) applied to the double-dot lateral gated quantum dot device was presented by our group in Ref. [43]. Here we shall summarize its most important results. Let us start by considering one electron with spin down confined in the double dot potential and at zero magnetic field. Here the electron can be placed can be placed in dot 1, forming a configuration $|1\rangle = c^+_{1\downarrow}|0\rangle$, or in dot 2, forming a configuration $|2\rangle = c^+_{2\downarrow}|0\rangle$. In this case the only relevant terms of the Hamiltonian $\hat{H}_{DD}$ are the first term, introducing onsite orbital energies, and the third term, providing a tunnel coupling between configurations. In the basis $\{|1\rangle, |2\rangle\}$ this Hamiltonian takes the matrix form

$$\hat{H}^{(1e)}_{DD} = \begin{bmatrix} E_1 & -t_{12} \\ -t_{12} & E_2 \end{bmatrix}. \tag{1.10}$$

By diagonalizing the above matrix we obtain the double-dot molecular single-particle energies - bonding E_B and antibonding E_{AB} in the form $E_{B,AB} = \frac{1}{2}(E_1 + E_2) \mp \frac{1}{2}\sqrt{(E_1 - E_2)^2 + 4t^2_{12}}$. In the absence of the tunnel coupling, i.e., at $t_{12} = 0$, the two dots are independent, and these energies are simply equal to the lower and the higher of the two energies (E_1, E_2), respectively. On the other hand, a finite interdot tunneling leads to the formation of molecular orbitals in the form of linear combinations of states $|1\rangle$ and $|2\rangle$. This molecular character is strongest for the system close to resonance, i.e., when $E_1 \approx E_2$. Then the ground state energy $E_B = E_1 - t_{12}$ corresponds to the symmetric bonding molecular orbital $|B\rangle = \frac{1}{\sqrt{2}}(|1\rangle + |2\rangle)$, and the energy of the excited state $E_{AB} = E_1 + t_{12}$ corresponds to the antisymmetric antibonding molecular orbital $|AB\rangle = \frac{1}{\sqrt{2}}(|1\rangle - |2\rangle)$. The two energies are separated by a gap of $2t_{12}$, i.e., proportional to the tunneling matrix element. As we detune the two dots, we still deal with molecular orbitals, although with asymmetric contributions of each configuration.

Let us now add the second electron to the system in such a way that we deal with a spin-polarized triplet configuration (both spins down). Due to the Pauli principle we can generate only one such configuration, $|T_-\rangle = c^+_{2\downarrow}c^+_{1\downarrow}|0\rangle$, with energy $E_{T-} = E_1 + E_2 - E_Z + V_{12}$. The other spin-polarized configuration, $|T_+\rangle = c^+_{2\uparrow}c^+_{1\uparrow}|0\rangle$, contains both electrons spin up. Its energy, $E_{T+} = E_1 + E_2 + E_Z + V_{12}$, is renormalized by the positive Zeeman term.

However, if we take the two electrons with antiparallel spins, we can distribute them in four ways. Two configurations, $|a\rangle = c^+_{2\uparrow}c^+_{1\downarrow}|0\rangle$ and $|b\rangle = c^+_{1\uparrow}c^+_{2\downarrow}|0\rangle$, involve single occupancy of the dots. In the remaining two states, $|c\rangle = c^+_{1\uparrow}c^+_{1\downarrow}|0\rangle$ and $|d\rangle = c^+_{2\uparrow}c^+_{2\downarrow}|0\rangle$, the two electrons form a spin-singlet pair on one of the dots. Further, we can rotate the two singly-occupied configurations into eigenstates of total spin. One of these states is the unpolarized spin triplet, $|T_0\rangle = \frac{1}{\sqrt{2}}(|a\rangle - |b\rangle)$, with the energy $E_{T0} = E_1 + E_2 + V_{12}$, while the other state is a spin singlet $|S\rangle = \frac{1}{\sqrt{2}}(|a\rangle + |b\rangle)$. The singlet states $|S\rangle$, $|c\rangle$, and $|d\rangle$ are coupled by the tunneling matrix elements, and therefore in order to find the singlet energy levels of the system we need to form a Hamiltonian matrix in the basis of these configurations. We have

$$\hat{H}_{DD}^{S} = \begin{bmatrix} E_1 + E_2 + V_{12} & -\sqrt{2}t_{12} & -\sqrt{2}t_{12} \\ -\sqrt{2}t_{12} & 2E_1 + U_1 & -\sqrt{2}t_{12} \\ -\sqrt{2}t_{12} & -\sqrt{2}t_{12} & 2E_2 + U_2 \end{bmatrix}. \tag{1.11}$$

Note that the energies of the unpolarized triplet $|T_0\rangle$ and the singly-occupied singlet configuration $|S\rangle$ are equal. However, coupling with the doubly-occupied configurations additionally lowers the singlet energy. Indeed, assuming that we deal with identical dots on resonance, i.e., $E_1 = E_2 = E$ and $U_1 = U_2 = U$, and in the weak coupling regime ($t_{12} \ll U - V_{12}$), the energy of the lowest singlet state can be approximated as $E_S \approx 2E + V_{12} - 4t_{12}^2/(U - V_{12})$. It is renormalized by the superexchange correction proportional to the square of the tunneling element. As a result, in the absence of the magnetic field the ground state of our two-electron double-dot molecule is a spin singlet.

Petta and co-workers [39] manipulated their two-electron double-dot device by adjusting the onsite energy E_2 relative to the energy E_1 from a large detuning, $E_1 \gg E_2$ to the resonance condition ($E_1 = E_2$). This tuning was accomplished electrostatically by means of gate voltages. Also, in order to separate the three triplets with different spin polarization, the system was placed in a constant, uniform magnetic field. In Fig. 1.4 (a) we show the energies of the three triplets and the lowest-energy singlet state as a function of the detuning $\varepsilon = E_1 - E_2$, treating E_2 as the reference energy. We calculate these energies with model Hubbard parameters $t_{12} = 0.1$, $U_1 = U_2 = 2$, and $V_{12} = 1$, representing typical energy relations in our system. Also, for better visibility, we apply a very large Zeeman energy of $E_Z = 0.5$. As we can see, the energies of each triplet change linearly with the detuning. This is due to the fact that for each spin polarization the triplet is realized by only one, singly occupied configuration, whose energy depends linearly on E_1. Additionally, the three triplets are split by the Zeeman energy.

In contrast, the energy of the singlet changes nonlinearly with the detuning. To understand this, let us first consider $\varepsilon \approx 0$, i.e., the case of two dots near resonance. Here the lowest-energy singlet configuration is $|S\rangle$, because its energy is renormalized by a smaller, interdot Coulomb term V. On the other hand, the electrons in doubly-occupied configurations $|c\rangle$ and $|d\rangle$, repel more strongly, and therefore these configurations are higher in energy.

As a result, the lowest singlet state contains only small admixtures of the doubly-occupied configurations. The gap between the triplet $|T_0\rangle$ and the lowest singlet is due only to the superexchange correction $4t_{12}^2/(U - V_{12})$. However, as we increase the detuning, the increase of the onsite energy characteristic for the singly-occupied configurations begins to compensate the excess Coulomb energy of the configuration $|d\rangle$. As a result, the lowest singlet evolves from a predominantly singly-occupied to a predominantly doubly-occupied configuration. At very large detunings the singlet energy saturates, because the dominant configuration $|d\rangle$ with two electrons on dot 2 is insensitive to the shifts of the onsite energy of dot 1.

Let us now recast the behavior of this system into the language of the quantum-computing Hamiltonian (1.7). In section 1.3.1 we have defined the two-spin computational space in terms of the basis configurations $|\downarrow\uparrow\rangle$ and $|\uparrow\downarrow\rangle$. Here, however, it is more convenient to encode the coded qubit states as the linear combinations of these two-spin configurations: the spin singlet $|S\rangle = \frac{1}{\sqrt{2}}(|\downarrow\uparrow\rangle - |\uparrow\downarrow\rangle)$ with energy $E_S = -3\hbar^2 J_{12}/4$ and the unpolarized spin triplet $|T_0\rangle = \frac{1}{\sqrt{2}}(|\downarrow\uparrow\rangle + |\uparrow\downarrow\rangle)$ with energy $E_{T0} = +\hbar^2 J_{12}/4$. In the absence of g-factor anizotropy these states are eigenstates of the two-spin Heisenberg Hamiltonian (1.7), and are separated by the energy gap $\hbar^2 J_{12}$. In the implementation of the coded qubit we identify the spin triplet state as our singly-occupied configuration $|T_0\rangle$, and the spin singlet state as the lowest eigenstate of the Hamiltonian (1.11). Hence, the value of the effective exchange constant J is found by calculating the energy gap between these states. In Fig. 1.4(b) we plot this gap as a function of the

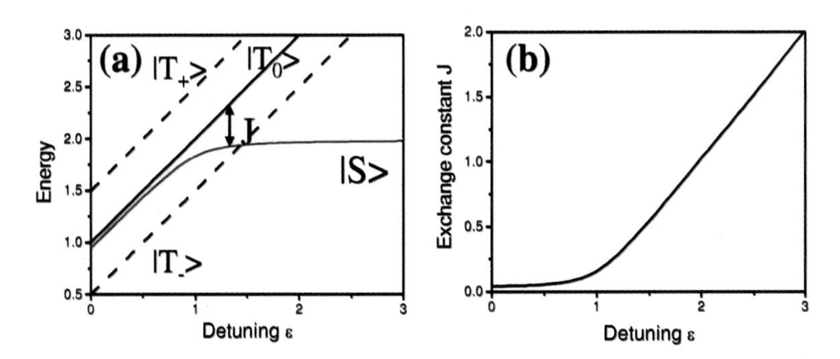

Fig. 1.4. (a) Energies of the three triplet states and the lowest-energy singlet state of two electrons on a double-dot gated lateral device, measured as a function of detuning ε between the onsite dot energies. (b) The exchange parameter $\hbar^2 J$ as a function of the detuning calculated as the width of the energy gap between the singlet and unpolarized triplet states.

detuning ε. As we can see, we are able to tune the exchange constant entirely by electrostatic means. Note that this ability is brought about by the doubly-occupied configurations mixed into in the singlet eigenstate. Although they play a key role in our coded qubit, these admixtures do not belong to the computational space defined by the configurations $|a\rangle$ and $|b\rangle$, and therefore are a source of computational error.

Petta and co-workers [39] manipulated the coded qubit in the following manner. First, they applied a large detuning to the two confined electrons, placing the system at the right-hand edge of Fig. 1.4(a). Here the ground state of the system is the spin singlet configuration composed predominantly of the configuration $|d\rangle$, i.e., both electrons are placed on dot 2 forming a singlet pair. Next, the detuning ε was tuned to zero. Assuming that this manipulation is accomplished adiabatically, and that there are no decoherence mechanisms outside of the Hamiltonian (1.9), the state of the molecule should evolve to a "separated singlet" configuration, i.e., one with the dominant contribution of the singly-occupied singlet $|S\rangle$. The system is then held in this configuration for time τ, after which the detuning is again increased. If this is done adiabatically as well, the ideal system should return to the starting configuration $|d\rangle$. We can measure the fidelity of this operation by measuring the charge density in the dot 2. The perfect fidelity of the system is confirmed when after the two steps we always recover the charge of $2e$, irrespectively of the length of time τ.

However, Petta and co-workers found that the fidelity decreased with the increase of τ, and the random magnetic field caused by the nuclear spins of atoms composing the sample was identified as the source of this decoherence. The amplitude of this field, $\Delta B_{nuc} \sim 1 - 5$ mT, defined an energy scale ΔE_{nuc} which was small compared to the Zeeman energy, but large compared to the exchange splitting $\hbar^2 J(\varepsilon \approx 0)$ close to resonance, and this is why it mixed only spin-unpolarized configurations. In the basis $\{|T_0\rangle, |S\rangle\}$ we can model this hyperfine coupling with the Hamiltonian

$$\hat{H}_{nuc} = \begin{bmatrix} \hbar J(\varepsilon) & \Delta E_{nuc} \\ \Delta E_{nuc} & 0 \end{bmatrix}, \tag{1.12}$$

with the energy of the singlet taken as the reference level. In the context of the two-spin coded qubit Hamiltonian (1.7) the nuclear hyperfine field can be viewed as an asymmetry of the system analogous to the difference in local g factors of the two quantum dots. Therefore, the random nuclear field leads to a rotation of the coded qubit.

Note that this rotation cannot be turned off due to the built-in character of the nuclear field. Let us assume that in our system the hyperfine coupling is *coherent*, and that at zero detuning the exchange gap is negligible. If we prepare our coded qubit state $|\Psi(t=0)\rangle$ in the "separated singlet" configuration, the nuclear field will lead to its time evolution in the form

$$|\Psi(t)\rangle = \cos(\Omega_{nuc}t)|S\rangle - i\sin(\Omega_{nuc}t)|T_0\rangle,$$

with the characteristic frequency $\Omega_{nuc} = \Delta E_{nuc}/\hbar$. In the above coherent evolution the state periodically returns to the singlet configuration. Therefore, in the projection experiment described above the charge on dot 2 measured as a function of the delay time τ should oscillate with a period set by the magnitude of the nuclear field.

However, Petta and co-workers found that the measured probabilities decayed without oscillations. This behavior indicated that the nuclear field coupled the separated singlet and triplet states in an *incoherent* manner. The impact of the decoherence mechanism could, however, be minimized by using the spin echo error correction scheme. Using this technique it was possible to demonstrate the Rabi oscillations of the coded qubit indicating coherent qubit operations.

The effects of the hyperfine coupling can be eliminated by adjusting the exchange gap so that it becomes larger than the energy ΔE_{nuc}. In this process, however, the qubit needs to remain close to zero detuning in order to minimize the admixture of doubly-occupied configurations, whose presence leads to computational errors. Since the superexchange coupling separating the singlet and triplet on resonance is proportional to t_{12}^2, the suppression of decoherence can be accomplished by increasing the tunnel coupling between dots. This scheme was demonstrated by Koppens and co-workers [38], who manipulated the magnitude of the element t_{12} electrostatically by adjusting the voltage on gates defining the lateral double dot device.

The next technological step, involving a demonstration of a network of coupled two-spin coded qubits, still awaits demonstration. As we have mentioned in section 1.3.1, the coupling between neighboring qubits can be realized by tuning the exchange interactions between the edge spins. However, the two-qubit operations could also be realized by taking advantage of the mixing of the separated singlet with the doubly-occupied configurations. As described by Taylor and co-workers [46], this scheme would involve detuning one of the coded qubits from the $\varepsilon = 0$ configuration. If the state of our qubit is a singlet, such a detuning will increase the contribution of the appropriate doubly-occupied configuration. At this point the charge density in the two dots constituting the qubit is no longer equal, and a charge dipole is formed. This does not take place if the qubit is in the triplet state. Taylor and co-workers show that the neighboring qubits can be coherently coupled via this spin-selective charge dipole.

1.4. Implementation of the Three-Spin Coded Qubit

Let us now summarize the progress made towards the realization of the voltage-controlled coded qubit formed with three electronic spins. To construct such a system we need to confine three electrons in a gated lateral triple dot device. In the inset to Fig. 1.5 we show a schematic picture of the triple dot created in our group superimposed on an SEM image of the gates generating it [47]. The same gate layout was used earlier to define the confinement of the double dot. However, in this case the stability diagram measured as a function of V_{1B} and V_{5B} and shown in the main panel of Fig. 1.5, exhibits three sets of addition lines, of which two have similar slopes. These two lines mark the voltages at which an electron is added to dot A or B, respectively, because they exhibit a greater sensitivity to the voltage V_{1B}. The third set of lines is almost vertical, indicating greater

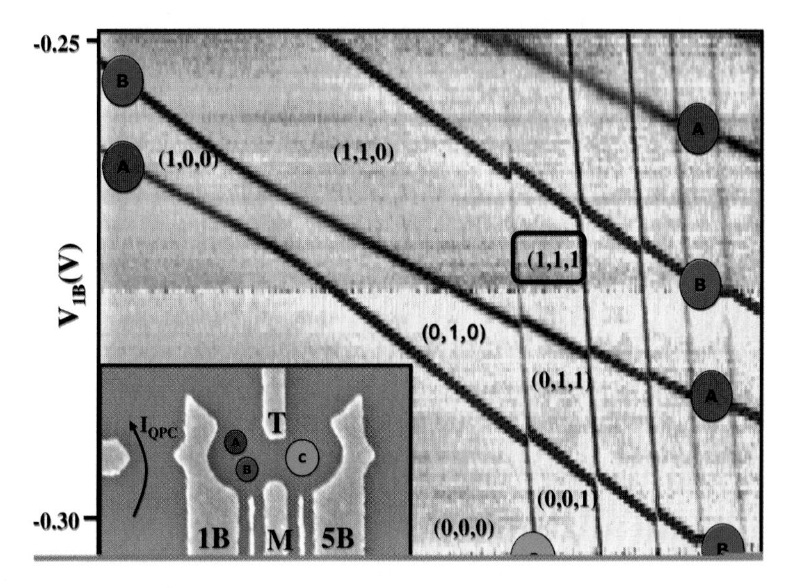

Fig. 1.5. Charging diagram of the lateral triple dot device. Inset shows an SEM image of the gate layout and the schematic positions of the three dots [47].

sensitivity to the voltage V_{5B}, and thus marking the addition events involving dot C. This triangular arrangement of dots results most likely from a mesoscopic fluctuation (an impurity) modifying the electrostatic confinement on the left-hand side of the device.

The three sets of addition lines appearing in Fig. 1.5 define stability regions with characteristic distribution of electrons (N_1, N_2, N_3), with N_i being the number of electrons in the ith dot. As we can see, in this experiment it was possible to empty the system completely (the configuration $(0, 0, 0)$ in the lower left-hand corner), and then fill it controllably with electrons to achieve the configuration $(1, 1, 1)$, i.e., with one electron in each dot. This is the operational regime of our three-spin coded qubit.

Before we can realize the coded qubit, we have to demonstrate that we can control the state of the three confined electrons in a coherent manner. This is yet to be achieved, however Gaudreau and co-workers [47] have demonstrated a great deal of control over the properties of the system by tuning the voltages on the gates defining the structure. In particular, a formation of the triple-dot molecule was achieved by bringing the three dots into simultaneous resonance.

Let us now turn to the mapping of the properties of the proposed triple-dot gated lateral device with three electrons onto those of the coded qubit based on three spins [30]. The layout of the model triple-dot system is shown in Fig. 1.6(a). Here, the gray rectangular gate, containing three circular openings, is responsible for creating the triple-dot confinement. The openings, placed in the corners of an equilateral triangle, translate into minima of the electrostatic confinement of electrons. The remaining gates are used to tune the properties of the system. The red gate influences the potential of dots 2 and 3, and so its action is analogous to that of the gate V_{1B} in Fig. 1.5. Similarly, the green gate is equivalent to the gate V_{5B} and controls the potential of dot 1. Finally, the blue gate can be used to tune the tunneling barrier between dots 1 and 3, just as the gate M, defining the tunneling between the dot B and C in the experimental setup.

In Ref. [48] we have shown that the properties of our system can be understood qualitatively in the frame of the three-site Hubbard model, whose Hamiltonian takes the form:

$$\hat{H}_{3QD} = \sum_{\sigma,i=1}^{3} E_i n_{i\sigma} - \sum_{\sigma,i,j=1,i\neq j}^{3} t_{ij} c_{i\sigma}^{+} c_{j\sigma} + \sum_{i=1}^{3} U_i n_{i\downarrow} n_{i\uparrow} + \frac{1}{2} \sum_{i,j=1,i\neq j}^{3} V_{ij} \varrho_i \varrho_j. \tag{1.13}$$

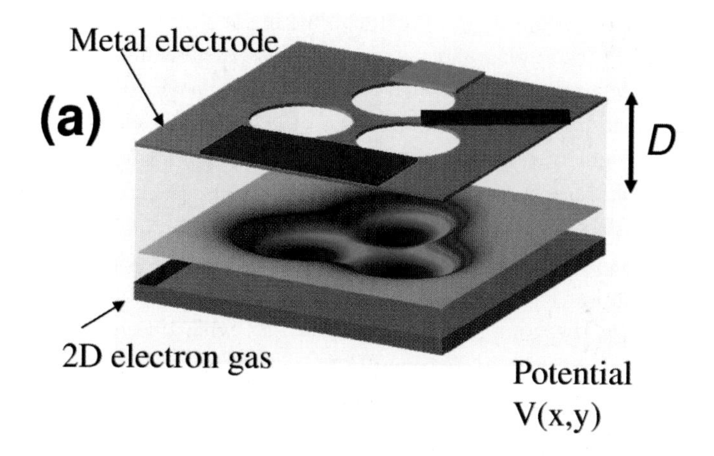

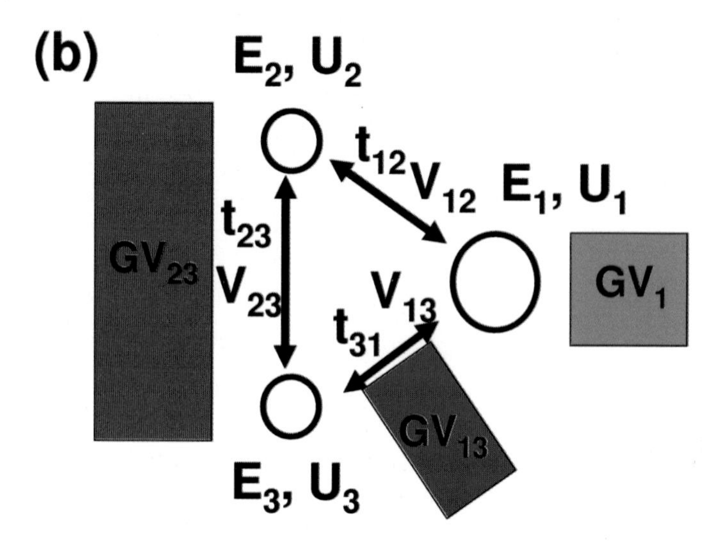

Fig. 1.6. (a) Cross-sectional view of the model triple-dot lateral gated device. (b) Schematic representation of the system described by the Hubbard parameters [48].

In the above formula we have used the notation analogous to that of the two-site Hamiltonian (1.9). Additionally, the above Hamiltonian is characterized by the energy levels of the i-th quantum dot E_i, the tunneling matrix elements t_{ij} between dots i and j, the on-site Hubbard repulsion U_i, and the direct Coulomb matrix elements V_{ij} between dots i and j. These Hubbard parameters are schematically shown in Fig. 1.6(b).

Let us first describe the formation of triple-dot molecular states by confining a single electron in our system. In this case it is convenient to assume the basis of non-overlapping orbitals $\{|1\rangle, |2\rangle, |3\rangle\}$, where $|i\rangle = c_{i\downarrow}^{+}|0\rangle$ and $|0\rangle$ denotes the vacuum. In this basis the diagonal Hamiltonian matrix elements are $\langle i|\hat{H}_{3QD}|i\rangle = E_i$ and the off-diagonal elements $\langle i|\hat{H}_{3QD}|j\rangle = -t_{ij}$. In general, we diagonalize this 3×3 matrix numerically to obtain quantum molecular orbitals and corresponding energies. However, with the three dots on resonance, i.e., with $E_1 = E_2 = E_3 = E$ and $t_{12} = t_{23} = t_{13} = t$, the calculation can be carried out analytically. For the typical case $t > 0$ we obtain the nondegenerate ground state with energy $E_A = E - 2t$, and the doubly-degenerate excited state with energy $E_B = E_C = E + t$. The corresponding molecular orbitals can be written as $|M_A\rangle = \frac{1}{\sqrt{3}}(|1\rangle + |2\rangle + |3\rangle)$, $|M_B\rangle = \frac{1}{\sqrt{2}}(|1\rangle - |2\rangle)$ and $|M_C\rangle = \frac{1}{\sqrt{6}}(|1\rangle + |2\rangle - 2 \cdot |3\rangle)$. The states $|M_B\rangle$ and $|M_C\rangle$ were chosen to be symmetric with respect to a mirror plane passing through the dot 1 and intersecting the $(2-3)$ base of the triangle at its midpoint. However, due to the degeneracy of the two levels, any pair of orthogonal states created as linear combinations of $|M_B\rangle$ and $|M_C\rangle$ will be viable as eigenstates. The degeneracy of the excited states is a direct consequence of the symmetry of the triangular molecule. Changing its topology, e.g., by increasing the tunneling barrier between dots 1 and 3, will remove the degeneracy. In the limit of an infinite barrier, i.e., $t_{13} = 0$, we deal with a linear triple-dot molecule, whose single-particle energy spectrum consists of three equally spaced levels: $(E - \sqrt{2}t, E, E + \sqrt{2}t)$. Thus, the triangular triple dot design makes it possible to engineer the degeneracy of states solely by electrostatic means, a property critical for our coded qubit design.

Let us now populate our triple dot with three electrons such that two of them have spin down, and one has spin up. This choice is dictated by the degree of polarization of the three spins necessary to form a coded qubit. For the system on resonance we expect that two of these electrons will form a spin-singlet pair on the lowest single-particle molecular state $|M_A\rangle$, while the third electron will be placed on the excited state. Due to the double degeneracy of that excited molecular orbital we expect that the ground state of the three-electron system will also be doubly degenerate. This is precisely what we found in section 1.2 for the three-spin system, in which the resonance condition was realized for all exchange parameters being equal. Any breaking of the triangular symmetry results in the removal of the degeneracy of the excited molecular state, which is expected to translate into a splitting of the two lowest three-electron states.

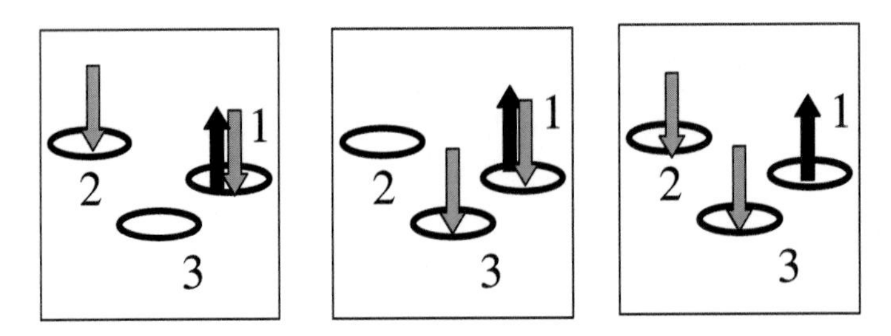

Fig. 1.7. Configurations of the three-electron system with $S_z = -1/2$ with the spin-up electron occupying dot 1 [48].

In a more quantitative analysis we enumerate all possible configurations of the three electrons with $S_z = 1/2$ on the three Hubbard sites. The sole spin-up electron can be placed on any orbital, and with each specific placement the remaining two spin-down electrons can be distributed in three ways. For example, Fig. 1.7 shows the three configurations with the spin-up electron occupying the dot 1. Thus, altogether we can generate nine different configurations. Three of these configurations involve single occupancy of the orbitals. They can be written as $|a\rangle = c^+_{3\downarrow}c^+_{2\downarrow}c^+_{1\uparrow}|0\rangle$, $|b\rangle = c^+_{1\downarrow}c^+_{3\downarrow}c^+_{2\uparrow}|0\rangle$, and $|c\rangle = c^+_{2\downarrow}c^+_{1\downarrow}c^+_{3\uparrow}|0\rangle$. These configurations are identified with the three-spin basis considered in section 1.2 for the coded qubit. The remaining six configurations involve doubly-occupied orbitals. They are $|d\rangle = c^+_{2\downarrow}c^+_{1\downarrow}c^+_{1\uparrow}|0\rangle$, $|e\rangle = c^+_{3\downarrow}c^+_{1\downarrow}c^+_{1\uparrow}|0\rangle$, $|f\rangle = c^+_{3\downarrow}c^+_{2\downarrow}c^+_{2\uparrow}|0\rangle$, $|g\rangle = c^+_{1\downarrow}c^+_{2\downarrow}c^+_{2\uparrow}|0\rangle$, $|h\rangle = c^+_{1\downarrow}c^+_{3\downarrow}c^+_{3\uparrow}|0\rangle$, $|j\rangle = c^+_{2\downarrow}c^+_{3\downarrow}c^+_{3\uparrow}|0\rangle$. In the basis of these nine configurations we construct the Hamiltonian matrix by dividing 9 configurations into three groups, each containing one of the singly-occupied configurations $|a\rangle$, $|b\rangle$, and $|c\rangle$. By labeling each group with the index of the spin-up electron, the Hamiltonian takes the form of a 3×3 matrix:

$$\hat{H}_{1/2} = \begin{bmatrix} \hat{H}_1 & \hat{T}_{12} & \hat{T}^+_{13} \\ \hat{T}^+_{12} & \hat{H}_2 & \hat{T}_{23} \\ \hat{T}_{31} & \hat{T}^+_{23} & \hat{H}_3 \end{bmatrix}. \tag{1.14}$$

The diagonal matrix, e.g.,

$$\hat{H}_1 = \begin{bmatrix} 2E_1 + E_2 + 2V_{12} + U_1 & t_{23} & -t_{13} \\ t_{23} & 2E_1 + E_3 + 2V_{13} + U_1 & t_{12} \\ -t_{13} & t_{12} & \begin{matrix} E_1 + E_2 + E_3 \\ +V_{12} + V_{13} + V_{23} \end{matrix} \end{bmatrix}$$

describes the interaction of three configurations which contain spin-up electron on site 1, i.e., two doubly-occupied configurations $|d\rangle$ and $|e\rangle$, and a singly-occupied configuration $|a\rangle$ (in this order). The configurations with double occupancy acquire the diagonal interaction term U. The three configurations involve a pair of spin-polarized electrons (spin triplet) moving on a triangular plaquette in the presence of a "spectator" spin-up electron. The remaining matrices corresponding to spin-up electrons localized on sites 2 and 3 can be constructed in a similar fashion. The interaction between them is given in terms of effective hopping matrix

$$\hat{T}_{ij} = \begin{bmatrix} 0 & -t_{ij} & 0 \\ 0 & 0 & -t_{ij} \\ +t_{ij} & 0 & 0 \end{bmatrix}.$$

There is no direct interaction between the configurations with single occupancy, since such scattering process would have to involve two electrons, one with spin up and one with spin down. This cannot be accomplished by the single-particle tunneling. These states are coupled only indirectly, involving the configurations with double occupancy.

Before we move on to describe the coded qubit operations on our system, we first need to parametrize the Hubbard Hamiltonian. We do so by mapping its single-electron and three-electron spectra onto those obtained with more microscopic methods. One of such methods, a real-space (RSP) approach presented in Ref. [30], involves numerical diagonalization of a discretized single-particle Hamiltonian. In the following we shall use the effective Rydberg, $1\mathcal{R} = m^* e^4/2\varepsilon^2\hbar^2$, as the unit of energy, and the effective Bohr radius, $1a_B = \varepsilon\hbar^2/m^* e^2$, as the unit of length, with e and m^* being the electronic charge and effective mass, respectively, and ε being the dielectric constant of the material. For GaAs parameters, $m^* = 0.067m_0$ and $\varepsilon = 12.4$, we have $1\mathcal{R} = 5.93$ meV and $1a_B = 9.79$ nm. In these units, we take the main gray gate to be a square with the side length of $22.4a_B$. The diameter of each circular opening is $4.2a_B$, the distance between the centers of each pair of the holes is $4.85a_B$. The gate is positioned $14a_B$ above the two-dimensional electron gas and a voltage of $-|e|V = 10\mathcal{R}$ is applied to it to create the symmetric triangular triple quantum dot. By comparing the spectra of the discretized RSP Hamiltonian and the Hubbard model for one confined electron we find that for the system on resonance the tunneling matrix elements $t_{12} = t_{13} = t_{23} = -0.0636\mathcal{R}$. With the single-particle orbitals and their energies obtained in the RSP calculation, in the next step we consider a system of three confined electrons. To this end we use the configuration interaction approach (CI), in which we create all possible configurations of our electrons on nine lowest molecular orbitals. The configurations are coupled by the Coulomb scattering matrix elements, which we also calculate numerically. In the basis of these configurations we write the Hamiltonian of our interacting system and diagonalize it numerically. By comparing the Hubbard and the RSP-CI spectra for the sysytem on resonance we find the Hubbard interaction parameters $V = 0.479\mathcal{R}$ and $U = 1.539\mathcal{R}$.

This parametrization allows us to examine how the system can be manipulated by adjusting the voltage on the gates. Let us start with the voltage V_{13}, applied to the gate shown in Fig. 1.6 in blue. As we make this voltage more negative, we suppress the tunneling t_{13} between the dots 1 and 3, assuming that all other Hubbard parameters remain unchanged. In Fig. 1.8 we present the spectra of the three electrons calculated in the RSP-CI (a) and the Hubbard approach (b), in each case treating the ground state energy as the reference level. Let us start our description of these spectra

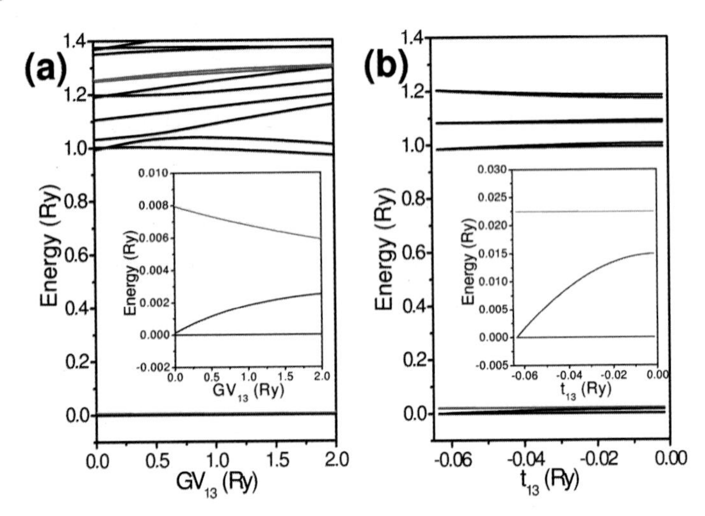

Fig. 1.8. (a) Three-electron energy levels calculated using the RSP-CI approach as a function of the gate voltage GV_{13} controlling the tunneling between dots 1 and 3. (b) A similar spectrum calculated in the Hubbard model as a function of the tunneling parameter t_{13}. Insets show the magnification of the low-energy segment of the spectra [48].

by analyzing the properties of the system on resonance. In both cases we find that the spectra are composed of two parts. We have three low-lying states, built primarily out of the singly-occupied configurations $|a\rangle$, $|b\rangle$, and

$|c\rangle$. The higher-lying states, on the other hand, are built primarily of the doubly-occupied configurations. Their energy is thus renormalized by the onsite Hubbard parameter U, which accounts for the gap separating the two segments of the spectrum.

Let us now look at the low-energy segments of Fig. 1.8(a), (b), magnified in the insets to both graphs. For the system on resonance this spectrum consists of a doubly-degenerate ground state with total spin $S = 1/2$ (drawn in black) and a nondegenerate excited state with total spin $S = 3/2$ (drawn in red). We can map this spectrum onto that of the three-spin coded qubit, calculated using the Hamiltonian (1.4), which allows us to identify our coded qubit states as $|0_L\rangle = \alpha_0 \frac{1}{\sqrt{2}}(|a\rangle - |b\rangle) + \beta_0|\Delta_0\rangle$ and $|1_L\rangle = \alpha_1 \frac{1}{\sqrt{6}}(|a\rangle + |b\rangle - 2|c\rangle) + \beta_1|\Delta_1\rangle$. Here $|\Delta_0\rangle$, $|\Delta_1\rangle$ are contributions of the doubly-occupied configurations. Note that, similarly to the two-spin coded qubit, the admixture of the doubly-occupied configurations is crucial for the coded qubit, as the resulting energy gap separates the low-spin states from the high-spin one. On the other hand, the additional terms do not belong to our computational space, and therefore may lead to computational errors.

As we make the voltage V_{13} more negative (or equivalently, decrease the tunneling element t_{13}), we see that the two coded qubit levels split. This behavior can be reproduced with the exchange Hamiltonian (1.4) by assuming the parameters $J_{12} = J_{23} = J$ to be constant, and decreasing the magnitude of the parameter J_{13}. As a result, both diagonal and offdiagonal elements of the exchange Hamiltonian change, and therefore we deal here with a superposition of the σ_x and σ_z rotations. The 2×2 part of that Hamiltonian can be readily diagonalized, giving the energy splitting between the two levels in the form $\Delta E_{13} = \hbar^2(J - J_{13})$. As the parameter J_{13} is tuned to zero, the gap approaches the value $\hbar^2 J$. By measuring this gap in the asymptotic limit we can establish the effective parameter J, by which we find a direct contact between our microscopic calculation and the Heisenberg three-spin model.

The second coded qubit rotation is carried out by tuning the voltage on the green gate. We assume that by doing so we renormalize the potential of the dot 1 without perturbing the dots 2 and 3. We have considered this case in Ref. [31] using an alternative microscopic approach based on the linear combination of atomic orbitals (LCAO). In this treatment we approximate the numerical triple-dot lateral confinement, obtained as a solution of the

Poisson equation, with a sum of three Gaussians:

$$V(x,y) = -\sum_{i=1}^{3} V_0^{(i)} \exp\left(-\frac{(x-x_i)^2+(y-y_i)^2}{d_i^2}\right). \tag{1.15}$$

The pairs (x_i, y_i) are coordinates of the center of each dot. In order to emphasize the evolution of the spectra under bias, we take the system parameters to be slightly different than those used up to now: for the resonant case we assume $V_0^{(1)} = V_0^{(2)} = V_0^{(3)} = 10\mathcal{R}$, $d_1 = d_2 = d_3 = 2.3a_B$ and the distances between dot centers of $4a_B$. We seek the quantum-molecular single-particle states in the form of linear combinations of single-dot orbitals localized on each dot. To simplify the calculations, we take these orbitals to be harmonic-oscillator (HO) wave functions of a two-dimensional parabolic potential, obtained by extracting the second-order component from each Gaussian. We take one s-type HO orbital per dot, and solve the generalized single-particle eigenproblem formulated in this nonorthogonal basis set. From the resulting spectrum we extract three quantum-molecular levels: a non-degenerate ground state and a doubly degenerate excited state, identified as our ground and excited molecular states $|M_A\rangle$, $|M_B\rangle$, and $|M_C\rangle$, respectively. The next step involves the CI calculation of the three-electron spectra in a manner analogous to that described earlier for the RSP-CI case. We focus specifically on the behavior of the three lowest three-electron levels as we tune the depth $V_0^{(1)}$ of the dot 1. In Fig. 1.9(a) we plot the energies corresponding to these states as a function of the bias ΔV_1 measured with respect to the depth of the dot potential at resonance. This bias is expressed in terms of the gap $\Delta\epsilon_S = 0.095\mathcal{R}$ appearing in the single-particle quantum molecular spectrum. In the Heisenberg model of the three-spin coded qubit we model this bias by decreasing the exchange parameters $J_{12} = J_{13} = J$ while keeping the parameter J_{23} constant. As a result, the diagonal matrix element of the Hamiltonian (1.4) are no longer equal, while the offdiagonal terms remain zero. Therefore, biasing of the dot 1 leads to the σ_z rotation of the coded qubit. Note that in this case the ground state energy of the system is $-3\hbar^2 J_{23}/4$, and remains constant under bias. The energy gap separating it from the first excited state is $\hbar^2(J_{23} - J)$ and increases linearly with the bias. In Fig. 1.9(b) we plot the two energies together with that of the state with total spin $S = 3/2$ as a function of the exchange parameter J. The very good qualitative agreement between the two approaches confirms that by biasing one of the dots we indeed perform the σ_z coded qubit rotation.

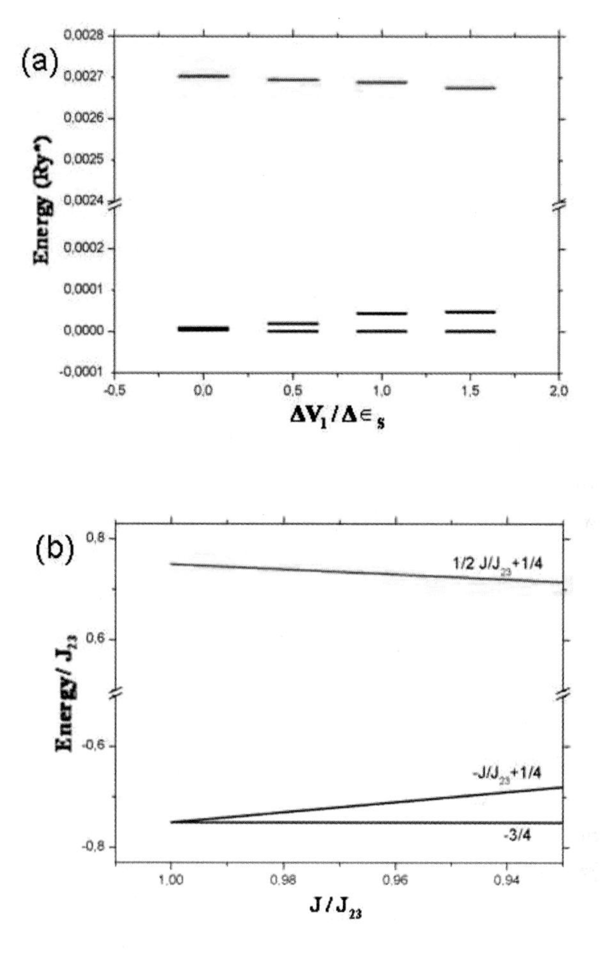

Fig. 1.9. (a) Three lowest energy levels of three electrons with $S_z = -1/2$ confined in the triple quantum dot calculated using the LCHO-CI approach as a function of the bias on dot 1. Here the ground state energy is used as a reference level. (b) The energy levels of a three-spin coded qubit as a function of the exchange parameter $J_{12} = J_{13} = J$ assuming constant parameter J_{23} [31].

We have demonstrated that the behavior of three electrons confined in a triple-dot gated lateral device can be mapped onto the three-spin Heisenberg Hamiltonian of a coded qubit. The effects of Coulomb electron-electron interactions, Fermi statistics and topology of the system can be translated into the exchange parameters J_{ij}, while tuning the triple-dot confinement with voltages translates into selective tuning of these parameters.

Suppressing the tunneling between two of the dots is equivalent to modifying one of the three parameters, which acts as a combination of the σ_x and σ_z rotations of the coded qubit state. Changing the depth of one of the dots, on the other hand, modifies two exchage parameters in the same way, which is equivalent to performing the σ_z qubit rotation. All the single-qubit operations are performed by voltage only, and therefore such a coded qubit does not suffer from the scaling problems inherent in the systems driven by local magnetic fields.

1.5. Summary

To summarize, we have presented a review of recent progress towards the realization of coded qubits based on electron spin. We discussed current approaches utilizing a two-spin system in a double-dot molecule. Further, we have shown that encoding the qubit states in the two lowest $S = 1/2$ states of a three-spin system allows to perform single-qubit operations by controlling the exchange interaction, without the need for local magnetic field or g-factor engineering. We have also described how the control over the exchange parameters can be achieved in lateral gated quantum dot devices by tuning the voltage on the gates defining the system.

Acknowledgments

We thank Irene Puerto-Gimenez, Andy Sachrajda, Louis Gaudreau, Michel Pioro-Ladriere, and Sergei Studenikin for collaboration. We thank Andy Sachrajda for the kind permission to include experimental data.

References

[1] Bennet, C. H., and DiVincenzo, D. P., Nature **404,** 247 (2000).
[2] Feynman, R. P., Foundations of Physics **16,** 507 (1986).
[3] Nielsen, M. A, and Chuang, I. L., *Quantum Computation and Quantum Information*, Cambridge University Press, 2000.
[4] Awschalom, D. D., and Kikkawa, J. M., Phys. Today **6,** 33 (1999).
[5] Fujisawa, T., Austing, D. G., Tokura, Y., Hirayama, Y., and Tarucha, S., Nature **419,** 278 (2002).
[6] Hanson, R., Witkamp, B., Vandersypen, L. M., Willems van Beveren, L. H., Elzerman, J. M., and Kouwenhoven, L. P., Phys. Rev. Lett. **91,** 196802 (2003).
[7] Brum, J. A., and Hawrylak, P., Superlatt. Microstruct. **22,** 431 (1997).
[8] Loss, D., and DiVincenzo, D. P., Phys. Rev. A **57,** 120 (1998).

[9] DiVincenzo, D. P., Bacon, D., Kempe, J., Burkard, G., and Whaley, K. B., Nature **408,** 339 (2000).
[10] Jacak, L., Hawrylak, P., and Wojs, A., *Quantum dots* (Springer, Berlin, 1998).
[11] Ashoori, R., Nature **379,** 413 (1996).
[12] Hawrylak, P., Phys. Rev. Lett. **71,** 3347 (1993).
[13] Tarucha, S., Austing, D. G., Honda, T., van der Hage, R. J., and Kouwenhoven, L. P., Phys. Rev. Lett. **77,** 3613 (1996).
[14] Wojs, A. and Hawrylak, P., Phys. Rev. B **53,** 10841 (1996).
[15] Pi, M., Emperador, A., Barranco, M., Garcias, F., Muraki, K., Tarucha, S., and Austing, D. G., Phys. Rev. Lett. **87,** 066801 (2001).
[16] Thornton, T. J., Pepper, M., Ahmed, H., Andrews, D., and Davies, G. J., Phys. Rev. Lett. **56,** 1198 (1986).
[17] Ciorga, M., Sachrajda, A. S., Hawrylak, P., Gould, C., Zawadzki, P., Jullian, S., Feng, Y., and Wasilewski, Z., Phys. Rev. B **61,** 16315 (2000).
[18] Klein, O., de Chamon, C., Tang, D., Abusch-Magder, D. M., Meirav, U., Wen, X.-G., Kastner, M. A., and Wind, S. J., Phys. Rev. Lett. **74,** 785 (1995).
[19] McEuen, P. L., Foxman, E. B., Kinaret, J. M., Meirav, U., Kastner, M. A., Wingreen, N. S., and Wind, S. J., Phys. Rev. B **45,** 11419 (1992).
[20] McEuen, P. L., Foxman, E. B., Meirav, U., Kastner, M. A., Meir, Y., Wingreen, N. S., and Wind, S. J., Phys. Rev. Lett. **66,** 1926 (1991).
[21] Scott-Thomas, J. H. F., Field, S. B., Kastner, M. A., Smith, H. I., and Antoniadis, D. A., Phys. Rev. Lett. **62,** 583 (1989).
[22] Hawrylak, P., Gould, C., Sachrajda, A., Feng, Y., and Wasilewski, Z., Phys. Rev. B **59,** 2801 (1999).
[23] Sachrajda, A. S., Hawrylak, P., and Ciorga, M., *Nano-spintronics with lateral quantum dots,* in: *Electronic Transport in Quantum Dots,* J. P. Bird, Editor (Kluwer, Boston, 2003).
[24] Ciorga, M., Pioro-Ladriere, M., Zawadzki, P., Hawrylak, P., and Sachrajda, A. S., Appl. Phys. Lett. **80,** 2177 (2002).
[25] Ciorga, M., Wensauer, A., Pioro-Ladriere, M., Korkusinski, M., Kyriakidis, J., Sachrajda, A. S., and Hawrylak, P., Phys. Rev. Lett. **88,** 256804 (2002).
[26] Sachrajda, A. S., Hawrylak, P., Ciorga, M., Gould, C., and Zawadzki, P., Physica E **10,** 493 (2001).
[27] Lidar, D. A. and Thywissen, J. H., J. Appl. Phys. **96,** 754 (2004).
[28] Kato, Y., Myers, R. C., Driscoll, D. C., Gossard, A. C., Levy, J., and Awschalom, D. D., Science **299,** 1201 (2003).
[29] Salis, G., Kato, Y., Ensslin, K., Driscoll, D. C., Gossard, A. C., and Awschalom, D. D., Nature **414,** 619 (2001).
[30] Hawrylak, P., and Korkusinski, M., Solid State Commun. **136,** 508 (2005).
[31] Puerto-Gimenez, I., Korkusinski, M., and Hawrylak, P., Phys. Rev. B 76, 075336 (2007).
[32] Bacon, D., Kempe, J., Lidar, D. A., and Whaley, K. B., Phys. Rev. Lett. **85,** 1758 (2000).
[33] Levy, J., Phys. Rev. Lett. **89,** 147902 (2002).

[34] Holleitner, A. W.,Blick, R. H., Hüttel, A. K., Eberl, K., and Kotthaus, J. P., Science **297,** 70 (2002).

[35] Bayer, M., Hawrylak, P., Hinzer, K., Fafard, S., Korkusinski, M., Wasilewski, Z. R., Stern, O., and Forchel, A., Science **291,** 451 (2001).

[36] Hatano, T., Stopa, M., and Tarucha, S., Science **309,** 268 (2005).

[37] Koppens, F. H. L., Buizert, C., Tielrooij, K. J., Vink, I. T., Nowack, K. C., Meunier, T., Kouwenhoven, L. P., and Vandersypen L. M. K., Nature **442,** 766 (2006).

[38] Koppens, F. H., Folk, J. A., Elzerman, J. M., Hanson, R., van Beveren, L. H. W., Vink, I. T., Tranitz, H. P., Wegscheider, W., Kouwenhoven, L. P., and Vandersypen, L. M. K., Science **309,** 1346 (2005).

[39] Petta, J. R., Johnson, A. C., Taylor, J. M., Laird, E. A., Yacoby, A., Lukin, M. D., Marcus, C. M., Hanson, M. P., and Gossard, A. C., Science **309,** 2180 (2005).

[40] Pioro-Ladriere, M., Abolfath, R., Zawadzki, P., Lapointe, J., Studenikin, S., Sachrajda, A. S., and Hawrylak, P., Phys. Rev. B **72,** 125307 (2005).

[41] Field, M., Smith, C. G., Pepper, M., Ritchie, D. A., Frost, J. E. F., Jones, G. A. C., and Hasko, D. G., Phys. Rev. Lett. **70,** 1311 (1993).

[42] Pioro-Ladriere, M., Ciorga, M., Lapointe, J., Zawadzki, P., Korkusinski, M., Hawrylak, P., and Sachrajda, A. S., Phys. Rev. Lett. **91,** 026803 (2003).

[43] Pioro-Ladriere, M., Sachrajda, A. S., Hawrylak, P., Abolfath, R., Lapointe, J., Zawadzki, P., and Studenikin, S., Physica E **34,** 437 (2006).

[44] Abolfath, R., Dybalski, W., and Hawrylak, P., Phys. Rev. B **73,** 075314 (2006).

[45] Dybalski, W. and Hawrylak, P., Phys. Rev. B **72,** 205432 (2005).

[46] Taylor, J. M., Engel, H.-A., Dür, W., Yacoby, A., Marcus, C. M., Zoller, P., and Lukin, M. D., Nature Physics **1,** 177 (2005).

[47] Gaudreau, L., Studenikin, S. A., Sachrajda, A. S., Zawadzki, P., Kam, A., Lapointe, J., Korkusinski, M., and Hawrylak, P., Phys. Rev. Lett. **97,** 036807 (2006).

[48] Korkusinski, M., Puerto Gimenez, I., Hawrylak, P., Gaudreau, L., Studenikin, S. A., and Sachrajda, A. S., Phys. Rev. B **75,** 115301 (2007).

Chapter 2

Quantum Optical Studies of Single Coupled Quantum Dot Pairs

Gareth J. Beirne, Michael Jetter, and Peter Michler

Institut für Halbleiteroptik und Funktionelle Grenzflächen, Universität Stuttgart, Allmandring 3, 70569 Stuttgart, Germany

In this chapter, we investigate the fundamental physical properties of both on and off resonantly coupled semiconductor quantum dots. The on resonant case could be potentially used to produce a quantum gate capable of performing quantum operations, whereas switching to the off resonant case could then be used to stop such operations. Lateral symmetric InGaAs quantum dots have been shown to be resonantly coupled via second-order cross-correlation function micro-photoluminescence measurements and have subsequently been used to study the on resonant case. Control of the degree of inter-dot electron tunnel coupling and an anomalous Stark shift of the exciton energies have also been observed by applying a variable lateral electric field along the coupling axis. On the other hand, vertical asymmetric InP coupled quantum dots have been employed to examine the stability of the off resonant situation as a function of inter-dot barrier distance. With decreasing barrier width distinct regimes ranging from no tunneling, to electron tunneling, and finally to electron and hole tunneling have been clearly manifest in time-integrated and time-resolved micro-photoluminescence experiments. We propose that this off resonant tunneling occurs via phonon-assisted scattering processes.

Contents

2.1. Introduction

2.1.1. *Quantum computing*

In principle, a quantum computer is any computing device that makes direct use of characteristically quantum mechanical phenomena, such as entanglement and superposition, to perform operations on data. In a classical computer, data is handled using bits; in a quantum computer, the data is handled using quantum bits or qubits for short. A qubit may be realized using a two-level system which can be represented by an atom or a quantum dot (QD). Quantum computation relies on the premise that the quantum properties of particles can be used to represent data, and that quantum mechanisms can be implemented to execute operations with these data.

For quantum computation to proceed a series of quantum gates, each composed of a pair of coherently interacting qubits, is needed [1]. The purpose of the quantum gate is to entangle the two qubits, as the realization of entanglement is the main prerequisite for quantum algorithms outperforming classical ones.

To date, quantum gates have been realized using a number of approaches, however, an epitaxially grown semiconductor alternative would help to enable future systems to be scaled up to large numbers of coherently interacting qubits using well established semiconductor technologies. Furthermore, subsequent integration into conventional electronic systems would be made easier. To this end, a number of theoretical studies have proposed that quantum operations in semiconductors may be performed using a pair of electronically coupled self-assembled semiconductor QDs as the basic scalable element [2–4].

2.1.2. *Coupled quantum dots*

The discrete nature of the electronic structure of quantum dots has led to their labeling as artificial atoms [5, 6] and the observation of a clear shell structure has been reported in the literature [7–11]. QDs have a number of distinct advantages over real atoms, for example, notwithstanding the fact that a single QD can be spatially controlled much more easily than a single atom, the discrete quantum states (electron and hole energy levels) in QDs can be tuned over a large range of values by simply varying the dot composition, dimensions, etc.

In an analogous way to atoms, electronically coupled QDs can also interact in ways that are intermediates between the ionic and covalent limits. The particular interaction may be defined with respect to the degree of electron delocalization across both dots. If the excited electrons are delocalized (shared) between the dots, the interaction can be thought of as being analogous to a covalent bond between two atoms. If, on the other hand, the electron is localized on one or other of the dots, the situation is analogous to an ionic bond between two atoms [12]. However, unlike real atoms, that only exhibit electron tunnel coupling, QDs can exhibit electron, hole, and even exciton tunnel coupling when the associated levels are energetically in resonance and the inter-dot separation is small enough.

More specifically, quantum coupling between QDs can occur (1) via a Coulomb related interaction, such as the dipole-dipole interaction or the resonant Förster transfer process [13–16], or (2) via a particle tunneling process, whereby the electron or hole or both can move from one dot to the other in either a uni-directional or bi-directional manner. Naturally, the actual coupling process that takes place for a given system of QDs primarily depends on (1) the individual dot parameters, (2) the similarity of the dots, and (3) the inter-dot barrier potential properties, such as, the material band gap and thickness [17].

The greatest hurdle toward realizing the full potential of such quantum dot molecules (QDMs), however, is the fact that the current fabrication techniques, for instance, Stranski-Krastanow growth, that are used to produce the quantum dot constituents still lack the degree of control needed at the single dot level. As a result the quantum dot molecules produced to date are intrinsically more asymmetric than desired with the two dots having somewhat different structural and compositional properties. This characteristic can either be suppressed or enhanced in order to produce nominally symmetric QDMs or to produce intentionally asymmetric QDMs, respectively.

If, however, this process could be significantly improved or if a superior process is discovered, and quantum dots could be engineered in terms of both size and position with precision on the quantum level, artificial molecules and even solids with a vast array of predetermined properties could be produced [18]. This would free one from the relatively restrictive nature of real atoms that have more or less fixed electronic structures and would also open up a route towards many new applications, such as quantum computation.

In this article, we present and discuss the results of optical measurements on both symmetric and asymmetric QDMs. The symmetric QDMs are comprised of dots that have very similar sizes and consequently have almost the same ground state photoluminescence (PL) emission energy within a range of less than two meV. As expected, such QDMs exhibit inter-dot electron tunnel coupling when the ground state energy levels of the two constituent dots are automatically in resonance or when they are brought into resonance by way of a static electric field. In terms of quantum computation, the latter resonant situation would be required by an operator that wishes to entangle two qubits in order to perform a quantum operation. Following this operation, however, the operator would naturally wish to suppress tunneling to stop the operation. This suppression can also be achieved by using a static electric field to bring the ground state energy levels of the two dots out of resonance. Such symmetric QDMs could therefore act as potential semiconductor quantum gates, as predicted. An open point still remains, however, namely, how long the off resonant case is stable as, for example, phonon assisted inter-dot tunneling could occur. In order to answer this question, we have also examined intentionally asymmetric QDMs containing two dots with considerably different sizes and consequently ground state emission energies that are out of resonance by tens of meVs.

2.2. Micro-photoluminescence Experimental Procedure

The micro-PL measurements were performed on samples that were cooled to a temperature of 4 K using a liquid helium flow cryostat. The cryostat was mounted on two nano-positioners in order to scan the laser excitation spot over the sample without misaligning the excitation and detection optical beam paths. Movements of less than 50 nm could be reliably achieved. A 95/5 beamsplitter was used to ensure that the PL signal (95% reflected to the detectors) was collected from the sample position onto which the laser (5% transmitted to the sample) was focussed using a 100× or 50× long-working-distance microscope objective. The position and subsequently the focus of the objective could be optimized in order to obtain a one to two micron diameter focus. A continuous wave laser providing 10 W of emission at 532 nm was used to pump a Ti:sapphire laser system that was in turn used to excite the samples. The Ti:sapphire laser can be operated at any wavelength between 700 and 1000 nm and can be run in continuous wave (cw) or pulsed mode. In the latter case providing a choice of either 120 fs pulses or 2 ps pulses at a repetition rate of 76.2 MHz. The sample PL was collected using the same objective and sent to a 0.75 m spectrometer for dispersion prior to detection. PL spectra were then acquired using a liquid-N_2-cooled charge coupled device camera. Single-channel avalanche photodiode detectors were used to perform photon statistics experiments in which case two spectrometers were used as monochromators in order to select the two narrow wavelength ranges to be cross-correlated using a Hanbury-Brown and Twiss type arrangement [19].

2.3. Symmetric InGaAs/GaAs Lateral Quantum Dot Pairs

For identical QDs, at relatively large inter-dot separations, the exciton states of individual QDs will remain as localized states. However, by decreasing the inter-dot separation, at a particular nano-scale separation, the electron states of the exciton will delocalize across both dots as a result of quantum mechanical inter-dot electron tunnel coupling. The heavy hole states, on the other hand, will remain localized on the individual dots unless the inter-dot separation is further reduced as a result of their larger effective mass. When such a tunneling process leads to a redistribution of the electron wave function across both dots the coupling is roughly analogous to the case of a covalent bond between atoms forming a covalent molecule. Covalent coupling leads to the formation of delocalized symmetric (bonding) and anti-symmetric (anti-bonding) states across both dots with the latter molecular states split by an energy that depends on the strength of the inter-dot coupling.

On the other hand, if the delocalized electron wave function becomes strongly localized on an individual dot the coupling is more analogous to the case of an ionic bond between atoms. Therefore, by introducing a perturbation, for instance, via an electric field, the inter-dot coupling can be controllably transformed between covalent and ionic type character.

As mentioned previously, using single QDs as the individual qubits, the coupling of two such QDs with similar sizes could be used to produce a quantum gate. With a view to creating a quantum gate early work concentrated on coupling vertically stacked QD pairs due to the somewhat easier fabrication process, whereby the upper dot preferentially self-assembles over the lower dot [20–23]. In this way, the inter-dot tunnel-barrier can be accurately varied by simply changing the barrier material thickness between

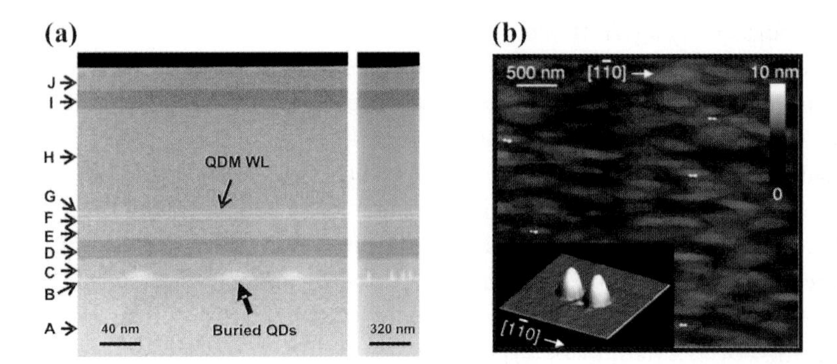

Fig. 2.1. (a) Cross-sectional transmission electron microscope image displaying the layer sequence used to produce the QDMs. Please note that the QDMs are not observed because the associated areal density is very low. (b) Atomic force microscopy studies have shown that each QDM (five in this image) consists of two structurally distinct QDs that are aligned along the [1-10] crystal direction. The inset is a 200 nm^2 image of a single QDM [32].

the upper and lower dot layers. We have used this technique to produce the asymmetric quantum dot pairs that are discussed later. In this Section, we discuss work in which we successfully demonstrate the presence of lateral quantum coupling between two QDs [24]. In principle, the coupling of QDs in the two lateral dimensions implies that coupling in all three dimensions is possible when combined with vertical coupling.

2.3.1. *Sample fabrication and structural characterization*

The lateral QDMs were self-assembled on GaAs(001) substrates by a combination of molecular beam epitaxy and atomic layer precise *in situ* etching. Referring to the cross-sectional transmission electron microscope (TEM) image shown in Fig. 2.1(a), the layer sequence may be outlined as follows; after substrate deoxidation and GaAs buffer growth (A), a high surface density layer of self-assembled InAs QDs is grown (B), this is followed by 20 nm of GaAs (C), 20 nm of $Al_{0.4}Ga_{0.6}As$ (D), and a further 20 nm of GaAs (E). The buried QD layer, which emits at about 1.02 eV, slightly roughens the GaAs surface, improving the QDM growth. This roughness may be seen more clearly in the right panel of Fig. 2.1(a), in which the width of the neighboring TEM image has been reduced by a factor of eight.

The QDM fabrication process is composed of four main parts: (1) a low surface density layer of InAs QDs is grown (F), (2) a 10 nm GaAs capping layer is deposited over the dots, (3) an *in situ* etching process is performed using $AsBr_3$ to a nominal depth of 5 nm with the strained material over the underlying dots preferentially etched, this leads to the formation of elongated nano-holes that are aligned along the [110]-direction, (4) another layer of QD material is deposited (G) with preferential QD self-assembly at the two short edges of the nano-holes, and hence, the QDMs are automatically aligned along the [1-10]-direction. Please note that the QDMs are not visible in this image because the QDM surface density is very low.

As the detectors used to study the QDMs are silicon-based, the QDM PL emission needed to be blue-shifted by partially capping them with 2 nm of GaAs and annealing the sample for 4 minutes. This *in situ* annealing step was also necessary to spectrally separate the QDM PL from that of the buried single QD layer (B) which remains unaffected by this procedure. Finally, the QDMs were capped by 100 nm of GaAs (H), 20 nm of $Al_{0.4}Ga_{0.6}As$ (I), and a further 20 nm of GaAs (J). For more information on the QDM formation process please refer to Refs. [25–28].

In the case of lateral QDMs both dots can also be structurally characterized using atomic force microscopy (AFM) measurements. A 3.75 μm^2 AFM image of uncapped QDMs is presented in Fig. 2.1(b). A remarkably low density of aligned QDMs, each comprised of two QDs, can be readily obtained (approx. 5×10^7 cm^{-2}). A 200 nm^2 image of a typical QDM is displayed as an inset in Fig. 2.1(b). The QDM is comprised of two structurally distinct QDs with an average height of 15 nm and an average lateral size of 35 nm. The dots are separated by an 8 ± 4 nm thick GaAs barrier which allows for a significant electron tunneling probability and implies that the center-to-center distance between the two QDs is relatively large at approximately 43 ± 5 nm. The height of the dots investigated via micro-PL is about 3 nm as a result of the *in situ* annealing step that is used to blue-shift the QDM PL emission.

2.3.2. *Micro-photoluminescence spectra*

The PL spectrum of the lateral QDMs almost invariably consists of six intense peaks [24], four of which are apparent even at very low power densities, X1, X2, Y, and Z, as shown in Fig. 2.2(a). This observation coupled with the fact that each exhibits a linear excitation power density dependence with respect to integrated PL intensity indicates the excitonic nature of each. When observed simultaneously in time-integrated PL spectra, X1 and X2 are thought to be due to the recombination of a delocalized (across both dots) electron with a hole in the larger or the smaller dot, respectively.

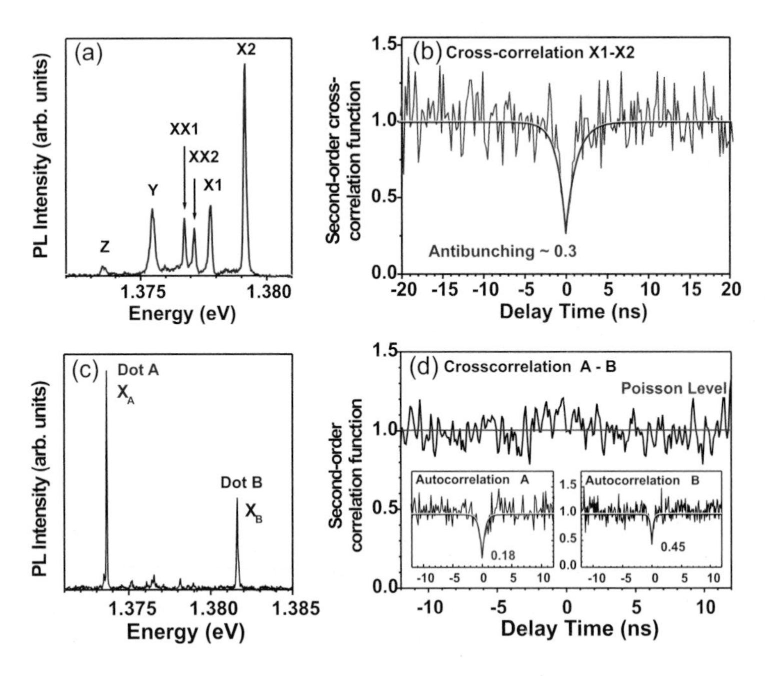

Fig. 2.2. (a) A representative 4 K PL spectrum from a single QDM taken under non-resonant cw excitation at 800 nm (1.55 eV). Four exciton PL lines are observed at low power density, namely, X1, X2, Y and Z. Two biexciton peaks, XX1 and XX2, emerge at high power density. (b) Second-order cross-correlation function measurements of the X1 and X2 emission lines, which clearly exhibits antibunching implying that the two dots are quantum coupled. (c) A PL spectrum from a different single-dot sample of two neighboring but nevertheless independent dots. (d) As expected, no antibunching is observed in the cross-correlation experiment between these two dots verifying that they are not coupled.

The energy difference between X1 and X2 has been observed to vary between as little as 0.5 and 1.7 meV. This implies that it should be possible to shift the associated energy levels into resonance using a small perturbation such as a static electric field. The other two less intense lines, Y and Z, are typically observed approximately 4.5 ± 1.5 meV and 5.5 ± 1.5 meV below the associated X2 peak, respectively [Fig. 2.2(a)]. These lines are thought to be related to charged excitonic transitions [24, 32]. Two additional biexciton PL lines, XX1 and XX2, emerge super-linearly at higher power densities (110 W/cm^2) and are thought to originate from transitions associated with two holes confined in the larger or the smaller dot, respectively. The biexcitons are typically located 1.0 to 2.5 meV [Fig. 2.2(a)] below the corresponding X1 and X2 lines.

2.3.3. *Photon statistics measurements*

Second-order intensity cross-correlation measurements have been carried out in order to determine whether the QDs are coupled or not. In such experiments, if the two dots under investigation are independent of each other, the second-order cross-correlation function, $g^{(2)}(\tau)$, will be equal to one for all delay times, i.e., independent dots will give a flat line response exhibiting no antibunching feature at zero time delay.

In order to emphasize that independent dots would behave in this way, we have performed cross-correlation measurements between two independent dots on a different sample, namely, an InGaAs single-QD sample. The PL spectrum of the two dots in question, labeled A and B, is presented in Fig. 2.2(c). The two dots are necessarily located within less than a few microns of each other as this is a clear prerequisite for their simultaneous optical excitation and detection using our micro-PL setup. As shown in the insets of Fig. 2.2(d), the single dot nature of dot A and B was verified prior to the cross-correlation measurement with both dots exhibiting a strong antibunching feature at zero time delay in autocorrelation measurements [29, 30]. However, in the subsequent cross-correlation measurement, between dot A and B [Fig. 2.2(d)], no antibunching feature is observed at zero time delay as expected for independent dots.

It is therefore very significant that pronounced antibunching has been observed between the two dots that make up the QDM i.e. in cross-correlation measurements between X1 and X2 [Fig. 2.2(b)]. In fact, antibunching has been observed between all six QDM emission lines implying that all of these transitions originate from the same quantum coupled double dot system. Furthermore, cascaded emission [31] has been observed in the cross-correlation experiments between the XX1 biexciton and the X1 exciton, and also between the XX2 biexciton and the X2 exciton [24].

2.3.4. *Electric-field dependent measurements*

The inter-dot coupling can be controlled by applying a parallel electric field to the QDM, that is, along the molecular (coupling) axis. The capacitor-like contact geometry was applied using inter-digital gate electrodes with 30 μm wide Schottky contacts separated by a spacing of 15 μm. The contacts consist of a 20 nm thick layer of Titanium followed by a 100 nm thick layer of gold.

In a time-integrated PL experiment, we propose that the X1 and X2 PL lines of a QDM are only observed simultaneously, when the two QD

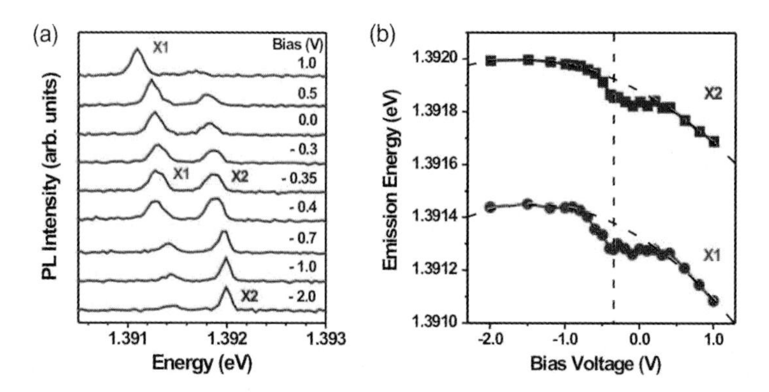

Fig. 2.3. A parallel electric field was applied to the QDMs using 30 μm wide Schottky contacts separated by 15 μm. (a) PL spectra recorded at different applied bias voltages from a QDM at 4 K. A clear switching of the PL intensities of the X1 and X2 PL lines can be observed. (b) The emission energy of the X1 and X2 PL lines as a function of applied voltage (bias). The voltage at which both peaks have approximately the same PL intensity is referred to as the alignment bias, and it is at this bias that an anomalous Stark shift is apparent. This shift is thought to indicate a redistribution of the electron probability density across both dots of the QDM [24].

electronic ground states of the constituent QDs are near to degeneracy and are consequently coupled. Degeneracy can be achieved using an electric field perturbation to tune the two QD electronic ground state energy levels into resonance via the quantum confined Stark effect.

Typical PL spectra obtained at a power density of 200 W/cm^2 from a QDM as a function of electric field are presented in Fig. 2.3(a). A clear anti-correlated behavior of the relative PL intensities can be observed, further demonstrating the presence of inter-dot quantum coupling. We refer to the particular voltage at which X1 and X2 possess approximately the same integrated PL intensity as the alignment bias. It is thought that maximum tunnel coupling is achieved at this field strength and that the electron probability density is equally distributed over both dots as a result. The exact value of this bias naturally depends on the individual QDM under observation. Significantly, when the E-field is applied perpendicular to the molecule axis, no such switching behavior is observed, as expected for such lateral QDMs. It is also worth noting that the relative intensities of the QDM X1 and X2 transitions do not exhibit any polarity preference with respect to electric field [32].

With respect to the peak emission energies of X1 and X2 as a function of applied electric field, both peaks are found to exhibit a combined linear and quadratic shift, with the former shift due to the permanent dipole moment of the QDM and the latter due to the polarizability of the QDM. The measured Stark shift is somewhat weaker than expected and is found to be comparable to that observed from QDs in vertical electric fields. Interestingly, at the alignment bias an anomalous Stark shift of both X1 and X2 is apparent in comparison to the interpolated peak energy that would normally be expected considering the Stark shift that is observed at both lower and higher field strengths (dashed curves in Fig. 2.3(b). This effect is also thought to occur when the two electronic ground states in the laterally coupled QDs are brought into resonance and reflects the presence of tunneling processes that result in a redistribution of the electronic wave function across both dots. Similar effects have also been reported for vertically coupled QDs [33].

The observation of this anomalous red-shift tends to imply that the X1 and X2 PL peaks observed at the alignment bias are primarily associated with the QDM electron symmetric state, that is, assuming that this state has a lower energy than the associated anti-symmetric state. Subsequently, an estimate of the coupling energy between the dots can be made and is determined to be 160 to 180 μeV from the 80 to 90 μeV red-shift of the X1 and X2 peaks at resonance.

Calculations [24, 32] indicate that while holes remain localized in the dots, the electrons become almost entirely delocalized between the two dots at the alignment bias. The different behavior of the electrons and holes can be explained when one considers the effective masses, 0.067 and 0.48 m_e (where m_e is the free electron mass), respectively, of these carriers in the inter-dot GaAs barrier material.

2.4. Asymmetric InP/GaInP Vertical Quantum Dot Pairs

Asymmetrically sized quantum dot pairs are also of interest and recent studies have shown that clear anti-crossings can be observed when the electron and hole levels of single vertical InAs/GaAs pairs are tuned into resonance [34, 35]. In the case of intentionally asymmetric QD pairs the situation is somewhat analogous to the case of ionic bonding between atoms with the electrons mainly localized on individual QDs. In other words, none of the electrons are shared by the QDs.

As the larger dot has a lower ground state transition energy, unidirectional carrier tunneling is energetically favored from the smaller dot to the larger dot. Subsequently, with the correct choice of inter-dot bar-

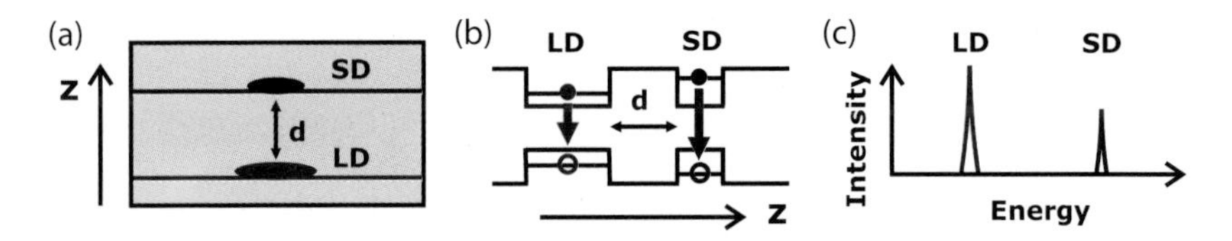

Fig. 2.4. (a) A schematic diagram of the asymmetric double InP QD arrangement in the samples, where Z represents the sample growth direction. (b) An illustration of the electronic structure of the dot pair, and (c) a schematic PL spectrum from such a pair illustrating that LD emits at a lower energy than SD.

rier spacer thickness one can attain significant inter-dot electron tunneling without also having hole tunneling. Therefore, when the small dot is excited the holes tend to be retained whereas the electrons are lost as a result of energetically favorable tunneling to the larger dot.

Single pairs of vertically stacked asymmetric InP/GaInP QDs have been investigated as a function of inter-dot spacer thickness [36]. PL measurements were performed and show a change of the intensity ratio between the two dots for different inter-dot separations. In addition, time-resolved PL experiments display an ever increasing difference in the PL decay time of the two dots with decreasing spacer thickness.

A schematic diagram of the dot arrangement within the samples is presented in Fig. 2.4(a). The larger of the two dots is labeled LD and is produced first during epitaxial growth. Following the deposition of a GaInP layer of thickness, d, which serves to separate the dots, the smaller dot, labeled SD, can be routinely grown over LD, due to the preferential strain situation. The value of d dictates the width of the tunnel barrier in the double dot system and consequently the degree of coupling between them. As LD is the larger dot, its ground state exciton emission energy is lower than that of the SD as illustrated in Fig. 2.4(b). Therefore, if d is sufficiently small and the dots are coupled, the electrons and holes will tend to be automatically captured at a higher rate by LD. This occurs simply because the LD represents the lowest energy state available to carriers in the local environment. As a result, we only expect to observe a peak originating from this dot in the associated PL spectrum sketched in Fig. 2.4(c) when the dots are coupled.

The QD size asymmetry primarily enables uni-directional tunneling of carriers from one dot to the other, as the ground state energy is different for the two dots. Therefore, the electronic levels are usually not aligned and resonant tunneling is prevented. Nevertheless, it is possible for carriers to tunnel out of the energetically higher-lying levels of the small dot into the lower-lying states of the large dot via phonon emission [37, 38].

2.4.1. *Sample growth and post-growth structuring*

The samples were grown by metal-organic vapor phase epitaxy on (100) GaAs substrates oriented by 6° toward the [111]A direction. The lower (larger) dots were grown using the Stranski-Krastanow growth mode by depositing 2.1 monolayers of InP on a 430 nm $Ga_{0.51}In_{0.49}P$ barrier layer. These dots were overgrown by a GaInP spacer layer of different thickness, namely, 2.5 nm, 10 nm, and 20 nm depending on the sample. The upper (smaller) QD layer was then grown using only 75% of the material that was used to produce the larger QD layer [39]. A 30 nm GaInP capping layer was also deposited for PL measurements. Finally, the samples were structured into mesas each with a diameter of 200 nm using electron-beam lithography and dry etching. This enabled QD pairs to be individually addressed via micro-PL measurements in which the mesas were excited using frequency-doubled Ti-Sapphire laser emission at 410 nm (3.02 eV) and a repetition rate of 76.2 MHz.

2.4.2. *Micro-photoluminescence spectra for various spacer thicknesses*

In Fig. 2.5, the PL spectra of dot pairs separated by different spacer widths are displayed as a function of excitation power density. At least five different pairs were examined for each available spacer thickness so as to ensure that the observed results could be reliably observed. Starting with the 20 nm spacer dot pair [Fig. 2.5(a)] an independent behavior of both dots is apparent, with both dots exhibiting similar intensities with respect to the dominant biexciton PL line from each. The weakness of the exciton related transition has been almost invariably observed and we have good reason to believe that this is due to the influence of an energetically lower-lying dark exciton state in such InP/GaInP QDs. This topic will be discussed in detail in a future publication. The peaks are energetically separated by 29.4 meV. The fact that two peaks can be observed, one from each dot implies that the asymmetric QDs are not coupled and that as expected the barrier spacing of 20 nm is too large to allow this process to occur to any significant degree. At high powers the additional PL line observed on the high energy side of the LD spectrum is thought to be due to p-shell emission.

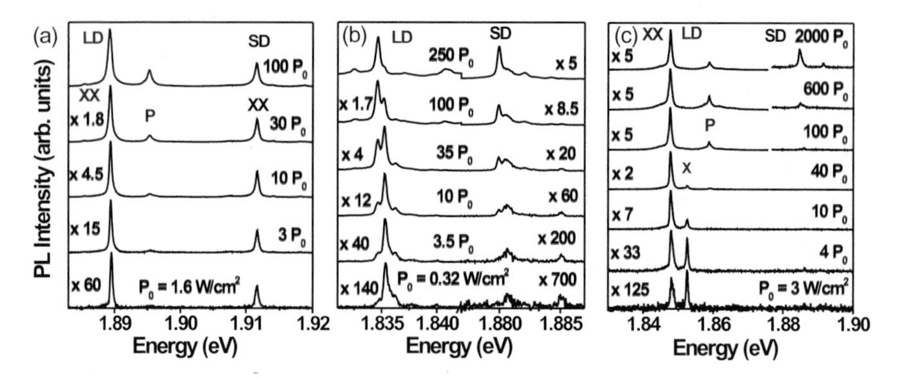

Fig. 2.5. Variation of the excitonic PL emission from the LD and the SD when separated by (a) a 20 nm spacer with no significant inter-dot tunneling, (b) a 10 nm spacer with significant inter-dot electron tunneling, and (c) a 2.5 nm spacer with significant inter-dot electron and hole tunneling. This parasitic coupling (from the viewpoint of the SD) is only clearly apparent when emission from the SD emerges at high power densities as a direct result of state filling in the LD.

For a 10 nm spacer layer [Fig. 2.5(b)], however, the situation changes considerably and the LD emission is now fifteen times more intense than that of SD at the lowest investigated power. We expect that this spacer thickness should allow for a uni-directional tunneling of electrons from SD to LD. At low power densities, the LD emitting at about 1.835 eV dominates the spectrum, with the associated emission consisting primarily of three transitions. The emission from the SD exhibits two emission lines at about 1.88 eV. The energy separation of the lines is typically between 0.7 and 2.5 meV. As these values are clearly smaller than the typical exciton-biexciton separation (4.5 - 5.5 meV) for this material system, such lines may result from charged exciton transitions. Alternatively, the observation of extra peaks may be due to the quantum confined Stark effect caused by the electric field that is created when an electron tunnels from the SD to the LD. We propose that the 10 nm barrier width is thin enough to allow electron tunneling from the SD to the LD, while hole tunneling is prevented due to the larger hole effective mass (m_e/m_h ($Ga_{0.5}In_{0.5}P$)= 0.22). This electron tunneling tends to leave holes or positively charged excitons in the SD, while at the same time leading to electrons or negatively charged excitons in the LD. It is therefore only at higher powers, when the LD ground state becomes saturated due to Pauli-blocking and electron tunneling is suppressed, that the neutral transitions from both dots clearly emerge.

In Fig. 2.5(c), a 2.5 nm thick GaInP spacer layer was used to separate the two dots and single lines are observed once again. For small and medium excitation powers, only the exciton (X) and biexciton (XX) emission from LD is observed, and it is only at very high excitation powers that the emission from the SD can be observed. We propose that this behavior may be explained by electron and hole tunneling times from SD to LD that are faster than the SD excitonic radiative recombination time. Hence, it is only at high excitation powers, when the LD electron and hole ground states are full, that emission from the SD emerges.

2.4.3. *Time-resolved photoluminescence measurements*

The time-resolved PL measurements are presented in Fig. 2.6. For the 20 nm spacing [Fig. 2.6(a)] the LD and the SD transients are very similar. In contrast, for the 10 nm spacer [Fig. 2.6(b)], although both dots still reach their PL maximum at a similar delay time, the decay of the SD PL signal is somewhat faster (290 ps) than that of LD (390 ps). It is thought that this decrease in the case of SD indicates the presence of a second decay channel for this dot, namely, electron tunneling to the LD. The change of decay time is even more pronounced for the 2.5 nm spacer (LD = 570 ps, SD = 140 ps) [Fig. 2.6(c)]. Furthermore, in this case, the PL emission from the SD clearly reaches a maximum before that of the LD. This is probably due to delayed carrier filling of the LD by the SD.

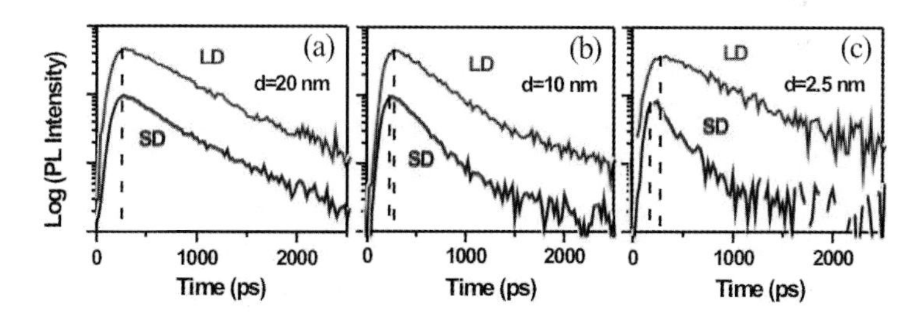

Fig. 2.6. Time-resolved PL measurements of the LD and the SD for (a) a 20 nm spacer, (b) a 10 nm spacer, and (c) a 2.5 nm spacer. The dashed lines are guides to the eye indicating the time at which the PL maxima occur.

Hence, we suggest transitions from vanishing tunnel-coupling to electron tunneling and finally to electron and hole tunneling for decreasing barrier widths. The results clearly show the nonresonant character of the tunneling process as a result of the different ground state energies of the unequally sized dots. The carrier tunneling times have been simulated using a rate equation model (not presented here) and are found to be in good agreement with those extracted from the time-resolved PL measurements [36]. When present, this non-resonant tunneling process from the SD to the LD is thought to occur via phonon scattering.

2.5. Conclusions

The results of optical measurements on both symmetric and asymmetric QD pairs have been presented. The lateral symmetric InGaAs QDMs exhibit inter-dot electron tunnel coupling leading to a splitting of the two former single-dot ground state energy levels into symmetric and anti-symmetric QDM levels. Strong evidence for coupling is provided by the observation of antibunching in the second-order cross-correlation function measurements between the various emission lines from the two dots that make up the QDM. The covalent coupling energy, or in other words, the degree of inter-dot electron tunnel coupling can be controlled by simply applying a lateral electric field along the molecular axis. This field can be used to bring the single-dot energy levels into or out of resonance with such control clearly manifest by a switching of the relative PL intensities of the neutral exciton PL lines from each dot. Furthermore, an anomalous Stark shift of the peak emission energies of these lines as a function of parallel electric field is observed. This anomaly is thought to be due to a delocalization of the electron wave function across both dots as a result of electron tunnel coupling. Furthermore, the value of this shift allows for a rough estimate of the coupling energy.

In relation to quantum computing, the on resonant (symmetric) case could be used to entangle two qubits to produce a quantum gate capable of performing a quantum operation, while the off resonant case, on the other hand, could be used to stop an operation by suppressing inter-dot electron tunneling. The stability of the latter off resonant situation has also been examined by investigating carrier tunneling in vertical asymmetric InP QD pairs as a function of inter-dot barrier distance. Integrated and time-resolved PL measurements have revealed transitions from regimes of no tunneling, to electron tunneling, and ultimately to both electron and hole tunneling, with decreasing barrier width. We propose that this nonresonant tunneling occurs via a phonon-assisted tunneling process.

Acknowledgments

The authors would like to gratefully acknowledge the invaluable contribution that the following people have made to the work outlined in this chapter; Claus Hermannstädter, Matthias Reischle, and Robert Roßbach from the Institut für Halbleiteroptik und Funktionelle Grenzflächen, at the University of Stuttgart; Lijuan Wang from the Max-Planck-Institut für Festkörperforschung, Stuttgart; Armando Rastelli and Oliver G. Schmidt from the Leibniz-Institut für Festkörper- und Werkstoffforschung, Dresden; Moritz Vogel from the Institut für Strahlwerkzeuge, at the University of Stuttgart; Elisabeth Müller from the Labor für Festkörperphysik, at the Swiss Federal Institute of Technology-Hönggerberg, Zürich; and Heinz Schweizer from the 4. Physikalisches Institut, at the University of Stuttgart. The authors also wish to acknowledge financial support from the Alexander von Humboldt Foundation and the Deutsche Forschungsgemeinschaft via Transregio grant no. SFB/TR21 and the DFG Forschergruppe 730 "Positioning of single nanostructures".

References

[1] Loss, D., and DiVincenzo, D. P., *Phys. Rev. A* **57**, 120 (1998).
[2] Biolatti, E., Iotti, R. C., Zanardi, P., and Rossi, F., *Phys. Rev. Lett.* **85**, 5647 (2000).
[3] Li, X-Q., and Arakawa, Y., *Phys. Rev. A* **63**, 012302 (2000).
[4] Lovett, B. W., Reina, J. H., Nazir, A., and Briggs, G. A. D., *Phys. Rev. B* **68**, 205319 (2003).
[5] Ashoori, R. C., *Nature* **379**, 413 (1996).
[6] Michler, P., (Ed.), *Single Quantum Dots*, Topics in Applied Physics Vol. **90**, Springer-Verlag, Berlin (2003).
[7] Roussignol, Ph., Heller, W., Filoramo, A., and Bockelmann, U., *Physica E (Amsterdam)* **2**, 588 (1998).
[8] Ono, M., Matsuda, K., Saiki, T., Nishi, K., Mukaiyama, T., and Kuwata-Gonokami, M., *Jpn. J. Appl. Phys., Part 2*, **38**, L1460 (1999).
[9] Zwiller, V., Pistol, M., Hessman, D., Coderstrom, R., Seifert, W., and Samuelson, L., *Phys. Rev. B* **59**, 5021 (1999).
[10] Dekel, E., Regelman, D. V., Gershoni, D., Ehrenfreund, E., Schoenfeld, W. V., and Petroff, P. M., *Phys. Rev. B* **62**, 11038 (2000).
[11] Beirne, G. J., Reischle, M., Roßbach, R., Jetter, M., Schulz, W. M., Seebeck, J., Gartner, P., Gies, C., Jahnke, F., and Michler, P., *Phys. Rev. B* **75**, 195302 (2007).
[12] Kouwenhoven, L. P., *Science*, **268**, 1440 (1995).
[13] Förster, T., *Ann. Phys.* **2**, 55 (1948).

[14] Berglund, A. J., Doherty, A. C., and Mabuchi, H., *Phys. Rev. Lett.* **89**, 068101 (2002).
[15] Nazir, A., Lovett, B. W., Barrett, S. D., Reina, J. H., and Briggs, G. A. D., *Phys. Rev. B* **71**, 045334 (2005).
[16] Govorov, A. O., *Phys. Rev. B* **71**, 155323 (2005).
[17] Bester, G., Zunger, A., and Shumway, J., *Phys. Rev. B* **71**, 075325 (2005).
[18] Schedelbeck, G., Wegscheider, W., Bichler, M., and Abstreiter, G., *Science* **278**, 1792 (1997).
[19] Hanbury-Brown, R., and Twiss, R. Q., *Nature (London)* **178**, 1447 (1956).
[20] Bayer, M., Hawrylak, P., Hinzer, K., Fafard, S., Korkusinski, M., Wasilewski, Z. R., Stern, O., and Forchel, A., *Science* **291**, 451 (2001).
[21] Krenner, H. J., Sabathil, M., Clark, E. C., Kress, A., Schuh, D., Bichler, M., Abstreiter, G., and Finley, J. J., *Phys. Rev. Lett.* **94**, 057402 (2005).
[22] Ortner, G., Bayer, M., Lyanda-Geller, Y., Reinecke, T. L., Kress, A., Reithmaier, J. P., and Forchel, A., *Phys. Rev. Lett.* **94**, 157401 (2005).
[23] Gerardot, B. D., Strauf, S., de Dood, M. J. A., Bychkov, A. M., Badolato, A., Hennessy, K., Hu, E. L., Bouwmeester, D., and Petroff, P. M., *Phys. Rev. Lett.* **95**, 137403 (2005).
[24] Beirne, G. J., Hermannstädter, C., Wang, L., Rastelli, A., Schmidt, O. G., and Michler, P., *Phys. Rev. Lett.*, **96**, 137401 (2006).
[25] Schmidt, O. G., Deneke, Ch., Kiravittaya, S., Songmuang, R., Heidemeyer, H., Nakamura, Y., Zapf-Gottwick, R., Müller, C., and Jin-Phillipp, N. Y., *IEEE J. Sel. Top. Quantum Electron* **8**, 1025 (2002).
[26] Songmuang, R., Kiravittaya, S., and Schmidt, O. G., *Appl. Phys. Lett.* **82**, 2892 (2003).
[27] Krause, B., Metzger, T. H., Rastelli, A., Songmuang, R., Kiravittaya, S., and Schmidt, O. G., *Phys. Rev. B* **72**, 085339 (2005).
[28] Wang, L., Rastelli, A., and Schmidt, O. G., *J. Appl. Phys.* **100**, 064313 (2006).
[29] Kimble, H. J., Dagenais, M., and Mandel, L., *Phys. Rev. Lett.* **39**, 691 (1977).
[30] Michler, P., Kiraz, A., Becher, C., Schoenfeld, W. V., Petroff, P. M., Zhang, Lidong Hu, E., and Imamoglu, A., *Science* **290**, 2282 (2000).
[31] Moreau, E., Robert, I., Manin, L., Thierry-Mieg, V., Gérard, J. M., and Abram, I., *Phys. Rev. Lett.* **87**, 183601 (2001).
[32] Beirne, G. J., Hermannstädter, C., Wang, L., Rastelli, A., Müller, E., Schmidt, O. G., and Michler, P., *Proceedings of SPIE Photonics West Conference*, **6471**, 647104 (2007).
[33] Oulton, R., Tartakovskii, A. I., Ebbens, A., Finley, J. J., Mowbray, D. J., Skolnick, M. S., and Hopkinson, M., *Physica E* **26**, 302 (2005).
[34] Stinaff, E. A., Scheibner, M., Bracker, A. S., Ponomarev, I. V., Korenev, V. L., Ware, M. E., Doty, M. F., Reinecke, T. L., and Gammon, D., *Science* **311**, 636 (2006).
[35] Bracker, A. S., Scheibner, M., Doty, M. F., Stinaff, E. A., Ponomarev, I. V., Kim, J. C., Whitman, L. J., Reinecke, T. L., and Gammon, D., *Appl. Phys.*

Lett. **89**, 233110 (2006).
[36] Reischle, M., Beirne, G. J., Roßbach, R., Jetter, M., Schweizer, H., and Michler, P., *Phys. Rev. B* **76**, 085338 (2007).
[37] Tackeuchi, A., Kuroda, T., Mase, K., Nakata, Y., and Yokoyama, N., *Phys. Rev. B* **62**, 1568 (2000).
[38] Rambach, M., Seufert, J., Obert, M., Bacher, G., Forchel, A., Leonardi, K., Passow, T., and Hommel, D., *Phys. Stat. Sol. (b)* **229**, 503 (2002).
[39] Roßbach, R., Reischle, M., Beirne, G. J., Schweizer, H., Jetter, M., Michler, P., *J. Cryst. Growth*, **298**, 603 (2007).

Chapter 3

Nuclear Spin Bi-stability in Semiconductor Quantum Dots

Alexander I. Tartakovskii and Maurice S. Skolnick
Department of Physics and Astronomy, University of Sheffield, Sheffield, UK

Alan Russell and Vladimir I. Fal'ko
Department of Physics, University of Lancaster, Lancaster, UK

The confinement of electrons on the nano-scale and the control of their spin is leading to both new physics and to novel devices with functionality at the quantum level. In III-V semiconductors where manipulation of a single electron spin has been possible in nano-structures, new challenges and opportunities arise due to the hyperfine interaction with the spin reservoir of lattice nuclei. In this work we show that by illuminating an InGaAs/GaAs self-assembled quantum dot with circularly polarized light, the nuclei of the tens of thousands of atoms constituting an InGaAs island can be driven into a bistable regime, in which either a threshold-like enhancement or reduction of the local nuclear field by up to 3 Tesla can be generated by varying the intensity of light. We show that such a nuclear spin "switch" can be controlled by both external magnetic and electric fields. The switch is shown to arise from the strong feedback of the nuclear spin polarization on the dynamics of spin transfer from electrons to the nuclei of the dot. A comprehensive theory describing the nuclear bi-stability phenomenon in optically pumped dots is presented.

Contents

3.1. Introduction

Recent achievements in condensed matter science enable control of quantum mechanical degrees of freedom in nano-structured solids. This eventually may lead to devices with operation based on a single electron charge and/or spin [1]. The major advantage of implementing the spin-based systems is in a much weaker coupling of spin to the solid-state environment when compared to charge [1]. In addition, localization of the electron spin in nano-structures is known to dramatically prolong its quantum phase coherence time when compared with structures with higher dimensionality [2–6]. Here an important example is given by semiconductor quantum dots (QD), providing three-dimensional electron confinement and thus the full motional quantization, which effectively decouples the electron spin from the orbital motion [2–6]. This property in turn suppresses the spin decoherence rate due to scattering processes (phonons, charge fluctuations) which couple to the spin via the spin-orbit interaction. Thus quantum dot systems permit accurate preparation on the nano-scale of quantum states with robust spin properties.

Currently of great research interest is another important type of spin states accessible on the nano-scale in semiconductors: spins of various nuclear isotopes [8–32]. Nuclei offer extended spin life-times and robust spin coherence properties [9, 31]. A single quantum dot provides access to these desirable properties through the nuclear spin interaction with accurately prepared QD electron spin states [12–23], well-controlled by external magnetic and electric fields and optical excitation.

In III-V semiconductors all nuclei possess non-zero spin opening a range of opportunities for addressing this quantum degree of freedom, both coherently and incoherently, using resonant techniques such as nuclear magnetic resonance (NMR), or through optical [12–23] or electrical [3–5, 25] pumping of spin polarized electrons. It has also been found that, especially in III-Vs, the collective state of the nuclear spin reservoir in a quantum dot strongly influences the spin properties of the confined electrons, including the spin life-time [21], tunneling probabilities in coupled dots [3–5, 25], spin polarization degree [6, 22] etc. It has also been shown that the coherence of the electron spin is reduced by fluctuations of the net nuclear spin [4, 5, 26–28]. This emphasises further the need for a complex approach to control of the spin in III-V semiconductor nano-structures, including addressing of both nuclear and electron spin.

In this Chapter we present the observation of the strong non-linearities arising due to interaction between an electron confined in a III-V self-assembled dot and an ensemble of 10^4 nuclei. The strong electron confinement results in enhanced interaction with the nuclei covered by the electron wave-function. We show that by illuminating an InGaAs/GaAs self-assembled quantum dot placed in magnetic field with circularly polarized light, the nuclei constituting the dot can be driven into a bistable regime, in which either a threshold-like enhancement or reduction of the local nuclear field by up to 3 Tesla can be generated by varying the intensity of light. We show that such nuclear spin 'switch' can be controlled by both external magnetic and electric fields. The 'switch' is shown to arise from the strong feedback of the nuclear spin polarization on the dynamics of spin transfer from electrons to the nuclei of the dot. Both experimental results and theoretical calculations are presented.

3.2. Hyper-fine Electron-nuclear Spin Interaction

The hyperfine interaction in solids arises from the coupling between the magnetic dipole moments of nuclear and electron spins [8, 9]. Two major effects arising from the hyperfine interaction are (i) inelastic relaxation of electron spin via the 'flip-flop' process (Fig. 3.1) and (ii) the Overhauser shift of the electron energy induced by the net magnetic field of the polarized nuclei [8, 9].

The dominant contribution to the coupling between the electron- and the nuclear-spin systems originates from the Fermi contact hyperfine interaction. It is notable that the nuclei interact only with electrons, the contact hyperfine interaction being negligible for holes due to the p-like nature of the Bloch functions [9]. In first order perturbation theory the electron-nuclear interaction in a dot can be expressed as:

$$\hat{H}_{hf} = \frac{v_0}{8} \sum_i A_i |\psi(\mathbf{R_i})|^2 \hat{\mathbf{S}} \cdot \hat{\mathbf{I}}^i \qquad (3.1)$$

where v_0 is the volume of the InAs-crystal unit cell containing eight nuclei, $\hat{\mathbf{S}}$ is the dimensionless electron spin operator, $\psi(\mathbf{r})$ is the electron envelope wave function and $\hat{\mathbf{I}}^i$ and $\mathbf{R_i}$ are the spin and location of the i-th nucleus, respectively. $A_i = \frac{2}{3}\mu_0 g_0 \mu_B \mu_i I^i |u(\mathbf{R_i})|^2$ is the hyperfine coupling constant, which depends on the nuclear magnetic moment μ_i, the nuclear spin I^i and on the value of the electron Bloch function $u(\mathbf{R_i})$ at the nuclear site.

μ_B is the Bohr magneton, g_0 is the free electron g-factor and μ_0 is the permeability of free space. A_i is of the order of 50 μeV for all nuclei in our InGaAs/GaAs dot system.

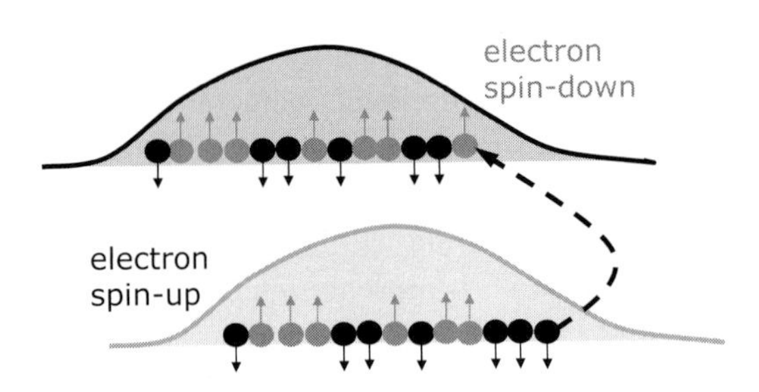

Fig. 3.1. Schematic diagram of the electron-to-nuclei spin-transfer mechanism often referred to as spin "flip-flop". The electron envelope wave-function "covers" $N = 10^4$ nuclei in a typical InGaAs dot. The electron-nuclear spin exchange process is described by the dynamical part in Eq. 3.1. The spin from one electron can be transferred to a single nucleus only.

$\hat{\mathbf{S}} \cdot \hat{\mathbf{I}}^i$ can be rewritten as $1/2(I^i_+ S_- + I^i_- S_+) + I^i_z S_z$ where $I^i_\pm$ and $S_\pm$ are the nuclear and electron spin raising and lowering operators, respectively. Eq. 3.1 thus contains a dynamical part, $1/2(I^i_+ S_- + I^i_- S_+)$ allowing for electron-nuclear spin exchange and a static part, $I^i_z S_z$, which simply acts to modify the energies of the two spin systems. The dynamical part is responsible for the transfer of angular momentum from spin-polarized electrons, which can be excited optically on the dot, to the nuclear spin reservoir of $N \approx 10^4$ nuclei in a typical InGaAs dot (see Fig. 3.1). The static part can be considered as an effective magnetic field acting either on electrons due to spin polarized nuclei (Overhauser field) [12, 14–20, 23] or on the nuclei due to a spin polarized electron (Knight field [16]).

In the absence of electrons, nuclei in the dot are unpolarized even at very low lattice temperatures and high magnetic fields. This is mainly caused by very small nuclear magnetic moments and hence Zeeman splittings, leading to negligible polarization due to thermalization. Indeed, in magnetic field of 2 Tesla, the Zeeman splitting for Ga^{69} nucleus is 70 neV only. Such small splittings result also in suppression of the thermalization process due to a very small probability of phonon scattering. On the other hand, rather large fluctuations of the nuclear polarization are possible leading to the fluctuations of the nuclear magnetic field, B_N, of order of $A/(\sqrt{(N)} g_e \mu_B) =$ 30mT [28]. Such fluctuations exist even at very low temperatures and their effect is further enhanced in a few nanometer dot due to a reduced number of nuclei. The fluctuations lead to decoherence and relaxation of the electron spin [26–29] and are detrimental for possible electron spin applications in quantum devices.

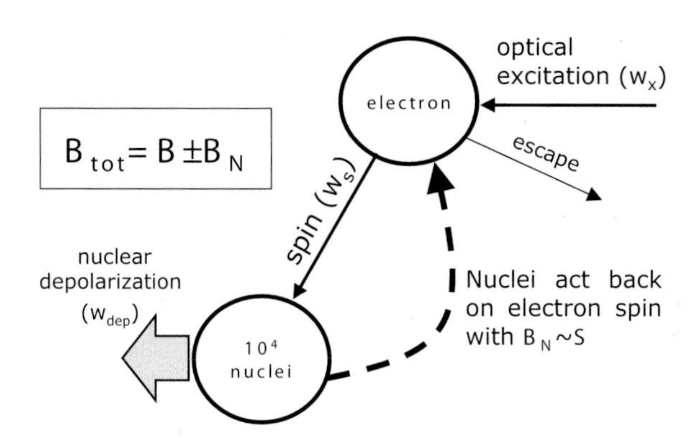

Fig. 3.2. Schematic diagram of the dynamical nuclear polarization in a quantum dot containing 10^4 nuclei. Spin polarized electrons are introduced to the dot by optical circularly polarized excitation with the rate w_x. They then escape from the dot either due to the radiative recombination with the holes or tunneling in the applied electric field. The electrons transfer their spin to the nuclear system with the rate w_s, which competes with the nuclear spin dissipation due to the diffusion into the bulk with the rate w_d. The three rates define the magnitude of the effective nuclear magnetic field, B_N, acting back on electrons and proportional to the steady state nuclear polarization, S. B_{tot} is the total magnetic field seen by the electron spin, which depends on the sign of the nuclear polarization and the external field B.

One of the possibilities to overcome the fluctuations of the nuclear field is to artificially maintain very high degree of nuclear polarization [29], which can be achieved by optical pumping. Eq. 3.1 predicts that a single electron can transfer its spin to a single nucleus only (see Fig. 3.1), leading to a necessity to excite a large number of electrons on the dot. Populating the dot with a single electron at a time will allow flexibility and control of the electron polarization: multi-charge population may lead to additional polarization dissipation channels such as the formation of spin-singlets and spin relaxation without spin transfer to the nuclei. Thus the requirement is to provide a flow of spin polarized electrons through the dot, where the electrons are introduced into the dot one-by-one.

Fig. 3.2 shows the detailed procedure of the nuclear pumping in an InGaAs dot used in our experiments carried out at low sample temperature of 10-20K. The electrons and holes are excited using circularly polarized light resonant with the wetting layer states (about 120 meV above the ground state of the dot) for photoluminescence (PL) measurements. After excitation at high energy the electrons relax to occupy the lowest energy state. Such energy relaxation is accompanied with partial loss of the spin polarization. Nevertheless, this excitation method leads to a high degree of spin polarization of the electrons in the ground state of the dot and hence results in efficient nuclear pumping.

Most of the electrons excited in the dot will escape from the dot either via the radiative recombination or tunneling in applied electric field before they transfer spin to the nuclei. However, if a high enough pumping rate is established the nuclear polarization will build up. This is a dynamical process: the nuclear spin pumping competes with the spin diffusion into the bulk of the crystal. The electron pumping, spin transfer and spin diffusion rates introduced in Fig. 3.2 will be explained in more detail below in Sections 3.3 and 3.4 where we present the study of the electron-nuclear interactions in individual InGaAs dots.

3.3. Nuclear Spin Bi-stability in External Magnetic Fields up to 3T: Experimental Results

In this Section we focus on fundamental properties of the coupling between the nuclear spin and the spin of a strongly localized electron. We show that in a few nanometre-scale QD a large local magnetic field (up to 3T) experienced by the spin of a single electron can be controllably switched on/off by manipulating the nuclear spin configuration. The origin of the threshold-like nuclear spin switching discussed in this Chapter is optically induced bistability of the nuclear spin polarization. We show that it arises due to a feedback between the optically induced Overhauser field B_N and the dynamics of the spin transfer between the confined electron and nuclei.

Non-linearities in the nuclear spin system and in particular bi-stability phenomenon, the first observation of which for optically active quantum dots is described in this Chapter, is a rather general observation for III-V semiconductors. Hysteretic behavior, characteristic for nuclear spin phenomena and bi-stability, has been reported in optical pumping experiments for a variety of systems including bulk GaAs [9] and GaAs/AlGaAs quantum wells [38] and in transport measurements on a 2D electron gas in n-doped GaAs in the quantum Hall regime [33–37]. Most recently hints of fast switching between several stable nuclear spin configurations has been found in transport measurements on shallow double-dots, electrostatically defined in a 2D electron gas in n-doped GaAs [3]. Oscillatory currents due to the ambient nuclear magnetic field have been observed in quantum dot structures based on vertical double dots lithographically fabricated from a double $In_{0.05}Ga_{0.95}As$/AlGaAs quantum well structure [39]. The origin of non-linearities and bi-stability is usually in the dynamic nature of the nuclear spin pumping and the occurrence of a feedback of the net nuclear spin polarization on the efficiency of the electron-to-nuclei spin transfer. The feedback is mainly based on the dynamic modification of the electron energy spectrum occurring due to the Overhauser magnetic field acting on the electron spin. This effect may result in either enhancement or suppression of the electron spin flip required for the spin transfer to the nuclear system. In this Chapter we describe such feedback mechanisms in detail for optically pumped InGaAs quantum dot system, where non-linear spin effects in an isolated system of 10000 nuclei are observed. In this class of self-assembled quantum dots the bi-stable behavior of the nuclear spin polarization is particularly pronounced and has now been reported by several groups [18–20].

The nuclear spin switch has been observed in several different structures containing self-assembled InGaAs/GaAs QD with ~3x20x20 nm size. Here we present results obtained at a temperature of 15K for two GaAs/AlGaAs Schottky diodes, where dots are grown in the intrinsic region of the device. The p- and n-type structures investigated were grown using molecular beam epitaxy on a semi-insulating GaAs substrate and are similar to those reported in Ref. [41] and Ref. [40]. The bottom contact was provided by a 50nm thick doped layer (Si and Be for n and p-type sample, respectively). The dots were then grown on top of a 25 nm tunneling GaAs barrier deposited on the doped layer. The structure was finished with deposition of the following undoped layers: 125 nm GaAs, 75 nm $Al_{0.3}Ga_{0.7}As$, 5 nm undoped GaAs cap. In both types of structures bias can be applied permitting control of the vertical electric field, F. For photoluminescence (PL) experiments, individual dots are isolated using 400 and 800 nm apertures in a gold shadow mask on the sample surface. In our experiments, samples containing InGaAs/GaAs self-assembled QDs are placed in an external magnetic field of $B = 1 - 5$T and low electric field so that the non-radiative carrier escape through tunneling is negligible from the ground state of the dot. Spin-polarized electrons are introduced one-by-one into a single 20nm

InGaAs dot at a rate w_x (see Fig. 3.2) by the circularly polarized optical excitation of electron-hole pairs 120 meV above the lowest QD energy states. The samples were placed on a cold finger in a continuous flow cryostat at a temperature of 15K. PL from the dots was detected using a double spectrometer and a CCD.

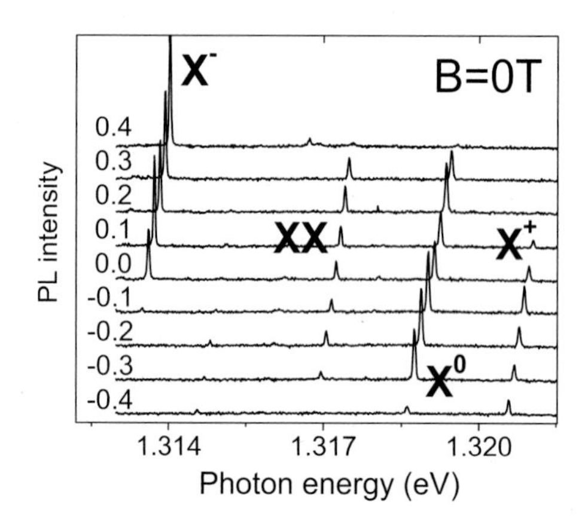

Fig. 3.3. Bias-dependent PL spectra recorded at T=15K and B=0T for a single dot in an n-type Schottky diode. The combination of optical excitation and electrical field tuning enables us to excite controllably the following configurations of the charges in the dots (e and h denotes an electron and a hole, respectively): ehh (positively charged exciton, X^+), eh (neutral exciton, X^0), eehh (bi-exciton, XX), eeh (negatively charged exciton, X^-).

Fig. 3.3 shows time-averaged (60s) PL spectra measured at $B = 0$. It is seen that by varying the bias applied to the sample it is possible to gradually tune the e-h population of the dot [41]. Even for a relatively low excitation power as in Fig. 3.3, at high reverse bias, just above the PL quenching threshold, we observe a peak corresponding to the positively charged trion X^+ (ehh configuration, where e and h represent an electron and a hole on the dot). This exciton complex is formed due to the slow tunneling of the holes in the applied electric field caused by the asymmetry of the diode structure and larger effective mass of the hole. The slow tunneling results in a higher hole occupancy of the dot ground state as compared to that of electrons. As seen in Fig. 3.3 as the bias is changed toward forward direction, X^0 (eh configuration), XX (eehh) and finally X^- (eeh, formed on the dot due to the electron tunneling from the n-contact) is observed. At high bias, where XX peak is weak, the X^+ intensity grows quicker than that of X^0 with the increasing excitation intensity (not shown), leading to domination of X^+ at high bias and excitation.

Note, that in the time averaged spectra as in Fig. 3.3 the relative intensities of the PL peaks represent probabilities to find the dot with a certain e-h population forming a bright exciton complex. The identification of the peaks in the measured spectra is usually performed following a standard procedure for dots embedded in Schottky diodes of both types [41] and is based on the bias- and power-dependence (not shown here) of their intensities and spectral positions. Note, that in the case of p-type structure, X^+, X^0 and XX will dominate the emission spectrum, while X^- is unlikely to be observed. In what follows, for both types of diode structures we will focus mainly on the regime where X^+, X^0 and XX states dominate.

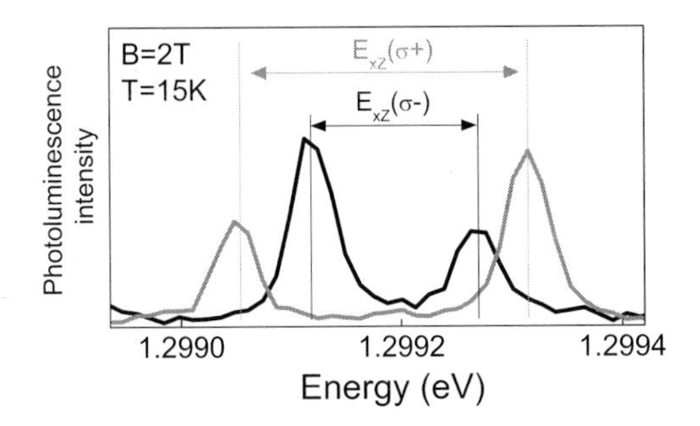

Fig. 3.4. X^0 photoluminescence spectra recorded for an individual InGaAs QD in an external magnetic field $B = 2$T at $T = 15$K. The spectrum excited with σ^+ (σ^-) light resonant with the wetting layer is plotted in gray (black). The horizontal arrows show the corresponding exciton Zeeman splittings.

Fig. 3.4 shows a time-averaged (60s) PL spectra recorded for a single QD in the n-type diode at a bias of -0.4V. The spectra are measured at an external magnetic field of 2T. Circularly polarized laser excitation in the low energy tail of the wetting layer (at 1.425eV) is employed and unpolarized PL is detected. This excitation/detection configuration is used for all data presented here. The portions of the spectra with peaks corresponding to the neutral exciton (X^0 or eh) PL are shown. For each excitation polarization the spectrum consists of an exciton Zeeman doublet, with the high (low) energy component dominating when σ^+ (σ^-) polarized excitation is used.

A strong dependence of the exciton Zeeman splitting (E_{xZ}) on the polarization of the excitation is observed in Fig. 3.4: $E_{xZ}(\sigma^+) = 260\mu$eV and $E_{xZ}(\sigma^-) = 150\mu$eV. Such a dependence is a signature of dynamic nuclear polarization [12–20, 23], which gives rise to a nuclear field B_N aligned parallel or anti-parallel to B for σ^+ or σ^- excitation, respectively. The difference between $E_{xZ}(\sigma^-)$ and $E_{xZ}(\sigma^+)$ is usually referred to as the Overhauser shift and is given by $E_{Ovh} = |g_e|\mu_B(B_N(\sigma^+)+B_N(\sigma^-))$, where g_e is the electron g-factor, μ_B is the Bohr magneton, and $B_N(\sigma^+)$ and $B_N(\sigma^-)$ are the nuclear fields for the two excitation polarizations. It is notable that in effect the shift is due to modification of the electron Zeeman splitting ϵ_{eZ} [9].

We now investigate the dependence of $B_N(\sigma^{\pm})$ on the pumping power (or w_x). The grey-scale plot in Fig. 3.5a presents PL spectra at $B = 2.5$T for a neutral exciton as a function of the power of σ^- excitation. At low power the Zeeman splitting $E_{xZ} = 310\mu$eV. As the power increases an abrupt decrease of the Zeeman splitting to $E_{xZ} = 225\mu$eV is observed, indicating the sudden appearance of a large nuclear field. Below we will refer to this effect as the nuclear spin switch.

E_{xZ} power dependences measured at $B = 2$T for both σ^+ and σ^- excitation polarizations are shown in Fig. 3.5b. For σ^- excitation even below the switch, the magnitude of E_{xZ} decreases gradually, whereas above the threshold a very weak power dependence is observed. Very different behavior is observed for σ^+ excitation, with a weak monotonic increase of E_{xZ} found similar to that previously reported [13]. The Zeeman splitting dependence in Fig. 3.5 reflects the variation of the nuclear field B_N. At low excitation $B_N \approx 0$ and $E_{xZ} = |g_e + g_h|\mu_B B$. As the excitation power is increased and non-zero B_N is generated, $E_{xZ}(\sigma^{\pm}) = |g_e + g_h|\mu_B B \pm |g_e|\mu_B B_N(\sigma^{\pm})$.

In order to present this in terms of the degree of nuclear polarization we will refer to Eq. 3.1. Assuming a homogeneous electron wave function $\psi(\mathbf{r}) \propto \sqrt{8/v_0 N}$ which is constant within the QD volume and zero outside, we will obtain for the shift of the exciton Zeeman splitting: $\Delta E_{xZ} = 2\Sigma_\alpha x_\alpha \langle I_z^\alpha \rangle A^\alpha$. Here x_α is the fraction of a given nuclear isotope in the composition of the dot, and I and A are the nuclear spin and hyperfine constant introduced in Section 3.2. Assuming further a typical dot composition of $In_{0.3}Ga_{0.2}As$, we will obtain the maximum change of E_{xZ} (or, in fact, ϵ_{eZ}) of $\approx 245\mu$eV. Such shift would be obtained when 100% nuclear spins in the dot are polarized.

We thus can estimate the degree of the pumped nuclear polarization above the threshold as shown in Fig. 3.5: it is 35% and 32% in Fig. 3.5a and Fig. 3.5b, respectively. The dependence of the nuclear polarization on the external magnetic field is notable. This trend is more evident in Fig. 3.6,

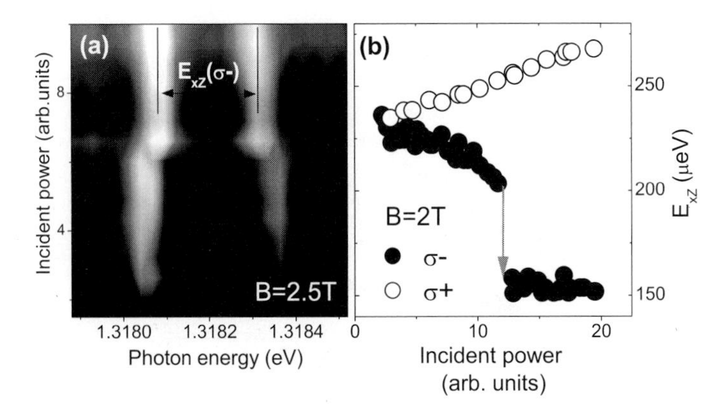

Fig. 3.5. (a) Grey-scale plot showing exciton PL spectra recorded for an individual InGaAs dot. The spectra are recorded at B=2.5T using unpolarized detection and σ^- excitation into the wetting layer at 1.425 eV. The spectra are displaced along the vertical axis according to the excitation power at which they are measured. (b) E_{xZ} power dependences measured at $B = 2$T for σ^+ and σ^- excitation polarizations.

where data of the type shown in Fig. 3.5 is presented as a function of B. The E_{xZ} data are divided into two regimes, below threshold (low power) and above threshold (high power). At all fields in Fig. 3.6 the nuclear spin switch occurs, and there is a weak E_{xZ} power dependence above threshold (as in Fig. 3.5), motivating the division of the power dependence into two regimes with $B_N = 0$ at low power and a high B_N above the threshold.

The nuclear field (and the degree of the nuclear polarization) generated as a result of the spin switch is in its turn dependent on B. The difference between the low and high power dependences $\Delta E_N = |g_e|\mu_B B_N(\sigma^-)$ is shown by solid circles in Fig. 3.6. ΔE_N increases linearly with B at low fields and then saturates at $B \approx 2.5 - 3$T. The inset in Fig. 3.6 shows that the threshold-power for the switch increases nearly linearly with B. No switch could be observed at $B > 3$T in the range of powers employed in our studies. The corresponding range of the nuclear polarization is shown on the right axis. The polarization changes from 20% to 40% as B is varied from 1 to 3 T. This estimate of the polarization degree shows that a majority of the nuclear spins on the dot are not sensitive to the optical pumping. This may be due to the fast nuclear spin diffusion from the dot into the bulk. This also raises the question about the involvement of different nuclear isotopes in the polarization process, which can only be answered from experiments employing resonant techniques such as NMR [12].

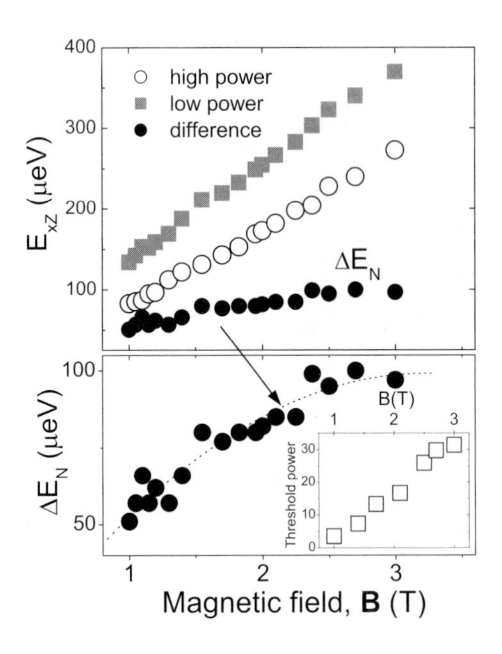

Fig. 3.6. Dependence of the QD exciton Zeeman splitting $E_{xZ}(\sigma^-)$ on the external magnetic field. Squares and open circles show high and low power data, respectively, and solid circles show their difference, ΔE_N, which is proportional to the nuclear magnetic field B_N. For all B shown in the figure the nuclear switch threshold was observed with the threshold power shown in the inset as a function of B.

For $B < 3$T, when the excitation power was gradually reduced from powers above the switching threshold, E_{xZ} was found to vary weakly with power until another threshold was reached, where the magnitude of the exciton Zeeman splitting abruptly increased, as shown in Fig. 3.7. This increase of E_{xZ} corresponds to depolarization of the nuclei and hence reduction of B_N. The observed hysteresis of the nuclear polarization shows that two significantly different and stable nuclear spin configurations can exist for the same external parameters of magnetic field and excitation power. We find that high nuclear polarization persists at low excitation powers for more than 15 min, this time most likely being determined by the stability of the experimental set-up. In what follows we will distinguish two thresholds at which (i) B_N increases (E_{xZ} decreases) and (ii) B_N decreases (E_{xZ} increases) with threshold powers P_{up} and P_{down}, respectively.

We also show in Figs.3.7a, b that the size of the hysteresis loop can be varied by changing either external magnetic or electric fields (the electric

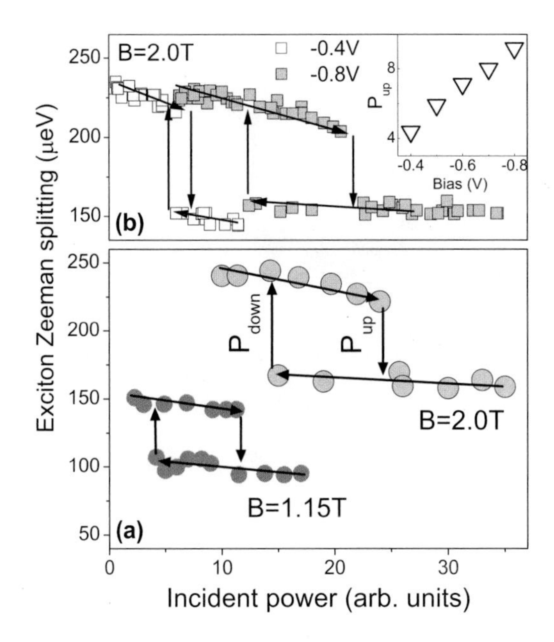

Fig. 3.7. (a) Power dependence of $E_{xZ}(\sigma^-)$ measured at $B = 2$T and 1.15T. The arrows show the direction in which the hysteresis loop is measured with two thresholds P_{up} and P_{down} at which $E_{xZ}(\sigma^-)$ abruptly decreases and increases, respectively. (b) $E_{xZ}(\sigma^-)$ power dependence measured at $B = 2.0$T for a p-type Schottky diode. The two hysteresis loops are measured at -0.4 and -0.8V applied bias. The inset shows the P_{up} dependence on the bias applied to the diode.

field is given by $F = (-V + 0.7)/d$, where V is the applied bias and $d = 230nm$ is the width of the undoped region of the device). The inset in Fig. 3.7b shows the P_{up} bias-dependence for a p-type diode. In general, both P_{up} and P_{down} increase with B and reverse bias, but also the difference between the two thresholds increases, leading to a greater range of incident powers in which the bistability is observed. The strong dependence of P_{up} and P_{down} on both B and F and, importantly, the dependence of B_N on B (as in Fig. 3.6) provides a flexible way to control precisely the degree of polarization of the nuclei in the dot.

In order to demonstrate that further, we show in Fig. 3.8a the bias-dependence of the X^+ Zeeman splitting measured for a single dot in an n-type Schottky diode for $B = 1.5$T. The dot is excited with σ^- polarized light of high intensity, which corresponds to the power above the nuclear spin switch threshold measured in Fig. 3.5b. During the bias scan the

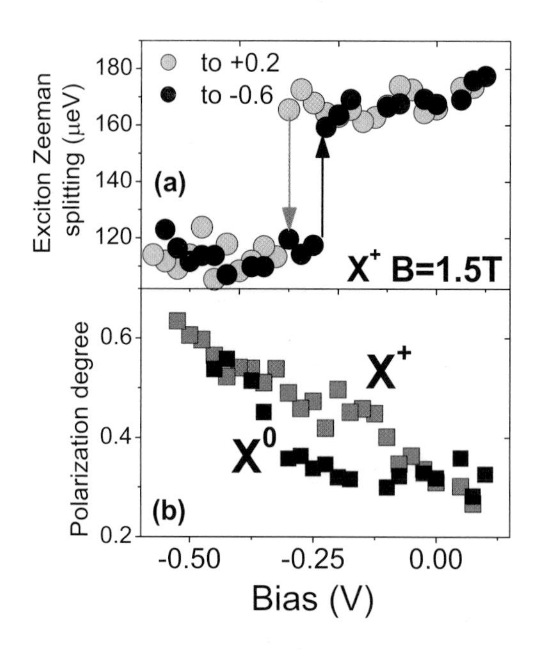

Fig. 3.8. (a) Positively charged exciton (X^+) Zeeman splitting measured for a single dot in an n-type Schottky diode as a function of applied bias. The measurement is carried out at $B = 1.5$T under σ^- polarized excitation of high power. The bias is scanned from forward to reverse for the curve shown with gray circles and in the opposite direction for the curve shown with black symbols. (b) Bias dependence of the degree of circular polarization of the X^+ and X^0 lines.

excitation power is kept constant. Similar bias-dependences are observed for all PL peaks belonging to this particular dot. We will first discuss the dependence where the bias is scanned from forward to reverse (gray symbols). For positive biases, where a high probability of the eeh (X^-) and eehh (XX) configurations of the dot population is observed, the exciton Zeeman splitting is large (170 μeV) and is practically not affected by the nuclear polarization accumulated in the dot, i.e. the nuclear spin pumping is inefficient. As the bias is decreased towards the reverse regime, where X^+ and X^0 dominate the spectrum, an abrupt decrease of the Zeeman splitting to 113 μeV is observed at -0.3V. The switch corresponds to the abrupt increase of the nuclear polarization by 23%. After the threshold-like change, the splitting remains nearly constant in the whole range of biases where PL is observed. If now the bias is tuned from reverse to forward regime (black symbols), the Zeeman splitting is small initially and then switches to a higher magnitude of 170 μeV at -0.23V. As seen from Fig. 3.8, the thresholds for the Zeeman splitting switching are observed at slightly different magnitudes of the bias for the bias scans in opposite directions and thus a hysteresis loop is formed, corresponding to the hysteresis in the large nuclear polarization switched on/off at the thresholds.

The observed bias-tuning represents an interesting tool for controllable aligning of the nuclear spins. A possible origin of the effect is in the increasing average electron spin on the dot, when the bias is tuned away from the regime where the ground state of the dot is occupied with two electrons with opposite spins (XX and X^- regime) and nuclear spin pumping is suppressed. Indeed, this is observed in Fig. 3.8b where the bias-dependences of the polarization degrees of X^+ and X^0 PL are shown. Both increase from 0.25 to 0.6 when the bias is tuned from 0.1 to -0.45. At high reverse bias, the electron population in the X^+ and X^0 complexes is most likely due to the direct relaxation of optically generated electron, possessing a relatively high polarization degree. Note, that electrons and holes are captured in the dot independently. This results in a more complex dynamics at low bias (see the theory part below), where various paths are possible for generation of a state with a single electron (X^0 and X^+) including initial generation of an electron left after XX and X^- recombination. In the latter case, there is a probability that the information about the initial dominating electron spin polarization excited in the wetting layer will be completely lost.

Note, that as seen from the inset to Fig. 3.7 for the p-type diode, the spin switch threshold power increases with reverse bias, the opposite trend to what is in effect observed for the n-type structure in Fig. 3.8. The difference arises from the opposite bias-dependence of the dot charge state in the two devices. Indeed, for both of them the nuclear spin pumping is most efficient when X^+ and X^0 are excited on the dot, which is possible for high reverse biases in the n-type structure and for forward biases in the p-type diode.

Before presenting an accurate model of the optical nuclear spin pumping in the next section, we will first consider semi-qualitative explanation of the observed effects. In order to explain the nuclear switching and the bistability, we employ a model based on spin-flip assisted electron-hole radiative recombination. We will consider that the nuclear spin is pumped due to the spin flip-flop in which a single electron and nucleus exchange spins. In external magnetic field, B, this process is prohibited due to the energy conservation requirement. We thus consider a process where the electron virtually occupies an optically active state with the opposite spin and the same energy and then recombines with a spin $\pm\frac{3}{2}$ heavy hole. The rate of this process involving a single nucleus [26] is,

$$w_s = |u|^2 w_r/(\epsilon_{eZ}^2 + \tfrac{1}{4}\gamma^2). \tag{3.2}$$

Here u is a typical energy of the hyperfine interaction with a single nucleus, γ is broadening of the electron energy level, and w_r is the rate at which the bright exciton recombines on the dot. $\epsilon_{eZ} = |g_e|\mu_B[B \pm B_N(\sigma^{\pm})]$ is the electron Zeeman splitting, modified due to the Overhauser field B_N. The form of the equation 3.2 implies a feedback due to the dependence of w_s on B_N. As shown below this feedback is key to the nuclear spin switching behavior, a manifestation of the nuclear spin bistability.

For the sake of simplicity, we will first describe the situation where the electron spin is 100% preserved during the energy relaxation and in addition, only one electron and one hole can be generated on the dot. This approach will well describe the general trends observed in our experiments, whereas the accurate theory presented in the next Section will describe a more general case with various charge and spin populations of the dot. As briefly described above, circularly polarized excitation generates e-h pairs with a well defined spin. The spin of the hole is partially randomized during energy relaxation, whereas the electron has a high probability of retaining its initial spin orientation due to its weaker spin-orbit coupling to the lattice. Both dark and bright excitons can be formed on the dot with rates αw_x and $(1-\alpha)w_x$, respectively. A bright exciton recombines at a rate w_r without spin transfer to the nuclei. In contrast, a dark exciton can recombine with the electron simultaneously flipping its spin due to the hyperfine interaction: the electron virtually occupies an optically active state with the opposite spin and the same energy and then recombines with the hole. In this process, the probability of which is described by Eq. 3.2 [26], spin will be transferred to a nucleus.

The spin pumping rate $w_p \propto \alpha w_x w_s$, is thus dependent on the dark exciton generation rate, and will compete with the nuclear depolarization rate $w_d \approx 1-10$ sec^{-1} [11] due to spin diffusion away from the dot into the surrounding GaAs (see Fig. 3.2). At the same time, Eq. 3.2 shows that w_s varies with the electron Zeeman splitting ϵ_{eZ}, given by a linear function of $B_N \propto S$, the degree of nuclear polarization. For the case of σ^- excitation, polarization of the nuclei leads to a decrease of ϵ_{eZ}, and thus positive feedback and speeding up of the spin transfer process: the more spin is transferred to the nuclear system the faster becomes the spin transfer rate. By contrast for σ^+ excitation, spin transfer leads to an increase of ϵ_{eZ}, leading to a saturation of S (and B_N) at high power.

For σ^- excitation the spin-pumping rate at high w_x may exceed the depolarization rate w_d, and thus triggers a stimulated polarization process leading to an abrupt increase of the nuclear spin at the threshold P_{up}. The stimulation stops when either (i) ϵ_{eZ} starts increasing again since $B_N > B$, causing reduction of w_s or (ii) the maximum achievable $B_N = B_N^{max}$ in the given dot is reached. This explains the dependence in Fig. 3.6, where ΔE_N, and hence the nuclear field, increases at low B and saturates at high fields, from which $B_N^{max} \approx$2.5-3T can be estimated. Note that a very similar magnitude of B_N is deduced from the dependence of $\Delta E_N(B)$ using $|g_e| \approx 0.5$.

When reducing the power from beyond the P_{up} threshold, the spin pumping rate w_p first falls due to the decrease of the excitation rate, although w_s remains high. When the condition $w_p < w_d$ is reached at sufficiently low w_x, strong negative feedback is expected: further nuclear depolarization will lead to even lower w_p due to the increase in the electron Zeeman energy ϵ_{eZ}. Thus, an abrupt nuclear depolarization will take place (at the threshold P_{down}). This explains the observed hysteresis behavior in Fig. 3.7, and also accounts for the existence of a bistable state in the nuclear polarization at intermediate powers, $P_{down} < P < P_{up}$.

3.4. Theory of Nuclear Spin Bi-stability in Semiconductor Quantum Dots

In this section we develop a general theory of the nuclear spin bistability in QDs in the regime of the non-resonant optical pumping. We consider only the spin and population dynamics of the lowest states of the dot. In this case the maximum charge population on the dot is two electrons and two holes. We study the dynamics of nuclear spins in a dot populated by electrons (el) and holes (h) arriving to the ground state of the dot with the independent rates w (el) and $\tilde{w}$ (h) and polarization degrees σ (el) and $\tilde{\sigma}$ (h). The theory considers a general case which also includes a special case of $w = \tilde{w}$, $\sigma = 1$ and $\tilde{\sigma} = 0$ presented in the qualitative description of the phenomenon in the previous section. Now, by employing a comprehensive theory, we demonstrate that bistability observed in experiments on self-assembled InGaAs dots [18–20] is indeed a general phenomenon possible in a wide range of experimental conditions. Importantly, the current theory also predicts a new phenomenon arising due to the bistable behavior of the nuclear polarization: the bistability of the average charge on the dot.

Similar to the semi-qualitative description in Section3.3, we employ a model based on spin-flip assisted electron-hole radiative recombination, i.e. a process where the electron virtually occupies an optically active state with the opposite spin and the same energy and then recombines with a heavy hole. The core element of this process is described by the spin transfer rate, w_s, in Eq. 3.2. However, in this Section we will consider other possibilities for the spin-flip assisted recombination, which in addition to dark neutral excitons will also involve all spin states of positively charged excitons. The kinetic model describing the carrier population in the ground state of the dot is based on the analysis of the probabilities of its 16 allowed configurations constructed from the two electron and two hole spin states. We solve the rate equations for the populations of these states and find the steady-state magnitude of the nuclear spin polarization S.

Below we denote the probability that the dot is empty by n, and use n_μ (n^μ) for the probabilities of the dot occupation by a single electron (hole), with the index $\mu = +/-$ representing the spin state of the particle. We refer to these states as D, D_μ and D^μ, respectively. The probabilities for the dot to be occupied with two electrons or two holes (states D_{+-} and D^{+-}) are n_{+-} and n^{+-}. The probability to find the dot in a dark exciton state (X^μ_μ) is n^μ_μ, and in a bright exciton state ($X^{-\mu}_\mu$) is $n^{-\mu}_\mu$. The probability to find the dot in a negative (positive) trion state labelled as X^μ_{+-} (X^{+-}_μ) is n^μ_{+-} (n^{+-}_μ) and, finally, n^{+-}_{+-} represents the dot in the biexciton state, X^{+-}_{+-}.

We now present the balance equations for the dot population. The two equations below describe the probability of the dot occupation by a single carrier.

$$\begin{aligned}
\dot{n}_\mu &= \tfrac{1}{2}(1+\mu\sigma)wn + w_r n^\mu_{+-} - \left[\tilde{w} + \tfrac{1}{2}(1-\mu\sigma)w\right] n_\mu; \\
\dot{n}^\mu &= \tfrac{1}{2}(1+\mu\tilde{\sigma})\tilde{w}n + w_r n^{+-}_\mu + \tfrac{1}{2}(1+\mu S)Nw_s n^{+-}_{-\mu} \\
&\quad - \left[w + \tfrac{1}{2}(1-\mu\tilde{\sigma})\tilde{w}\right] n^\mu.
\end{aligned} \tag{3.3}$$

Both include gains due to the arrivals of an electron/hole into the empty dot and the recombination of a charged bright exciton, and losses due to the arrival of an electron or a hole. The second equation also has a gain due to a possible spin-flip-assisted recombination from a positive trion ($X^{+-}_\mu \to D^{-\mu}$) in which the spin is transferred to a nucleus. The process is impossible for a negative trion since in the lowest orbital state the flip-flop is blocked by the presence of the second electron. The probability for an e-h pair to recombine via spin-flip depends on the number of nuclei available (maximum 10000 in a typical InGaAs/GaAs dot), which leads to a factor

$(1+\mu S)Nw_s n_{-\mu}^{+-}$ in Eq. 3.2, where S is the degree of nuclear polarization and N is the total number of nuclei covered by the electron wave function.

The following two equations describe the dynamics of D_{+-} and D^{+-}:

$$\dot{n}_{+-} = \tfrac{1}{2}\sum_{\mu}(1-\mu\sigma)wn_{\mu} - \tilde{w}n_{+-};$$
$$\dot{n}^{+-} = \tfrac{1}{2}\sum_{\mu}(1-\mu\tilde{\sigma})\tilde{w}n^{\mu} - wn^{+-}. \quad (3.4)$$

The dynamics of the exciton states X_{μ}^{μ} and $X_{\mu}^{-\mu}$ are described by

$$\begin{aligned}\dot{n}_{\mu}^{\mu} =& \tfrac{1}{2}(1+\mu\tilde{\sigma})\tilde{w}n_{\mu} + \tfrac{1}{2}(1+\mu\sigma)wn^{\mu}\\ &-\tfrac{1}{2}\left[(1-\mu S)Nw_s + (1-\mu\sigma)w + (1-\mu\tilde{\sigma})\tilde{w}\right]n_{\mu}^{\mu};\\ \dot{n}_{\mu}^{-\mu} =& \tfrac{1}{2}(1-\mu\tilde{\sigma})\tilde{w}n_{\mu} + \tfrac{1}{2}(1+\mu\sigma)wn^{-\mu} + w_r n_{+-}^{+-}\\ &-\left[w_r + \tfrac{1}{2}(1-\mu\sigma)w + \tfrac{1}{2}(1+\mu\tilde{\sigma})\tilde{w}\right]n_{\mu}^{-\mu}.\end{aligned} \quad (3.5)$$

Both neutral bright and dark exciton populations decrease when more carriers arrive onto the dot. The neutral bright exciton can also be created and removed due to the e-h pair recombination in the processes $X_{+-}^{+-} \to X_{\mu}^{-\mu}$ and $X_{\mu}^{-\mu} \to D$, respectively. The dark exciton can decay due to the spin-flip-assisted recombination ($X_{\mu}^{\mu} \to D$) leading to spin transfer to a nucleus.

The occupation of the trion states X_{+-}^{μ} and X_{μ}^{+-} is described by

$$\begin{aligned}\dot{n}_{+-}^{\mu} =& \tfrac{1}{2}(1+\mu\tilde{\sigma})\tilde{w}n_{+-} + \tfrac{1}{2}\sum_{\nu=\pm}(1-\nu\sigma)wn_{\nu}^{\mu}\\ &-\left[w_r + \tfrac{1}{2}(1-\mu\tilde{\sigma})\tilde{w}\right]n_{+-}^{\mu};\\ \dot{n}_{\mu}^{+-} =& \tfrac{1}{2}(1+\mu\sigma)wn^{+-} + \tfrac{1}{2}\sum_{\nu=\pm}(1-\nu\sigma)wn_{\mu}^{\nu}\\ &-\left[w_r + \tfrac{1}{2}(1-\mu S)Nw_s + \tfrac{1}{2}(1-\mu\sigma)w\right]n_{\mu}^{+-}.\end{aligned} \quad (3.6)$$

Both trion populations decrease due to the recombination ($X_{\mu}^{+-} \to D^{\mu}, X_{+-}^{\mu} \to D_{\mu}$) or the arrival of a single additional charge (the ground states of the dot permit maximum four carriers). The positive trions can also recombine in the spin-flip-assisted process ($X_{\mu}^{+-} \to X^{-\mu}$), forbidden for the negative trions. Finally, the biexciton state X_{+-}^{+-} does not contribute to the nuclear spin pumping since it decays without the spin-flip ($X_{+-}^{+-} \to X_{+}^{-}, X_{-}^{+}$):

$$\begin{aligned}\dot{n}_{+-}^{+-} =& \tfrac{1}{2}\sum_{\mu}\left[(1-\mu\tilde{\sigma})\tilde{w}n_{+-}^{\mu} + (1-\mu\sigma)wn_{\mu}^{+-}\right]\\ &-2w_r n_{+-}^{+-}.\end{aligned} \quad (3.7)$$

The 15 equations above are complemented by the normalization condition $1 = n + n_{+-} + n^{+-} + n_{+-}^{+-} + \sum_\mu n_\mu + n^\mu + n_\mu^\mu + n_\mu^{-\mu} + n_{+-}^\mu + n_\mu^{+-}$. We write these equations in the form $\hat{M}\vec{n} = (1, 0, ..., 0)^T$, where the components of $\vec{n}$ are the occupation numbers and $\hat{M}$ is a 16×16 matrix with elements determined by the coefficients in Eqs. (3.3-3.7) and the normalization condition. The solutions for components of $\vec{n}$ are given by $C_{i,1}/detM$ where $C_{i,1}$ is the relevant cofactor of $\hat{M}$.

For simplicity, we consider spin $\pm\frac{1}{2}$ nuclei, since higher spins will result only in re-parameterization of N. Combining the balanced equations for the occupation numbers of the nuclei with spin up ($f_\Uparrow$) and down ($f_\Downarrow$) we will obtain the steady-state value for the nuclear polarization S (defined as $S = f_\Uparrow - f_\Downarrow$):

$$\dot{S} = I \equiv \sum_\mu \mu \left(1 - \mu S\right) \left(n_\mu^\mu + n_\mu^{+-}\right) w_s - 2Sw_d. \tag{3.8}$$

Eq. 3.8 summarizes the processes leading to the nuclear spin pumping: S (as defined above) is increased as a result of the spin-flip-assisted recombination of X_+^+ and X_+^{+-} and reduced due to a similar recombination process involving X_-^- and X_-^{+-}. Thus the balance between the populations of X_+^+ and X_+^{+-} on one hand and X_-^- and X_-^{+-} on the other will eventually define the sign of the net nuclear polarization. However, an additional important contribution to the depolarization of the nuclei has to be taken into account. It arises from their mutual dipole-dipole interaction effectively leading to the nuclear spin diffusion from the dot into the bulk semiconductor, described in our model by the rate w_d (last term in Eq. 3.8) [11]. For a dot with radius $r \approx 5nm$, we approximate $w_d \approx D_N/r^2 \approx 1 - 10s^{-1}$, where $D_N \approx \mu_n^2/\hbar a$ is the coefficient of polarization diffusion due to the dipole-dipole interaction between magnetic moments μ_n of neighbouring nuclei and $a = 0.56nm$ is the lattice constant.

We now present the analysis of the above equations, where the following important parameters will be employed:

$$x = \frac{B}{B_N^{max}}, \qquad z = 2N^2\frac{w_d}{w_r}, \qquad P = \frac{\tilde{w}}{zw_r}, \tag{3.9}$$

Here B_N^{max} is defined through the Overhauser field B_N as $B_N = B_N^{max}S$. Now Eq. 3.2 can be presented in the form:

$$w_s \equiv \frac{w_r}{N^2\left(x - S\right)^2 + \frac{1}{4}\alpha^2}, \tag{3.10}$$

where $\alpha = \gamma / g_e \mu_B B_N^{max}$. For InGaAs quantum dots used in Refs. [16–22], $w_r \approx 10^9 s^{-1}, N \approx 10^4, B_N^{max} \approx 2.5 - 3T, \alpha \approx 0.01$.

The steady-state values of S are given by the solutions of the equation $I(S) = 0$, satisfying the condition $\frac{dI}{dS} < 0$ (solutions with $\frac{dI}{dS} > 0$ are unstable). Figure 3.9 demonstrates that for a fixed external magnetic field the number of stable solutions for the nuclear spin polarization varies: it can be one or two depending on the incident power and other experimental parameters such as $w_d, \sigma, \tilde{\sigma}$ and the ratio $w/\tilde{w}$. At small powers only a single low value of S is possible. At high powers when two stable solutions become possible, including one with a large S, the dot enters the regime of the nuclear spin bistability.

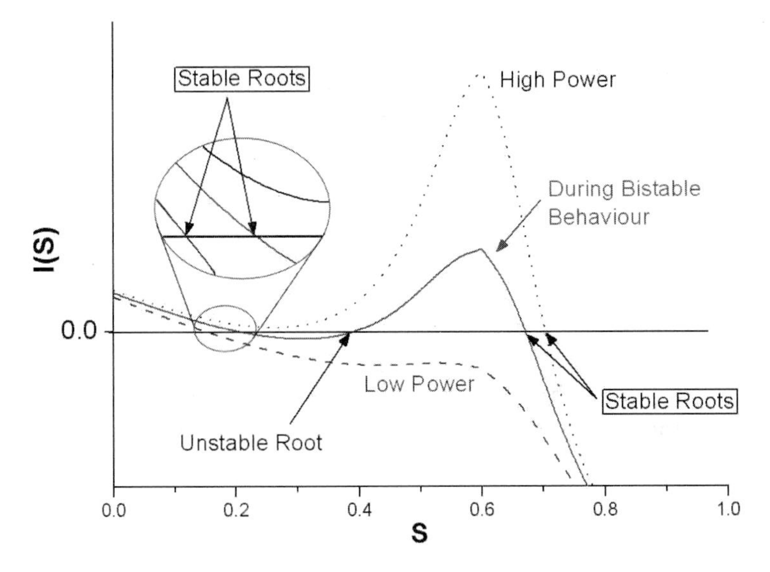

Fig. 3.9. The function $I(S)$ for the situation where $w = \tilde{w}, \sigma = 0.9, \tilde{\sigma} = -0.2, x = 0.6$ and $z = 8$ for three different powers: $P = 0.0001, 0.0003, 0.0005$. Stable roots correspond to the solutions of $I(S) = 0$ where $\frac{dI}{dS} < 0$.

The bottom parts of Figs. 3.10(a) and (b) show the calculated evolution of the nuclear polarization in a dot for realistic magnitudes of z in the regime where electrons have a high degree of spin memory and arrive with the same rate as the holes. Fig. 3.10(a) contains a large hysteresis loop in the power dependence of S for a fixed magnetic field (we consider $x = 0.6$), similar to those observed in experiment [18, 19]. The bistable behavior occurs for a wide range of z: $5 \lesssim z \lesssim 14$. Experimentally, the evolution of S with power

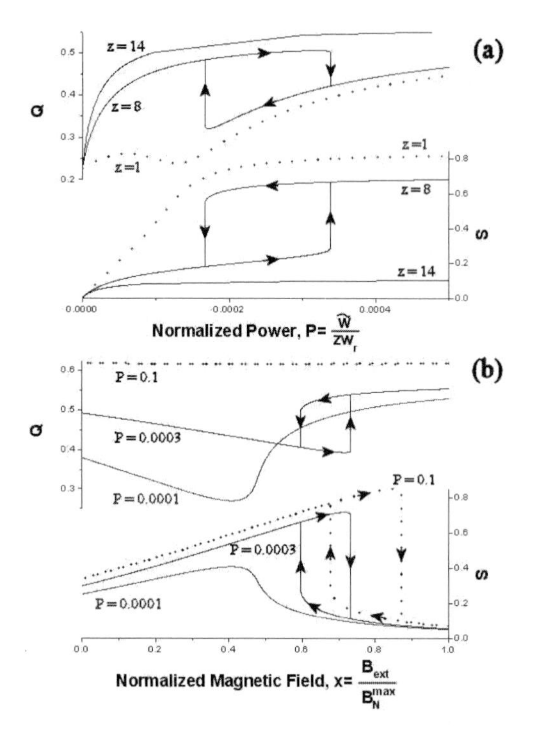

Fig. 3.10. Evolution of nuclear polarization (S) and the average charging state of the dot (Q) for $w = \tilde{w}, \sigma = 0.9$ and $\tilde{\sigma} = -0.2$: (a) as a function of power for $x = 0.6$ and various values of z, with the arrows indicating a forwards or backwards sweep. Although not shown in the figure, at high powers $P \approx 1$, both Q and S start to decrease due to the dot being dominated by the biexciton (for which the spin-flip process is blocked); (b) as a function of magnetic field for $z = 8$ and various power values.

can be detected in polarization-resolved PL experiments on individual self-assembled InGaAS/GaAs quantum dots, where S can be deduced from the measured exciton Zeeman splitting.

We also find that the bistability in S leads to a novel phenomenon: a nuclear-spin induced hysteresis in the average dot charge, Q (see top parts of Fig. 3.10a,b). This occurs in the case when the electrons arriving to the dot have a high degree of spin polarization, permitting their recombination with one spin orientation of weakly polarized holes only. Thus an extra hole with the opposite spin is likely to remain on the dot, leading to on average positive dot charge. The enhancement of the spin-flip-assisted recombination for large S, removing such holes, will result in reduction of the charge. Therefore the hysteresis in S will be reflected as a hysteresis in the average dot charge. A similar bistable behaviour in both S and Q can also be found if the external magnetic field is varied [20] at a fixed optical pump power, as shown in Fig. 3.10(b) for $z = 8$.

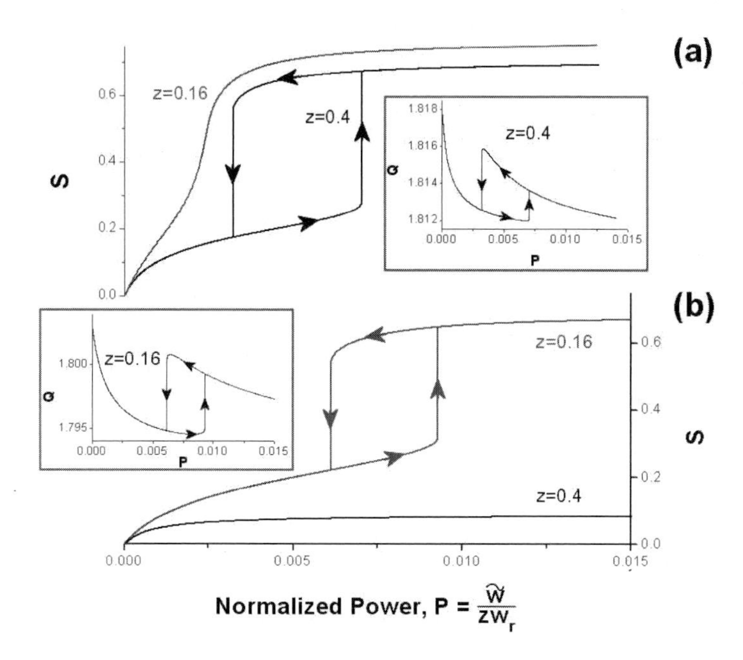

Fig. 3.11. Evolution of S with P in the regime where $\tilde{w} = 10w$ for $z = 0.05, 0.4$ and different polarizations of arriving electrons/holes. (a) $\sigma = 0.9, \tilde{\sigma} = -0.2$. The inset shows the evolution of the charging state of the dot for $z = 0.4$ (with a very small hysteresis loop). (b) Same for $\sigma = 0.45, \tilde{\sigma} = -0.1$, showing that bistability in the nuclear polarization can occur even in unfavourable regimes.

Figure 3.11 illustrates that the range of parameters for which the bistability can occur strongly depends on the ratio between the arrival rates of electrons and holes, w and $\tilde{w}$, as well as on their polarizations, σ and $\tilde{\sigma}$. In experiment the ratio $w/\tilde{w}$ can be varied by applying electric field in a diode containing QDs in the intrinsic region [16–20]: Because of a light effective mass, electrons can tunnel to the contacts before relaxing to the dot ground state, which in effect reduces their arrival rate as compared to that of the holes. Fig. 3.11(a) shows the evolution of $S(P)$ for $w = 0.1\tilde{w}$. The dot is dominated by the state D^{+-} so that the average charge of the dot is large ($\approx +1.8$) and exhibits a weak power-dependence with a negligible hysteresis loop (see inset), despite a pronounced hysteresis loop in the nuclear polarization. As seen from the figure, for such low values of

$w/\tilde{w}$ higher powers are required to pump a significant nuclear polarization and the spin bistability is observed for smaller values of z ($0.05 \lesssim z \lesssim 0.5$). Figure 3.11(b) illustrates that when the polarizations of both electrons and holes is reduced by 50% the bistability can still be observed but in a smaller range of parameter z $0.01 \lesssim z \lesssim 0.05$.

3.5. Conclusions

To summarize, we have observed a strong optically induced bistability of the nuclear spin polarization in self-assembled InGaAs QDs and have developed a general theory for this bistability phenomenon in optically active quantum dots. In our experiments, the nuclear spin bistability has been observed for optically pumped InGaAs/GaAs QDs at temperatures $T = 15 - 30$K and external magnetic fields $B = 1 \div 3$T. The phenomenon leads to a characteristic threshold-like switching of the nuclear spin polarization in a dot pumped with circularly polarized light, which can be described as an abrupt change of up to 3T of the local magnetic field seen by the confined electron. The effect arises due to the strong feedback of the nuclear spin polarization on the dynamics of the electron-nuclear spin transfer accompanying the radiative recombination process, which is accelerated when the Overhauser and external magnetic fields cancel each other. We show that nuclear magnetic fields of a few Tesla can be switched on and off in individual dots by varying one of three external controlling parameters: electric and magnetic fields and intensity of circularly polarized excitation. The presented theory predicts that such bistable nuclear spin behavior can be observed in non-resonantly optically pumped semiconductor quantum dots with a wide range of dot properties, including the number of nuclei, e-h radiative recombination life-time and nuclear spin diffusion rate. We therefore suggest that the nuclear spin bistability is a general phenomenon for dots pumped with circularly polarized light. In addition, we find that the nuclear spin polarization can also strongly influence the charge dynamics in the dot leading to bistability of the average dot charge.

Acknowledgments

Many people directly contributed to this work including Maxim Makhonin, Tim Wright, Ilias Drouzas, Alexander Van'kov, Joanna Skiba-Szymanska, Roman Kolodka, Hui-Yun Liu, Paul Fry, Abbes Tahraoui, and Mark Hopkinson. We are grateful for their enthusiasm and skill. This work has been supported by the Sheffield EPSRC Programme grant GR/S76076, the Lancaster-EPSRC Portfolio Partnership EP/C511743, the EPSRC IRC for Quantum Information Processing, ESF-EPSRC network EP/D062918, and by the Royal Society. AIT was supported by the EPSRC Advanced Research Fellowship EP/C54563X/1 and research grant EP/C545648/1.

References

[1] S. A. Wolf, D. D. Awschalom, R. A. Buhrman, J. M. Daughton, S. von Molnar, M. L. Roukes, A. Y. Chtchelkanova, D. M. Treger, Science **294**, 1488 (2001).
[2] A. Greilich, D. R. Yakovlev, A. Shabaev, Al. L. Efros, I. A. Yugova, R. Oulton, V. Stavarache, D. Reuter, A. Wieck, M. Bayer, Science **313**, 341 (2006).
[3] F. H. L. Koppens, J. A. Folk, J. M. Elzerman, R. Hanson, L. H. Willems van Beveren, I. T. Vink, H. P. Tranitz, W. Wegscheider, L. P. Kouwenhoven, L. M. K. Vandersypen, Science **309**, 1346-1350 (2005).
[4] F. H. L. Koppens, C. Buizert, K. J. Tielrooij, I. T. Vink, K. C. Nowack, T. Meunier, L. P. Kouwenhoven, L. M. K. Vandersypen, Nature **442**, 766 (2006).
[5] J. R. Petta, A. C. Johnson, J. M. Taylor, E. A. Laird, A. Yacoby, M. D. Lukin, C. M. Marcus, M. P. Hanson, A. C. Gossard, Science **309**, 2180-2184 (2005).
[6] M. Atature, J. Dreiser, A. Badolato, A. Hogele, K. Karrai, A. Imamoglu, Science **312** 551 (2006).
[7] M. Krouvtar, Y. Ducommun, D. Heiss, M. Bichler, D. Schuh, G. Abstreiter, J. J. Finley, Nature **432**, 81 (2004).
[8] A. W. Overhauser, Phys. Rev. **92**, 411 (1953).
[9] F. Meier, B. P. Zakarchenya, *Optical Orientation.* (Elsevier, New York, 1984).
[10] D. Paget, G. Lampel, B. Sapoval, V. I. Safarov, Phys. Rev. **B 15**, 5780 (1977).
[11] D. Paget, Phys. Rev. **B 25**, 4444 (1982).
[12] D. Gammon, S. W. Brown, E. S. Snow, T. A. Kennedy, D. S. Katzer, D. Park, Science **277**, 85 (1997).
[13] S. W. Brown, T. A. Kennedy, D. Gammon, E. S. Snow, Phys. Rev. **B 54**, 17339 (1996).
[14] T. Yokoi, S. Adachi, H. Sasakura, S. Muto, H. Z. Song, T. Usuki, S. Hirose, Phys. Rev. **B 71**, 041307(R) (2005).
[15] A. S. Bracker, E. A. Stinaff, D. Gammon, M. E. Ware, J. G. Tischler, A. Shabaev, Al. L. Efros, D. Park, D. Gershoni, V. L. Korenev, I. A. Merkulov, Phys. Rev. Lett. **94**, 047402 (2005).
[16] C. W. Lai, P. Maletinsky, A. Badolato, A. Imamoglu, Phys. Rev. Lett. **96**, 167403 (2006).

[17] B. Eble, O. Krebs, A. Lemaitre, K. Kowalik, A. Kudelski, P. Voisin, B. Urbaszek, X. Marie, T. Amand, Phys. Rev. **B 74**, 081306(R) (2006).
[18] P.-F. Braun, B. Urbaszek, T. Amand, X. Marie, O. Krebs, B. Eble, A. Lemaitre, P. Voisin, Phys. Rev. **B 74**, 245306 (2006).
[19] A. I. Tartakovskii, T. Wright, A. Russell, V. I. Falko, A. B. Vankov, J. Skiba-Szymanska, I. Drouzas, R. S. Kolodka, M. S. Skolnick, P. W. Fry, A. Tahraoui, H.-Y. Liu, M. Hopkinson, Phys. Rev. Lett. **B 98**, 026806 (2006).
[20] P. Maletinsky, P. Maletinsky, C. W. Lai, A. Badolato, A. Imamoglu, Phys. Rev. **B 75**, 035409 (2007).
[21] P.-F. Braun, X. Marie, L. Lombez, B. Urbaszek, T. Amand, P. Renucci, V. K. Kalevich, K.V. Kavokin, O. Krebs, P. Voisin, Y. Masumoto, Phys. Rev. Lett. **94**, 116601 (2005).
[22] R. Oulton, A. Greilich, S. Yu. Verbin, R. V. Cherbunin, T. Auer, D. R. Yakovlev, M. Bayer, I. A. Merkulov, V. Stavarache, D. Reuter, A. D. Wieck, Phys. Rev. Lett. **98**, 107401 (2007).
[23] I. Akimov, D. H. Feng, F. Henneberger, Phys. Rev. Lett. **97**, 056602 (2006).
[24] G. Yusa, K. Muraki, K. Takashina, K. Hashimoto, Y. Hirayama, Nature **434**, 1001 (2005).
[25] A. C. Johnson, J. R. Petta, J. M. Taylor, A. Yacoby, M. D. Lukin, C. M. Marcus, M. P. Hanson, A. C. Gossard, Nature **435**, 925 (2005).
[26] S. I. Erlingsson, Y. V. Nazarov, V. I. Fal'ko, Phys. Rev. **B 64**, 195306 (2001).
[27] A. V. Khaetskii, D. Loss, L. Glazman, Phys. Rev. Lett. **88**, 186802 (2002).
[28] I. A. Merkulov, A. L. Efros, M. Rosen, Phys. Rev. **B 65**, 205309 (2002).
[29] A. Imamoglu, E. Knill, L. Tian, P. Zoller, Phys. Rev. Lett. **91**, 017402 (2003).
[30] S. E. Barrett, R. Tycko, L. N. Pfeiffer, K. W. West, Phys. Rev. Lett. **72**, 1386 (1994).
[31] H. Sanada, S. Matsuzaka, K. Morita, C. Y. Hu, Y. Ohno, H. Ohno, Phys. Rev. Lett. **94**, 097601 (2005).
[32] G. Salis, D. T. Fuchs, J. M. Kikkawa, D. D. Awschalom, Y. Ohno, H. Ohno, Phys. Rev. Lett. **86**, 2677 (2001).
[33] M. Dobers, K. v. Klitzing, J. Schneider, G. Weimann, K. Ploog, Phys. Rev. Lett. **61**, 1650 (1988).
[34] D.C. Dixon, K. R. Wald, P. L. McEuen, M. R. Melloch, Phys. Rev. **B 56**, 4743 (1997).
[35] T. Machida, S. Ishizuka, T. Yamazaki, S. Komiyama, K. Muraki, and Y. Hirayama, Phys. Rev. **B 65**, 233304 (2002).
[36] J. H. Smet, R. A. Deutschmann, F. Ertl, W. Wegscheider, G. Abstreiter, K. von Klitzing, Nature **415**, 281 (2002).
[37] G. Yusa, K. Hashimoto, K. Muraki, T. Saku, Y. Hirayama, Phys. Rev. **B 69**, 161302(R) (2004).
[38] H. Sanada, S. Matsuzaka, K. Morita, C. Y. Hu, Y. Ohno, H. Ohno, Phys. Rev. **B 68**, 241303(R) (2003).
[39] K. Ono, S. Tarucha, Phys. Rev. Lett. **92**, 256803 (2004).
[40] R. Oulton, J. J. Finley, A. D. Ashmore, I. S. Gregory, D. J. Mowbray, M. S. Skolnick, M. J. Steer, San-Lin Liew, M. A. Migliorato, A. J. Cullis Phys.

Rev. **B 66**, 045313 (2002).
[41] R. J. Warburton, C. Schäflein, D. Haft, F. Bickel, A. Lorke, K. Karrai, J. M. Garcia, W. Schoenfeld, P. M. Petroff, Nature **405**, 926 (2000).

Chapter 4

Nonequilibrium Optical Spin Cooling in Charged Quantum Dots

I. A. Akimov, D. H. Feng, and F. Henneberger

Institut für Physik, Humboldt Universität zu Berlin, Newtonstr.15, 12489 Berlin, Germany

Optical spin pumping and the hyperfine dynamics is investigated in the situation where the spin of a localized electron interacts only with a few hundred nuclear moments in the surrounding lattice. Both pumping and read-out of the spin state is accomplished via the trion feature. The formation of a dynamical nuclear polarization as well as its subsequent decay by the dipole-dipole interaction is directly resolved in time. Because not limited by intrinsic nonlinearities, polarization degrees as large as 50 % are achieved, even at elevated temperatures. The data signify a nonequilibrium mode of nuclear polarization, distinctly different from the standard spin temperature concept.

Contents

4.1. Introduction

The electron spin is a fundamental digital unit that can be utilized to store and to process information. Practical applications require large spin polarizations without the need of strong external magnetic fields and at temperatures as high as possible. This demands concepts that work far from thermal equilibrium, generally summarized under the term "spin cooling".

In this context, a resident electron in a charged quantum dot (QD) is a promising candidate. Since largely isolated from the environment, processes that limit the electron spin life-time in bulk semiconductors are strongly suppressed. Moreover, a QD can be selectively addressed by optical or electrical injection, allowing thus for spin manipulation on the nanometer length scale. The electron spin is coupled by the contact hyperfine interaction to the nuclear moments of the lattice atoms. This coupling has been extensively studied many years ago on bulk semiconductors [1, 2]. A prominent effect is the formation of a dynamical nuclear polarization (DNP). Strongly polarized nuclei increase the electron spin life-time. In addition, the hyperfine interaction offers the possibility to manipulate the state of the nuclear ensemble and to use it as a long-lived quantum memory as well.

Table 4.1. Hyperfine parameters of CdSe and InAs/GaAs QD structures. I, nuclear spin quantum number; a, isotope abundance; μ_I, nuclear magnetic moments in units of nuclear magneton μ_N; A, hyperfine coupling constant; N_L, number of lattice atom.

		I	$a(\%)$	μ_I	$A(\mu eV)$	N_L
CdSe	Cd	1/2	25.01	-0.608	~ -10	10^3
	Se	1/2	7.58	0.534		
InAs GaAs	In	9/2	100	5.533	~ 100	$10^5 - 10^6$
	As	3/2	100	1.439		
	Ga	3/2	100	2.232		

The hyperfine interaction is controlled by the coupling constant A of the unit cell, the number of lattice atoms N_{L} seen by the electron, and the isotope abundance a [2]. Regarding these parameters, prototype II-VI QD structures stand for a situation quite different from their III-V counterparts (see Table 4.1). Weaker coupling, a 1/2 spin of the dominant isotope, low abundance, and, in particular, an orders of magnitude smaller number of atoms per dot create a specific scenario that is the subject of this contribution.

Optical excitation of charged QDs generates three-particle states called trions, consisting of two electrons and one hole. Their electronic structure is essential for optical spin control and will be thus considered in some detail first. Based on this knowledge, optical pumping of the resident electron spin by circular excitation photons is addressed next. Finally, the dynamics of the hyperfine interaction is summarized. The formation of a DNP as well as its subsequent decay by the dipole-dipole interaction is directly resolved in time. Because not limited by intrinsic nonlinearities related to the Overhauser field, large nuclear polarizations are achieved, even at elevated lattice temperatures. The data demonstrate a nonequilibrium mode of DNP, distinctly different from the standard spin temperature concept exploited on bulk semiconductors.

4.2. Optical Excitations in Charged Quantum Dots

The CdSe/ZnSe QD structures are grown by molecular beam epitaxy [3]. Making use of the native n-type of these materials, dots charged with a resident electron are formed under appropriate stoichiometry conditions. The dot height is below 2 nm and the lateral extension < 10 nm [4]. In ensemble measurements, finestructures of the optical features are generally hidden by the inhomogeneous broadening resulting from size, shape, composition, and strain fluctuations. The number of QDs under study can be systematically reduced down to a single-dot level by fabricating mesa structures of different sizes (2 μm - 100 nm) addressed optically in a confocal arrangement. The propagation direction of both excitation and detected photons is always along the growth direction of the structure, for more experimental details see, e.g., Ref. 5.

The fundamental excitation of a charged QD is the trion. Unlike the exciton, the trion has half integer spin. According to the Kramers theorem, its eigenstates are degenerate with respect to the sign of the spin projection. The energy states and optical transitions of relevance in the present context are schematized in Fig. 4.1. The photoluminescence (PL) emerges form the singlet-type trion ground-state consisting of two electrons with anti-parallel spins in the lowest electron shell and one hole in the lowest heavy-hole shell. The total spin projection F_z is thus defined by the hole with $J_z = \pm 3/2$. In the radiative recombination, a single electron with spin $S_z = \pm 1/2$ is left behind. Out of the four possibilities, only the two transitions $|3/2\rangle \rightarrow \sigma^+ + |1/2\rangle$ and $|-3/2\rangle \rightarrow \sigma^- + |-1/2\rangle$ are allowed by angular momentum conservation where $\sigma^\pm$ denotes the circular polarization orientation of the photons emitted.

Fig. 4.2 depicts typical single-dot trion PL spectra, also in presence of an external magnetic field. The exciton state in uncharged QDs exhibits a characteristic finestructure generated by the electron-hole (eh) exchange interaction. It shows most prominently up by a zero-field splitting of the emission feature. In contrast, such splitting is absent for the trion PL which appears as a single line labeled by X^-. The exchange of the hole with each

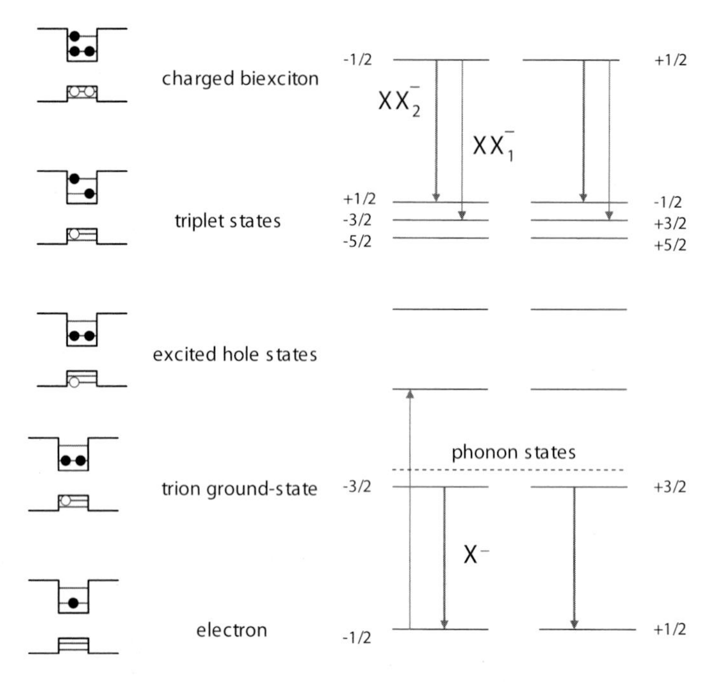

Fig. 4.1. Schematics of energy levels and optical transitions in a QD charged with a single electron. For explanations see text.

of the two electrons compensates to zero because of their opposite spin projections. The p-character of the hole states imposes a spin quantization axis oriented along the [001] growth direction of the QD structures. A longitudinal magnetic field (Faraday geometry) reveals the two recombination pathways of the trion degenerate at zero field and their opposite circular polarization. A transverse magnetic field (Voigt geometry) creates mixtures of the pure spin states so that all four possible transitions become visible. The smaller splitting within each of the two main PL components signifies a noticeable in-plane g factor of the hole, the origin of which is addressed below.

Excited trion states are built-up by holes and/or electrons occupying energetically higher single-particle shells. In view of the larger hole mass, states with an excited hole appear next to the ground-state. Their spin is again defined by the hole. In case of electron excitation, new spin configurations are formed. The electron-electron exchange interaction with typical energies in the 10-meV range splits the excited-electron trion into a singlet state with total electron spin $S_{\text{tot}} = 0$ and a triplet state $S_{\text{tot}} =$

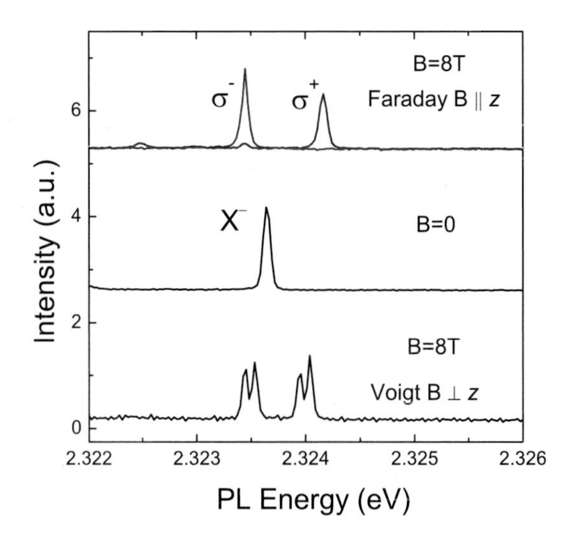

Fig. 4.2. Single-dot PL of the trion (X^-). The data are taken under non-resonant excitation at a temperature $T = 2$ K.

1 ($S_z^{tot} = 0, \pm 1$) situated at lower energies [6]. Accounting for the hole, the triplet state comprises the spin configurations $F_z = \pm 1/2, \pm 3/2, \pm 5/2$. While absent in the ground-state, the eh exchange creates a finestructure of the triplet state [7, 8]. The isotropic part is diagonal in F_z and splits the three doublets equidistantly apart by an energy $\tilde{\Delta}_0$.

The emission from the charged biexciton state appearing at higher excitation levels directly reflects the finestructure of the triplet state. In the ground-state of this five-particle complex, the third electron must occupy the second shell to comply with the Pauli principle. The spins of the two other electrons as well as the two holes are compensated so that the total spin of the charged biexciton is defined by the extra electron and has thus the projections $\pm 1/2$. The most probable optical transition is recombination of an electron and a hole from the lowest shells, i.e., the trion triplet state is the final state of the charged biexciton recombination. Angular momentum conservation can be only fulfilled for $F_z = \pm 1/2, \pm 3/2$. Note that the trion transitions themselves obey the same selection rules, since the final-state electron has also 1/2 spin. Indeed, as shown in the left part of Fig. 4.3, a double feature XX_1^- and XX_2^- emerges energetically below X^- [5]. The separation of about 2 meV compares well with the isotropic exchange energy of the exciton in uncharged QDs [9].

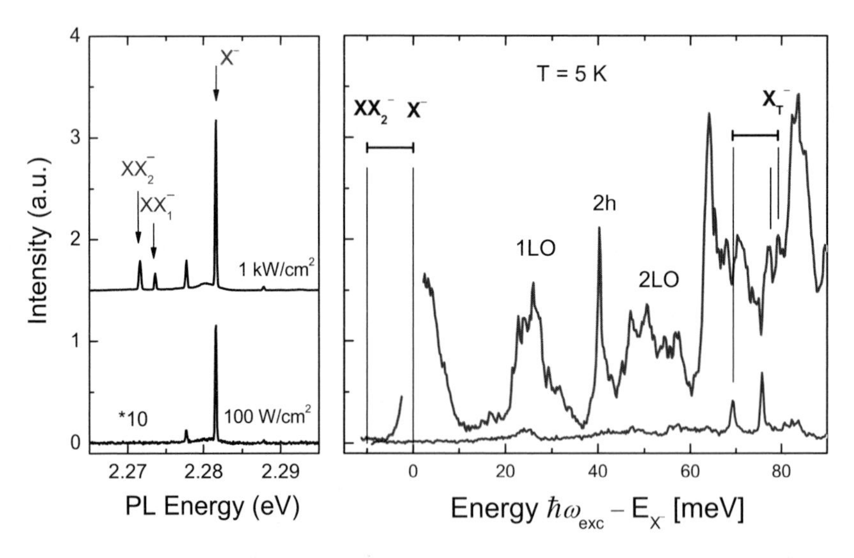

Fig. 4.3. Left: Emergence of the charged biexciton emission XX_1^- and XX_2^- at higher excitation density. Right: Excitation spectra of the trion (blue) and charged biexciton (red) PL. The position of X^- is taken as origin of the energy scale. The horizontal bars indicate the separation of XX_1^- and XX_2^- in PL. All data are single-dot.

Optical spin pumping requires knowledge of the absolute energy positions of the states addressed. This goal can be achieved by photoluminescence excitation (PLE) spectroscopy. The right part of Fig. 4.3 depicts PLE spectra recorded at the position of the trion (X^-) and the charged biexciton (XX_2^-). The trion spectrum is rich in structure. Broad features related to acoustic phonons in close proximity of the PL as well as to LO-phonons are present. These combined trion-phonon states allow for selective excitation of a particular QD with minimum excess energy. This type of excitation is preferentially used in the spin pumping measurements and will be denoted by "quasi-resonant". The narrow peak labeled "2h" is associated with the first excited hole state. At higher energy, the spectrum becomes increasingly complex, making a reliable assignment difficult. The PLE spectra taken at the charged biexciton features XX_1^- and XX_2^- are virtually identical. There is a marked energy gap with practically no absorption. The first distinct peak results from cascaded two-photon absorption [10]: The first photon creates an excited trion that rapidly relaxes into the groundstate where the second photon is resonantly absorbed generating a charged biexciton. The next higher peak is due to an excited two-exciton state.

The charged biexciton PL is low-energy shifted from the trion ground-state PL line by $W = W_{\mathrm{XX}^-} - W^{\mathrm{T}}_{\mathrm{X}^-} - W^{\mathrm{G}}_{\mathrm{X}^-}$ comprising the Coulomb energies of the eh complexes involved in the transitions (W_{XX^-}: charged biexciton, $W^{\mathrm{T}}_{\mathrm{X}^-}$: trion triplet, $W^{\mathrm{G}}_{\mathrm{X}^-}$: trion ground-state). The trion triplet state is high-energy shifted by the same W with respect to first peak of the charged biexciton PLE [10, 11]. Indeed, two weak but clearly resolvable peaks with the energy separation of XX^-_1 and XX^-_2 are present at this position in the trion PLE spectrum of Fig. 4.3. The data uncover hence a singlet-triplet separation of about 80 meV. The oscillator strength of the direct triplet transition is small as electron and hole are excited in shells of different parity.

A further point important for optical spin pumping is the circular polarization selectivity of the excitation. The strict circular selection rules described so far are undermined by in-plane size and strain anisotropies in the QD structures reducing the symmetry from $\mathrm{D}_{2\mathrm{d}}$ to $\mathrm{C}_{2\mathrm{v}}$ or even below. In this case, the anisotropic eh exchange comes into play. It couples the radiative triplet states according to $\Delta F_z = \pm 2$ providing mixed states of the type $a|\pm\frac{3}{2}\rangle + b|\mp\frac{1}{2}\rangle$ [7, 8]. The mixing gives rise to elliptically polarized transitions with a linear degree ρ_{l} determined by the weights a and b. The charged biexciton lines XX^-_1 and XX^-_2 exhibit indeed considerable linear polarization (Fig. 4.4a). The weights are functions of the ratio $\tilde{\Delta}_1/\tilde{\Delta}_0$ where $\tilde{\Delta}_1$ is the anisotropic eh exchange energy. A fit to the data yields $\tilde{\Delta}_1 \sim$ 300-600 μeV. The anisotropic exchange coupling is equivalent to a combined eh spin flip in the trion state and, as will be discussed below, essentially determines the relaxation pathway of optically excited eh pairs in charged QDs.

A further consequence of the in-plane anisotropy is an enhanced heavy-light hole coupling [12]. This coupling is responsible for the transverse hole g factor seen in Fig. 4.2. It produces mixed spin states $|\pm 3/2\rangle + (\gamma^{\pm}/\Delta E_{\text{l-h}})|\mp 1/2\rangle$ where $\gamma^{\pm}$ is the complex coupling parameter and $E_{\text{l-h}}$ the energy separation between the heavy and light hole shell. In this way, trion transitions below the triplet energy become also partially linearly polarized. Again, this is observed experimentally. The linear degree of the trion ground-state can become as large as 0.4, but scatters considerably across the ensemble. In Fig. 4.4a, a QD with a low degree of only 0.05 is selected. The linear polarization of the first excited hole state is generally more pronounced and can come close to $\rho_{\mathrm{l}} = 1$. Even for the weakly anisotropic QD of Fig. 4.4a, it reaches a value of 0.4. The polarization axes are not identical with those of the charged biexciton PL demonstrating that

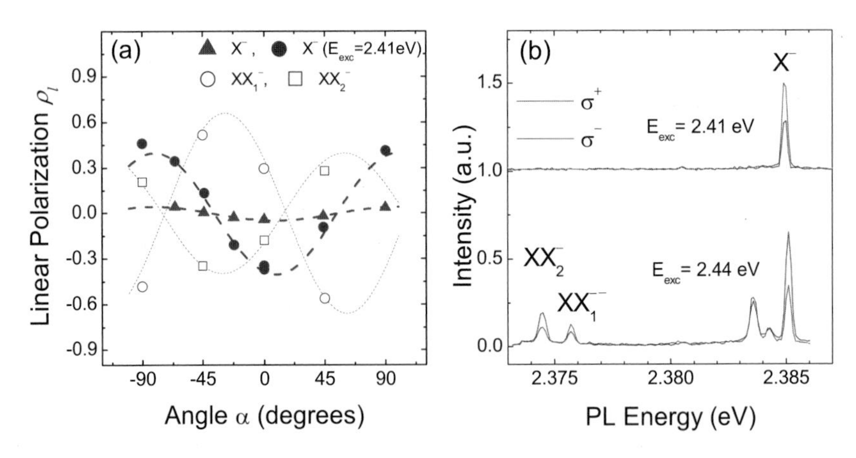

Fig. 4.4. Polarization properties of the PL features of charged QDs. (a) Linear degree $\rho_l = [I(\alpha) - I(\alpha + 90°)]/[I(\alpha) + I(\alpha + 90°)]$, α is the angle between the polarization axis probed and the [110] direction. Black open symbols: Charged biexciton components, blue triangles: trion ground-state, red dots: excited 2h trion state. In the last case, α is related to the excitation polarization, while it refers to detection otherwise. (b) Circular polarization. The spectra are taken in σ^+ (blue) σ^- (red) detection under σ^+ excitation. Top: Quasi-resonant excitation in the trion-LO-phonon state ($\rho_c \approx 0.3$). Bottom: Excitation at higher energies (about 65 meV above X^-) in the biexciton resonance.

the origin of the linear polarization is indeed different.

The circular polarization degree of the PL directly reflects the populations of the trion spin states and is thus of immediate relevance for the spin pumping. It is defined by $\rho_c = (I_+ - I_-)/(I_+ + I_-)$ where $I_\pm$ refers to the line intensity in $\sigma^\pm$ polarization. Fig. 4.4b shows that ρ_c can be both positive and negative with respect to the circular excitation polarization depending on the excitation photon energy. Pumping with σ^+ polarization via the 1-LO feature clearly below the triplet states, it holds $I_+ > I_-$ or $\rho_c > 0$. At higher excitation energies, when the trion triplet becomes involved, ρ_c changes sign. The role of the triplet states in the formation of a negative circular polarization has been addressed in various studies [14–16]. An interesting finding in this regard is that direct excitation of the charged biexciton by cascaded two-photon absorption (Fig. 4.4b) provides also $\rho_c < 0$ at the trion PL. Excitation by σ^+ photons creates selectively a spin-down biexciton state which decays subsequently in the -3/2 or 1/2 triplet state by the emission of a σ^- or σ^+ photon, respectively. Exactly these polarizations are observed for XX_1^- and XX_2^-.

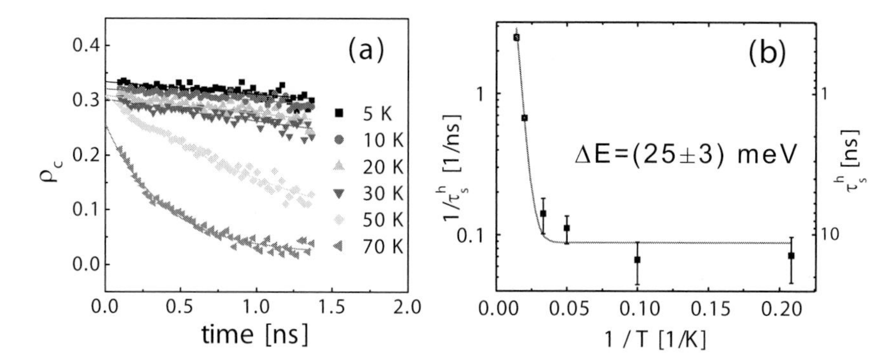

Fig. 4.5. Decay of circular polarization degree after short-pulse quasi-resonant σ^+ excitation of X^- at various temperatures. The solid lines are exponential fits to deduce the longitudinal hole spin life-time τ_s^h. (b) Temperature dependence of the spin-flip rate $1/\tau_s^h$. Solid curve is a fit with $[0.1 + 84 \cdot \exp(\Delta E/k_B T)]$ ns^{-1}.

The spin dynamics of the hole in the trion ground-state can be directly pursued by time-resolving the circular-polarization degree after short pulse excitation [17]. Quasi-resonant excitation ensures a sufficiently large initial degree. As seen from Fig. 4.5, almost no decay of ρ_c is observed at low temperature. The longitudinal life-time of the hole spin is beyond the detection window where a processible PL signal can be observed and thus longer than 10 ns. Data recorded in a longitudinal magnetic field confirm this result. Increasing temperature causes the spin-flip rate to grow with an activation energy corresponding to the energy of the LO-phonon. Already at 70 K, the spin life-time is as short as 400 ps. The hole spin-flip in the LO-phonon-trion state is obviously fast, probably a result of its large disorder-induced broadening (8 meV), see Ref. 17.

4.3. Optical Electron Spin Pumping

The finestructure and polarization properties of the trion states reviewed above make it possible to orient the spin of the resident electron by optical excitation. The transition scheme underlying the spin pumping is depicted in Fig. 4.6. Unlike the V-type exciton transition starting from a common ground-state, it consists of two arms corresponding to the two possible spin projections of the electron.

We neglect heavy-light hole coupling and ignore momentarily the hyperfine interaction. The elements of the spin density matrix are denoted by ϱ_{ij} where the indices refer to the projections of either the electron $(\downarrow, \uparrow)$ or the

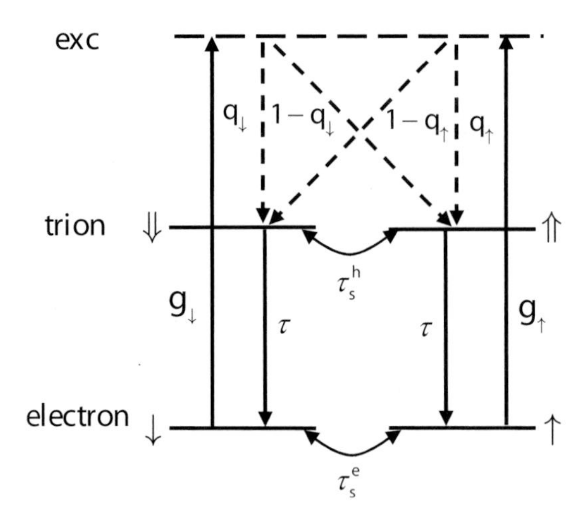

Fig. 4.6. Schematics of optical spin pumping of a negatively charged QD. For explanations see text.

hole $(\Downarrow, \Uparrow)$ in the trion ground-state, respectively. The master equations for the electron spin populations read as

$$\begin{aligned}\dot{\varrho}_{\uparrow\uparrow} &= \frac{\varrho_{\Uparrow\Uparrow}}{\tau} - g_{\uparrow}\varrho_{\uparrow\uparrow} - \frac{\varrho_{\uparrow\uparrow} - \varrho_{\downarrow\downarrow}}{2\tau_{\mathrm{s}}^{\mathrm{e}}}, \\ \dot{\varrho}_{\downarrow\downarrow} &= \frac{\varrho_{\Downarrow\Downarrow}}{\tau} - g_{\downarrow}\varrho_{\downarrow\downarrow} + \frac{\varrho_{\uparrow\uparrow} - \varrho_{\downarrow\downarrow}}{2\tau_{\mathrm{s}}^{\mathrm{e}}}.\end{aligned} \tag{4.1}$$

Here, g_j is the rate at which the electron spin level j is depopulated by optical excitation, τ the radiative life-time of the trion, and $\tau_{\mathrm{s}}^{\mathrm{e}}$ summarizes all spin relaxation processes of the electron aside from the hyperfine interaction. The populations $\varrho_{\Uparrow\Uparrow}$ and $\varrho_{\Downarrow\Downarrow}$ depend on the relaxation processes taking the trion from the state of excitation labeled by "exc" in Fig. 4.6 down to its ground-state. If these processes are sufficiently fast, one can condense the whole relaxation in two parameters $q_{\uparrow}$ and $q_{\downarrow}$ which are the probabilities that the trion arrives at the ground-state of the same arm, $\Uparrow$ or $\Downarrow$, where the electron was de-excited, while $(1 - q_{\uparrow})$ and $(1 - q_{\downarrow})$ are the probabilities of cross-relaxation. It follows then

$$\begin{aligned}\dot{\varrho}_{\Uparrow\Uparrow} &= q_{\uparrow}g_{\uparrow}\varrho_{\uparrow\uparrow} + (1 - q_{\downarrow})g_{\downarrow}\varrho_{\downarrow\downarrow} - \frac{\varrho_{\Uparrow\Uparrow}}{\tau} - \frac{\varrho_{\Uparrow\Uparrow} - \varrho_{\Downarrow\Downarrow}}{2\tau_{\mathrm{s}}^{\mathrm{h}}}, \\ \dot{\varrho}_{\Downarrow\Downarrow} &= (1 - q_{\uparrow})g_{\uparrow}\varrho_{\uparrow\uparrow} + q_{\downarrow}g_{\downarrow}\varrho_{\downarrow\downarrow} - \frac{\varrho_{\Downarrow\Downarrow}}{\tau} + \frac{\varrho_{\Uparrow\Uparrow} - \varrho_{\Downarrow\Downarrow}}{2\tau_{\mathrm{s}}^{\mathrm{h}}},\end{aligned} \tag{4.2}$$

where $\tau_{\mathrm{s}}^{\mathrm{h}}$ stands for the hole spin-flip time. The radiative life-time of $\tau \sim$

500 ps is, as will be seen below, much shorter than the time-scale of the electron spin dynamics. It is thus justified to treat the trion populations adiabatically, i.e., they follow instantaneous the slow change of the electron spin populations so that always $\dot{\varrho}_{\Uparrow\Uparrow} = \dot{\varrho}_{\Downarrow\Downarrow} = 0$. For weak excitation, inversion of the electron-trion transition can be neglected ($\varrho_{\Uparrow\Uparrow} + \varrho_{\Downarrow\Downarrow} \ll \varrho_{\uparrow\uparrow} + \varrho_{\downarrow\downarrow} \approx 1$). The equation of motion of the electron spin defined through $S = (\varrho_{\uparrow\uparrow} - \varrho_{\downarrow\downarrow})/2$ is then

$$\dot{S} = P - \frac{S}{\tau_{\mathrm{R}}} \tag{4.3}$$

with the total spin pumping and relaxation rate

$$P = \frac{1}{4}\left[-(1-Q_{\uparrow})g_{\uparrow} + (1-Q_{\downarrow})g_{\downarrow}\right], \tag{4.4}$$
$$\frac{1}{\tau_{\mathrm{R}}} = \frac{1}{\tau_{\mathrm{s}}^{\mathrm{e}}} + \frac{1}{2}\left[(1-Q_{\uparrow})g_{\uparrow} + (1-Q_{\downarrow})g_{\downarrow}\right],$$

respectively. The relaxation rate is intrinsically excitation-dependent, as the pumping process itself limits the electron life-time. The new parameters $Q_j = (2q_j - 1)Q_{\mathrm{s}}^{\mathrm{h}}$ include the final degree of the hole spin relaxation $Q_{\mathrm{s}}^{\mathrm{h}} = \tau_{\mathrm{s}}^{\mathrm{h}}/(\tau + \tau_{\mathrm{s}}^{\mathrm{h}})$ reached in the trion ground-state prior to recombination. The PL signal is proportional to the trion population and strictly circularly polarized, namely $I_+ \propto \varrho_{\Uparrow\Uparrow}$ and $I_- \propto \varrho_{\Downarrow\Downarrow}$. Choosing the unimportant factor appropriately, one finds from the steady-state solutions of Eqs. (4.1) and (4.2)

$$I_{\pm} = (1 \pm Q_{\uparrow})g_{\uparrow}(1/2 + S) + (1 \mp Q_{\downarrow})g_{\downarrow}(1/2 - S). \tag{4.5}$$

That is, the optically induced electron spin can be directly read-out by the secondary emission of the QD. Next, we discuss two limiting cases.

(i) Resonant excitation: When the trion ground-state is directly accessed, excitation with circular polarization is fully arm-selective, i.e., $g_{\uparrow} = g$, $g_{\downarrow} = 0$ for σ^+ and $g_{\uparrow} = 0$, $g_{\downarrow} = g$ for σ^-, with a single relaxation parameter $Q_{\uparrow/\downarrow} = Q_{\mathrm{s}}^{\mathrm{h}}$. Denoting by $p = \pm$ and $d = \pm$ the helicities of pumping and detection, respectively, it follows

$$P = -p\frac{1}{4}(1 - Q_{\mathrm{s}}^{\mathrm{h}})g, \quad \frac{1}{\tau_{\mathrm{R}}} = \frac{1}{\tau_{\mathrm{s}}^{\mathrm{e}}} + \frac{1}{2}(1 - Q_{\mathrm{s}}^{\mathrm{h}})g, \tag{4.6}$$

and

$$I_d^p = (1 + pd\, Q_{\mathrm{s}}^{\mathrm{h}})(1/2 - |S|)g. \tag{4.7}$$

Excitation, say with σ^+ photons, removes selectively a spin-up electron. When the hole spin-flips, a spin-down electron is left behind and an average spin $S < 0$ just opposite to the excitation photon momentum is formed.

(i) Continuum excitation: Circular selectivity is lost when excited electron shells become involved. E.g., the triplet states can be excited along both arms for a given circular polarization. Eventually, for sufficiently large photon energies, it holds $g_\uparrow = g_\downarrow = g$ regardless of σ^+ or σ^- excitation providing

$$P = \frac{1}{4}(Q_\uparrow - Q_\downarrow)g, \quad \frac{1}{\tau_\mathrm{R}} = \frac{1}{\tau_\mathrm{s}^\mathrm{e}} + \frac{1}{2}(2 - Q_\uparrow - Q_\downarrow)g, \tag{4.8}$$

and

$$I_\pm = [1 \pm (Q_\uparrow - Q_\downarrow)/2 \pm (Q_\uparrow + Q_\downarrow)S]g. \tag{4.9}$$

Orientation of the electron spin is only possible when the relaxation is arm-selective ($q_\uparrow \neq q_\downarrow$). This can indeed happen through the finestructure of the triplet state. A σ^+ photon creates an eh pair $(\downarrow, \Uparrow)$. It is reasonable to assume that the hole spin very quickly randomizes in the continuum [15]. Then, effectively, pairs $(\downarrow, \Uparrow)$ and $(\downarrow, \Downarrow)$ are generated with equal probability. The resultant configurations $(\uparrow, \downarrow, \Uparrow)$ and $(\uparrow, \downarrow, \Downarrow)$ excited along the spin-up arm can can relax without any spin-flip directly to the 3/2 and -3/2 trion ground-state, respectively. From the states $(\downarrow, \downarrow, \Uparrow)$ and $(\downarrow, \downarrow, \Downarrow)$ created along the spin-down arm, the first one can reach the 1/2 triplet state without spin-flip. Cross-relaxation down to the 3/2 ground-state requires a single-electron spin-flip which is very unlikely. The combined eh spin-flip mediated by the anisotropic exchange (Section 2) serves as shortcut, leading to the -3/2 triplet state followed by further relaxation along the spin-down arm. The second configuration returns also predominantly along this arm though an electron spin-flip has to occur after it has reached the -5/2 triplet state. Therefore, $q_\uparrow \sim 1/2, q_\downarrow \sim 1$ or $Q_\uparrow \sim 0, q_\downarrow \sim Q_\mathrm{s}^\mathrm{h}$ and, as a consequence, the electron spin is oriented downwards.

While both pumping mechanisms provide a spin polarization in the same direction, there are specific differences. Under resonant excitation, the electron spin is probed in absorption. The depletion of the electron spin population in the arm selected by the incident photon polarization decreases the excitation rate so that the total PL signal disappears like $I_+ + I_- = g(1 - 2|S|)$. The circular polarization degree $\rho_\mathrm{c} = pQ_\mathrm{s}^\mathrm{h}$ is constant and the spin pumping rate P proportional to $(1 - \rho_\mathrm{c})$. That is, the smaller the circular degree, the more efficient is the pumping. Complete spin polarization ($S = -p/2$) is approached at sufficiently high excitation rate g. In contrast, the total PL signal is independent on S for off-resonant excitation ($I_+ + I_- = 2g$). Here, the electron spin is probed by the circular emission degree $\rho_\mathrm{c} \sim$

$-pQ_s^h(1/2 + |S|)$. There is no spin pumping for $\rho_c = 0$ and the saturation spin is only $S = -pQ_s^h/2(2 - Q_s^h)$.

The presence of a negative circular polarization does not necessarily imply that orientation of the electron spin is indeed accomplished. The most simple example is excitation of a pure light-hole state. A σ^+ photon removes then a spin-down electron and creates a 1/2-hole trion state which relaxes preferably in the -3/2 trion ground-state, since parity breaking is not required in this channel. Here, a σ^- photon is emitted, however, the spin-down electron is recovered. Another source of negative circular polarization is the charged biexciton. As already argued, after selective excitation of a spin-down state, the subsequent radiative decay addresses directly the trion triplet states 1/2 and -3/2. A combined eh spin flip transforms 1/2 to -3/2 so that the relaxation runs along the spin-down arm leading finally to the emission of a σ^- photon. The data in Fig. 4.4 where the biexciton is resonantly excited by cascaded two-photon absorption confirm hence the presence of the exchange-induced eh spin-flip. Whether a at least partial spin selection is maintained under nonresonant excitation of the biexciton and whether this might be used to improve spin pumping deserve further investigation.

In order to study experimentally the sole electron-spin pumping scenario, a longitudinal external field of 100 mT is applied which is sufficient to suppress the influence of the hyperfine interaction on the electron spin dynamics. The polarization of the excitation is periodically switched between σ^+ and σ^- providing rectangular pulses in a given polarization with duration t_{on} that can be varied over orders of magnitude. Fig. 4.7 summarizes single-dot PL transients generated for different photon-energies in excess to the trion ground-state. A QD with negligible heavy-light hole mixing in X^- is selected.

Resonant excitation of the trion ground-state is faced with invincible stray light problems. The energetically lowest excitation is thus quasi-resonant to the 1-LO phonon feature (Fig. 4.3b). The polaron-like state has the same spin structure as the trion ground-state and obeys hence the same optical selection rules. In the context of spin pumping, such excitation has even the advantage that the hole spin-flip in this state is faster than in the ground-state (Fig. 4.5b). Denoting by $\tau_s^{h,LO}$ the respective time constant and by τ_{LO} the life-time of the polaron state related to its relaxation in the ground-state, it is easy to show that $Q_{\uparrow/\downarrow} = \tau_s^{h,LO}\tau_s^h/(\tau + \tau_s^h)(\tau_{LO} + \tau_s^{h,LO})$. The experimental transients follow entirely the above predictions. At the

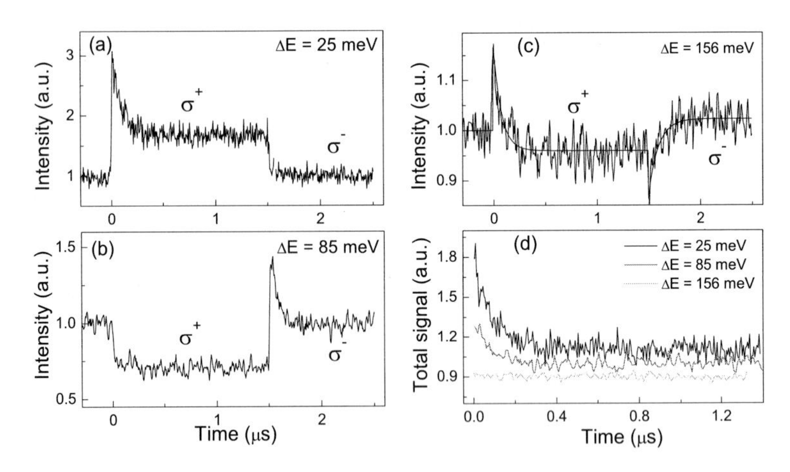

Fig. 4.7. (a)-(c) Single-dot electron spin pumping transients at different excitation photon energy ($T = 5$ K, $B = 100$ mT, $I_{\mathrm{exc}} = 200$ W/cm^2). The excitation is $\sigma^+ - \sigma^-$ modulated at a rate of 333 kHz using a Pockels cell with a rise time of 20 ns. Detection is in σ^+ mode. Detection in σ^- provides identical results, only with the role of σ^+ and σ^- in excitation reversed. (d) Total PL signal ($I_+^+ + I_+^-$). The excess energies ΔE are given relative to the PL position of X^-. Note the different vertical scales of the panels.

beginning of the σ^+ excitation period ($t = 0$), the electron starts with the steady-state spin $S_{\mathrm{st}} = P\tau_{\mathrm{R}} > 0$ built-up within the preceding σ^- period and reaches finally the opposite value $-S_{\mathrm{st}}$ at $t = t_{\mathrm{on}}$. The amplitude of the transients measures thus the spin polarization $2S_{\mathrm{st}}$ generated and the relaxation time τ_{R} is given by the $1/e$ point of the curves. In counter-polarized excitation-detection mode, the overall signal level is weak, indicating that $Q_{\mathrm{s}}^{\mathrm{h}}$ is noticeably larger than 0.5 at low temperature so that the spin switching is hardly seen on the intensity scale of Fig. 4.7. The absolute spin polarization can be directly extracted from the experimental data. Using Eq. (4.7) it follows

$$2S_{\mathrm{st}} = \frac{I_+^+(0) - I_+^-(0) - I_+^+(t_{\mathrm{on}}) + I_+^-(t_{\mathrm{on}})}{I_+^+(0) - I_+^-(0) + I_+^+(t_{\mathrm{on}}) - I_+^-(t_{\mathrm{on}})}, \tag{4.10}$$

where the unknown proportionality factor drops out. This yields a polarization as large as $2S_{\mathrm{st}} = 0.5$ for the transient in Fig. 4.7a.

At higher photon energies, the shape of the transients depend sensitively of the specific state addressed by the excitation. Fig. 4.7b shows an example where circular degree is clearly negative. However, the shape is indicative of a light-hole state. The fact that the same time constants occur as for quasi-resonant excitation assures that spin pumping is indeed involved. The total signal represented in Fig. 4.7d demonstrates that the circular excitation selectivity indeed increasingly desappears. In view of the number of unknown parameters, extraction of $2S$ is not straightforward in such cases. The circular degree declines when the excitation is shifted to still higher energies and is only $\rho_c = 0.04$ for the data in Fig. 4.7c. Obviously, $Q_\downarrow$ and $Q_\uparrow$ are not much different under real conditions. The spin generated is thus much smaller than for quasi-resonant excitation. A general disadvantage of spin pumping without circular excitation selectivity, as can be easily seen from Eq. (4.8), is a very low saturation spin if one is not close the idealized situation $Q_\downarrow = -Q_\uparrow$.

The same type of spin transients is observed in ensemble measurements (Fig. 4.8). While the amplitude of the transients provides now only a relative measure of the electron spin, the relaxation time τ_R is scarcely affected by the average. Consequently, the spin dynamics can be also studied on QD ensembles with considerably shorter integration times and better signal-to-noise ratio. In addition, excitation more closely to the trion ground-state via acoustic phonons (Fig. 4.8, bottom left) with only a few meV in excess becomes possible here. The observation of virtually identical transients in this case signifies that lattice heating is of no relevance in the spin pumping process.

A characteristic feature of the spin relaxation rate under optical pumping is its power dependence [see Eq. (4.6)]. Indeed, $1/\tau_R$ is over two orders of magnitude an almost linear function of the excitation intensity in the experiment (Fig. 4.9). However, as most clearly seen from single-dot data, the electron spin generated stays practically constant at about $S = 0.25$. The intrinsic life-time τ_s^e is thus also power-dependent. This conclusion reveals a characteristic spin-relaxation scenario where the QD captures an extra charge created in its environment as a concomitant of direct optical excitation. E.g., capture of a hole produces an exciton that radiatively recombines followed by recapture of another electron. Recharging processes of that type have been also exposed by photon-bunching studies [18]. They limit the life-time of the single-electron state itself, but also randomize the spin as the latter is undetermined when the QD recovers to single-electron occupation. The data in Fig. 4.9 provide $1/\tau_R = 2(P + \eta P^\gamma)$, $\eta \approx 1$, $\gamma \approx 0.8$, combining both the intrinsic and recharging contribution. Decreasing the excitation rate down to a level at which the PL signal disappears in the noise floor, relaxation times as long as 10 μs are found. However, τ_R does not approach an off-set value, demonstrating that the intrinsic spin-flip time due to spin-orbit coupling is significantly longer than

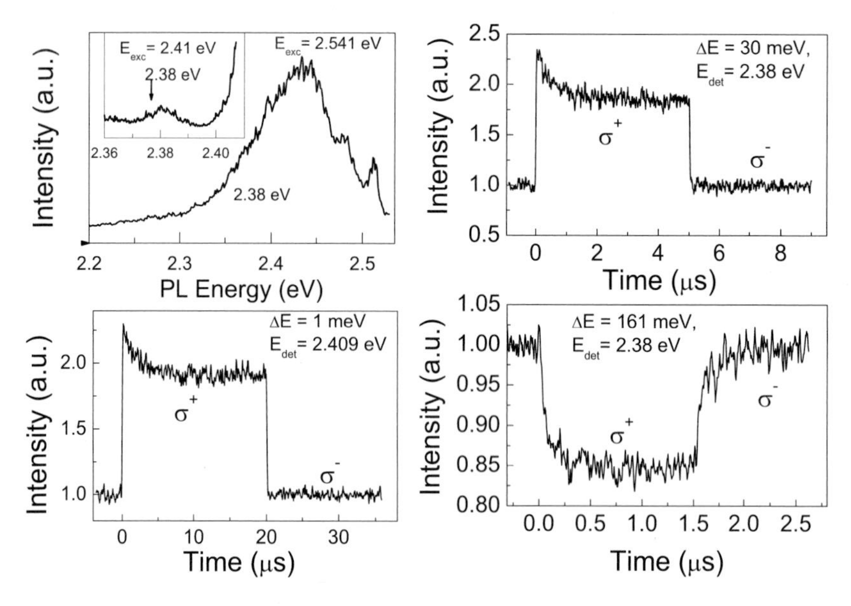

Fig. 4.8. Ensemble electron spin pumping transients. As depicted in the upper left panel, a certain fraction of QDs from the inhomogeneously broadened PL band is selected by the detection energy ($E_{\rm det}$) and excited with well-defined excess energy ΔE. The inset demonstrates spectral selectivity for 1-LO excitation ($T = 5$ K, $B = 100$ mT, $I_{\rm exc}$ $= 500$ W/cm^2). Detection is in σ^+ polarization.

10 μs.

At higher temperature, a spin amplitude becomes also visible in counter-polarized excitation-detection mode indicating, consistent with the data of Fig. 4.5b, speed-up of the hole spin-flip so that the relaxation parameter $Q_{\rm s}^{\rm h}$ decreases. The temperature dependence of S, $\tau_{\rm R}$, and $Q_{\rm s}^{\rm h}$ are summarized in Fig. 4.10. Though a shorter hole spin-flip results in a larger pumping rate P at given optical power, the spin amplitude declines with increasing temperature. The reason is that $\tau_{\rm R}$ and thus $1/\tau_s^e$ shorten in parallel, most probably caused by thermal activation of the recharging rate η. However, even up to 100 K, a residual spin amplitude survives.

In the absence of the external magnetic field, the hyperfine interaction comes into play. Modulation of the excitation at sufficiently high rates avoids formation of a DNP studied in the next Section. In this regime, the electron spin interacts with the randomly fluctuating nuclear field. This opens an extra channel through which a part of the spin decays on a time-scale defined by $\tau_2^* \sim \sqrt{aN_L}\hbar/(aA)$ [19]. This time is about 5 ns for the present QD structures and below the resolution of the setup used for record-

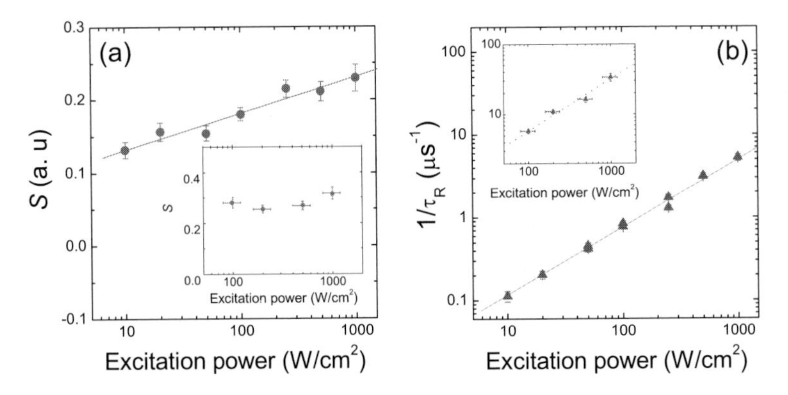

Fig. 4.9. Excitation power dependence of electron spin pumping ($T = 5$ K). (a) Spin amplitude. (b) Spin relaxation rate. Main panels: Ensemble under acoustic-phonon excitation ($B = 50$ mT). Insets: Single-dot under quasi-resonant excitation ($B = 100$ mT).

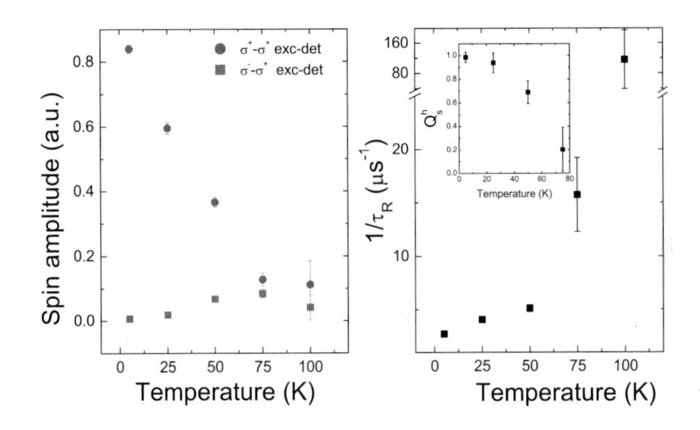

Fig. 4.10. Temperature dependence of electron spin pumping. (a) Spin amplitude. (b) Relaxation rate. Inset: Hole spin relaxation parameter. Data are taken from ensemble measurements under quasi-resonant excitation ($B = 100$ mT, $I_{\rm exc} = 500$ W/cm^2).

ing the transients. Fig.4.11 depicts the disappearance of the spin amplitude with decreasing magnetic field strength. However, the spin pumping relaxation rate $1/\tau_{\rm R}$ is independent of B demonstrating that the spin part surviving the initial rapid drop within τ_2^* follows the scenario described above.

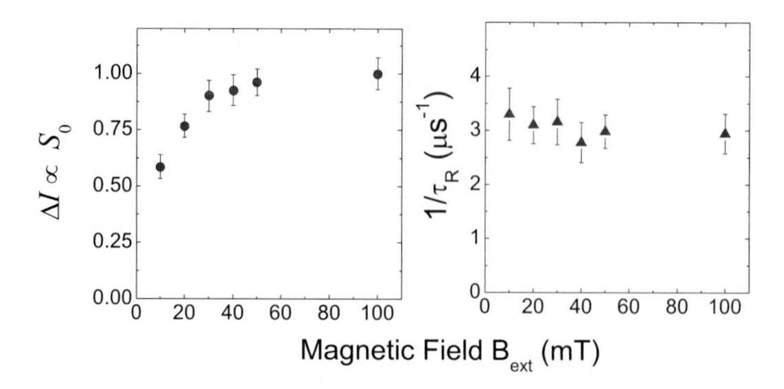

Fig. 4.11. Magnetic field dependence of electron spin pumping. Data are taken from ensemble transients under quasi-resonant excitation. (a) Spin amplitude. (b) Relaxation rate. ($t_{on} = 3$ μs, $T = 5$ K)

4.4. Hyperfine Nuclear-electron Spin Dynamics

After elaboration of the sole spin pumping scenario, the coupling of the electron to the nuclear moments by the hyperfine interaction is now being discussed. Fig. 4.12 represents spin transients recorded without external field and for much longer pumping periods than in the previous Section. While their shape is fully analog to the short transients, the spin response is now much slower, though one expects just the opposite: A shorter spin life-time in the absence of a longitudinal field. However, the transients are no longer related to the single electron spin life-time τ_R, but reflect the formation of a DNP under continuous spin pumping.

In what follows, we concentrate on the measurements with quasi-resonant excitation as the spin pumping is here most effective. The transverse part of the hyperfine interaction induces flip-flop processes by which the spin is exchanged between the electron and the nuclear system [20]. If one ignores inhomogeneity of the hyperfine coupling as well as collective nuclear excitations [21], the flip-flop processes are accounted for by extending Eq. (4.3) according to

$$\begin{aligned} \frac{\mathrm{d}S}{\mathrm{d}t} &= \pm P - \frac{S}{\tau_\mathrm{e}} + N\frac{J-S}{\tau_\mathrm{hf}}, \\ \frac{\mathrm{d}J}{\mathrm{d}t} &= \frac{S-J}{\tau_\mathrm{hf}} - \frac{J}{\tau_\mathrm{d\text{-}d}}. \end{aligned} \tag{4.11}$$

When only the hyperfine contribution, characterized by a single time constant τ_hf, is considered, these equations ensure total spin conservation $\dot{S} + N\dot{J} = 0$. $N = aN_\mathrm{L}$ is the number of nuclear moments interacting

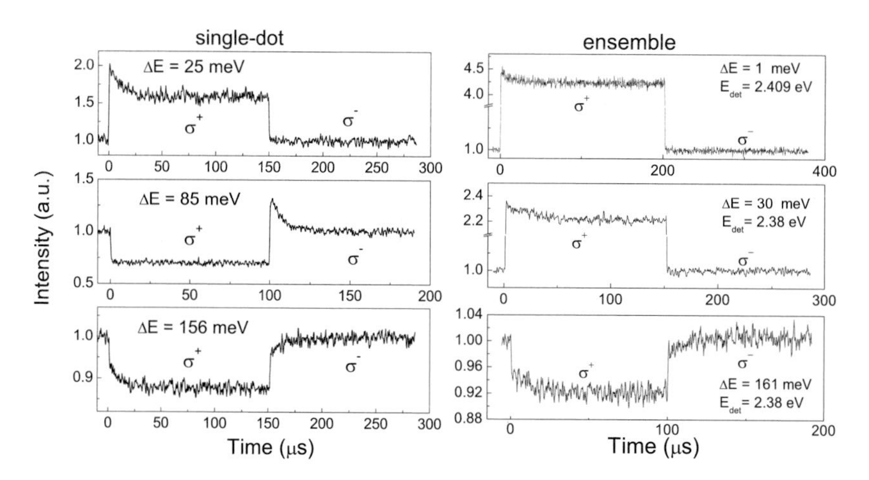

Fig. 4.12. Nuclear spin cooling transients. Left panel: Single-dot, $X^- = 2.385$ eV. Right panel: Ensemble. Same polarization configurations as in Fig. 4.7 and Fig. 4.8. ΔE: excitation photon energy in excess, E_{det}: detection energy. Note the enlarged time-scale ($T = 5$ K, $B = 0$, $I_{\mathrm{exc}} = 2$ kW/cm^2).

with the electron. The electron wavefunction φ_{e} obtained by calculating the energy position of the ensemble PL provides an average number $N_{\mathrm{L}} = 8/(v_0 \int \mathrm{d}V \varphi_{\mathrm{e}}^4(r)) \approx 1600$ (v_0: volume of unit cell) [22, 23]. Accounting for only the major contribution from the ^{111,113}Cd isotopes ($a = 0.125$), it follows $N = 200$. The nuclear spin J per isotope is defined (analog to S) through the diagonal elements of the nuclear spin density matrix. During spin pumping, the electron time τ_{e} is given by τ_{R}, while it equals the orders of magnitude longer spin-orbit time in the absence of optical excitation. $\tau_{\mathrm{d-d}}$ is the dipole-dipole time governing the decay of the DNP by the interaction between different nuclei. For the 1/2 isospin, a quadrupole contribution is absent [24]. The linear system is easily solved analytically yielding for $N \gg 1$ two time constants

$$\frac{1}{\tau_{1/2}} = \frac{1}{2}\left[\frac{1}{\tau_{\mathrm{e}}} + \frac{N}{\tau_{\mathrm{hf}}} + \frac{1}{\tau_{\mathrm{d-d}}} \pm \sqrt{\left(\frac{1}{\tau_{\mathrm{e}}} + \frac{N}{\tau_{\mathrm{hf}}} - \frac{1}{\tau_{\mathrm{d-d}}}\right)^2 + \frac{4}{\tau_{\mathrm{hf}}}\left(\frac{1}{\tau_{\mathrm{d-d}}} - \frac{1}{\tau_{\mathrm{e}}}\right)}\,\right] \tag{4.12}$$

by which the steady-state values

$$S_{\mathrm{st}} = \pm P \frac{\tau_{\mathrm{e}}(\tau_{\mathrm{hf}} + \tau_{\mathrm{d\text{-}d}})}{2(\tau_{\mathrm{hf}} + \tau_{\mathrm{d\text{-}d}} + N\tau_{\mathrm{e}})}, \quad J_{\mathrm{st}} = S_{\mathrm{st}} \frac{\tau_{\mathrm{d\text{-}d}}}{\tau_{\mathrm{hf}} + \tau_{\mathrm{d\text{-}d}}} \tag{4.13}$$

are approached.

Spin pumping: At the typical excitation levels used $\tau_e \sim 100$ ns. The dipole-dipole time is determined by the interaction energy of neighbored nuclei and estimates from NMR linewidths provide a time-scale of 10^{-4} s [25]. It holds thus $\tau_e \ll \tau_{\text{d-d}}$ which produces in lowest order $\tau_1 = \tau_e \tau_{\text{hf}}/(\tau_{\text{hf}} + N\tau_e)$ and $\tau_2 = (\tau_{\text{hf}} + N\tau_e)/(1 + \tau_{\text{hf}}/\tau_{\text{d-d}} + N\tau_e/\tau_{\text{d-d}})$. The meaning of these two times is more clearly seen from the time-dependent solutions. In terms of the deviations $\delta S = S - S_{\text{st}}$ and $\delta J = J - J_{\text{st}}$, those read as

$$\delta S(t) = \delta S_0 \mathrm{e}^{-t/\tau_1} - \delta J_0 K(\mathrm{e}^{-t/\tau_1} - \mathrm{e}^{-t/\tau_2}),\ \delta J(t) = \delta J_0 \mathrm{e}^{-t/\tau_2}, \quad (4.14)$$

Therefore, τ_1 describes the direct response of the electron spin on optical pumping and is too short to be resolved in Fig. 4.12, whereas τ_2 represents the formation time τ_F of the DNP. If the dipole-dipole decay can be neglected ($\tau_{\text{d-d}} \to \infty$), $\tau_F = \tau_{\text{hf}} + N\tau_e$ expressing that the electron spin has to be N times oriented in order to polarize the nuclei. $K = N\tau_e/(\tau_{\text{hf}} + N\tau_e)$ measures the degree by which the presence of the DNP increases the electron spin polarization. It is this quantity which makes it possible to pursue experimentally the nuclear dynamics by PL measurements. The formation time $\tau_F \sim 10$ μs revealed by the transients in Fig. 4.12 is orders of magnitude shorter than in bulk semiconductor where it lies in the second or even hour range [2]. We further find from fits to the data $\tau_{\text{hf}} = 25$ μs and $K = 0.3$, i.e, hyperfine flip-flop and electron life-time ($N\tau_e \simeq 10$ μs) contribute about equally to τ_F.

Spin relaxation: Without pumping ($P, S_{\text{st}}, J_{\text{st}} = 0$), the single-spin flip-flop defines the shortest time-scale ($\tau_{\text{hf}}/N \ll \tau_{\text{hf}}, \tau_{\text{d-d}}, \tau_e$) providing $\tau_1 = \tau_{\text{hf}}/N$, $\tau_2 = \tau_{\text{d-d}}$, and $K = 1$. The short τ_1 enforces $S \approx J$, while both spin polarizations decay slowly by the dipole-dipole time.

In order to uncover the spin decay, a dark period of duration t_{dark} is introduced between the excitation pulses, again either co- or counter-polarized. A difficulty of such measurements is an increasing number of dark counts when t_{dark} exceeds considerably t_{on}. The data presented below are selectively verified on a single-dot level, systematic studies at reasonable integration times are made on ensembles.

Application of periodic boundary conditions results in DNP-related spin amplitudes $\Delta S = \Delta S_\infty[1 \pm \exp(-t_{\text{dark}}/\tau_{\text{d-d}})]$ for alternating and co-polarization mode, respectively, where $\Delta S_\infty = -KJ_{\text{st}}$ ($t_{\text{on}} \gg \tau_F$). The experimental data in Fig. 4.13c follow closely this prediction yielding $\tau_{\text{d-d}} = 250$ μs. The formation time of the DNP is hence about one order of magnitude shorter than its decay by the dipole-dipole interaction. In marked contrast, just the opposite holds for bulk semiconductors because

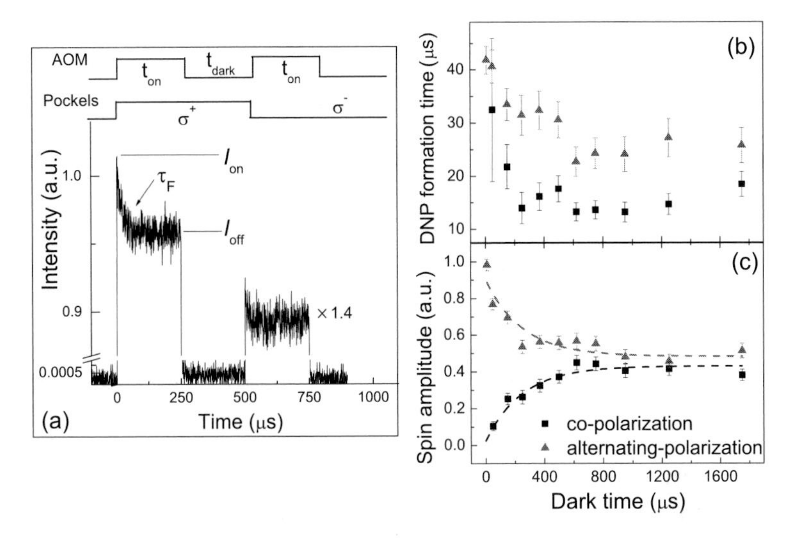

Fig. 4.13. Nuclear dynamics. (a) Optical excitation trains in alternating polarization mode (top) as well as PL transients (main panel) recorded at $T = 60$ K. (b) DNP formation time and (c) DNP related spin amplitude ($I_{\rm on} - I_{\rm off}$) as functions of the dark time in both polarization modes ($B = 0$, $I_{\rm exc} = 2$ kW/cm^2). The dashed curves are single-exponential fits with a time constant of 250 μs. Excitation is quasi-resonant to the 1-LO-phonon state, detection in σ^+ polarization. For more experimental details see Ref. 27.

of the huge number of nuclei seen by the electron. Here, spin temperature cooling in an external magnetic field can produce an equilibrium DNP. The field Zeeman-splits the nuclear states sufficiently up and selectively pumping one of the populations via the flip-flop process creates a nuclear spin temperature different to that of the lattice [1]. The decay of the equilibrium DNP requires dissipation of the Zeeman energy by spin-lattice interaction which takes place on time-scales of a second and beyond [2].

The linewidth broadening of the electron spin levels is determined by the life-time of the off-diagonal density matrix elements $\varrho_{\uparrow\downarrow}$. Denoting this time by $\tau_{\rm c}$ and assuming $(A/N_{\rm L})\tau_{\rm c} \ll 1$ leads to the standard expression [26]

$$1/\tau_{\rm hf} = 1/T_{1\rm e} = (A/N_{\rm L})^2\tau_{\rm c}/(1 + \Delta\omega_{\rm B}^2\tau_{\rm c}^2) \tag{4.15}$$

where $\Delta\omega_{\rm B} = g_{\rm e}\mu_{\rm B}B + aAJ$ represents the electron Zeeman splitting caused by the external field (B) as well as the Overhauser field ($B_{\rm N}$) associated with the DNP. The much smaller splitting of the nuclear levels can be neglected. The increase of the DNP formation time and the disappearance of the DNP-related spin amplitude anticipated from Eq. (4.15) are depicted

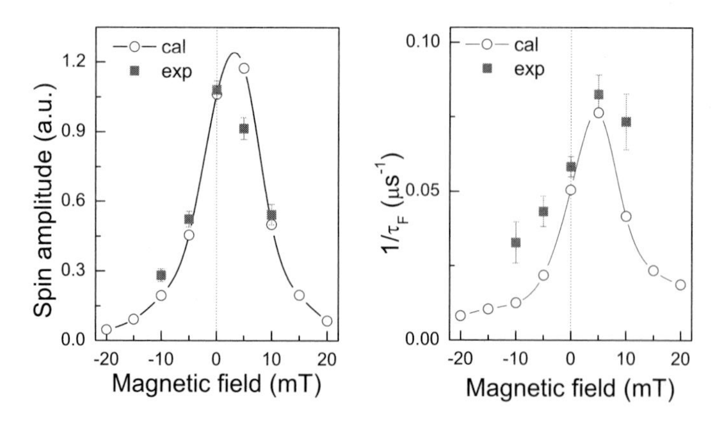

Fig. 4.14. Magnetic-field dependence of the spin amplitude and DNP formation time. Experimental data are taken in $\sigma^+ - \sigma^+$ excitation-detection mode ($T = 50$ K). Calculations are based on Eqs. (4.11) and (4.15) ($N = 200$, $\tau_e = 10$ ns, $\hbar A = -6$ μeV, $\tau_c = 2$ ns, $\tau_{\text{d-d}} = 250$ μs).

in Fig. 4.14. The electron g factor $g_e = 1.1$ [12] provides for the external contribution $\Delta\omega_B = 9$ ns$^{-1}B/100$ mT with which the estimate $\tau_c \sim 1$ ns follows for the coherence time. There is also a weak, but clearly experimentally resolvable asymmetry regarding the directions of the external field and the electron spin. Switching the excitation from σ^- to σ^+, as used to record the data in Fig. 4.14, orients the electron spin quickly down, while the nuclear spin is initially still up. The nuclear magnetic moment of the Cd isotopes is negative (μ_I^{Cd} = -0.6) and the Overhauser field points thus always in the direction opposite to J. That is, B and B_N partially compensate if the external field is applied in positive direction, while they add for $B < 0$ to a larger effective field making the decline of the spin amplitude stronger. Exactly this behavior is seen in the experiment. In view of the ensemble average, calculations based on Eqs. (4.11) and (4.15) reproduce the experimental data reasonably well. The estimate of the Overhauser field $|B_N| = a\hbar AJ/g_e\mu_B \simeq 3$ mT is consistent with the hyperfine coupling constant $\hbar A = -6\mu$eV [28].

For no external field, the Overhauser field introduces nonlinearities in the dynamics which limit the nuclear spin polarization by increasing the formation time. For III-V QDs, where B_N can reach the 1-Tesla range, strong nonlinearities have been described by various groups [29–31]. The DNP is here directly displayed by a zero-field splitting of the dot emission, but formation times estimated from indirect measurements are $\tau_F \sim 1$ s and longer [29, 32]. Contrary, the Overhauser feedback is weak in the present

CdSe/ZnSe QD structures defining a regime where the DNP-induced spin splitting does not exceed markedly the broadening of the Zeeman levels ($aA\tau_c \sim 1$). This fact is corroborated by the dark-time dependence of the DNP formation time (Fig. 4.13b). Indeed, τ_F is longer for smaller t_{dark} because of a stronger initial DNP, but the overall change is only about 30 %. Spin transients computed with the appropriate hyperfine parameters predict virtually the same nuclear and electron spin polarizations (Fig. 4.15). Indeed, single-dot measurements provide direct experimental evidence for an almost complete closing of the hyperfine flip-flop rate ($S \simeq J$). The slow DNP-related transients approach the same final level as the sole electron spin pumping transients at $B = 100$ mT (Fig. 4.16). Accordingly, the same electron spin polarization $| S_{st} |= P\tau_e$ is created in both situations. It may hence be concluded that the nuclear spin polarization becomes also close to $2J \approx 0.5$.

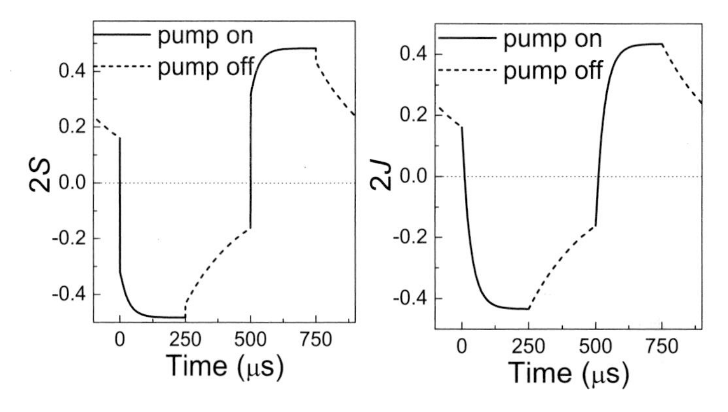

Fig. 4.15. Calculated electron and nuclear spin transients from Eqs. (4.11) and (4.15) ($N = 200$, $\tau_e = 50$ ns, $\hbar A = -6$ μeV, $\tau_c = 1.2$ ns, $\tau_{d\text{-}d} = 250$ μs).

The DNP formation time shortens monotonically with temperature, while the spin amplitude initially slightly increases reaching a maximum at about 50 K before it starts to decline (Fig. 4.17). A shorter τ_F at higher temperature is expected from the shortening of the electron life-time. A temperature increase of K is indicative of a shortening of τ_{hf}. This is reasonable since the shorter coherence time τ_c at higher temperature reduces the Overhauser contribution in the denominator of τ_{hf}. At higher temperature, the shortening of the electron life-time takes over. Single-dot measurements where contributions of nontrionic origin can be excluded confirm these tendencies. Note, however, that the ensemble data, neither of Fig. 4.10 nor

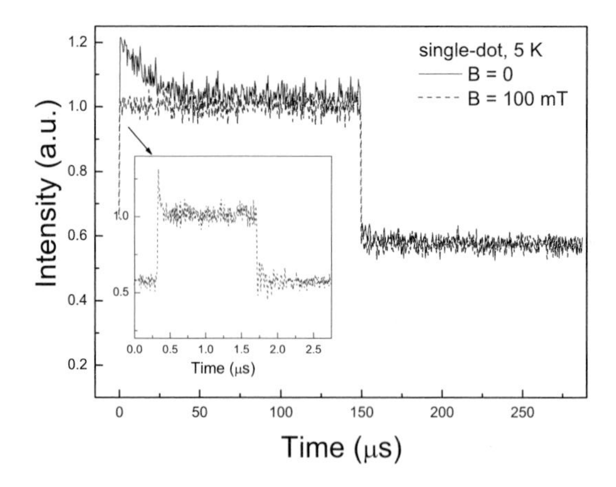

Fig. 4.16. Single-dot spin cooling transients at $B = 0$ and $B = 100$ mT.

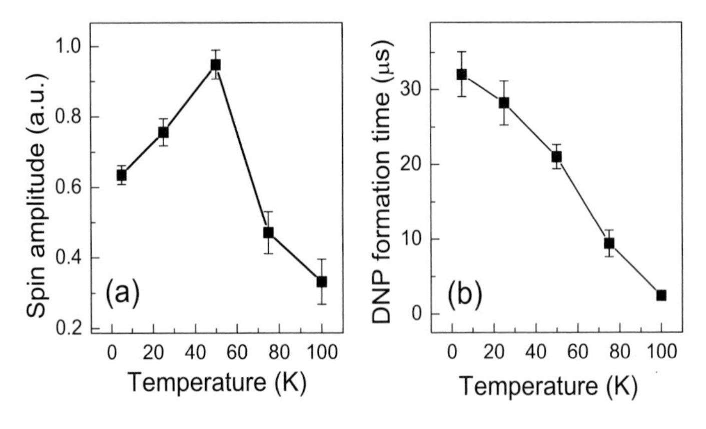

Fig. 4.17. Temperature dependence of (a) DNP spin amplitude and (b) formation time for $t_{\text{dark}} = 0$ taken in alternating polarization mode ($B = 0$).

Fig. 4.17, allow for conclusions about the absolute spin polarizations.

The 250 μs dipole-dipole decay time is in accord with the interaction energy of 10^{-5} μeV of neighboring nuclei. Standard spin cooling is described by a depolarization time $\propto B^2/\tilde{B}_{\text{L}}^2$ where $\tilde{B}_{\text{L}}$ is the effective local dipole field [33]. The hyperfine interaction can be also viewed as a so-called Knight field produced by the electron and seen by the nuclear spins. It has been argued that an external field is not required for QDs, as the hyperfine Knight field is strong enough to ensure spin cooling conditions [32, 34]. A substantial Knight field $B_{\text{e}} \propto S$ added to B would show up by an asym-

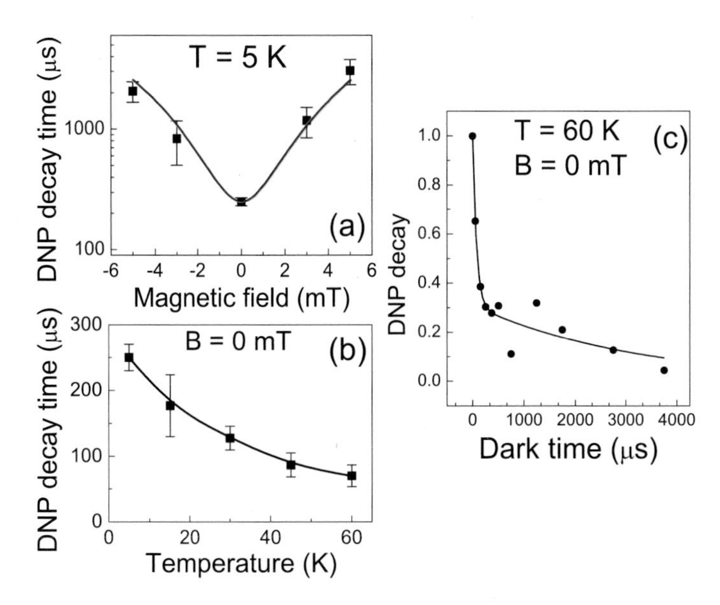

Fig. 4.18. Dynamics of the DNP decay. (a) Magnetic field dependence of the decay time obtained from single-exponential fits to the experimental data. The line represents $\tau_{\text{d-d}}(B) = 250\mu\text{s} + 91\ \mu\text{s/mT}\ B^2$. (b)Temperature dependence. (c) Decay transient at elevated temperature on a longer time-scale.

metry of the decay transients with respect to the directions of electron spin and external field. No such asymmetry is found beyond the experimental resolution. The magnetic-field dependence of the dipole-dipole time is depicted in Fig. 4.18a. It increases about one order of magnitude for modest field strength ($B = 5$ mT), clearly demonstrating the opening of a nuclear spin splitting suppressing dipole-dipole-mediated transitions between different nuclei. The zero-field time decreases only smoothly with temperature (Fig. 4.18b) so that the relation $\tau_{\text{F}} < \tau_{\text{d-d}}$ is maintained up to 100 K. Measurements on an extended time-scale (Fig. 4.18c) uncover in addition to the 100 μs component a second, clearly smaller part which persists up to the millisecond range. This part becomes increasingly visible at higher temperature and makes up a 0.2 portion of the total amplitude at 60 K. The origin of the long persisting DNP needs further investigations. It might be due to the Se isotopes with lower abundance or spin diffusion between different dots.

4.5. Conclusions

Optical spin cooling without the need of an external magnetic field can be efficiently accomplished by quasi-resonantly exciting the trion transition of charged QDs. Single-dot measurements demonstrate a spin orientation degree of about 50 % for both the electron as well as the nuclear system. The scenario uncovered by our experimental and theoretical analysis relies on two specific features realized by the CdSe/ZnSe QD structures under study. First, the splitting induced by the Overhauser field stays within the homogeneous width of the electron spin states avoiding nonlinearities that limit the available spin polarization. Note that the definition of a spin temperature is not meaningful in such case. Second, the number of nuclear spins per QD is small, i.e., in the 100-range. This leads to a situation where the formation of the nuclear polarization is an order of magnitude faster than its decay by the dipole-dipole coupling ($\tau_F < \tau_{d\text{-}d}$). Unlike the vast amount of previous work on semiconductors, this allows for the formation of a strong nonequilibrium DNP.

The spin polarizations are limited by a type of parasitic process where the electron disappears from the QD through recharging by carriers created in the background. Suppression of this loss channel by improving structure perfection may lead to higher polarizations. Theoretical work has shown that collective dark states in the nuclear system can impose quite strong restrictions for the spin cooling [21]. The present measurements suggest a very short electron spin coherence time $\tau_c \sim 1$ ns under optical excitation. Even shorter times ($\tau_c = 25$ ps) have been observed for electrons localized on very diluted donors in ultra-pure GaAs and assigned to spin exchange processes with carriers in the environment [35]. Such processes seem to play a role also in Stranski-Krastanov QD structures. We conjecture that the fast spin decoherence of the electron during spin pumping is imprinted in the nuclear system so that the nuclei interact indeed independently with the electron. Signatures of a DNP formation are seen up to a lattice temperature of about 100 K. At those temperatures, the optical spin pumping and read-out via the trion features brake down. Whether or not a DNP can be established at still higher temperatures can not be decided by the current experiments.

Acknowledgments

This work was supported by the Deutsche Forschungsgemeinschaft within Project No. He 1939/18-1.

References

[1] A. Abragam and W. G. Proctor, Phys. Rev. **109**, 1441 (1958).
[2] M. I. Dyakonov and V. I. Perel, in *Optical Orientation*, ed. F. Meier and B. P. Zakharchenya, (North Holland, Amsterdam, 1984) p. 11; V. G. Fleisher and I. A. Merkulov, ibid., p. 173.
[3] M. Rabe, M. Lowisch, and F. Henneberger, J. Cryst. Growth **184/185**, 248 (1998).
[4] D. Litvinov et al., Appl. Phys. Lett. **81**, 640 (2002).
[5] I. A. Akimov, A. Hundt, T. Flissikowski, and F. Henneberger, Appl. Phys. Lett. **81**, 4730 (2002).
[6] R. J. Warburton et al., Nature **405**, 926 (2000).
[7] K. V. Kavokin , Phys. Status Solidi (a)**195**, 592 (2003).
[8] I. A. Akimov, K. V. Kavokin, A. Hundt, and F. Henneberger, Phys. Rev. B **71**, 075326 (2005).
[9] J. Puls, M. Rabe, H.-J. Wünsche, and F. Henneberger, Phys. Rev. B **60**, R16303 (1999).
[10] I. A. Akimov, A. Hundt, T. Flissikowski, P. Kratzert, and F. Henneberger, Physica E (Amsterdam) **17**, 31 (2003).
[11] I. A. Akimov, T. Flissikowski, A. Hundt, and F. Henneberger, phys. stat. sol. (a) **201**, No. 3, 412420 (2004).
[12] A. V. Koudinov, I. A. Akimov, Yu. G. Kusrayev, and F. Henneberger, Phys. Rev. B **70**, 241305(R) (2004).
[13] R. I. Dzhioev, B. P. Zakharchenya, V. L. Korenev, P. E. Pak, D. A. Vinokurov, O. V. Kovalenkov, and I. S. Tarasov, Phys. Sol. State **40**, 1587 (1998).
[14] Michio Ikezawa, Bipul Pal, Yasuaki Masumoto, Ivan V. Ignatiev, Sergey Yu. Verbin, and Il'ya Ya. Gerlovin, Phys. Rev. B **72**, 153302 (2005).
[15] S. Laurent, M. Senes, O. Krebs, V. K. Kalevich, B. Urbaszek, X. Marie, T. Amand, and P. Voisin, Phys. Rev. B **73**, 235302 (2006).
[16] M. E. Ware, E. A. Stinaff, D. Gammon, M. F. Doty, A. S. Bracker, D. Gershoni, V. L. Korenev, S. C. Badescu, Y. Lyanda-Geller, and T. L. Reinecke, Phys. Rev. Lett. **95**, 177403 (2005).
[17] T. Flissikowski, I. A. Akimov, A. Hundt, and F. Henneberger, Phys. Rev. B **68**, 161309 (2003).
[18] C. Santori, D. Fattal, J. Vukovi, G. S. Solomon, E. Waks, and Y. Yamamoto, Phys. Rev. B **69**, 205324 (2004).
[19] I. A. Merkulov, Al. L. Efros, and M. Rosen, Phys. Rev. B **65**, 205309 (2002).
[20] D. Paget and V. L. Berkovits, in *Optical Orientation*, ed. F. Meier and B. P. Zakharchenya, (North Holland, Amsterdam, 1984) p. 381.

[21] Such effects are treated by H. Christ, J. I. Cirac, and G. Giedke, Phys. Rev. B **75**,155324 (2007).
[22] P. R. Kratzert et al, Appl. Phys. Lett. **79**, 2814 (2001).
[23] A. Hundt, J. Puls, and F. Henneberger, Phys. Rev. B**69**, R121309 (2004).
[24] R. I. Dzhioev et al., Phys. Rev. Lett.**99**, 037401 (2007).
[25] D. Paget, G. Lampel, B. Sapoval, and V. I. Safarov, Phys. Rev. B **15**, 5780 (1977).
[26] M. I. Dyakonov and V. I. Perel, Sov. Phys. JETP **38**, 177 (1974).
[27] D. H. Feng et al., Phys. Rev. Lett. **99**, 036604 (2007).
[28] We use the Bloch factor $|u(r_{\mathrm{Cd}})|^2 = 2200$ from CdTe: A. Nakamura et al., Solid State Commun. **30**, 411 (1979).
[29] P. F. Braun et al., Phys. Rev. B **74** 245306 (2006).
[30] P. Maletinsky et al., Phys. Rev. B**75** 035409(2007).
[31] A. I. Tartakovskii et al. ,Phys. Rev. Lett. **98**, 026806 (2007).
[32] C. W. Lai et al, Phys. Rev. Lett. **96**, 167403 (2006).
[33] D. Gammon et al., Phys. Rev. Lett. **86**, 5176 (2001).
[34] I. A. Akimov, D.H. Feng, and F. Henneberger, Phys. Rev. Lett. **97**, 056602 (2006).
[35] D. Paget, Phys. Rev. B **25**, 4444 (1982); see also: R. I. Dzhioev et al., Phys. Rev. Lett. **88**, 256801 (2002).

PART 2

Qubit Control, Readout, and Transfer

Chapter 5

Electron Spin Quantum Bits in Quantum Dots: Initialization, Decoherence, and Control

A. Greilich, D. R. Yakovlev and M. Bayer

Experimentelle Physik II, Universität Dortmund, D-44221 Dortmund, Germany

A. Shabaev and A. L. Efros

Naval Research Laboratory, Washington, DC 20375, USA

Electron spins confined in semiconductor quantum dots have been considered as promising quantum bit candidates. This proposal is based mostly on longitudinal spin relaxation times in the milliseconds range at cryogenic temperatures. Here we substantiate the prospective properties of such quantum bits by addressing both the static and dynamic properties of electron spins in (In,Ga)As/GaAs self-assembled quantum dots. In particular we discuss first the g-factor tensor and turn then to the creation of spin coherence by optical laser pulses. We also discuss the times during which the spin coherence is maintained, and how the spin coherence can be tailored by the details of the laser excitation protocol.

Contents

5.1. Introduction

Recently the coherent dynamics of elementary excitations in semiconductor heterostructures has attracted considerable interest for applications in quantum information processing such as cryptography or computing [1]. In the fields of atom quantum optics and nuclei magnetic resonance it is rather easy to identify well defined two-level systems which can be used as carriers of quantum information (quantum bits) and are well separated from the environment. Therefore the quantum information activities started to flourish in these fields due to the superior coherence properties of the elementary excitations. Quite a few proof-of-principle experiments have been demonstrated, such as few qu-bit entanglement, quantum gate operation and design of simple quantum processors [1]. However, currently these approaches appear to be limited due to the lack of scalability towards large numbers of involved qu-bits, which is less a problem for quantum communication but which is indispensable for quantum computing.

The potential to reach this goal has been attributed to solid state and in particular semiconductor systems [1], due to the proven level of system integration in conventional electronics. Therefore the quantum information ideas and concepts have been transferred to semiconductors, even though it was clear, for example that it is much more complicated to identify well isolated two-level systems, by which coherence and therefore quantum information can be retained for long enough times. This has consequently directed interest toward semiconductor quantum dots because of their discrete energy level structure, due to which they bear some resemblance to atoms found in nature. The limitations of this analogy have been, however, clearly worked out in the meantime.

The 'artificial atom' analogy has been studied a lot by optical spectroscopy, for which self-assembled quantum dot structures are very well suited due to their high quantum efficiency. For example, at cryogenic temperatures the linewidth of the radiative decay of electron-hole pairs (excitons) confined in quantum dots is limited by the radiative decay time T_1^X, corresponding to widths of a few μeV [2]. But at elevated temperatures the interaction with higher lying confined states in the dots and with continuum states of the dot environment becomes so important that the linewidth reaches a few meV [3]. Further, recent studies have also shown that the simple exponential decay laws which give a perfect description of radiative decays in atomic physics can typically not be applied for quantum dots. Only for strictly resonant excitation of the transition between the valence and conduction band ground states at low temperatures a two-level-scheme may be appropriate [4].

On the other hand, from ultrafast spectroscopy it is well established that coherent manipulation of excitons created by optical excitation can be done on a sub-picoseconds time scale, and therefore attraction was caught first by charge excitations. The manipulation time scale has to be compared, however, to the coherence time T_2^X. Long coherence times are required for performing a sufficient number of quantum manipulations before destruction of coherence occurs. The decoherence of charges such as electron and hole typically occurs very fast in semiconductors, but charge neutral complexes such as excitons show longer coherence. Non-linear optical studies on quantum dot excitons have rendered T_2^X-values in the ns-range, which are ultimately limited by the radiative lifetime [5]. This time might be extended, for example, by suppression of spontaneous emission which would require a tailoring of the photonic environment in which the quantum dots are located. This could be achieved by a photonic crystal, for example, requiring sophisticated nanopatterning technology [6]. Imperfections in this patterning could be, however, a source of decoherence. Alternately, by application of an electric field the electron-hole overlap may be reduced, but it is not clear yet, whether the required field variation can be done adiabatically. In any case it seems hard to increase the T_1^X and T_2^X-times by more than about an order of magnitude. This coherence time span might be too short for quantum computing but could turn out to be sufficient for application in quantum communication, requiring a rather limited number of involved qu-bits and operations on them. Further, when quantum dots are coupled to molecules, as required in certain quantum gate designs relying on direct electronic coupling of qu-bits, the coherence time may be reduced as compared to the quantum dot case, setting further limitations on their use.

Therefore interest has moved toward spin excitations in semiconductors [7–10], motivated in particular by the observation of very long electron spin coherence times T_2^S already for bulk semiconductors [11]. Further, it has been shown that the spin relaxation mechanisms which are effective in higher-dimensional systems are strongly suppressed in quantum dots. For electrons, for example, only the spin-orbit coupling and the interaction with the nuclear background is effective, while for holes also the interaction with the nuclei is suppressed. The interest in quantum dot spins was enhanced further by the demonstration of very long electron spin relaxation lifetimes, T_1^S, in the milliseconds-range [12]. This has raised hopes that T_2^S, which may theoretically last as long as $2T_1^S$ [13], could be similarly long, with encouraging indications to that effect found lately [14].

In this chapter we give some insight into the current status of coherent optical manipulation of electron spins in self-assembled (In,Ga)As/GaAs quantum dots. We point out that all studies have been done on quantum dot ensembles, raising on one hand the hope that robust quantum bits can be realized in that way, while on the other hand this approach may suffer from the unavoidable inhomogeneities of the ensembles.

In the next Section II we describe the samples as well as the experimental techniques used for studying them. Section III addresses the electron g-factor. In Section IV we describe how the spins can be oriented efficiently by coherent optical excitation, in Section V and VI we describe measurements of the spin coherence and all-optical manipulation of the spins. The chapter is concluded by a summary and an outlook on future work.

5.2. Experiment

The experiments were performed on self-assembled (In,Ga)As/GaAs quantum dots, which were fabricated by molecular beam epitaxy on a (001)-oriented GaAs substrate. Here we use the generic term (In,Ga)As for the quantum dot material as the precise composition is unknown. To obtain strong enough light-matter interaction, the sample contained 20 quantum dot layers separated by 60 nm wide barriers. The layer dot density is about $10^{10}\,\mathrm{cm}^{-2}$. For an average occupation by a single electron per dot, the structures were n-modulation doped 20 nm below each layer with a Si-dopant density roughly equal to the dot density. From the Faraday rotation studies presented below we estimate that about half of the dots is singly charged by an electron, while the other half is uncharged. The precise ratio varies with the position on the sample and also with the experimental conditions such as temperature, illumination etc.

The as-grown sample shows ground state emission at wavelengths around 1.2 μm, which is outside of sensitivity range of Silicon photodiodes. Therefore the structure was thermally annealed for 30 s at 945 °C so that its emission occurs around 1.396 eV, as seen from the luminescence spectrum in Fig. 5.1. This range is easily accessible for Si-detectors. The full width at half maximum of the emission is about 10 meV, demonstrating a rather good homogeneity, achieved through the annealing step. Further optical properties of these dots can be found in Refs. [15, 16].

Most of the experiments reported here were performed with the sample immersed in liquid helium at a temperature $T = 2\,\mathrm{K}$. In this variable temperature insert the temperature could, however, also be increased up to several tens of K. The sample chamber was placed between the coils of an optical split-coil magneto-cryostat for fields up to $B = 10\,\mathrm{T}$. For reference, we define the sample growth direction [001] as the z-axis, which gives also the direction of light propagation. The orientation of the sample could be varied relative to the magnetic field. Experiments were performed for longitudinal (Faraday geometry) or transverse (Voigt geometry) magnetic field orientation relative to the z-axis. In addition, the sample could be rotated about the growth axis.

For optical excitation, a Ti-sapphire laser emitting pulses with a duration of $\sim$1.5 ps (full width at half maximum of $\sim$1 meV) was used, hitting the sample along the z-axis, as mentioned before. The laser repetition rate was 75.6 MHz, corresponding to a period $T_R = 13.2$ ns between the pulses. The laser pulse separation could be increased to multiples of T_R by a pulse picker system. The emission energy was tuned to be in resonance with the ground state transition of the charged quantum dots (see Fig. 5.1).

This laser system was used as basis for implementation of two different optical techniques which allow us to study the electron spin dynamics, both of which can be categorized as time-resolved pump-probe Faraday rotation [17, 18]. The first technique exploits an intense circularly polarized pump pulse for inducing circular dichroism of the quantum dots by optical orientation of carrier spins. The second technique, optically induced linear dichroism, exploits a linearly polarized pump beam which results in optical alignment of excitons in the quantum dots. In both cases, the optical anisotropies due to the pump pulses were analyzed by measuring the rotation angle of the polarization plane of a linearly polarized probe pulse of rather weak intensity. For detecting the rotation angle of the linearly polarized probe beam, a homodyne technique based on phase-sensitive balanced detection was used. The pump beam hit the sample at time zero, and the probe beam could be delayed relative to the pump beam by a mechanical delay line.

5.3. Electron g-factor

The open circles in Fig. 5.1 show the variation of the electron g-factor across the inhomogeneously broadened emission of the quantum dot ensemble. To measure it, the energy of the exciting laser was shifted across the emission band. Details of the g-factor determination can be found below. The magnetic field was oriented perpendicular to the heterostructure growth direction along the [1-10] crystal direction (the y-direction). The g-factor modulus decreases with increasing emission energy from 0.57 on the low energy side to less than 0.50 on the high energy side, and therefore shows a variation of about 7.5 % about its mean value.

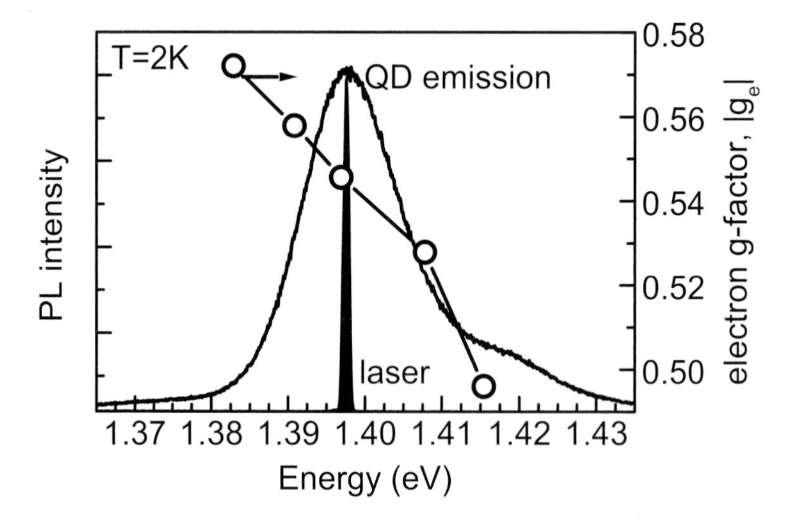

Fig. 5.1. Photoluminescence spectrum of the studied (In,Ga)As/GaAs quantum dot ensemble. The filled trace gives the spectrum of the excitation laser used in the Faraday rotation experiments, which could be tuned across the inhomogeneously broadened emission band. The symbols give the electron g-factor along the [1-10] direction in the dot plane across this band, for which the right scale is relevant. [19]

The g-factor of a conduction band electron typically differs considerably from its value of $g_0 = 2$ for free electrons. The reason is the strong spin-orbit interaction in semiconductors, leading to a strong mixing of bands. If one makes the assumption that the main effect of the quantum dot confinement is an increase of the band gap E_g between conduction and valence band as compared to bulk, but neglects all other effects such as changes of spin-orbit splittings etc., the deviation of the electron g-factor from 2, as determined by $\mathbf{k} \cdot \mathbf{p}$ perturbation theory, can be estimated by using the form for the g-factor in bulk [20]:

$$g_e = g_0 - \frac{4m_0 P^2}{3\hbar^2} \frac{\Delta}{E_g (E_g + \Delta)}. \tag{5.1}$$

Here m_0 is the free electron mass. P is the matrix element describing the coupling between valence and conduction band. Δ is the spin-orbit splitting in the valence band. For GaAs- or InAs-based semiconductors the coupling matrix element and the splitting are so large that the g-factor even becomes negative, for example it amounts -0.44 in GaAs bulk at cryogenic temperatures.

From our measurements we do not obtain direct access to the sign of the g-factor, but the systematic variation across the emission band allows us to trace it indirectly. Increasing emission energy corresponds to an increase of the band gap, leading to a reduction of the right hand side in equation 5.1. The decrease of the g-factor modulus with increasing emission energy in Fig. 5.1 can then be only explained if the g-factor is negative.

There is another striking difference between the g-factors of a free electron and a crystal electron. Due to the spatial anisotropy in a crystal it can in general no longer be described as a scalar quantity, but has to be described by a tensor of second order. In crystals with cubic symmetry this tensor can be reduced to a scalar, but for nanostructures this cannot be done in any case. Still for GaAs-based quantum wells, for example, the conduction band g-factor can often be taken as isotropic as the electron wave function is formed from s-type atomic orbitals. However, for self-assembled quantum dots this scalar approximation cannot be used, as shown in Fig. 5.2, for which the magnetic field orientation was varied in the quantum dot plane. The full circles give the electron g-factor at $B = 5\,\mathrm{T}$. Note that for the electron we do not find a g-factor dependence on field strength. For comparison also the exciton g-factor is shown by the full triangles. For both quantities a remarkable anisotropy is seen.

The anisotropy can be well described by a pattern with two-fold symmetry. Therefore, for an arbitrary direction, characterized by the angle φ relative to the x-axis, the electron g-factor can be written as

$$\sqrt{g_{e,x}^2 \cos^2 \varphi + g_{e,y}^2 \sin^2 \varphi} = g_{e\perp}, \tag{5.2}$$

where $g_{e,x}$ and $g_{e,y}$ are the g-factors along the x and y-axes, [110] and [1-10], respectively. The solid lines in Fig. 5.2 are fits to the data using equation 5.2. From these fits we obtain $g_{e,x} = 0.57$ and $g_{e,y} = 0.54$ for the electron. This corresponds to a relative variation of 2.7 % around the mean value.

We have also done measurements of the electron g-factor with the magnetic field aligned along the heterostructure growth direction, exploiting the linear dichroism in Faraday rotation measurements. From these studies (not shown here) we obtain an average g-factor of the electron along z to be -0.61 with a variation of about 10 % across the ensemble. Therefore it is considerably larger than the average g-factor in the dot plane.

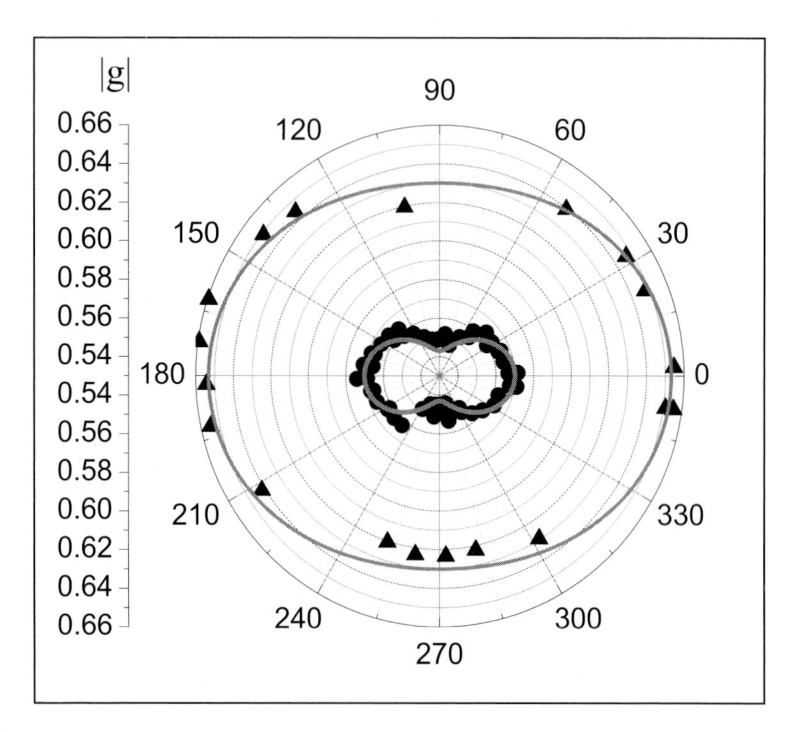

Fig. 5.2. In-plane angular dependence of the electron (circles) and exciton (triangles) g-factors obtained from the circular dichroism experiments. Solid lines are fits to data, see text for details. $B = 5\,\text{T}$. Angle zero corresponds to field orientation along the x-axis which coincides with the [110] crystal axis. [19]

5.4. Creation of Spin Coherence by Spin Initialization

For addressing electron spin coherence, the quantum dot sample was studied by Faraday rotation spectroscopy. The pump beam was circularly polarized and directed along the heterostructure growth direction. Since it was resonant with the ground state, it may inject an electron and a hole into the conduction and valence band ground states of the quantum dots. These carriers will have well defined spin orientations due to the optical selection rules. For example, for σ^+ (σ^-)-excitation the electron have a spin projection, $S_z = -1/2$ ($S_z = +1/2$), along z, while the total angular momentum of the hole (being the sum of the orbital moment and the spin moment coupled due to the spin-orbit interaction) are $J_z = +3/2$ ($J_z = -3/2$). Injection of such an electron-hole combination will of course only be possible if it is allowed by the Pauli principle, since there is already an electron in the quantum dot due to the doping: The spin orientation of this residual electron has to be opposite to the one of the optically injected electron in order to allow for optical excitation.

Any optically created spin imbalance along z corresponds to a net spin polarization. If one assumes a spin polarization in the ensemble such that, for example, the quantum dots contain more electrons with spin-up ($S_z = +1/2$ in state $|\uparrow\rangle$) than with spin-down ($S_z = -1/2$ in state $|\downarrow\rangle$) this will be reflected by the transmitted probe beam (propagating under a slight angle relative to z, to avoid interference with the pump). Its linear polarization can be decomposed into two counter-circularly polarized components of equal weight. Due to the spin imbalance, the interaction of the σ^+-polarized part will be smaller than that of the σ^--polarized one, leading to different propagation speeds. Adding the two components behind the sample will result again in linear polarization but due to their different propagation times a phase shift has occurred, reflected by a rotation of the polarization plane. This angle of rotation is measured in our experiments.

This is the description for a quasi-static situation. In the following the carrier spins are injected in a transient fashion, as after some time the electron-hole pair will recombine radiatively. In addition, a static magnetic field is applied normal to the spin orientation so that the carrier spins precess about this field, which is oriented along x. Due to this precession also the spin polarization oscillates, which can be mapped through the oscillating rotation angle of the probe beam's polarization. An example for the experimental data which can be obtained in that way is given in Figure 5.3(a) showing the Faraday rotation signal of the (In,Ga)As/GaAs quantum dots versus the delay between pump and probe for different magnetic field strengths. Pronounced electron spin quantum beats are observed with some additional modulation at high B.

In a quantum mechanical language the precession corresponds to a quantum beating between two spin-split levels. For the electron for example, the two Zeeman-split eigenstates in transverse magnetic field are spin parallel and spin antiparallel to the field, i.e. the spin points either along the $+x$ or the $-x$ direction. Using the S_z-basis, these states can be written as: $|\pm x\rangle = (|\uparrow\rangle \pm |\downarrow\rangle)/\sqrt{2}$, reflecting the zero spin polarization along z. Shining in with a laser pulse which is short enough so that its spectral width covers the energy separation between the split states, can lead to excitation of a superposition of the two split states. The time evolution of this superposition shows oscillations with a frequency corresponding to the splitting.

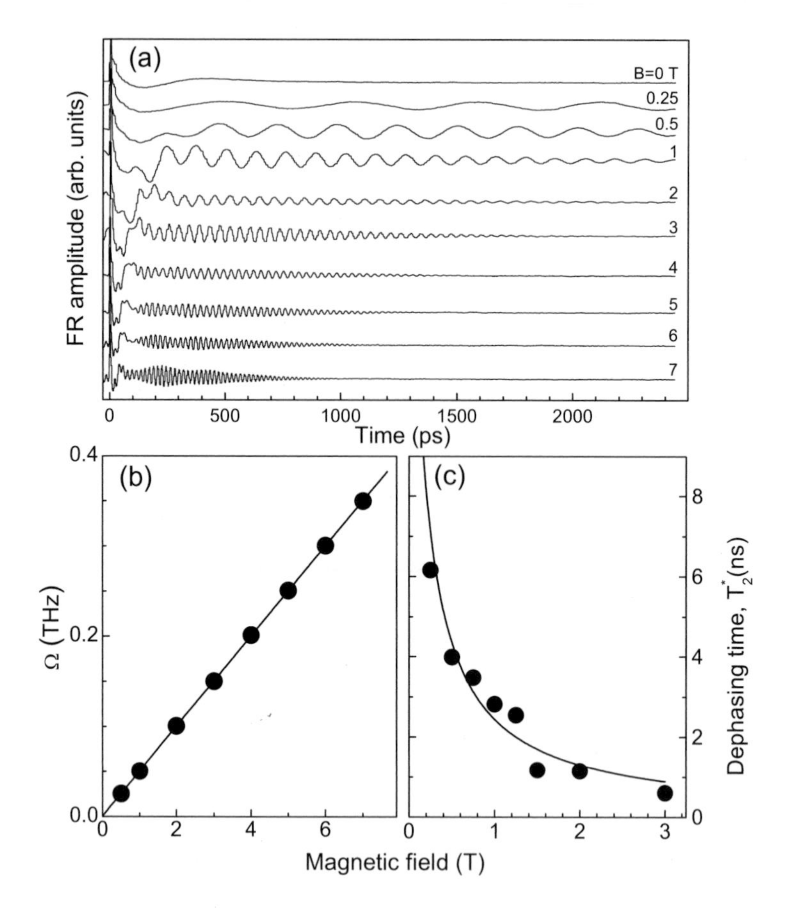

Fig. 5.3. (a): Faraday rotation traces of n-doped (In,Ga)As/GaAs quantum dots vs delay between pump and probe at different magnetic fields. The pump power was ~10 mW. (b): Field dependencies of the electron precession frequency. The line is a B-linear fit to the data, to determine the electron g-factor. (c): Spin dephasing time T_2^* versus B. The line is a $1/B$-fit to the data, to determine the electron g-factor variation in the ensemble. [15]

The modulations of the beats at strong applied fields are seen only at short delay times not exceeding 400 ps. This time corresponds to the lifetime of electron-hole pairs, as determined from time-resolved photoluminescence. As can be seen from the signal at weak fields, for longer delays the Faraday rotation signal contains oscillations with only a single frequency component, which are exponentially damped. The lifetime of these beats is as long as 4 ns at $B = 0.5\,\mathrm{T}$, for example, exceeding essentially the lifetime of optically excited carriers. Therefore we attribute these long-lived oscillations to residual electrons in the dots. The modulation at early delays apparently arises from interference of the long-lived oscillation with an oscillation related to optically excited carriers, and the two oscillation frequencies lie close to each other, so that the observed beating behavior occurs.

Three features are to be noted for the appearance of the oscillations:

- We have first analyzed the long lived precession component by the form:

$$\exp\left(-\frac{t}{T_2^\star}\right)\cos\left(\omega_e t\right), \tag{5.3}$$

  where $T_2^\star$ is the dephasing time and ω_e is the electron precession frequency given by the spin-splitting $\omega_e = g_{e,x}\mu_B B/\hbar$ with the Bohr magneton μ_B. $g_{e,x}$ is the electron g-factor along the field. From the field dependence of the precession frequency the g-factor can be determined, and this was the technique which was applied to gain the data shown in Section II.
  Fig. 5.3(b) shows the field dependence of the precession frequency ω_e (the circles) obtained from fitting our data by the form above. It is in agreement with a linear dependence on B, as expected. Note that in general also deviations from such a linear behavior might occur if the magnetic field modifies the band structure, leading to a change of the g-factor. This might be the case in particular for holes, but less so for electrons. From a B-linear fit [the black solid line in Fig. 5.3(b)] we obtain $\mid g_{e\perp} \mid = 0.57$.
- The spin beats become increasingly damped with increasing magnetic field, corresponding to a reduction of the ensemble spin dephasing time $T_2^\star$, plotted in Fig. 5.3(c). The damping arises from variations Δg_e of the electron g-factor within the quantum dot ensemble, which are translated into a spread of the precession frequency: $\Delta\Omega_e = \Delta g_e \mu_B B/\hbar$. The electron spins become oriented at the moment of pump pulse arrival, after which they start to precess about the field. Due to the frequency variations the precessions of the electrons run out of phase with increasing delay, so that the coherent signal is reduced. Note, however, that this is a destructive interference effect from the ensemble, but does not mean that the coherence of an individual spin in a quantum dot is lost.

Obviously the frequency spread increases linearly with increasing magnetic field, which in the time domain (as measured by $T_2^\star$) leads to a dependence inversely proportional to the magnetic field. Therefore the dephasing can be described by $[T_2^\star(B)]^{-1} = [T_2^\star(0)]^{-1} + \Delta g_e \mu_B B/\sqrt{2}\hbar$. The solid line in Fig. 5.3(c) shows a $1/B$-fit to the $T_2^\star$ data, by which a g-factor variation $\Delta g_e = 0.004$ has been extracted, which is only about 0.7 % of the mean value. This variation appears to be surprisingly small as we address an inhomogeneously broadened ensemble of millions of dots. However, one has to keep in mind that we select by our laser pulse a rather narrow energy range of about 1 meV of quantum dot ground state transition energies.
From the data one can also conclude that $T_2^\star(0)$ exceeds 6 ns in the limit of zero magnetic field, for which the g-factor variations are avoided. The zero-field dephasing is mainly caused by electron spin precession about the frozen, but random magnetic field of the nuclei [21]. The net nuclear orientation varies from dot to dot, and it is these variations that lead to ensemble spin dephasing. On a ns time scale the electron spin polarization drops to one third of the initial value, and on a longer time scale the polarization is completely lost due to flip-flop processes with the nuclei.

- As discussed, the additional modulation of the quantum beats at high fields can be assigned to photoexcited carriers, which precess about the field with a frequency close to that of the electron. This results in the beating from interference of the two signals. From the data in Fig. 5.2 the exciton has a similar g-factor as the electron, and therefore we attribute these short lived beats to the exciton spin precession in quantum dots which do not contain a resident electron. This precession persists only during the radiative decay of the excitons, which is in good accord with the beat lifetime. From the ratio of the amplitudes of the electron and exciton beats we have estimated the ratio of charge neutral and singly charged quantum dots, that has been given above. Dots with more than two electrons do not show a considerable spectroscopic response in resonant Faraday rotation, as Pauli-blocking by the spin singlet state of the two electrons prevents interaction with light, so that these dots do not show a laser induced spin polarization.

After this analysis of the static g-factor properties, we address now the question, why spin precession is observed at all. At least in high magnetic fields, for which the spin-splitting is quite large compared to the thermal energy, the system is in thermal equilibrium before photoexcitation. This means that the spin is either parallel or antiparallel to the magnetic field along x, corresponding to zero spin polarization along the light propagation axis, so that the probe pulse does not undergo a rotation of polarization. Through the optical excitation we are apparently able to rotate the electron spin by 90 deg., so that the observed spin precession can occur. To obtain further insight into the underlying mechanism, additional information is needed:

Figure 5.4(a) shows FR signals at $B = 1\,\mathrm{T}$ for different powers of the pump pulses. The corresponding FR amplitudes are plotted in Fig. 5.4(b) versus the laser pulse area Θ, which is defined as $\Theta = 2\int [\boldsymbol{d}\boldsymbol{E}(t)]\,dt/\hbar$ in dimensionless units. Here $\boldsymbol{d}$ is the dipole matrix element for the transition from the valence to the conduction band. For pulses of constant duration, but varying power, as used here, Θ is proportional to the square root of excitation power, and it is given in arbitrary units in Fig. 5.4(b). The Faraday rotation amplitude shows a non-monotonic behavior with increasing pulse area. It rises first to reach a maximum, then drops to about 60%. Thereafter it shows another strongly damped oscillation.

This behavior is similar to the one known from Rabi-oscillations of the Bloch vector describing an electron-hole excitation, whose z-component describes the corresponding population [22]. The laser pulse coherently drives this population, leading to coherent oscillations as function of the pulse area Θ. For $\Theta = 0$ (no pulse) it does not change the population, while for $\Theta = \pi$ the system inverted, leading to electron-hole pair population in an undoped quantum dot. For $\Theta = 2\pi$ the Bloch vector is rotated by 360° and so on. In our case, the Faraday rotation amplitude becomes maximum when applying a π-pulse as pump, and it becomes minimum for a 2π-pulse. However, the observed oscillations are strongly damped.

To observe periodic oscillations, this damping would have to be suppressed as much as possible, that is, the system has to be quite homogeneous and the driving laser pulse has to be shorter than any decoherence times. In our case, the damping of the oscillations most likely is due to ensemble inhomogeneities of quantum dot properties such as the dipole matrix element $\boldsymbol{d}$ [5].

Based on these observations, we can understand the mechanism of generating electron spin coherence in the quantum dots. For that purpose we

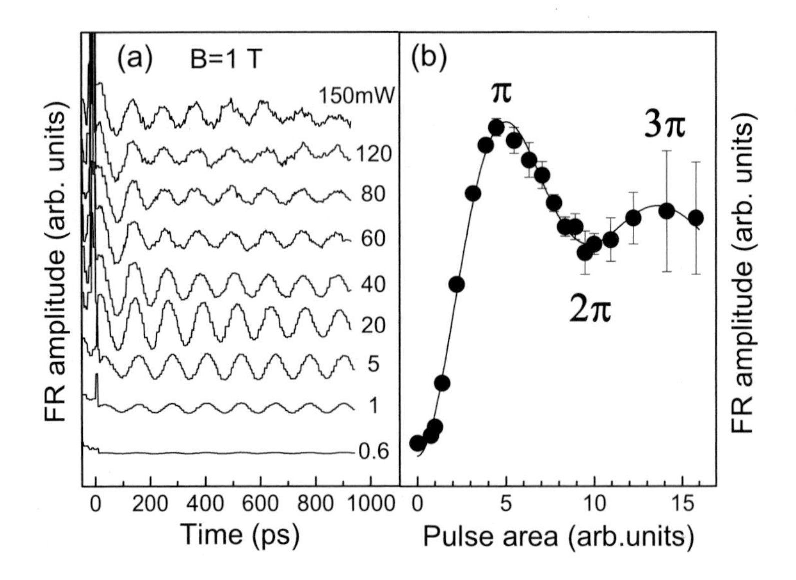

Fig. 5.4. (a): Short delay closeup of the Faraday rotation signal at $B = 1\,\mathrm{T}$ for different pump powers. (b): Faraday rotation amplitude, as determined from fitting the data in panel (a) by an exponentially damped harmonic, versus laser pulse area Θ. The line is a guide to the eye. [15]

discuss first charge neutral dots. A resonant optical pulse with σ^- polarization creates a superposition of vacuum and exciton:

$$\cos\left(\frac{\Theta}{2}\right) \mid 0\rangle - i \sin\left(\frac{\Theta}{2}\right) |\uparrow\Downarrow\rangle, \tag{5.4}$$

where $\mid 0\rangle$ describes the deexcited semiconductor. The hole spin orientations $J_{h,z} = \pm 3/2$ are symbolized by the arrows $\Uparrow$ and $\Downarrow$, respectively. The electron and hole spins in the exciton become reversed for σ^+-excitation. The exciton component in the superposition precesses about the magnetic field for a time, which cannot last longer than the exciton lifetime. The amplitude of the contribution to the ensemble Faraday rotation signal is given by the square of the exciton coefficient $\sin^2(\Theta/2)$. The Faraday rotation signal becomes reduced either by dephasing in the ensemble or by decoherence through spin-scattering of electron or hole.

Let us turn now to singly charged quantum dots, for which the resonant excitation leads to excitation of trions. We assume that the deexcited quantum dot state is given by an electron with arbitrary spin orientation:

$$\alpha \left|\uparrow\right\rangle + \beta \left|\downarrow\right\rangle, \tag{5.5}$$

with $\mid \alpha \mid^2 + \mid \beta \mid^2 = 1$. As seen above, a σ^--polarized laser pulse 'tries to place' an exciton with spin configuration $\left|\uparrow\Downarrow\right\rangle$ in the quantum dot. This action is, however, restricted by the Pauli-principle, due to which the optically excited electron must have a spin orientation opposite to the resident one. Therefore the pulse can excite only the second part of the initial electron state.

In consequence a coherent superposition state of an electron and a trion is created:

$$\alpha \left|\uparrow\right\rangle + \beta \cos\left(\frac{\Theta}{2}\right) \left|\downarrow\right\rangle - i\beta \sin\left(\frac{\Theta}{2}\right) \left|\downarrow\uparrow\Downarrow\right\rangle. \tag{5.6}$$

The trion consists of two electrons combining to a spin singlet and a hole in state $\mid \Downarrow\rangle$. Here we assume that decoherence does not occur during excitation, i.e., the pulse length is much shorter than the radiative decay and the carrier spin relaxation times. One sees that the electron-hole population oscillates with pulse area Θ. Restricting to rather low pump powers, where damping effects are quite small, the excitation is most efficient for $\Theta = \pi$, which gives the superposition state:

$$\alpha \left|\uparrow\right\rangle - i\beta \left|\downarrow\uparrow\Downarrow\right\rangle, \tag{5.7}$$

After some time the electron hole pair will relax, leaving a resident electron in the quantum dot. In the ensemble this occurs on a mean time scale given by the radiative lifetime. This ensemble averaging will wipe out any contribution from the second summand to the Faraday rotation signal. If before recombination hole spin relaxation occurs, the situation will not be changed, as the ensemble average will again nullify the contribution from the second part.

The efficiency of this protocol is obviously determined by the quality of the suppression of the pure $\left|\downarrow\right\rangle$-component which in effect reduces the electron spin polarization along z. The probability to excite it is given by $\cos^2(\Theta/2)$, or vice versa, the probability to avoid it is $1 - \cos^2(\Theta/2) = \sin^2(\Theta/2)$. Since the Faraday rotation signal is proportional to the electron spin polarization, we expect a dependence proportional to $\sin^2(\Theta/2)$, neglecting any damping. This is reflected by the observed Rabi-oscillations in Fig. 5.4.

Let us consider the problem more quantitatively: By variation of the area Θ not only the electron and trion state populations are changed periodically with period $\Theta = 2\pi$, but also the orientation of electron and trion spins $\boldsymbol{S}$ and $\boldsymbol{J}$ are controlled. The electron spin polarization can be described by a spin vector $\boldsymbol{S} = (S_x, S_y, S_z)$ defined by: $S_x = \mathrm{Re}(\alpha\beta^*)$, $S_y = -\mathrm{Im}(\alpha\beta^*)$, $S_z = (1/2)(|\alpha|^2 - |\beta|^2)$. Similarly, one can introduce the spin vector, $\boldsymbol{J} = (J_x, J_y, J_z)$, which represents the polarization of the trion, $|\bar{\psi}\rangle = \bar{\alpha}|\uparrow\downarrow\Uparrow\rangle + \bar{\beta}|\uparrow\downarrow\Downarrow\rangle$. The spin vectors $\boldsymbol{S}$ and $\boldsymbol{J}$ represent 6 of the 16 components of the four level density matrix, and their dynamics is given by the corresponding equations of motion [23].

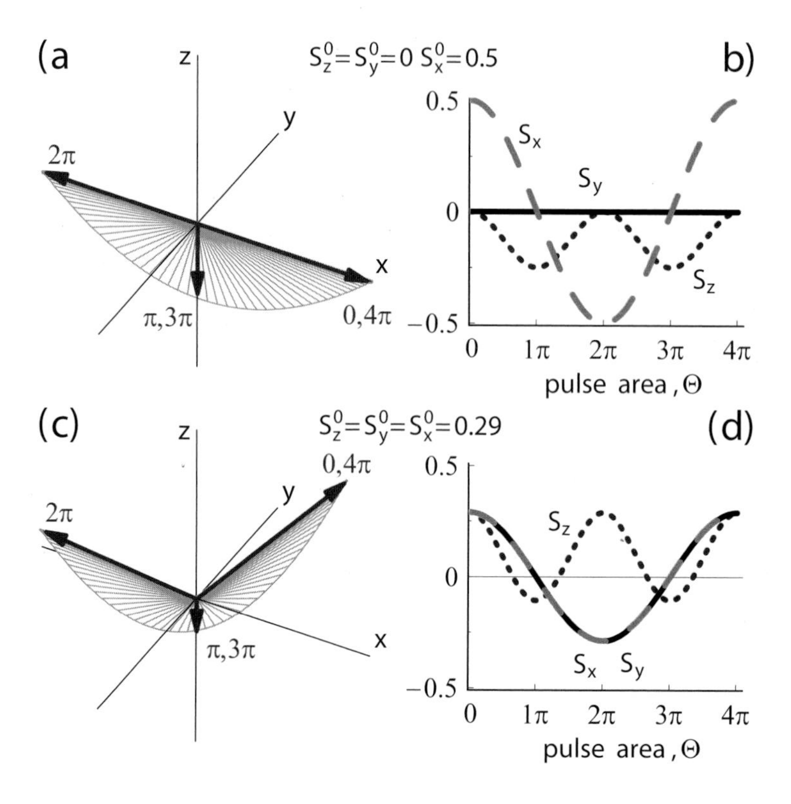

Fig. 5.5. (a) and (c): Reorientation of the electron spin polarization by application of a resonant optical pulse of varying area as denoted. Calculations have been done for two different initial values of spin polarization, S_x^0 and S_z^0. (b) and (d): Electron spin polarization components versus pulse area Θ. [15]

The electron spin vector evolution as function of Θ is shown in Fig. 5.5 for two initial orientations: one is parallel to the magnetic field and the other exemplifies an arbitrary direction (see figure caption for details). A short σ^+ polarized pulse excites the initial electron spin state, $\alpha|\uparrow\rangle + \beta|\downarrow\rangle$, into an electron-trion superposition state $\alpha\cos(\Theta/2)|\uparrow\rangle + \beta|\downarrow\rangle - i\alpha\sin(\Theta/2)|\uparrow\downarrow\Uparrow\rangle$. The light induced change of the S_z component, $|S_z - S_z^0| = |\alpha|^2\sin^2(\Theta/2)$ varies with the $|\uparrow\rangle$ state population, and independently of the initial conditions it reaches maximum for $\Theta = (2n+1)\pi$-pulses, for which the S_x and S_y components vanish. In particular, $S_z([2n+1]\pi) = -0.25$ for $S_z^0 = 0$ [24]. Unlike the S_z component, the electron spin components normal to z swing between their initial directions (S_x^0, S_y^0, S_z^0) and the opposite direction $(-S_x^0, -S_y^0, S_z^0)$ with a period of 4π, because the $S_{x,y}$ components are proportional to $\cos(\Theta/2)$. They describe the coherence of the electron spin state and both vary with the phase of the spin wave function.

The control of spin dynamics by an optical pulse allows for a fast spin alignment. In a quantum dot ensemble, a small area pulse, $\Theta \ll 1$, induces a coherent spin polarization proportional to Θ [25]. With increasing Θ, the total spin polarization oscillates with a period 2π as does the S_z component of each individual spin in the ensemble, explaining the Faraday rotation amplitude oscillations in Fig. 5.4. The long trion lifetimes in our quantum dots could enable realization of a regime in which a pulse of rather low power, but long duration can be used to reach a large pulse area without decoherence due to radiative decay. Further, the S_x and S_y components change sign with period 2π. This implies that $2n\pi$-pulses can be used for refocusing the precessing spins, similar to spin-echo techniques [26].

Let us turn now to the spin dynamics after initialization by a short pulse. Then the off-diagonal component of the density matrix, describing the optical coherence between electron and trion, is decoupled from the electron and trion spin vectors, which are governed independently by two vector equations:

$$\frac{d\boldsymbol{J}}{dt} = [\boldsymbol{\Omega}_h \times \boldsymbol{J}] - \frac{\boldsymbol{J}}{\tau_s^h} - \frac{\boldsymbol{J}}{\tau_r},$$

$$\frac{d\boldsymbol{S}}{dt} = [(\boldsymbol{\Omega}_e + \boldsymbol{\Omega}_N) \times \boldsymbol{S}] + \frac{\left(\boldsymbol{J}\hat{\boldsymbol{z}}\right)\hat{\boldsymbol{z}}}{\tau_r}, \tag{5.8}$$

where $\boldsymbol{\Omega}_{e,h} \parallel \boldsymbol{e}_x$ and $\boldsymbol{\Omega}_N = g_e\mu_B\boldsymbol{B}_N/\hbar$ is the electron precession frequency in the effective magnetic field, $\boldsymbol{B}_N$ due to the frozen nuclear configuration in a dot. In the second equation we do not include the electron spin relaxation time, τ_s^e, explicitly. At low temperatures, τ_s^e is on the order of μs and is mainly determined by fluctuations of the nuclear field $\mathbf{\Omega}_N$ in a single quantum dot [12, 16, 21, 27]. This long time scale is irrelevant for our problem. The spin relaxation of the hole in the trion, τ_s^h, is caused by phonon assisted processes and at low temperatures may be as long as τ_s^e [28, 29].

Solving equation (5.8) we obtain the time evolution of the spin vectors $\boldsymbol{S}$ and $\boldsymbol{J}$. After trion recombination ($t \gg \tau_r$), the amplitude of the long-lived electron spin polarization excited by a $(2n+1)\pi$-pulse is given by

$$S_z(t) = \mathrm{Re}\left\{\left(S_z(0) + \frac{0.5J_z(0)/\tau_r}{\gamma_T + i(\omega + \Omega_h)} + \frac{0.5J_z(0)/\tau_r}{\gamma_T + i(\omega - \Omega_h)}\right)\exp(i\omega t)\right\}, \tag{5.9}$$

where $S_z(0)$ and $J_z(0)$ are the electron and trion spin polarizations created by the pulse. $\omega = \Omega_e + \Omega_{N,x}$. $\gamma_T = 1/\tau_r + 1/\tau_s^h$ is the total trion decoherence rate. If the radiative relaxation is fast $\tau_r \ll \tau_s^h, \Omega_{e,h}^{-1}$, the induced spin polarization $S_z(t)$ is nullified on average by trion relaxation, as $S_z(0) = -J_z(0)$. In contrast, if the spin precession is fast, $\Omega_{e,h} \gg \tau_r^{-1}$, the electron spin polarization is maintained after trion decay [24, 30]. This is the situation that we find in our experiment.

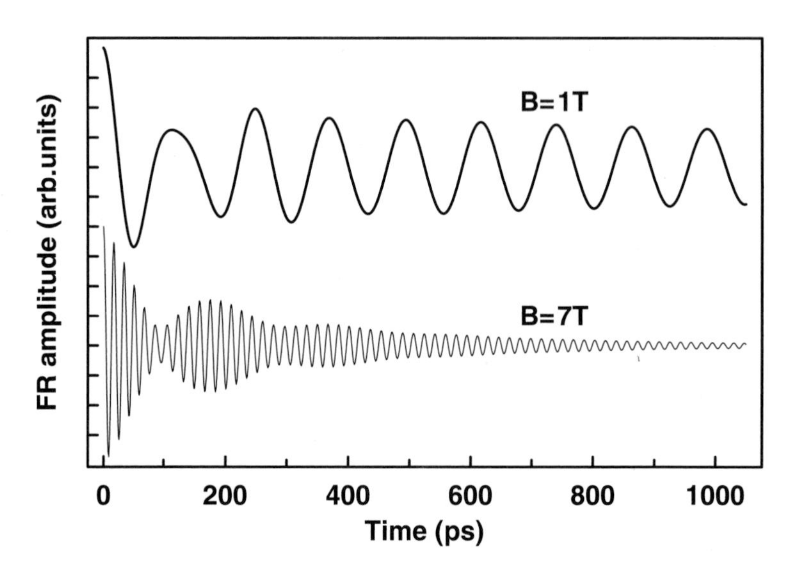

Fig. 5.6. Calculated time dependence of pump-probe Faraday rotation signal of n-doped quantum dots excited by a σ^+ polarized pulse. τ_r=400 ps, τ_s^h=170 ps, $|\, g_e \,| = 0.57$, and $\Delta g_e = 0.004$. [15]

For an ensemble of quantum dots, the electron spin polarization is obtained by averaging equation (5.8) over the distribution of g-factors and nuclear configurations. At low B, the random nuclear magnetic field becomes more important for the electron spin dephasing than the g-factor dispersion, leading to dephasing during several nanoseconds [21]. As discussed, the rotation of the linear probe polarization is due to the difference in scattering of its σ^+ and σ^- polarized components by one of the transitions $|\uparrow\rangle \rightarrow |\uparrow\downarrow\Uparrow\rangle$ and $|\downarrow\rangle \rightarrow |\uparrow\downarrow\Downarrow\rangle$. The scattering efficiency is proportional to the population difference of the states involved in these transitions $\Delta n_+ = n_\uparrow - n_\Uparrow$ or $\Delta n_- = n_\downarrow - n_\Downarrow$. The Faraday rotation angle is $\phi(t) \sim (\Delta n_+ - \Delta n_-)/2 = S_z(t) - J_z(t)$. Figure 5.6 shows the Faraday rotation signal after a σ^+-polarized excitation pulse, calculated with input parameters corresponding to the experimental situation. The traces are in good accord with the experimental situation. In particular, at $B = 7\,\mathrm{T}$, the Faraday rotation shows modulated beats resulting from interference of the electron and exciton precessions.

5.5. Electron Spin Coherence

For quantum information applications, the electron spin dynamics needs to be understood in detail. In particular, the time scales during which the information of a spin state is retained have to be addressed. Phenomenologically the spin dynamics can be described by two times, the longitudinal spin relaxation time T_1 and the transverse spin relaxation time T_2 . In a simple picture these time scales can be understood in the following way. A 'longitudinal' magnetic field leads to a spin-splitting. The T_1-time describes then the time scale on which the relaxation of a spin from the upper into the lower state occurs. If the magnetic field is, on the other hand, oriented normal to the spin, the spin precesses about the field. In this case, the T_2 time describes the time during which the precession is going on in an unperturbed way until the first scattering followed by a phase change occurs. It is this latter time scale which is the relevant one for quantum information processing.

In the previous chapter we had introduced an additional time constant, $T_2^\star$, to describe the decay of the ensemble coherent signal, called dephasing. As we had pointed out, one of the origins of this fast decay in the ns-range are the ensemble inhomogeneities which lead to a strong variation of the precession frequency. These variations are dominant at strong enough fields. Toward zero field the spin coherence lifetime is limited by dot-to-dot variations of the nuclear fields about which each electron precesses. Besides such momentary inhomogeneities in an ensemble, T_2^* might in general be also limited by variations of the experimental conditions during the measurement time, such as signal integration times which are much longer than the time during which the conditions can be kept stable. This is the typical case for single quantum dot measurements, for which one has to perform an experiment many times in order to get a statistically significant result.

Generally the dephasing time is much shorter than T_2. Note, however, that dephasing does not lead to a destruction of the coherence of an individual spin. But it does mask the duration of the single spin coherence due to the rapid loss of coherence among the phases of the spins. Theoretically, the single-spin coherence time may be as long as twice the spin relaxation time, which is on the scale of milliseconds [12], as recent experiments have demonstrated. The true spin coherence time may be obtained by sophisticated spin-echo techniques [31], which typically are quite laborious. In general, a less complicated and robust measurements scheme would be highly desirable, by which also the spin coherence could be preserved so that ultimately many of the operations critical to the processing of quantum information, including initialization, manipulation, and read-out of a coherent spin state, would become possible.

To address this point, we look again at Faraday rotation traces, recorded in a similar way as those discussed above. There we had shown the traces only for positive delays between pump and probe. Now we take a look also at the negative delay side. This is done for two different magnetic fields, 1 and 6 T, in Fig. 5.7, lower panel. Long-lived electron spin quantum beats are seen at positive delays, as discussed before. Surprisingly, also for negative delays strong spin beats with a frequency corresponding to the electron precession are observed. The amplitude of these quantum beats increases when approaching zero delay $t = 0$. Spin beats at negative delay have been reported for experimental situations in which the decay time exceeds the time interval between the pump pulses: $T_2^* \geq T_R$ [10]. This is clearly not the case here, where the Faraday rotation signal has fully vanished after 1.5 ns at $B = 6$ T, for example, so that the dephasing is much faster than the pulse repetition period. Within the experimental error, the rise time of the signal at negative delays is the same as the decay time at positive delays for all magnetic fields, suggesting that the negative delay signal also can be traced to electron spin precession.

The upper panel shows the signal when scanning the delay over a larger range in time, in which four pump pulses, separated by 13.2 ns from each

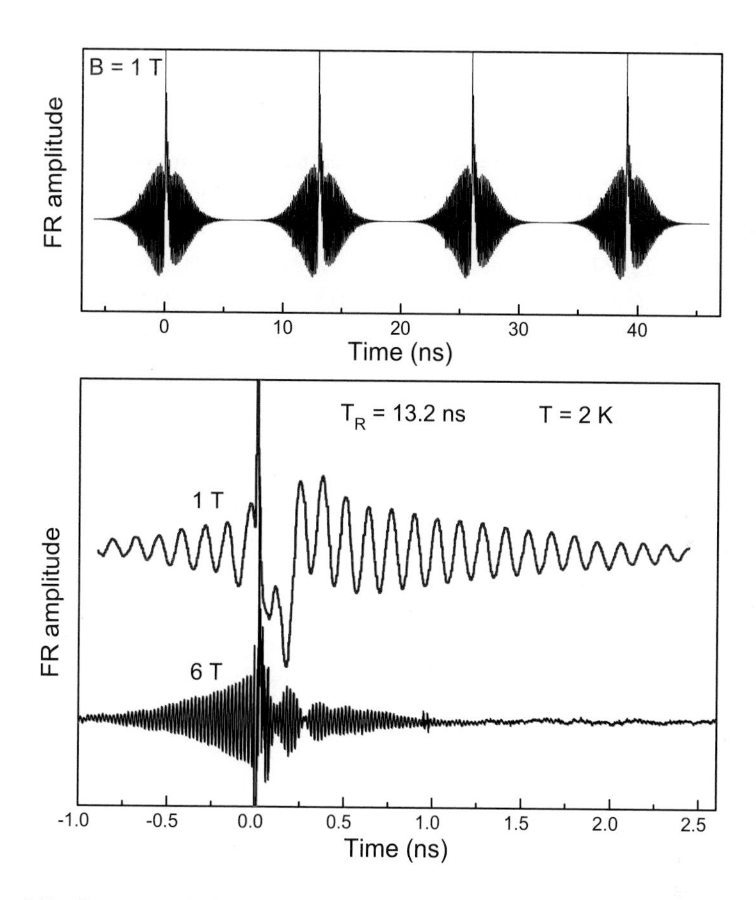

Fig. 5.7. Lower panel: Pump-probe Faraday rotation signal versus delay measured at $B = 1$ and 6 T on singly charged (In,Ga)As/GaAs quantum dots. The pump power density is $60\,\mathrm{W/cm^2}$, the probe density is $20\,\mathrm{W/cm^2}$. Upper panel: Faraday rotation signal recorded for a longer delay range in which four pump pulses were located. [34]

other, are located. At each pump arrival electron spin coherence is created, which after a few ns is quickly dephased. Before each pump arrival the coherent signal from electrons appears again. This negative delay precession can occur only if the coherence of the electron spin in each single dot prevails for much longer times than T_R. Independent of the origin of the coherent signal, this opens a pathway towards measuring the spin coherence time T_2: When increasing the pump pulse separation continuously until it becomes comparable with the T_2-time, the amplitude of the signal on the negative delay side should continuously decrease. Further, if the pump pulse separation is increased far above the coherence time, the signal should vanish completely.

Corresponding data at $B = 6\,\mathrm{T}$ measured for two pump densities differing by a factor of two are given in Fig. 5.8, showing the Faraday rotation amplitude on the negative delay side shortly before the next pump arrival as function of T_R. T_R was increased from 13.2 up to 990 ns by means of a laser pulse-picker. A significant Faraday rotation signal can be measured even for the longest pulse interval of a μs. For longer T_R it is difficult to measure a coherent signal, not because T_R would exceed the coherence time, but because by increasing T_R the duty cycle in the measurement protocol is too low, i.e., the effective time during which signal is collected is reduced too much. From the data which have been corrected for the real measurement times, we see, that a drop of Faraday rotation amplitude is seen, meaning that we scan a pump separation range comparable with T_2.

In order to understand why the single quantum dot coherence time can be seen at all in an ensemble measurement, let us consider excitation of a single quantum dot by a periodic train of π-pulses which all have the same circular polarization. The first impact of the pulse train is a synchronization of electron spin precession. For discussing this effect we define the degree of spin synchronization by $P(\omega_e) = 2|S_z(\omega_e)|$. Here the z-component of the electron spin vector, $S_z(\omega_e)$, is taken at the moment of pulse arrival. If the pulse period, T_R, is a multiple N of the electron spin precession period, $2\pi/\omega_e$, the subsequent action of such π-pulses leads to almost complete electron spin alignment along the light propagation direction z [24]. In general the synchronization is given by $P_\pi = \exp(-T_R/T_2)/[2 - \exp(-T_R/T_2)]$ for π-pulses [24]. In our case it reaches its largest value $P_\pi = 1$, corresponding to almost 100% synchronization, because for excitation with a high repetition rate (as in experiment) $T_R \ll T_2$ so that $\exp(-T_R/T_2) \approx 1$.

An ensemble contains quantum dots whose precession frequencies fulfill the synchronization relation with the laser which we term phase synchronization condition in the following:

$$\omega_e = 2\pi N/T_R \equiv N\Omega. \tag{5.10}$$

Since the electron spin precession frequency is typically much larger than the laser repetition rate for not too small magnetic fields, $N \gg 1$. As in addition the spread of precession frequencies is also much larger than the laser repetition rate, multiple subsets within the optically excited quantum dot ensemble satisfy the equation (5.10) for different N. This is illustrated by

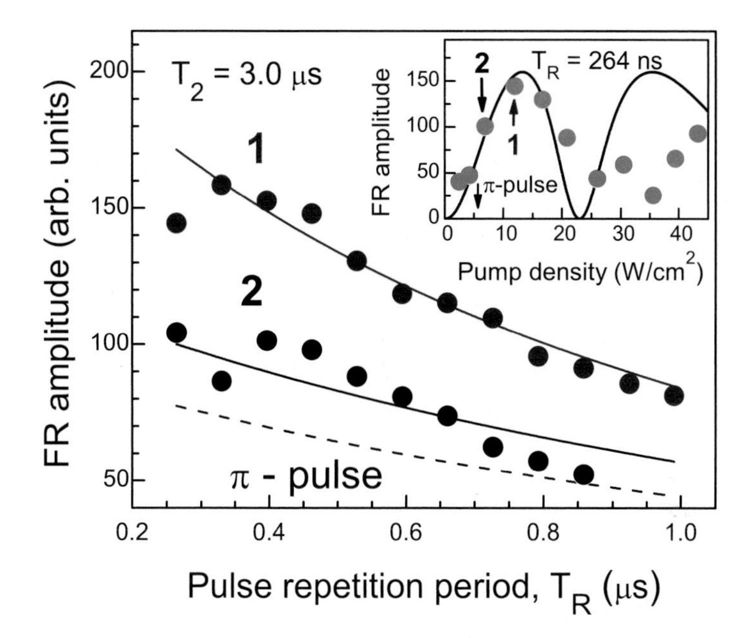

Fig. 5.8. Faraday rotation amplitude at negative delay as function of time interval between pump pulses. The experimental data were measured at $B = 6\,\mathrm{T}$ for two pump densities of 12 and $6\,\mathrm{W/cm^2}$ shown in the inset by the red and blue arrows. $T = 6\,\mathrm{K}$. The solid lines show the theoretical dependencies [34], which contains as single fit parameter $T_2 = 3.0\,\mu\mathrm{s}$. In the inset the Faraday rotation amplitude measured at $T_R = 264\,\mathrm{ns}$ is shown as function of pump density. The solid line shows again the theoretical dependence. The comparison of experiment and theory allows us to determine the pump density, which corresponds to a π-pulse (shown by the black arrow). The theoretical dependence of the Faraday rotation amplitude on T_R calculated for π-pulse excitation is shown by the dashed line. [34]

Fig. 5.9, where panel A sketches the precession for $N = K$ and $K + 1$, and panel B gives the spectrum of phase synchronized precession modes. The number of synchronized subsets, ΔN, can be estimated from the broadening of the electron spin precession frequencies, γ, by: $\Delta N \sim \gamma/\Omega$. It increases linearly with magnetic field, B, and pulse period, T_R. The spins in each subset precess between the pump pulses with frequency $N\Omega$, starting with an initial phase which is the same for all subsets. Their contribution to the spin polarization of the ensemble at a time t after the pulse is given by $-0.5\cos(N\Omega \cdot t)$. As sketched in Fig. 5.9, the sum of oscillating terms from all synchronized subsets leads constructive interference of their contributions to Faraday rotation signal around the times of pump pulse arrival. The rest of the quantum dots does not contribute to the average electron spin polarization $\overline{S}_z(t)$ at times $t \gg T_2^*$, due to dephasing. The synchronized spins therefore move on a background of dephased electrons, which, however, also still precess individually during the spin coherence time.

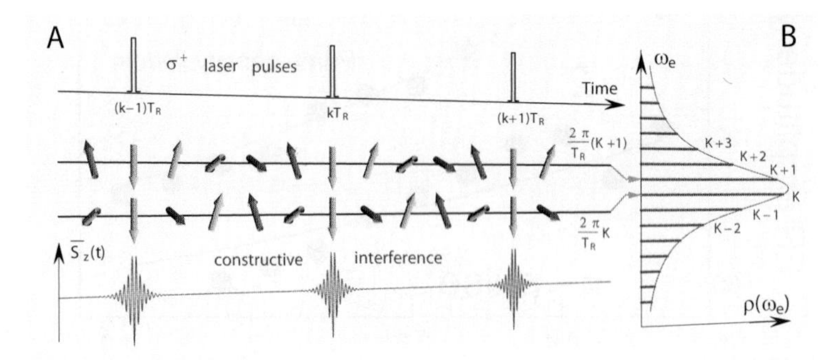

Fig. 5.9. Phase synchronization of electron spin precession by a train of π-pulses of circularly polarized light. The top panel shows the train of σ^+ polarized laser pulses with repetition period T_R. The train synchronizes the electron spin precession in quantum dots where the precession frequency is a multiple of $(2\pi/T_R)$: $\omega_e = N(2\pi/T_R)$. In these quantum dots, the spins are aligned at the moment of the pulse arrival: each spin is opposite to the light propagation direction. The two middle panels show the phase synchronization for two spins with precession frequencies differing by $2\pi/T_R$: $N = K$ and $N = K + 1$ (K is a large integer). The bottom panel shows a time evolution of the average spin polarization $\overline{S}_z(t)$, resulting from a constructive interference of the phase synchronized quantum dot subsets. (B) Spectrum of phase synchronized electron spin precession modes enveloped by the density of precession frequencies $\rho(\omega_e)$ in a quantum dot ensemble. Only those electron spins which are synchronized by the pulse train give a contribution to the spectrum, consisting of sharp peaks at the frequencies $\omega_e = N(2\pi/T_R)$ $(N = \ldots, K-1, K, K+1, \ldots)$ which satisfy the phase synchronization condition equation (5.10). [34]

The average spin polarization $\overline{S}_z(t) = -0.5\Omega \sum_{N=-\infty}^{\infty} \cos(N\Omega{\cdot}t)\rho(N\Omega)$, where $\rho(\omega_e)$ is the density of the quantum dot precession frequencies within the laser excitation profile. Assuming that this density has a Lorentzian shape [32] $\rho(\omega_e) = (\gamma/\pi)(1/[(\omega_e - \overline{\omega_e})^2 + \gamma^2])$, centered around the average frequency $\overline{\omega_e}$, an analytic result can be obtained:

$$\overline{S}_z(t) \approx \frac{\beta \cosh\{\beta[1 - 2\,\mathrm{mod}(t, T_R)]\} - \sinh\beta}{\beta \sinh\beta} \cos(\overline{\omega_e} t) \,, \qquad (5.11)$$

where $\beta = \gamma T_R/2$ and $\mathrm{mod}(x, y) = x - y\,[x/y]$ is the modul function, with $[x/y]$ defined as integer division. The resulting time dependence of $\overline{S}_z(t)$

(Fig. 5.9A) explains the appearance of Faraday rotation signal at negative delays.

Obviously, π-pulse excitation is not critical for the electron spin phase synchronization by the circularly polarized light pulse train. Any resonant pulses of arbitrary intensity create a coherent superposition of the trion and electron state in a quantum dot, leading to a long-lived coherence of resident electron spins, because the coherence is not affected by the radiative decay of the trion component. Each pulse of σ^+ polarized light changes the electron spin projection along the light propagation direction by $\Delta S_z = -(1 - 2|S_z(t \to t_n)|)W/2$, where $t_n = nT_R$ is the time of the n-th pulse arrival, and $W = \sin^2(\Theta/2)$ [15, 33]. Consequently, a train of such pulses orients the electron spin opposite to the light propagation direction, and it also increases the degree of electron spin synchronization P. Application of $\Theta = \pi$-pulses (corresponding to $W = 1$) leads to a 99% degree of electron spin synchronization already after a dozen of pulses. However, if the electron spin coherence time is long enough ($T_2 \gg T_R$), an extended train of pulses leads to a high degree of spin synchronization also for $\Theta \ll 1$ ($W \approx \Theta^2/4$).

The general consideration of this problem (see [34] and SOM to this paper) gives a degree of spin synchronization $P(\omega_e) = 2|S_z(\omega_e)|$:

$$P(\omega_e) = \frac{(W/2T_R)(W/2T_R + 1/T_2)}{(W/2T_R + 1/T_2)^2 + (\omega_e - N\Omega)^2} . \tag{5.12}$$

One sees that: (i) a train of pulses synchronizes the spin precession of quantum dot electrons with precession frequencies in a narrow range of width $W/2T_R + 1/T_2$ around the phase synchronization condition, (ii) the electron spin synchronization still reaches 100% if $W/2T_R \gg 1/T_2$.

The effect of the pump density (namely of the pump area) on the distribution of the spin polarization synchronized by the pulse train at moment of pulse arrival ($t = t_n$) for $\Theta = 0.4\pi$, π and 1.6π, is shown in Figs. 5.10A-C. Calculations were done for $T_R = 13.2$ ns (red) and 52.8 ns (blue). The density of the electron spin precession modes is shown by the black line, which gives the envelop of the spin polarization distribution. The quasi-discrete structure of the distribution created by the pulse train is the most important feature, which allows us to measure the long spin coherence time of a single quantum dot on an ensemble: A continuous density of spin precession modes would cause fast dephasing with a time inversely proportional to the total width of the frequency distribution: $T_2^* = \hbar/\gamma$. Only the gaps in the density of precession modes facilitate the constructive interference at negative delay times in Fig. 5.7. These gaps are created by mode locking of the electron spins with the periodic laser emission.

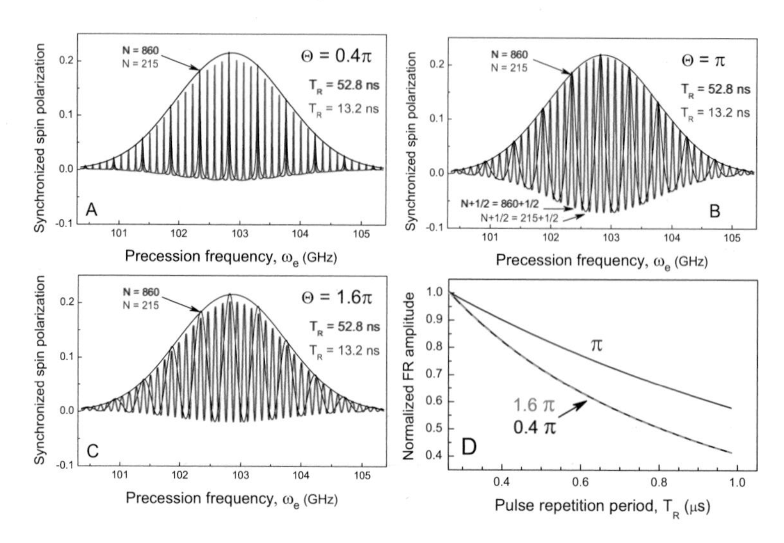

Fig. 5.10. Spectra of phase synchronized electron spin precession modes created by a train of circularly polarized pulses, $-\overline{S}_z(t_n)$, calculated for the pulse area $\Theta = 0.4\pi$, π, and 1.6π, respectively. The spectra have been calculated for two pump pulse repetition periods: $T_R = 13.2$ ns (red) and 52.8 ns (blue). At low pumping intensity (panel A) the pulse train synchronizes electron spin precession in a very narrow frequency range near the phase synchronization condition: $\omega_e = N(2\pi/T_R)$. The π-pulses (panel B) widen the range of synchronized precession frequencies. In addition, the electron spins with opposite polarization at frequencies between the phase synchronization condition become significantly synchronized. The degree of synchronization for these spins decreases at $\Theta > \pi$ (panel C). (D) Negative delay Faraday rotation amplitude dependence on pump pulse repetition period T_R calculated for the same three pulse areas. The amplitude is normalized to its value at $T_R = 264$ ns. All calculations have been done for a magnetic field of $B = 2$ T with $g_e = |0.57|$, $\Delta g_e = 0.005$ and $T_2 = 3.0\,\mu$s. [34]

Figure 5.10D shows the decay of the normalized Faraday rotation amplitudes as a function of pulse repetition time T_R. The decay depends on the pulse area and is minimized for a π-pulse. The decay rate for pulses with areas deviating from π are equal for $\Theta = \pi - \alpha$ and $\pi + \alpha$. We have fitted the experimental data of the Faraday rotation amplitude on T_R (Fig. 5.8) and its dependence on pump density (inset in Fig. 5.8) with theoretical dependence [34]. The fit allows us to determine the pump density corresponding to the π-pulse and the single quantum dot coherence time $T_2 = 3.0 \pm 0.3\,\mu$s, which is four orders of magnitude longer than the ensemble dephasing time $T_2^\star = 0.4$ ns at 6 T.

The Faraday rotation amplitude does not reach its largest value at π-pulse pumping (see inset of Fig. 5.8). This is because the train of π-pulses synchronizes the spin precession for a broad distribution of precession frequencies and not only for the ω_e satisfying the phase synchronization condition. For example, in the quantum dots with $\omega_e = (N + 1/2)\Omega$ the spin synchronization degree is 1/3. However, the S_z projection of electron spin in these quantum dots is opposite to the one for quantum dots which satisfy the phase synchronization condition ($\omega_e = N\Omega$), as seen in Fig. 5.10B. This leads to cancellation effects in the total Faraday rotation amplitude of the quantum dot ensemble. On the other hand, one can see in Fig. 5.10C that pulses with an area $\Theta > \pi$ are not so efficient in synchronizing the electron spin precession in quantum dots which do not satisfy the phase synchronization condition. This diminishes the "negative" contribution of such quantum dots to the electron spin polarization and increases the Faraday rotation amplitude. Generally the rise of the excitation intensity from zero to π-pulses increases the number of quantum dots contributing to the Faraday rotation signal at negative delays.

After having shown that a specific protocol for a laser pulse sequence can be used for selecting a subset of synchronized quantum dots with the single dot dephasing time, we turn to testing the degree of control that can be achieved by such sequences. For that purpose each pump is split into two pulses with a fixed delay $T_D < T_R$ between them. The results of measurements for $T_D = 1.84\,\text{ns}$ are shown in Fig. 5.11A. Both pumps were circularly co-polarized and had the same intensities. When the quantum dots are exposed to only one of the two pumps (the two upper traces), the Faraday rotation signals are identical except for a shift by T_D. The signal changes drastically under excitation by the two pulse train (the lower trace): Around the arrival of pump 1 the same Faraday rotation response is observed as before in the one-pump experiment. Also around pump 2 qualitatively the same Faraday rotation pattern is observed with considerably larger amplitude. This means that the coherent response of the synchronized quantum dot ensemble can be amplified by the second laser pulse. Even more remarkable are the echo-like responses showing up in the Faraday rotation signal before the first and after the second pump pulse. They have a symmetric shape with the same decay and rise times T_2^*. The temporal separation between them is a multiple of T_D. Note that these Faraday rotation bursts show no additional modulation traces as seen at positive delay times when a pump is applied. This is in accord with the assignment of the modulation to the photogenerated carriers [15].

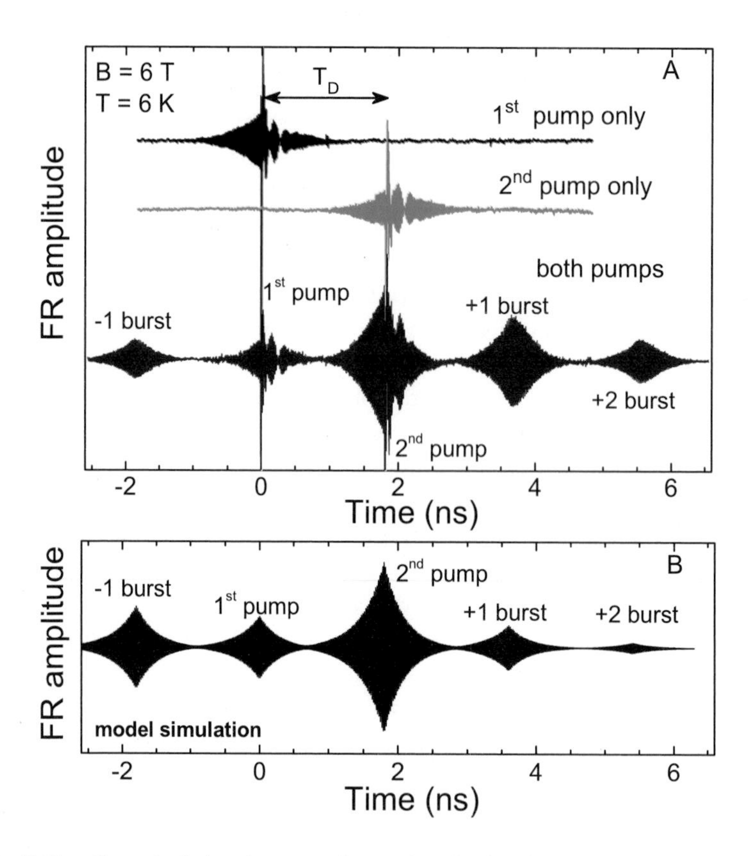

Fig. 5.11. Control of the electron spin synchronization by two trains of pump pulses with $T_R = 13.2\,\text{ns}$ shifted in time by $T_D = 1.84\,\text{ns}$. (A) Experimental Faraday rotation signal measured for separate action of the first or the second pump (the two upper curves) and for joint action of both pumps (the bottom curve). The pumps were co-polarized (σ^+). (B) Theoretical modelling of the spin echo-like signals in the two pulse experiment with the parameters: $\Theta = \pi$ and $\gamma = 3.2\,\text{GHz}$. [34]

Apparently, the electron spins in the quantum dot sub-ensemble, which is synchronized with the laser repetition rate, have been clocked by introducing a second frequency which is determined by the laser pulse separation. The clocking results in multiple bursts in the Faraday rotation response. This behavior can be explained by our theoretical model: The echo-like signal has the same origin as the Faraday rotation revival in the single pump experiment, which is constructive interference of the Faraday rotation amplitudes from quantum dot subsets with phase-synchronized electron spin precession. We have calculated the distribution of electron spin polarization created by a train of π-pulses in the two pump experiment, using a technique similar to the one described above for the single pump experiment. The resultant time dependence of the Faraday rotation signal reproduces the experimental burst signals well (Fig. 5.11B).

The mode locking mechanism in an ensemble of quantum dots with inhomogeneously broadened precession frequencies rises the question what properties should have a quantum dot ensemble for use in various quantum coherent devices. In general quantum dot ensembles whose spin states are only homogeneously broadened would be optimal for quantum information processing. However, fabrication of such ensembles cannot be foreseen based on current state-of-the-art techniques, which always lead to sizeable inhomogeneities. Under these circumstances, a sizable distribution of the electron g-factor is good for mode locking, as the phase synchronization condition is fulfilled by many quantum dot subsets, leading to strong spectroscopic response. Further, it gives some flexibility when changing, for example, the laser protocol (e.g. wavelength, pulse duration and repetition rate) by which the quantum dots are addressed, and therefore changing the phase synchronization condition, as the ensemble involves other quantum dot subsets for which the single dot coherency can be recovered. However, a very broad distribution of electron g-factors would lead to a very fast dephasing in the ensemble, making it difficult to observe the Faraday rotation both after and before pulse arrival. In this case the phase synchronization can be exploited only during a very short period of time.

5.6. Control of Ensemble Spin Precession

Figure 5.12 shows FR traces excited by the two-pulse train with a repetition period $T_R = 13.2\,\text{ns}$, in which both pulses have the same intensity and polarization, and the delay between these pulses T_D was varied between $\sim T_R/7$ and $\sim T_R/2$. The FR pattern varies strongly for the case when the delay time T_D is commensurate with the repetition period T_R: $T_D = T_R/i$ with $i = 2, 3, 4, ...$, and for the case $T_D \neq T_R/i$. For commensurability $T_D = T_R/i$, the FR signal shows strong periodic bursts of quantum oscillations only at times equal to multiples of T_D, as seen in the left panel for $T_D = 1.86\,\text{ns} \approx T_R/7$. Commensurability is also given to a good approximation for delays $T_D = T_R/4 \approx 3.26\,\text{ns}$ and $T_D = T_R/3 \approx 4.26\,\text{ns}$.

For incommensurability of T_D and T_R, $T_D \neq T_R/i$, the FR signal shows bursts of quantum oscillations between the two pulses of each pump doublet,

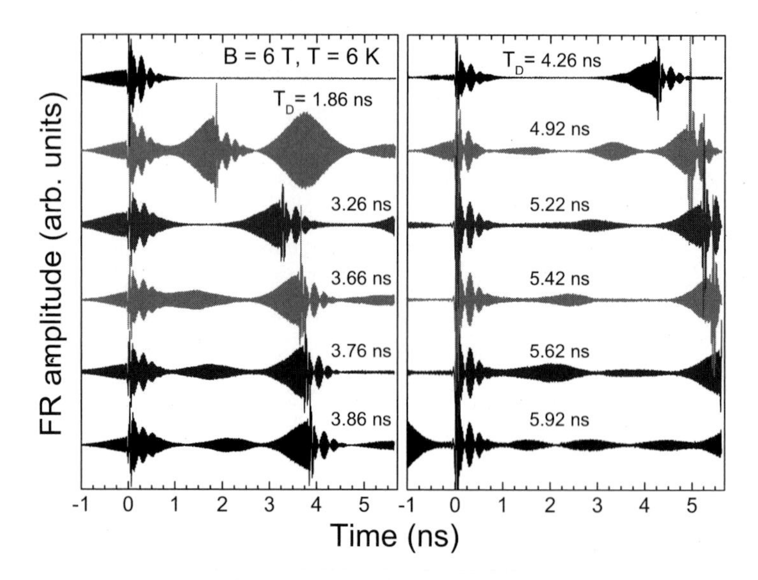

Fig. 5.12. Faraday rotation traces measured as function of delay between probe and first pump pulse at time zero. A second pump pulse was applied, delayed relative to the first one by T_D, indicated at each trace. The top left trace gives the FR without second pump. [36]

in addition to the bursts outside of the doublet. For example, one can see a single burst in the mid between the pumps for $T_D = 3.76$ and $5.22\,\text{ns}$. Two bursts, each equidistant from the closest pump and also equidistant from one another, appear at $T_D = 4.92$ and $5.62\,\text{ns}$. Three equidistant bursts occur at $T_D = 5.92\,\text{ns}$. Note also that the FR amplitude before the second pump arrival is always significantly larger than before the first pump for any T_D.

Although the time dependencies of the FR signal look very different for commensurate and incommensurate T_D and T_R, in both cases they can be fully controlled by designing the synchronization of electron spin precession modes in order to reach constructive interference of their contributions to the FR signal [34]. For a train of pump pulse doublets the phase synchronization condition accounts for the intervals T_D and $T_R - T_D$ in the laser excitation protocol

$$\omega_e = 2\pi NK/T_D = 2\pi NL/(T_R - T_D) \ , \tag{5.13}$$

where K and L are integers. On first glance this condition imposes severe limitations on the T_D values, for which synchronization is obtained:

$$T_D = [K/(K+L)]T_R \ , \tag{5.14}$$

which for $T_D < T_R/2$ leads to $K < L$. When equation (5.14) is satisfied, the contribution of synchronized precession modes to the average electron spin polarization $\overline{S}_z(t)$ is proportional to $-0.5\cos[N(2\pi Kt/T_D)]$. Summing over all relevant oscillations leads to constructive interference of their contributions with a period T_D/K in time [34]. The rest of quantum dots does not contribute to $\overline{S}_z(t)$ at times longer than the ensemble dephasing time. The phase synchronization condition explains the position of all bursts in the FR signal for commensurate and incommensurate ratios of T_D and T_R. For commensurability, $K \equiv 1$ and $T_D = T_R/(1+L)$ according to equation (5.14). In this case constructive interferences should occur with period T_D as seen in Fig. 5.12 for $T_D = 1.86$ ns ($L = 6$).

For incommensurability of T_D and T_R the number of Faraday rotation bursts between the two pulses within a pump doublet and the delays at which they appear can be tailored. There should be just one burst between the pulses, when $K \equiv 2$, because then the constructive interference must have a period $T_D/2$. A single burst is seen in Fig. 5.12 for $T_D = 3.76$ and 5.22 ns. The corresponding ratios T_D/T_R are 0.285 and 0.395, respectively. At the same time equation (5.14) gives a ratio $T_D/T_R = 2/(L+2)$, which is equal to 0.285 and 0.4 for L=5 and 3, respectively, in good accord with experiment.

Next, two Faraday rotation bursts are seen for $T_D = 4.92$ and 5.62 ns, corresponding to $T_D/T_R \approx 0.372$ and 0.426. The corresponding constructive interference period $T_D/3$ is reached for $K \equiv 3$. Then Faraday rotation equation (5.14) $T_D/T_R = 3/(L+3)$, giving 0.375 and 0.429 for L= 5 and 4, respectively. Finally, the Faraday rotation signal with $T_D = 5.92$ ns ($T_D/T_R \approx 0.448$) shows three Faraday rotation bursts between the two pumps. The constructive interference period $T_D/4$ is obtained for $K \equiv 4$, for which eq. (5.14) gives $T_D/T_R = 4/(L+4) \approx 0.444$ with L= 5. Obviously good general agreement between experiment and theory is established, highlighting the high flexibility of the laser protocol. In turn, this understanding can be used to induce Faraday rotation bursts at wanted delays T_D/K, so that at these times further coherent manipulation of all electron spins involved in the burst is facilitated.

However, the question arises how accurate condition equation (5.14) for the T_D/T_R ratio must be fulfilled to reach phase synchronization. Formally, one can find for any arbitrary T_D/T_R large K and L values such that equation (5.14) is satisfied with high accuracy. But the above analysis shows, that only the smallest of all available L leads to PSC matching. Experimentally, the facilities to address this point are limited, as the largest T_D for which Faraday rotation signal can be measured are delays around 5 ns between the two pumps. For larger delays the Faraday rotation bursts shift out of the scanning range. For short T_D, on the other hand, the bursts are overlapping with the Faraday rotation signal from the pump pulses.

To answer this question, we have modeled the Faraday rotation signal for commensurate and incommensurate ratios of T_D and T_R. Figure 5.13 shows the results together with spectra of synchronized spin precession modes (SSPM) at the moment of the first and second pump pulse arrival. The SSPM were calculated similar those induced by a single pulse train [34]. Figure 5.13(a) gives the SSPM for commensurate $T_D = T_R/3$ superimposed on the SSPM created by a single pulse train with the same T_R. Panel (c) shows the Faraday rotation signal created by such a two pulse train. The SSPM for the considered strong excitation are considerably broadened and contain modes for which $\omega_e = 2\pi M/T_D = 2\pi 3M/T_R$ with integer M, which coincide with each third mode created by a single pulse train. However, the SSPM given by $\omega_e = 2\pi N/T_R$, which do not satisfy the phase synchronization condition for a two pulse train, are not completely suppressed, because the train synchronizes the electron spin precession in some frequency range around the PSC. One sees also, that at $t = 0$ the two pulse train leads to a significant alignment of electron spins opposite to the direction of spins satisfying the PSC. This "negative" alignment decreases the constructive interference magnitude and therefore the Faraday rotation signal before the first pulse arrival, and is also responsible for a significantly larger magnitude of the Faraday rotation signal before the second pulse arrival [see Figs. 5.12 and 5.13(c)].

For incommensurate ratios of T_D and T_R the SSPM become much more complex. Still we are able to recover the modes which satisfy the phase synchronization condition at the pulse arrival times. In Fig. 5.13(b) we show the SSPM at $t = 0$ and $t = T_D$ for $T_D = 2T_R/7$ ($K = 2$, $L = 5$), where the arrows indicate the frequencies which satisfy the phase synchronization condition for the two pulse train. Only a small number of such modes fall within the average distribution of electron spin precession modes, because the distance between the phase synchronization condition modes is proportional to $2\pi K/T_D = 2\pi(K + L)/T_R$. The diluted spectra of PSC modes for incommensurability decrease the magnitude of the Faraday rotation bursts between the pump pulses, in accord with experiment. This shows, that

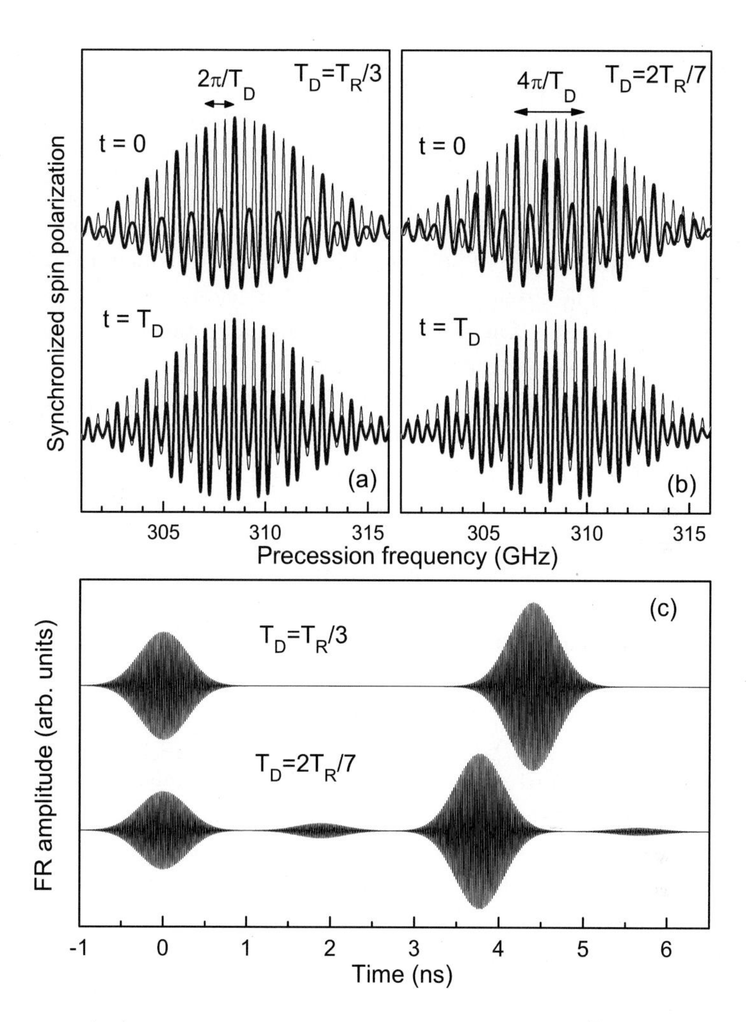

Fig. 5.13. (a,b): Spectra of electron spin precession modes, $-\overline{S}_z(t)$, which are phase synchronized by the two-pulse train calculated for $T_D = T_R/3$ and $T_D = 2T_R/7$ at the moments of first ($t = 0$) and second ($t = T_D$) pulse arrival (red). Single-pump spectra are shown in blue. (c): Faraday rotation traces calculated for two ratios of T_D/T_R. Laser pulse area $\Theta = \pi$. $T_R = 13.2\,\text{ns}$. Electron g-factor dispersion $\Delta g = 0.005$. [36]

although any ratio of T_D/T_R can be satisfied by large K and L, the Faraday rotation signal between the pulses should be negligibly small in this case. Consequently, not any ratio of T_D/T_R leads to pronounced Faraday rotation bursts.

To obtain further insight into the tailoring of electron spin coherence, which can be reached by a two-pulse train, we have turned from co- to counter-circularly polarized pumps. The delay between pumps T_D was fixed at $T_R/6 \approx 2.2\,\text{ns}$. The time dependences of the corresponding Faraday rotation signals are similar, as shown in Fig. 5.14. Besides the two Faraday rotation bursts directly connected to the pump pulses, one sees bursts due to constructive interference of spin synchronized modes which are periodic in time with T_D. The insets in Fig. 5.14(a) show closeups of the different Faraday rotation bursts. The sign, κ, of the Faraday rotation amplitude for the counter-circular configuration undergoes T_D-periodic changes in time relative to the co-circular case, as seen in Fig. 5.14(b), which demonstrates optical switching of the electron spin precession phase by π in an *ensemble* of quantum dots.

The observed effect of sign reversal is well described by our model. Let us consider first a two-pulse train with delay time $T_D = T_R/2$ for which the two pumps are counter-circularly polarized. In this case an electron spin can be synchronized only if at the moment of pulse arrival it has an orientation opposite to the orientation at the previous pulse. This leads to the phase synchronization condition $\omega_e = 2\pi(N + 1/2)/T_D$. The contribution of such precession modes to the electron spin polarization is proportional to $\cos[2\pi(N + 1/2)t/T_D)] = \cos(2\pi Nt/T_D)\cos(\pi t/T_D) - \sin(2\pi Nt/T_D)\sin(\pi t/T_D)$. Summing these contributions, only the first term gives a constructive interference, whose modulus has period T_D, while the sign of $\cos(\pi t/T_D)$ changes with period $2T_D$. Only each third of the precession frequencies can be synchronized by a counter-circularly polarized two pulse train when the delay time is $T_R/6$ as in our experiment. The corresponding phase synchronization condition is: $\omega_e = 6 \times 2\pi(N + 1/2)/T_R$. A consideration similar to the previous shows that the constructive interference modulus has period $T_R/6$ and changes its sign with period $2T_R/6$. The relative sign of the Faraday rotation amplitude for the counter- and co-circularly case, $\kappa = \text{sign}\{\cos[\pi t/(T_R/6)]\}$, is in accord with the experimental dependence in Fig. 5.14(b).

The constructive interferences of the electron spin contributions can be seen only as long as the coherence of the electron spins is maintained. In this respect the temperature stability of the constructive interference is especially important. Fig. 5.14(c) shows Faraday rotation traces in a two-pump-pulse configuration with $T_D = 1.88$ ns at different temperatures. For both positive and negative delays, the Faraday rotation amplitude at a fixed delay is about constant for temperatures up to 25 K, irrespective of slight

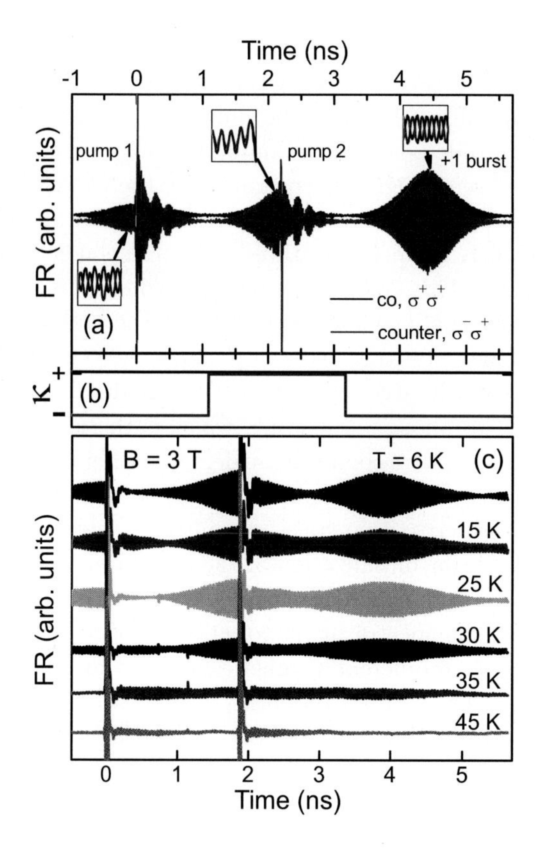

Fig. 5.14. (a): Faraday rotation traces in the co-circularly (blue traces) or counter-circularly (red traces) polarized two pump pulse experiments measured for $T_D = 2.2\,\text{ns}$ and $B = 6\,\text{T}$. Insets give close-ups showing the relative sign, κ, of the Faraday rotation amplitude between the two traces. κ is plotted in (b) vs time. (c): Effect of temperature on the Faraday rotation amplitude in two-pump-pulse experiment. $T_D = 1.88\,\text{ns}$. [36]

variations which might arise from changes in the phase synchronization of quantum dot subsets. Above 30 K a sharp drop occurs, which can be explained by thermally activated destruction of the spin coherence.

The electron spin coherence in charged quantum dots is initiated by generation of a superposition of an electron and a charged exciton state by resonant pump pulses [15, 33]. The simultaneous decrease of the Faraday rotation magnitude before each pump pulse and afterwards (when the constructive interference signal is controlled by the excitation pulse) suggests that the coherence at elevated temperatures is lost already during its generation. The 30 K temperature threshold corresponds to an activation energy of ~2.5 meV. This energy may be assigned only to the splitting between the two lowest confined hole levels, because the electron level splitting dominates the 20 meV splitting between p- and s-shell emission in photoluminescence and is much larger than 2.5 meV. The decoherence of the hole spin results from two phonon scattering, which is thermally activated and should occur on a sub-picosecond time scale, i.e. within the laser pulse [35]. The fast decoherence of the hole spin at $T > 30$ K suppresses formation of the electron-trion superposition state. ps-pulses as used here are therefore not sufficiently short for initialization of the superposition and creation of a long-lived electron spin coherence.

Note also, that the spectra in the lower and upper panels of Fig. 5.14 have been recorded at different magnetic fields (3 and 6 T). While the appearance of the Faraday rotation bursts changes with the field strength, the times at which the bursts appear remain unchanged. This shows the stability of the control of electron spin precession against magnetic field variations.

5.7. Summary

In summary we have performed detailed studies of the electron g-factor in quantum dots. The spin is described by a complex g-factor tensor with pronounced anisotropies. With this knowledge we have addressed the coherent manipulation of the spin. We have shown first a very efficient technique by which the spin can be oriented (initialized) by circularly polarized laser pulses. By such pulses the spin orientation can be controlled. We have then shown that the electron can be phase synchronized with the periodic laser protocol. As a first trade-off this technique allowed as an electron spin coherence time of 3 μs at cryogenic temperatures. We have then succeeded with a first demonstration that this method also allows a far-reaching coherent control of the spins: A two-pulse protocol allowed us to clock the electron spin such that periodic bursts appear in the Faraday rotation signal. Further, by tailoring the laser protocol we can control the pattern of

Faraday rotation bursts, and we can also induce π phase changes.

This result shows that the deficits which are typically attributed to quantum dot spin ensembles may be overcome when combining them with elaborated laser excitation protocols, with all the related advantages due to the robustness of the phase synchronization of the quantum dot ensemble: (i) a strong detection signal with relatively small noise; (ii) changes of external parameters like repetition rate and magnetic field strength can be accommodated for in the phase synchronization condition due to the broad distribution of electron spin precession frequencies in the ensemble and the large number of involved quantum dots. This should be elaborated further in future studies.

Acknowledgments

This work was supported by the Bundesministerium für Bildung und Forschung (program 'nanoquit'), the Defense Advanced Research Program Agency (program 'QuIST'), the Office of Naval Research, and the Deutsche Forschungsgemeinschaft (Forschergruppe 'Quantum Optics in Semiconductor Nanostructures'). Parts of this work were done in collaboration with R. Oulton, E. A. Zhukov, I. A. Yugova. Samples, on which the described work was done are grown in Bochum University by D. Reuter, and A. D. Wieck.

References

[1] See, for example, *The Physics of Quantum Information*, Eds. D. Bouwmeester, A. Ekert and A. Zeilinger (Springer, Berlin, 2000); Yamamoto Y., Quantum Information Procesing **5** (5): 299 (2006); Zoller P., Beth T., Binosi D., *et al.*, European Physical Journal D **36** (2): 203-228 (2005).
[2] W. Langbein, P. Borri, U. Woggon, V. Stavarache, D. Reuter, and A. D. Wieck, Phys. Rev. B **70**, 033301 (2004).
[3] M. Bayer and A. Forchel, Phys. Rev. B **65**, 041308 (2002).
[4] M. Schwab, H. Kurtze, T. Auer, T. Berstermann, M. Bayer, J. Wiersig, N. Baer, C. Gies, F. Jahnke, J. P. Reithmaier, A. Forchel, M. Benyoucef, P. Michler, Phys. Rev. B **74**, 045323 (2006).
[5] P. Borri, W. Langbein, S. Schneider, U. Woggon, R. L. Sellin, D. Ouyang, and D. Bimberg, Phys. Rev. Lett. **87**, 157401 (2001).
[6] J. Vuckovic and Yammamoto, App. Phys. Lett. **82** (15), 2374 (2003).
[7] V. Cerletti, W. A. Coish, O. Gywat and D. Loss, *Recipes for spin-based quantum computing*, Nanotechnology **16** (2005) R27.
[8] D. Loss and D. P. DiVincenzo, Phys. Rev. A **57**, 120 (1998).

[9] A. Imamoglu, D. D. Awschalom, G. Burkard, D. P. DiVincenzo, D. Loss, M. Sherwin, and A. Small, Phys. Rev. Lett. **83**, 4204 (1999).

[10] *Semiconductor and Quantum Computation*, ed. by D. D. Awschalom, D. Loss, and N. Samarth (Springer-Verlag, Heidelberg 2002).

[11] J. M. Kikkawa and D. D. Awschalom, Science **287**, 473 (2000).

[12] J. M. Elzerman, R. Hanson, L. H. Willems Van Beveren, B. Witkamp, L. M. K. Vandersypen, and L. P. Kouwenhoven, Nature **430**, 431 (2004); M. Kroutvar, Y. Ducommun, D. Heiss, M. Bichler, D. Schuh, G. Abstreiter, and J. J. Finley, Nature **432**, 81 (2004).

[13] W. A. Coish and D. Loss, Phys. Rev. B **70**, 195340 (2004).

[14] J. R. Petta, A. C. Johnson, J. M. Taylor, E. A. Laird, A. Yacoby, M. D. Lukin, C. M. Marcus, M. P. Hanson, and A. C. Gossard, Science **309**, 2180 (2005).

[15] A. Greilich, R. Oulton, E. A. Zhukov, I. A. Yugova, D. R. Yakovlev, M. Bayer, A. Shabaev, Al. L. Efros, I. A. Merkulov, V. Stavarache, D. Reuter, and A. Wieck, Phys. Rev. Lett. **96**, 227401 (2006).

[16] R. Oulton, A. Greilich, S. Yu. Verbin, R. V. Cherbunin, T. Auer, D. R. Yakovlev, M. Bayer, V. Stavarache, D. Reuter, and A. Wieck, Physical Review Letters **98**, 107401 (2007).

[17] S. A. Crooker, D. D. Awschalom, J. J. Baumberg, F. Flack, and N. Samarth, Phys. Rev. B **56**, 7574 (1997).

[18] R. E. Worsley, N. J. Trainor, T. Grevatt, and R. T. Harley, Phys. Rev. Lett. **76**, 3224 (1996).

[19] I.A. Yugova, A. Greilich, E.A. Zhukov, D.R. Yakovlev, and M. Bayer, V. Stavarache, D. Reuter, and A.D. Wieck, Physical Review B **75**, 195325 (2007).

[20] P. Y. Yu and M. Cordona, *Fundamentals of Semiconductors*, (Springer, Berlin, 1996).

[21] I. A. Merkulov, Al. L. Efros, and M. Rosen, Phys. Rev. B **65**, 205309 (2002).

[22] see, e.g., T. H. Stievater, X. Li, D. G. Steel, D. S. Katzer, D. Park, C. Piermarocchi, and L. Sham, Phys. Rev. Lett. **87**, 133603 (2001); A. Zrenner, E. Beham, S. Stufler, F. Findeis, M. Bichler, and G. Abstreiter, Nature **418**, 612 (2002).

[23] A. Shabaev and Al. L. Efros, unpublished.

[24] A. Shabaev, Al. L. Efros, D. Gammon, I. A. Merkulov, Phys. Rev. B **68**, 201305(R) (2003).

[25] M. V. Gurudev Dutt, J. Cheng, Bo Li, Xiaodong Xu, Xiaoqin Li, P. R. Berman, D. G. Steel, A. S. Bracker, D. Gammon, Sophia E. Economou, Ren-Bao Liu, and L. J. Sham, Phys. Rev. Lett. **94**, 227403 (2005).

[26] G. Morigi, E. Solano, B.-G. Englert, and Herbert Walther, Phys. Rev. A **65**, 040102(R) (2002).

[27] A. V. Khaetskii, D. Loss, and L. Glazman, Phys. Rev. Lett. **88**, 186802 (2002).

[28] D. V. Bulaev and D. Loss, Phys. Rev. Lett. **95**, 076805 (2005).

[29] T. Flissikowski, I. A. Akimov, A. Hundt, and F. Henneberger, Phys. Rev. B **68**, 161309(R) (2003).

[30] S. E. Economou, Ren-Bao Liu, L. J. Sham, and D. G. Steel, Phys. Rev. B **71**, 195327 (2005).
[31] C. P. Slichter, *Principles of Magnetic Resonance* (Springer-Verlag, Berlin 1996).
[32] We chose a Lorentzian profile for the quantum dot precession frequencies in the consideration because it leads to the closed form for $\overline{S}_z(t)$ in equation 2. Generally our numerical calculations do not show any of qualitative or quantitative differences for the both Gaussian or Lorentzian profiles as long as the distribution $\rho(\omega_e)$ is smoothly going to zero on the scale of its width.
[33] T. A. Kennedy, A. Shabaev, M. Scheibner, Al. L. Efros, A. S. Bracker, D. Gammon, Phys. Rev. B **73**, 045307 (2006).
[34] A. Greilich, D. R. Yakovlev, A. Shabaev, Al. L. Efros, I. A. Yugova, R. Oulton, V. Stavarache, D. Reuter, A. D. Wieck, and M. Bayer, Science **313**, 341 (2006).
[35] T. Takagahara, Phys. Rev. B **62**, 16840 (2000).
[36] A. Greilich, M. Wiemann, D.R. Yakovlev, I.A. Yugova, M. Bayer, A. Shabaev, Al.L. Efros, V. Stavarache, D. Reuter, and A.D. Wieck, Physical Review B **75**, 233301 (2007).

Chapter 6

Optical Control of Quantum-Dot Spin States

M. Atatüre

Cavendish Laboratory, University of Cambridge, Cambridge CB3 0HE, United Kingdom

J. Dreiser, A. Badolato & A. Imamoğlu

Institute of Quantum Electronics, ETH Zurich, HPT G10, Zurich 8093, Switzerland

Among many stringent prerequisites of quantum information processing (QIP), the ability to prepare and measure a state is the essential first step before an experimental realization of gate operations can be pursued. Initialization of the qubits to a simple fiducial state is in fact one of the widely used diVincenzo criteria for QIP. Readout is a projective measurement of the gate-modified state in the computational basis: in this sense an assessment of the fidelity of a quantum manipulation can only be tested if a realiable readout scheme exists. In this Chapter, we present the state-of-the-art in optical state preparation and readout of spins confined in self-assembled quantum dots.

Contents

6.1. Introduction

In the last ten years, self-assembled semiconductor quantum dots (QDs) have emerged as a model system for solid-state quantum optics and QIP.

Photoluminescence (PL) studies of nonresonantly excited single QDs have led to the generation of single photons [1, 2] and entangled photons [3] as well as the observation of cavity-quantum electrodynamics (QED) effects both in the weak-coupling [4–6] and strong-coupling [7–9] regimes. Similarly, resonant excitation has enabled the observation of Rabi oscillations [10] and coherent manipulation of excitons and biexcitons [11]: all indicators of an interesting quantum optical system. These milestones have further strengthened various QIP-motivated proposals, including those regarding optical access to spins in QDs [12].

To that end, we will first discuss spin-state preparation of a single electron trapped in an Indium Arsenide (InAs) quantum dot. The underlying principle of optical pumping is profoundly simple and was first put to use by J. Brossel *et al.* in the context of spin polarizing an atomic ensemble [13]. In this scheme, the optical selection rules allow for spin-selective transitions leading to a final dark state. Same principle can be applied to a single quantum-dot spin: Optical coupling of electronic spin states can be achieved using resonant excitation of the charged excitonic transitions (trions) along with the spin-state mixing, which leads to nonzero rates for spin-flip Raman transitions. Using this mechanism it has been shown that an electron can indeed be pumped into one of the spin states with a high fidelity despite being in a solid-state environment. The figure of merit for QIP, i.e. the state-preparation fidelity exceeds 0.998 for this candidate system.

The second key prerequisite for applications in QIP, the ability to read-out the state of a single confined spin, is as essential as the ability to prepare it in the first place. The first demonstration of single semiconductor spin read-out was performed using transport measurements based on spin-charge conversion. Here, we will discuss an all-optical dispersive measurement of the electron spin state, virtually the observation of Faraday effect from the smallest possible magnet in Nature. Access to the spin-state information is obtained through conditional Faraday rotation of a spectrally detuned optical field, induced by the polarization- and spin-selective trion transitions.

6.2. Resonant Scattering from Charge-Tunable QDs

Photoluminescence is arguably the most commonly used optical spectroscopy technique for single QDs to-date. An excitation laser is used for generating electrons and holes in the surrounding host matrix such as Gallium Arsenide (GaAs) followed by the relaxation and entrapment of these carriers into the lower lying discrete QD states. Recombination of an electron-hole pair, namely an exciton generates a photon with energy corresponding to that of the QD transition. This technique allows for mapping out the optical transitions, yet it still renders any systematic access to individual spins near impossible. An alternative technique is the resonant scattering spectroscopy, where a laser of single optical frequency addresses a transition of interest directly and selectively allowing us to study features such as optical selection rules and oscillator strengths by observing the scattered laser light [14]. Another advantage of this technique is the superior spectral resolution yielding around 10^3-fold improvement over photoluminescence.

In parallel, a single electron has to be trapped in a QD with certainty before its spin can be studied. Incorporating a QD layer in between a highly doped GaAs layer and a surface gate electrode thus forming a Schottky-diode heterostructure [Fig. 6.1] is probably the most convincing solution to-date [15]. By choosing an appropriate voltage applied between the two leads of the device we can guarantee one and only one electron to occupy a single QD. In such a situation, the ground state is a single electron in the conduction band, and the excited state is an electron pair in a singlet state with zero total spin in the presence of a hole in the valence band. The two ground states are determined by electron spin states with $S_z = \pm 1/2$, while the two excited states are determined by the hole spin states with $J_z = \pm 3/2$. The experimental results discussed here were achievable thanks to the combination of these approaches.

Figure 6.1 illustrates the device structure and the experimental setup for resonant scattering experiments where the transmitted laser field is measured by photodetectors rather than the emitted photons being detected by a spectrometer. The key elements are the quarter-wave plate (QWP) which maps circular polarization states into linear ones and the polarizing beam splitter (PBS) which distributes each linear polarization to one designated detector. This system can measure the signal from both circularly-polarized

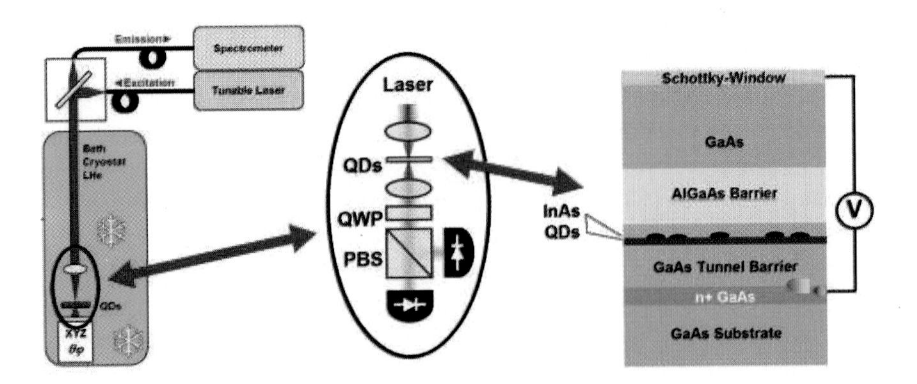

Fig. 6.1. Resonant scattering setup and device structure.

QD transitions simultaneously. When the laser field is in resonance with a QD transition, the optical field scattered by the QD interferes with the background optical field [16]. Therefore, the total field observed upon transmission through the quantum dot includes the signature of the quantum dot response to the light field. In short, the ratio of the scattered to incident laser power can be expressed as

$$\frac{P_{\text{Scattered}}}{P_{\text{Laser}}} = \eta \cdot \frac{3\lambda^2}{2\pi A_{\text{Laser}}} \cdot \frac{\Gamma^2}{4\delta^2 + \Gamma^2 + 2\Omega^2_{\text{Laser}}}, \tag{6.1}$$

where the symbols P_i, η, λ, A_{Laser}, Γ, δ, and Ω stand for incident and scattered laser power, mode matching efficiency, QD transition wavelength, laser focus area, QD transition linewidth, spectral detuning of the laser from the QD transition, and the Rabi frequency. This Lorentzian response can be mapped out in two ways: Either by sweeping the frequency of the laser with respect to the QD transition or by shifting the quantum dot emission energy via DC Stark effect. In our experiments both methods are employed, and Fig. 6.2 displays an example laser scan over the trion transition with a Rabi frequency comparable to the QD transition linewidth Γ. The PL data is also displayed for completeness. It is worth noting that a typical value for Eq. (6.1) is below 1% for a singly charged QD due to the small η factor. Therefore, signal-to-noise improving protocols such as lock-in detection are typically employed.

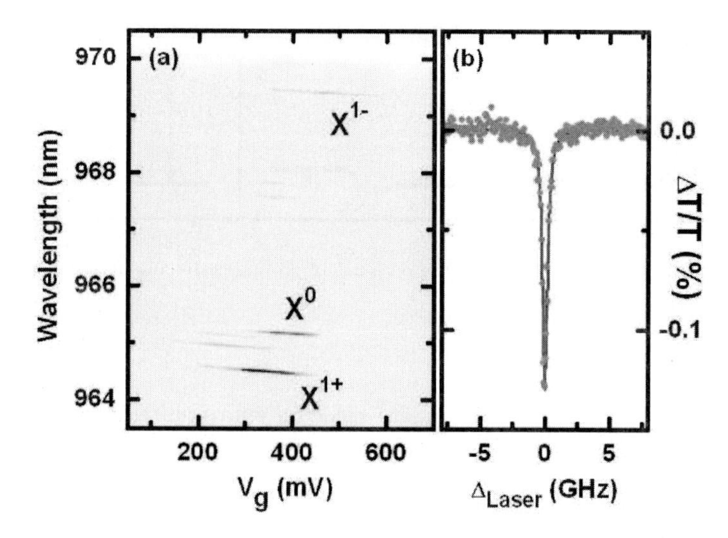

Fig. 6.2. The voltage-dependent PL data for a typical QD in a Schottky heterostructure along with resonant scattering data from the degenerate trionic transition in linear polarization basis.

6.3. Spin-State Preparation

A single electron spin in a solid-state environment has numerous channels for spin relaxation. Fundamental interactions that determine spin dynamics in QDs are hyperfine coupling to QD nuclear spin ensembles, spin-phonon coupling and exchange-type interactions with a nearby Fermi sea of electrons. Further, optical access necessitates careful consideration of heavy-light hole mixing and its effects on electron spin dynamics in the presence of a driving laser. For the goal of spin-state preparation, we have to operate in a regime that brings forth the mechanisms that help with spin pumping and minimize those that lead to spin relaxation. The strength of Pauli blockade of the corresponding optical transitions works as a means to infer the electron spin state. The physical mechanisms for state mixing and spin heating are discussed in detail elsewhere [17].

A single-electron-charged QD in the trion picture is analogous to the four-level system illustrated in Fig. 6.3, where state $|\uparrow\downarrow,\Downarrow\rangle$ ($|\uparrow\downarrow,\Uparrow\rangle$) corresponds to two ground-state electrons forming a singlet and a ground-state hole with angular momentum projection $J_z = -\frac{3}{2}$ $(+\frac{3}{2})$ along the quantization axis. The strong trion transitions, $|\uparrow\downarrow,\Downarrow\rangle \longleftrightarrow |\downarrow\rangle$ and

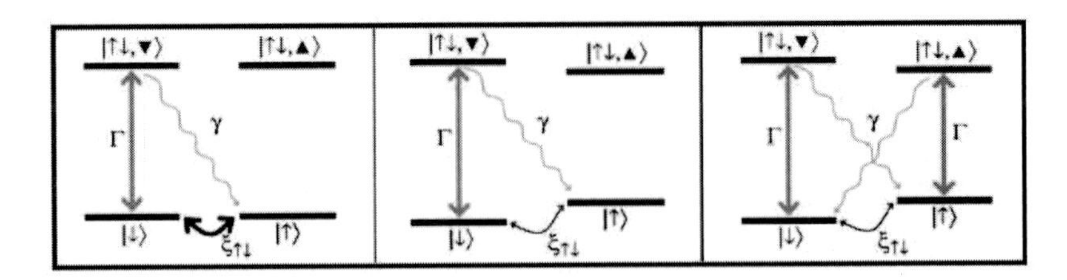

Fig. 6.3. Electron level scheme in the presence of ground-state coupling and branching of optical transitions.

$| \uparrow\downarrow, \Uparrow\rangle \longleftrightarrow | \uparrow\rangle$, leave the resident electron spin unaltered, whereas the weak transitions, $| \uparrow\downarrow, \Downarrow\rangle \longleftrightarrow | \uparrow\rangle$ and $| \uparrow\downarrow, \Uparrow\rangle \longleftrightarrow | \downarrow\rangle$, lead to a net spin-flip of the resident electron. The latter transitions are ideally forbidden by the optical selection rules, however the various interaction mechanisms mentioned above allows for $\Gamma \gg \gamma \neq 0$, where Γ and γ are the allowed and forbidden spontaneous emission rates, respectively, as indicated in Fig. 6.3 (left). In addition to these optical transitions, the ground states for the electron spin can be coupled at a rate $\xi_{\uparrow\downarrow}$, the main culprits for this channel being hyperfine interaction, cotunneling, and phonon-assisted spin-orbit coupling. For the case of hyperfine interaction, previous studies on similar structures have shown that $\xi_{\uparrow\downarrow}^{(B=0)} \gg \Gamma$ in the absence of an external magnetic field B but is strongly suppressed even under relatively weak magnetic fields ($B \sim 0.1$ T) because of incommensurate electron and nuclear Zeeman energies [18, 19]. This is the case depicted in Fig. 6.3 (middle), where $\xi_{\uparrow\downarrow}^{(B>0)} \ll \Gamma$. In this regime, driving any one of the two optical transitions will naturally lead to a shelving of the electron into the dark spin state. How well the electron can be kept in that spin state is indeed determined by the relative strengths of all the involved transition rates. The third scenario that will be considered is a control experiment where a second re-pump laser is resonant with the other Zeeman-split trion transition $| \uparrow\downarrow, \Uparrow\rangle \longleftrightarrow | \uparrow\rangle$ simultaneously. Then, the $| \uparrow\rangle \rightarrow | \downarrow\rangle$ transition will also take place via spontaneous spin-flip Raman scattering, with a rate proportional to the ratio of the two laser intensities, as long as we are not driving the transitions to saturation. Figure 6.3 (right) illustrates this scenario also known as the double-Λ scheme in quantum optics.

Figure 6.4 is the experimental realization of Fig. 6.3, showing the expected X^{1-} plateau at $B = 0$ T, where the quantum dot is single-electron-

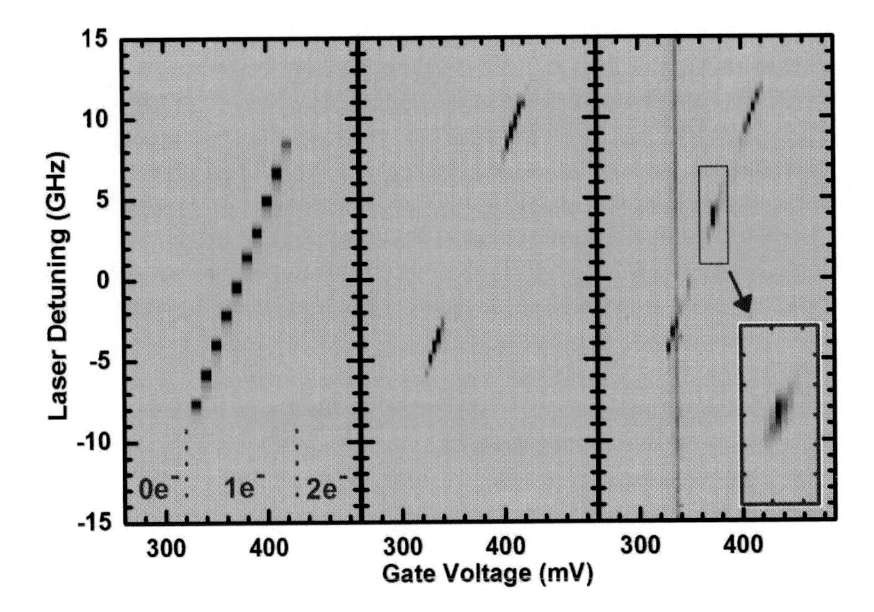

Fig. 6.4. Absorption plots as a function of magnetic field and optical pumping.

charged for gate voltages (V_{gate}) in the 320- to 424-mV range (left plot). The probe laser was scanned across the $|\uparrow\downarrow, \Downarrow\rangle \longleftrightarrow |\downarrow\rangle$ transition and had the corresponding circular polarization $[\sigma^{(-)}]$ as determined by the optical selection rules. Voltage-dependent shift of the QD transition frequency is the DC Stark shift due to the permanent trionic dipole moment. The edges of the charging plateau are the strong cotunneling regimes where $\xi_{\uparrow\downarrow}$ remains very strong and prohibits spin pumping. In fact, the strength of spin relaxation generated by the previously listed fundamental interactions can be changed by up to five orders of magnitude upon varying the external electric and magnetic fields [17]. Through such strong dependency, the QD system can be tuned from the regime of an isolated artificial atom to that of a quantum-confined solid-state system coupled either to a charge or a spin reservoir. Therefore, we will only consider the center of the charging plateau as the relevant regime for QIP. The middle portion of the charging plateau in Fig. 6.4 shows a suppression of the signal at a 0.2-T magnetic field: The quantum dot becomes transparent for gate voltages in the 344- to 396-mV range. This strong suppression of the signal in the X^{1-} plateau center is a signature of optical electron-spin pumping into the $|\uparrow\rangle$ state due to the unidirectional spontaneous Raman scattering process. From the surviving signal strength we can conclude that at 0.2 Tesla this leads to a 98.5% electron spin-state preparation fidelity. When we reach 0.3 T the electron is in the $|\uparrow\rangle$ state more than 99.8% of the measurement time, a value limited by our signal-to-noise level. This value remains to be the highest state-preparation fidelity reported in a solid-state system and can be reached for both spin states [20]. An alternative figure of merit is the spin temperature. We can say that the electron spin can be cooled optically from 4.2 K down to 20 mK, i.e. $(\frac{1}{210})^{\text{th}}$ of the ambient temperature of 4.2 K.

Reduction of the light scattering signal is the result of Pauli blockade, but it can alternatively come from QD ionization, i.e. exciting the electron to higher energy levels from where it may escape from the QD. To prove that the electron was only shelved in the $|\uparrow\rangle$ dark state by the probe laser at the $|\uparrow\downarrow,\Downarrow\rangle \longleftrightarrow |\downarrow\rangle$ transition, we simultaneously applied a repump laser on the $|\uparrow\downarrow,\Uparrow\rangle \longleftrightarrow |\uparrow\rangle$ transition with orthogonal circular polarization $[\sigma^{(+)}]$. Figure 6.4 (right) shows the resulting gate sweep, where an absorption peak at $V_{\text{gate}} = 372$ mV appears. The linewidth of this peak is equal to that shown in Fig. 6.2, and it was observed when the re-pump laser was detuned from the probe laser by exactly 6 GHz; that is, the independently measured trionic Zeeman splitting. The fact that we fully recovered the probe laser absorption only when both lasers were resonant with the corresponding trion transitions indicates that the system is a realization of the last (right) illustration in Fig. 6.3.

The results discussed in this Section rely on spontaneous spin-flip Raman transition and constitute the first step toward stimulated Raman transition [21] on a single QD electron for coherent preparation of an arbitrary superposition of spin states, as well as cavity-assisted spin-flip Raman transitions as a source of indistinguishable single photons with near-unity collection efficiency [22].

6.4. Spin-State Measurement

We will next try to answer the second question that is essential for the suitability of optically accessed quantum dot spin states as qubits, i.e. can we measure the spin states after a particular operation? Numerous proposals have considered resonance fluorescence as a way to measure electron spin state, where each emitted photon upon an excitation already carries the needed electron spin-state information. This in principle yields nanosecond measurement times limited only by the excited state lifetime, however the current practical limitations keep us far from this regime: Single photon detection approach suffers from limited extraction and detection efficiencies while resonant scattering approach suffers from minute $\eta \frac{\sigma_0}{A_{\text{Laser}}}$ values as discussed before. Unfortunately, both limitations currently push the possible measurement times well into the milliseconds regime. Considering the spin-flip rates we have discussed in the previous Section, spin measurement with resonant scattering is rendered impossible without altering the spin state itself. Therefore, we have to look for alternative routes for spin measurements.

In our quest for optical access to QD spins so far we have only considered the absorptive response of an optical transition, however dispersive response can also be observed with a modified measurement scheme. Ensemble spin measurements using Faraday rotation of light polarization have been implemented in semiconductors [23], but single spin sensitivity has only been achieved in the last year [24, 25]. Once again the underlying principle is rather simple: A detuned optical field experiences a phase shift in the vicinity of an optical transition. If the transition and the laser have circular and linear polarizations, respectively, then this acquired phase shift becomes detectable as a rotation of the linear polarization plane for the laser by a corresponding angle. This is referred to as the Faraday rotation (FR) angle. As previously discussed, we have two optical transitions available for a single electron charged QD. The $[\sigma^{(-)}]$ ($[\sigma^{(+)}]$) is linked to the $|\uparrow\rangle$ ($|\downarrow\rangle$) spin ground state. If a linearly polarized laser, e.g. $[\pi^{(+)}]$, feels this transition, the $[\sigma^{(-)}]$ ($[\sigma^{(+)}]$) polarization component acquires a phase shift rotating the laser's polarization by an angle Θ ($-\Theta$) in the linear basis. Owing to Pauli Blockade, only one of these transitions is available at any given time and the laser polarization is rotated in positive or negative angular direction conditional on the electron's spin state. Spin measurement via Faraday rotation exploits precisely this spin-state-dependence of the laser polarization. This alternative route to spin detection, however, comes at a cost: We have to remain far detuned from the transition in order to sufficiently suppress light scattering due to the absorptive response. Fortunately, dispersive response dominates over the absorptive part as the detuning is increased due to the linear versus quadratic dependence on spectral detuning. In this Section, we will discuss a measurement of a QD spin by detecting this dispersive response through spin-dependent Faraday rotation angle of a far-detuned linearly polarized laser.

In order to observe this polarization rotation, the measurement scheme of Fig. 6.1 has to be modified as follows: A polarized laser propagates through a gated heterostructure incorporating quantum dots. In the absence of the quarter-wave plate, the polarizing beam splitter distributes the transmitted light into two linear polarizations in the rectilinear basis of (X, Y) and directs each arm to a photodiode. Along with each detector's output (T_{x} and T_{y}), such a configuration allows us to measure their sum and difference simultaneously. When normalized to the total incident laser intensity their sum (T_{sum}) indicates a change of photon number in the total light, yielding the absorptive response, while a similarly normalized difference of the two signals (T_{diff}) is linked to the polarization rotation arising from the dispersive response. Under an external magnetic field, the Zeeman splitting (E_{Zeeman}) lifts the spectral degeneracy of the two spin-selective trionic transitions. Consequently, an antisymmetric lineshape due to the dispersive response is expected to appear along with the Lorentzian absorptive part in the spectral vicinity ($\delta \ll E_{\mathrm{Zeeman}}$) of a transition in the form

$$T_{\mathrm{diff}} = \frac{1}{2}\Theta[\mathrm{radians}] = \eta \cdot \frac{3\lambda^2}{4\pi A_{\mathrm{Laser}}} \cdot \frac{\Gamma\delta}{4\delta^2 + \Gamma^2 + 2\Omega_{\mathrm{Laser}}^2}, \tag{6.2}$$

where Θ is the Faraday rotation angle.

Figure 6.5 shows how the sum (open circles) and difference (filled circles) signals from one of the Zeeman-split transitions behave as a function of laser polarization basis, when an external magnetic field of 1 Tesla is applied along the strong confinement axis of the quantum dot. The data presented here is obtained in the cotunneling regime to avoid spin pumping. The transitions are split by 26 GHz, which is about 60 times larger than the total transition linewidth Γ, and the laser's response in the near vicinity of the $[\sigma^{(+)}]$-polarized transition linked to spin-up electronic state is displayed only. When the laser is also circularly polarized (left plot), it acquires an overall phase that can not be detected leading to a purely absorptive signal (open circles). Same scan with a linearly polarized laser (right plot) displays the dispersive response (filled circles) alongside the absorptive (open

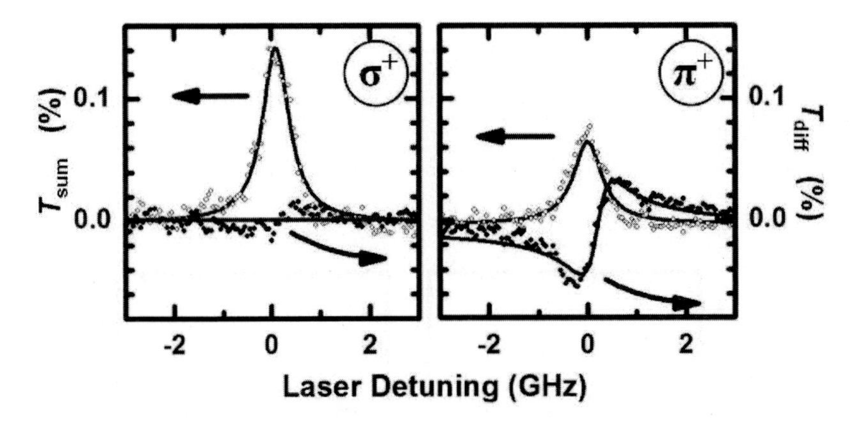

Fig. 6.5. The absorptive and dispersive signals from a quantum dot transition conditional on laser polarization.

circles): since the acquired phase is now relative, it leads to a linear polarization rotation, which can be detected by the difference detection system. At such small detunings with respect to Zeeman splitting, the laser experiences Faraday rotation primarily due to one Zeeman transition nearby. Following this initial demonstration of mapping out the dispersive response with single QD-spin sensitivity, we now proceed to the QIP-relevant operation regime, i.e. probe-laser detunings that are considerably larger than Zeeman splitting ($\delta \gg E_{\text{Zeeman}}$), where the difference signal arises from a competition between the two transitions. The top two plots of Figure 6.6 display the theoretical prediction and experimental realization for a 4-GHz Zeeman splitting case. Clearly, the combination of the two dispersive responses leads to an overall time-averaged signal that is enhanced between the transitions while suppressed as the laser detuning becomes large. The bottom left plot of Fig. 6.6 shows what we expect as the remaining tails of the dispersive response for a laser detuning that is considerably large with respect to the transition linewidth or the 26-GHz Zeeman splitting. The red (blue) curve is for a spin-up (spin-down) electron, while the black is the time-averaged signal for a random spin state. The key features are the strongly reduced signal levels and the spin-dependent change of sign. The bottom right plot of the same Figure shows what can be observed for a 100-msec effective measurement timescale per data point, plotted as a function of probe-laser detuning when the QD spin is prepared to state $|\uparrow\rangle$ (red data set), prepared to state $|\downarrow\rangle$ (blue data set) and left to randomize (black data set). The signal from the optically prepared spin switches sign as expected and the curves are in accordance with the expected inverse detuning (δ^{-1}) dependence. This plot further shows that the signal-to-noise ratio remains above unity all the way out to 90-GHz detuning from the center of the Zeeman-split transitions.

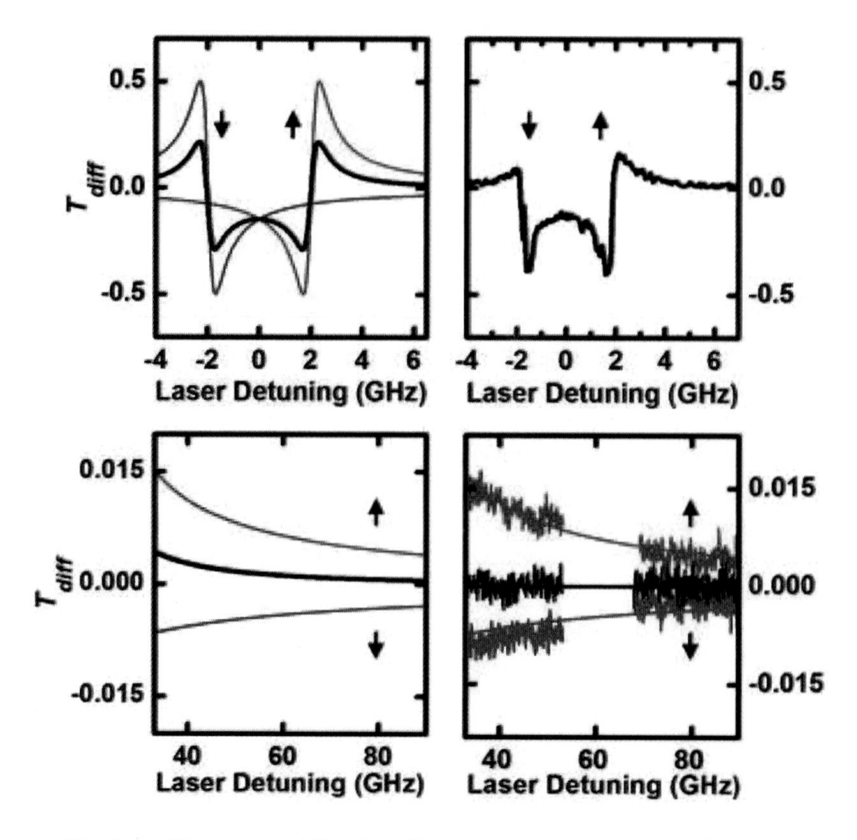

Fig. 6.6. Time-averaged Faraday effect from two transitions of a QD spin.

In order to tell the sign of the Faraday angle, and hence the spin of the electron within a limited measurement time, unity SNR value is sufficient. However, achieving spin read-out that is fast with respect to spin-flip dynamics, i.e. T_1 time, is not the only criteria. If, during this measurement time, a spin-flip event occurs due to unwanted optical spin pumping mechanism on a timescale still faster than the natural spin-flip times, then the measured dynamics will be distorted by this back-action. Therefore, while the measurement time can be reduced by choosing the reasonable detuning, it is essential that the probability of a back-action event in the form of spin-flip Raman scattering remains small within the T_1 time, i.e. $T_{\text{measure}} < T_1 < T_{\text{back-action}}$. Rigorously satisfying this criteria in fact opens the possibility for a quantum nondemolition or back-action evading measurement. The data of Fig. 6.6 is not in this regime, since during the required measurement time to obtain a unity SNR level (100 msec), around 5 back-action events occur (once every 20 msec for a branching ratio of 10^{-4}, as extracted from spin-relaxation studies [17]). The obvious solution to this last remaining challenge is to improve the $\eta\frac{\sigma_0}{A_{\text{Laser}}}$ factor scaling our signal: By using a combination of index-matched solid-immersion lenses ($\times n_{\text{GaAs}}^2 \sim 12$) and commercially available higher efficiency photo-detectors ($\times 8$), achieving a $\sim$100-fold improvement of this coefficient is plausible. It would then be possible to effectively eliminate spin-flip Raman scattering from the probe laser within the anticipated measurement time of $\sim$10 msec and resolve spin quantum jumps with these improvements. Alternatively, incorporating gated structures into photonic crystal nano-cavities [6] is demanding, but the existence of a far-detuned cavity mode can well be the way to obviate measurement back-action.

6.5. Conclusions and Outlook

We have covered recent advances on two key issues relevant to the utilization of self-assembled quantum dot spins for quantum information processing. First, we discussed how to prepare a single electron spin via optical spin pumping with fidelity beyond 99.8%. Optical coupling of electronic spin states is achieved using resonant excitation of the charged quantum dot (trion) transitions along with finite spin-flip Raman scattering rates. Second, we discussed a new approach to all-optical spin measurement with minimal back-action. The spin-state information is obtained through conditional Faraday rotation of a spectrally detuned laser, induced by the polarization- and spin-selective transitions. What still remains as a goal is time-resolved single spin read-out in the absence of back-action. This ability, kept from us only by technological shortcomings that should be not too difficult to overcome, will enable the observation of electron spin-flip quantum jumps in a continuous measurement scheme. This will be the first checkpoint for many other experiments to come. Another issue is the all-optical coherent manipulation of electron spin, although preparation and measurement suggest that achieving coherent manipulation is likely to follow. In parallel, sufficient increase of the SNR levels could obviate the need for lock-in detection scheme and real-time measurements can be realized [26]. Finally, the marriage of quantum-dot spins and solid-state cavity quantum electrodynamics was rather far fetched at the time of its proposal, since there was little achieved on either of these fronts at the time. Today, cavities with mode quality factors exceeding 10^4 and volumes on the order of a wavelength can routinely be fabricated. Further, positioning a single cavity mode on top of a single quantum dot has already been demonstrated. The advances on spins, partly covered in this work, and the current state of the art in nanotechnology may merge together for the realization of cavity-mediated spin-spin interactions in not-so-distant future.

References

[1] P. Michler, A. Imamoglu, M. D. Mason, P. J. Carson, G. F. Strouse, and S. K. Buratto, Quantum correlation among photons from a single quantum dot at room temperature, *Nature*. **406**(6799), 968–970, (2000).

[2] J. Kim, O. Benson, H. Kan, and Y. Yamamoto, A single-photon turnstile device, *Nature*. **397**, 500–503, (1999).

[3] R. M. Stevenson, R. J. Young, P. Atkinson, K. Cooper, D. A. Ritchie, and A. J. Shields, A semiconductor source of triggered entangled photon pairs, *Nature*. **439**, 179–182, (2006).

[4] A. Kiraz, P. Michler, C. Becher, B. Gayral, A. Imamoglu, L. Zhang, E. Hu, W. V. Schoenfeld, and P. M. Petroff, Cavity-quantum electrodynamics using a single inas quantum dot in a microdisk structure, *Applied Physics Letters*. **78**, 3932–3934, (2001).

[5] C. Santori, D. Fattal, J. Vuckovic, G. S. Solomon, and Y. Yamamoto, Indistinguishable photons from a single-photon device, *Nature*. **419**, 594–597, (2002).

[6] A. Badolato, K. Hennessy, M. Atatüre, J. Dreiser, E. Hu, P. M. Petroff, and A. Imamoglu, Deterministic coupling of single quantum dots to single nanocavity modes, *Science*. **308**, 1158–1161, (2005).

[7] J. P. Reithmaier, G. Sk, A. Lffler, C. Hofmann, S. Kuhn, S. Reitzenstein, L. V. Keldysh, V. D. Kulakovskii, T. L. Reinecke, and A. Forchel, Strong coupling in a single quantum dot-semiconductor microcavity system, *Nature*. **432**, 197–200, (2004).

[8] T. Yoshie, A. Scherer, J. Hendrickson, G. Khitrova, H. M. Gibbs, G. Rupper, C. Ell, O. B. Shchekin, and D. G. Deppe, Vacuum rabi splitting with a single quantum dot in a photonic crystal nanocavity, *Nature*. **432**, 200–203, (2004).

[9] K. Hennessy, A. Badolato, M. Winger, D. Gerace, M. Atatüre, S. Gulde, S. Faelt, E. Hu, and A. Imamoglu, Quantum nature of a strongly coupled single quantum dot-cavity system, *Nature.* **445**, 896–899, (2007).

[10] T. H. Stievater, X. Q. Li, D. G. Steel, D. Gammon, D. S. Katzer, D. Park, C. Piermarocchi, and L. J. Sham, Rabi oscillations of excitons in single quantum dots, *Physical Review Letters.* **87**, 133603, (2001).

[11] X. Q. Li, Y. W. Wu, D. Steel, D. Gammon, T. H. Stievater, D. S. Katzer, D. Park, C. Piermarocchi, and L. J. Sham, An all-optical quantum gate in a semiconductor quantum dot, *Science.* **301**, 809–811, (2003).

[12] A. Imamoglu, D. D. Awschalom, G. Burkard, D. P. DiVincenzo, D. Loss, M. Sherwin, and A. Small, Quantum information processing using quantum dot spins and cavity qed, *Physical Review Letters.* **83**, 4204–4207, (1999).

[13] J. Brossel, A. Kastler, and J. Winter, Creation optique dune inegalite de population entre les sous-niveaux zeeman de letat fondamental des atomes, *Journal de Physique et le Radium.* **13**, 668, (1952).

[14] B. Alen, F. Bickel, K. Karrai, R. J. Warburton, and P. M. Petroff, Stark-shift modulation absorption spectroscopy of single quantum dots, *Applied Physics Letters.* **83**, 2235–2237, (2003).

[15] R. J. Warburton, C. Schaflein, D. Haft, F. Bickel, A. Lorke, K. Karrai, J. M. Garcia, W. Schoenfeld, and P. M. Petroff, Optical emission from a charge-tunable quantum ring, *Nature.* **405**, 926–929, (2000).

[16] K. Karrai and R. J. Warburton, Optical transmission and reflection spectroscopy of single quantum dots, *Superlattices and Microstructures.* **33**, 311–337, (2003).

[17] J. Dreiser, M. Atatüre, C. Galland, T. Muller, A. Badolato, and A. Imamoglu, Optical investigations of quantum-dot spin dynamics, *arXiv:0705.3557v1.* (2007).

[18] J. M. Elzerman, R. Hanson, L. H. W. van Beveren, B. Witkamp, L. M. K. Vandersypen, and L. P. Kouwenhoven, Single-shot read-out of an individual electron spin in a quantum dot, *Nature.* **430**, 431–435, (2004).

[19] M. Kroutvar, Y. Ducommun, D. Heiss, M. Bichler, D. Schuh, G. Abstreiter, and J. Finley, Optically programmable electron spin memory using semiconductor quantum dots, *Nature.* **432**, 81–84, (2004).

[20] M. Atatüre, J. Dreiser, A. Badolato, A. Hogele, K. Karrai, and A. Imamoglu, Quantum-dot spin-state preparation with near-unity fidelity, *Science.* **312**, 551–553, (2006).

[21] M. V. G. Dutt, J. Cheng, B. Li, X. D. Xu, X. Q. Li, P. R. Berman, D. G. Steel, A. S. Bracker, D. Gammon, S. E. Economou, R. B. Liu, and L. J. Sham, Stimulated and spontaneous optical generation of electron spin coherence in charged GaAs quantum dots, *Physical Review Letters.* **94**, 227403, (2005).

[22] A. Kiraz, M. Atatüre, and A. Imamoglu, Quantum-dot single-photon sources: Prospects for applications in linear optics quantum-information processing, *Physical Review A.* **69**, 032305, (2004).

[23] D. D. Awschalom and J. M. Kikkawa, Electron spin and optical coherence in semiconductors, *Physics Today.* **52**, 33–38, (1999).

[24] J. Berezovsky, M. H. Mikkelsen, O. Gywat, N. G. Stoltz, L. A. Coldren, and D. D. Awschalom, Nondestructive optical measurements of a single electron spin in a quantum dot, *Science.* **314**(5807), 1916–1920, (2006).

[25] M. Atatüre, J. Dreiser, A. Badolato, and A. Imamoglu, Observation of faraday rotation from a single confined spin, *Nature Physics.* **3**(5807), 101–106, (2007).

[26] A. N. Vamivakas, M. Atatüre, J. Dreiser, S. T. Yilmaz, A. Badolato, A. K. Swan, B. B. Goldberg, A. Imamoglu, and M. S. Unlu, Strong extinction of a far-field laser beam by a single quantum dot, *Nano Letters.* **7**, 2892–2896, (2007).

Chapter 7

Optical Read-out of Single Carrier Spin in Semiconductor Quantum Dots

Filippo Troiani, Ignacio Wilson-Rae, and Carlos Tejedor

CNR-INFM National Research Center S3, c/o Dipartimento di Fisica, via G. Campi 213/A, 41100, Modena, Italy

The capability of manipulating and reading-out information at the single-spin level represents a key challenge for spintronics and spin-based quantum information [1, 2]. In semiconductor quantum dots, the strategies so far implemented for the single- spin read-out are based on spin-to-charge conversion [3, 4]. There, the dot is electrically manipulated in order to obtain a spin-dependent occupation number. The charge state is then measured with high sensitivity thanks to the dot's capacitive coupling to a quantum-point contact. These measurement schemes, however, imply the tunneling-out of the electron (at least in half of the cases), and therefore the destruction of the quantum state of interest. Projective measurements are instead repeatable, and, in the ideal case, the result of repetitions is completely predictable. This makes them also potential tools for the fast initialization of the qubits [5].

The optical manipulation represents an alternative route to spin-based quantum information [6, 7]. Also there, indirect strategies have been implemented for the initialization and read-out of carrier spins. These rely on the optical selection rules that build-up specific correlations between the spin state of the electron-hole pair and the circular polarization of the light emitted (absorbed) with the pair's annihilation (creation) [8, 9]. As in the previous case, however, the conversion from the spin degree of freedom to the auxiliary ones implies the loss of the spin state, and this impedes the measurement's repetition. In order to achieve a reliable single-spin measurement, an order-of-magnitude increase in the photon collection efficiency is needed. Most likely, this might not suffice, unless the possibility of repeatedly probing the same quantum state will also be demonstrated. Photon-collection efficiencies can be largely increased if the dot is optically coupled to an optical microcavity, which mediates its optical emission into free space [10]. The additional requirement that the measurement be non-destructive requires suitable devices and manipulation schemes. Hereafter, we illustrate one such proposals, specifically aimed at implementing a quantum non-demolition measurement of single carrier spin in a dot-cavity system.

Contents

7.1. Quantum Non-demolition Measurement

A quantum non-demolition (QND) measurement is a specific kind of indirect measurement [11]. It ideally consists of a sequence of two steps. In a first step, the interaction between the system (S) and a probe (P) is turned on, in such a way that the observable of interest (called *signal*, A_S) affects the value of a probe observable (A_P), without being perturbed itself by such interaction. This can be formally expressed as follows: $[A_S, H_{SP}] = 0$, $[A_P, H_{SP}] \neq 0$, and $\partial H_{SP}/\partial A_S \neq 0$, where H_{SP} accounts for the signal-probe coupling. Such interaction results in a reversible (i.e., unitary) evolution, and generates a correlation between the states of S and P. In a second step, A_P is measured, and the outcome is used to infer the eigenstate of A_S. In order for such sequence to be repeatable, the post-measurement evolution of A_S must not become unpredictable due to: (i) possible information that the measurement provides on its conjugated variables (A_S^c), and (ii) the uncertainty that the state projection induces in A_S^c. The latter condition calls for the observable to be a constant of motion of the system Hamiltonian: $\partial H_S/\partial A_S^c \neq 0$ [12].

In the QND measurement discussed hereafter, the signal and probe are represented by the hole spin and the circular polarization of the cavity photons, respectively. More specifically, $A_S = J_z$ and $A_P = n_+ - n_-$, where $n_\pm$ are the occupation numbers of the cavity modes corresponding to the $+$ and $-$ circular polarizations. Here, we restrict ourselves to the heavy-hole subband, corresponding to the states $|\uparrow\rangle \equiv |J_z = +3/2\rangle$ and $|\downarrow\rangle \equiv |J_z = -3/2\rangle$. The goal of the present scheme is to optically induce an effective coupling between carrier spin and cavity-photon polarization that builds up correlations between J_z and $n_+ - n_-$: $|\uparrow\rangle \otimes |0\rangle \longrightarrow |\uparrow\rangle \otimes |\sigma_+\rangle$,

$|\downarrow\rangle \otimes |0\rangle \longrightarrow |\downarrow\rangle \otimes |\sigma_-\rangle$, where σ_+ (σ_-) generically denotes a state with $n_+ > n_-$ ($n_+ < n_-$). This allows to finally infer the spin state by means of a polarization-resolved detection of the cavity photons.

7.2. Level Scheme

The mapping of the spin state onto the cavity-photon polarization requires a suitable level scheme. This can be provided by two vertically-coupled self-assembled QDs. Within the Stranski-Krastanov growth technique, the vertical alignment of the dots is strongly favored by the strain field that each dot induces in the upper layers. Besides, the width of the interdot barrier can be varied in a controlled manner, thus allowing to modulate the strength of the interdot coupling [13]. Due to the combined effect of strain and of effective mass asymmetry between valence and conduction bands, electrons and holes typically experience different coupling regimes. In fact, while the electronic wavefunctions correspond to delocalized bonding and antibonding states, the holes tend to be localized in either of the two dots [14, 15].

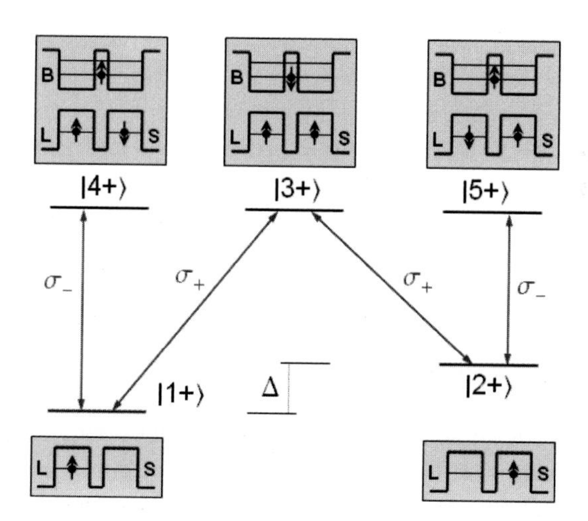

Fig. 7.1. Relevant level scheme of the positively-doped artificial molecule: "+" subspace (see the text). The two lowest hole states ($|n < 3+\rangle$), separated by an energy gap Δ, correspond to the hole being localized either in the larger (L) or smaller (S) dot. The three optically active trion states ($|n > 2+\rangle$) are schematically represented by the dominant single-particle configurations, all involving the bonding state (B) for the electron.

As a consequence, the low-energy level scheme of the artificial molecule (Fig. 7.1), doped with an excess hole, is typically characterized by the presence of two stable states for each spin orientation, corresponding to the carrier being localized either in the larger dot (L) or in the smaller one (S). The spin up states $|1+\rangle \equiv d^{\dagger}_{L,\uparrow}|0\rangle$ and $|2+\rangle \equiv d^{\dagger}_{S,\uparrow}|0\rangle$ are optically coupled to a common trion state, which for simplicity we identify with the configuration of lowest single-particle energy, namely one hole per dot, and the electron in the bonding state (B): $|3\pm\rangle \equiv c^{\dagger}_{B,\downarrow}d^{\dagger}_{L,\uparrow}d^{\dagger}_{S,\uparrow}|0\rangle$. Possible small corrections to this picture arising from the Coulomb interactions are irrelevant to our purposes, as far as the Λ-scheme is preserved [16]. Both the above transitions are induced by light with σ_+ circular polarization; their frequencies differ by an energy gap Δ, which reflects the asymmetry between the dots L and S. In addition to these states, that are functional to the scheme implementation, we consider the lowest trion states corresponding to the alternative spin configurations: $|4+\rangle \equiv c^{\dagger}_{B,\uparrow}d^{\dagger}_{L,\uparrow}d^{\dagger}_{S,\downarrow}|0\rangle$ and $|5+\rangle \equiv c^{\dagger}_{B,\uparrow}d^{\dagger}_{L,\downarrow}d^{\dagger}_{S,\uparrow}|0\rangle$. The $|1+\rangle \longleftrightarrow |4+\rangle$ and $|2+\rangle \longleftrightarrow |5+\rangle$ transitions are degenerate with $|1+\rangle \longleftrightarrow |3+\rangle$ and $|2+\rangle \longleftrightarrow |3+\rangle$, respectively, but they correspond to an opposite circular polarization (σ_-). Slight departures of the dot geometry from the circular symmetry might select linearly polarized excitons as a preferred basis. Even in this case, however, a magnetic field applied in the growth direction would allow to recover the required circularly polarized basis. By inverting all the spin orientations and the light polarizations in Fig. 7.1, one obtains a perfect replica of the level scheme: hereafter, we refer to the two as the "+" and "−" subspaces. Finally, we assume the dots to be coupled to an optical microcavity (MC) [17, 18] that supports an energetically resolved fundamental mode, doubly degenerate with respect to polarization and close to resonance with the excitonic transitions (see below).

7.3. Manipulation Scheme

Given the above level scheme, the correlation between the hole spin and the photon polarization is established by two cavity-assisted Raman transitions [19]: those coupling the states $|1\zeta\rangle$ and $|2\zeta\rangle$, through the interconnecting states $|3\zeta\rangle$, with $\zeta = \pm$. The first of these transitions, induced by a laser pulse of central frequency ω_1, results in the hole transfer from dot L to dot S, and in the creation of a σ_ζ cavity photon [Fig. 7.2(a)]. The second transition, induced by a pulse of frequency ω_2, transfers the hole from S to L, and also generates a σ_χ photon [Fig. 7.2(b)]. In both cases,

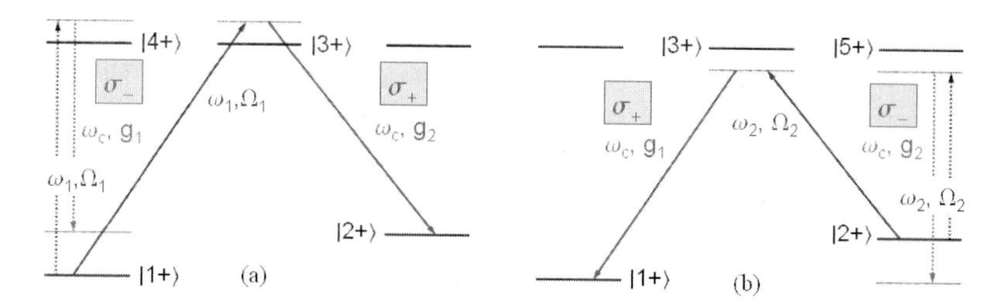

Fig. 7.2. Raman transitions that couple the two hole states in the "+" subspace, with initial state $|1+\rangle$ (a) or $|2+\rangle$ (b). The desired transitions (solid lines) result in the hole transfer from dot L to S, or viceversa, and in the emission of a σ_+ cavity photon. The off-resonant ones (see, e.g., dotted lines) can generate σ_- cavity photons. $g_{1,2}$ and $\Omega_{1,2}$ are the dot-cavity and dot-laser coupling constants, respectively (see the text). Specular schemes apply to the "−" subspace.

the spin state is left unaffected. In fact, in the absence of decoherence and imperfections, the two spin components evolve independently from one another, for the interaction of the molecule with light doesn't couple the "+" and "−" subspaces. The actual occupation of the trion states $|n > 2\,\chi\rangle$ can result in photon emission into the leaky modes, and therefore in a reduced detection efficiency. This suggests to keep the single-photon transitions sufficiently out of resonance. Also the Raman transitions involving states $|4\chi\rangle$ and $|5\chi\rangle$ are a potential sources of error, for they imply the generation of anti-correlated cavity photons (σ_- for $|\uparrow\rangle$ and σ_+ for $|\downarrow\rangle$), and therefore tend to reduce the overall correlation between spin state and photon polarization. Thanks to the presence of the energy gap Δ, the two-photon resonance condition for the desired Raman transitions can be fulfilled, while keeping out of resonance the unwanted ones. In particular, this is achieved by setting $\omega_1 = \omega_c + \Delta + \delta_1$, and $\omega_2 = \omega_c - \Delta + \delta_2$, where ω_c is the frequency of the cavity modes and $\delta_{1,2}$ are smaller corrections accounting for the AC Stark shifts. In particular, a reliable read-out of individual spins is due to the weakness of the signal. The possibility of repeating the measurement might therefore represent a crucial resource. Repeatability requires the measurement to be non-destructive. In the ideal case the result of the repetitions must be predictable. Measurement of the first kind, projective measurements. The importance of such measurements is also related to initialization, error correcting codes. [1]

The above equations show that Δ also gives the difference between the laser and the cavity frequencies; this has a practical relevance, for it allows to spectrally resolve the photons emitted by the cavity from the stray light.

Before discussing in more detail two possible implementations of the scheme, a few general comments are in order. The role played by the cavity here is twofold: on the one hand, it provides the optical coupling which is required for inducing the above Raman transitions; on the other hand, it makes the photon emission extremely more rapid and directional than it would be otherwise [20]. These are both crucial issues, and call for a fine tuning of the dot-cavity coupling (see below). It is also relevant that both the possible spin eigenstates give rise to photon emission: the fact that the measurement is ultimately decided depending on whether n_+ is larger or smaller than n_-, rather than on the presence versus absence of photons in a given mode, makes the scheme more robust with respect to photon loss or detection inefficiency.

7.4. Pulsed Implementation

One possible implementation of the measurement scheme corresponds to driving the system by a sequence of non-overlapping, alternate pulses from two lasers. These operate at the frequencies ω_1 and ω_2, and emit linearly-polarized light. Therefore, if the hole spin is initially in the $|\uparrow\rangle$ ($|\downarrow\rangle$) state, the σ_+ (σ_-) component of the laser light induces the required Raman transitions, whereas the σ_- (σ_+) is uneffective. Within the sequence, each pulse ideally triggers the emission of one cavity photon, thus allowing for a single detection. In Fig. 7.3 we report the simulation [21, 22] corresponding to the case of two pulses (shaded gray areas), one from each laser, and to the hole being initialized in state $|1+\rangle$. The first pulse, with frequency ω_1, drives the hole to dot S, while leaving its spin state unaffected [panel (a)]. The molecule's $|1+\rangle \longrightarrow |2+\rangle$ transition is accompanied by the creation of a single σ_+ cavity photon (figure inset). The second laser pulse, with frequency ω_2, drives the hole back to the initial state, and generates a second σ_+ cavity photon. We incidentally note that the photons are emitted from the cavity before being reabsorbed by the molecule: the measurement scheme would not profit from such a reversible dot-cavity dynamics. The dotted line in panel (a) represents the overall occupation of the excitonic manifold: this is evidently kept negligible throughout the process, and could be further reduced by applying some pulse chirping. The main source of error is represented by the hole-spin relaxation [23–25], which induces a population leakage to the subspace "$-$"; this results in a finite probability of emitting a σ_- photon (figure inset). Such population leakage sets a time constraint, for the overall measurement must be completed on a timescale

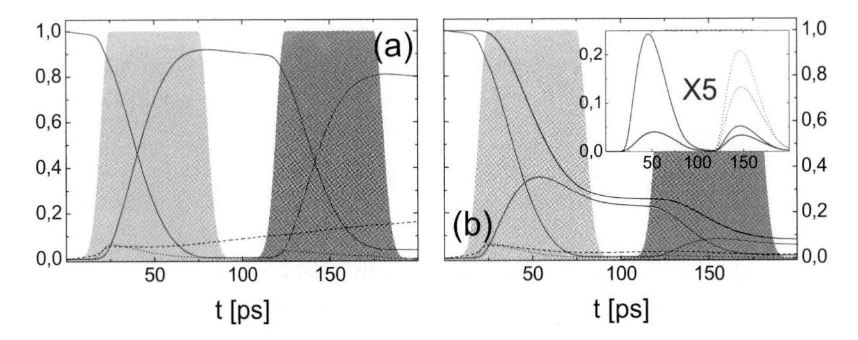

Fig. 7.3. Simulated time evolution of the dot-cavity system, driven by a sequence of two laser pulses (shaded gray areas). Panels (a) and (b) correspond to the unconditioned ($\eta = 0$) and conditional ($\eta = 0.75$) dynamics, respectively. The different lines give the occupations of: state $|1+\rangle$ (red), $|2+\rangle$ (blue), manifold $|n > 2+\rangle$ (black dotted), "$-$" subspace (black dashed). The solid black line in (b) is $\text{Tr}\{\rho_\eta(t)\}$. Inset: average occupation of the σ_+ (red) and σ_- (blue) cavity modes, for $\eta = 0$ (dotted) and $\eta = 0.75$ (solid).

shorter than those characterizing the spin relaxation (here we assume $\Gamma_s = 1\,\text{ns}^{-1}$). In other words, it limits the number of repetitions that can be performed before the initial state of the spin is lost.

Repeating the measurement allows in principle to overcome two possible shortcomings: the finite detection efficiency, and the imperfect correlation between signal and probe. The first limitation might leave the observer without any information at all on the state of the system; the second one reduces the reliability of the inference allowed by the measurement of the probe. In the present case the first problem is more severe than the second one. Therefore, it is reasonable to decide the measurement on the basis of the first detected photon. From the theoretical point of view, this translates into solving a conditional master equation, where the Liouvillian that accounts for the loss of the cavity photons now depends on the detection efficiency η [26]. The solution $\rho_\eta(t)$ of such equation corresponds to the state of the dot-cavity system conditioned upon not having detected any photon up to time t. The corresponding probability is given by $P_0(t) = \text{Tr}\{\rho_\eta(t)\}$, which is no longer preserved (solid brown line in the plot). The failure probability $\epsilon \equiv P_0(T)$ is clearly reduced by repeating the number of pulses within the duration T of the overall measurement. The probability of detecting one (and only one) photon with either polarization is proportional to the time integral of the cavity mode occupation: $P_\pm = \kappa\eta \int_0^T \langle a^\dagger(t)a(t)\rangle dt$, with κ the cavity-photon emission rate. The reliability of the information provided by the photons, is quantified by the measurement fidelity ($\mathcal{F}$) [27]. The results we obtain in the above illustrative case are: $P_+ = 0.71\,(0.87)$, $P_- = 0.027\,(0.048)$ after one (two pulses), resulting in a fidelity close to the optimum value of 1, namely $\mathcal{F} = 0.982\,(0.973)$; the failure probability, instead, is $\epsilon = 0.262\,(0.082)$. Here, the measurement duration ($T = 200\,\mathrm{ps}$) is comparable to the spin-flip time $1/\Gamma_s$; therefore, the measurement repetition, while decreasing the failure probability, tends to decrease the overall fidelity. The best tradeoff between these conflicting requirements depends on the value of η, and on the ratio between the measurement and the decoherence rates (see below).

7.5. Continuous Implementation

An alternative implementation of the scheme consists in driving the system in a continuous mode: the two lasers are simultaneously switched on, and the measurement ends when the predetermined number of photons is detected. In the virtual Raman regime, the excitonic manifold is prevented from being meaningfully populated during the process. The corresponding range of physical parameters is identified by the inequality $|\delta_T| \lesssim |\tilde{\Delta}| \sim \Delta$, with the latter much larger than all the other frequency scales in the problem ($2\tilde{\Delta} \equiv \omega_1 - \omega_2$ and $\delta_T \equiv \omega_1 + \omega_2 - 2\omega_T + \Delta$). There, the unitary part of the system dynamics can be described by means of an effective Hamiltonian H_{eff}, defined within the ground-state manifold. To lowest order in $1/\Delta$, H_{eff} takes the following form:

$$H_{\mathrm{eff}} = \sum_{\zeta=\pm} \left[\frac{g_2|\Omega_1^{(1)}(t)|}{\delta_T + \tilde{\Delta}} \sigma_+^{(\zeta)} + \frac{g_1|\Omega_2^{(2)}(t)|}{\delta_T - \tilde{\Delta}} \sigma_-^{(\zeta)} \right] a_\zeta^\dagger + \mathrm{H.c.}. \tag{7.1}$$

Here, $a_\pm$ are the annihilation operators for the $\sigma_\pm$ polarized cavity modes; $\sigma_+^{(\zeta)} \equiv |2\zeta\rangle\langle 1\zeta|$ and $\sigma_-^{(\zeta)} \equiv |1\zeta\rangle\langle 2\zeta|$ are the ladder operators connecting the two orbital hole states; $\Omega_{1/2}^{(1)}(t)$ $[\Omega_{1/2}^{(2)}(t)]$ are the Rabi frequencies for the transitions between $|1/2\pm\rangle$ and the trion states $|3\pm\rangle$ induced by laser 1 (2). In order to optimize the process, ω_1, ω_2, and the Rabi frequencies are chosen so that all the detunings and induced AC Stark shifts cancel out [22].

Under the combined effect of the H_{eff} and of the MC-photon loss, the dot-cavity system tends to a stationary state, where the $n_+ - n_-$ reflects and amplifies the initial carrier-spin state:

$$\rho_{\mathrm{ss}}^{\uparrow} = |\uparrow\rangle\langle\uparrow| \otimes \sum_{\xi=\pm} |\xi\rangle\langle\xi|\rho_h|\xi\rangle\langle\xi| \otimes |\lambda_\xi\rangle\langle\lambda_\xi|_+ \otimes |0\rangle\langle 0|_- \,. \tag{7.2}$$

Here $|0\rangle_\pm$ and $|\lambda_\xi\rangle_\pm$ are the vacuum and coherent states for the cavity modes, respectively, with $\lambda_\xi \equiv -\xi i\tilde{\Omega}_0/\kappa$ and $\tilde{\Omega}_0 = |\Omega_1^{(1)}(t_s)|(g_1^2 - g_2^2)/g_2\Delta$; besides $|\xi\rangle \equiv (|2\rangle + \xi|1\rangle)/\sqrt{2}$, and ρ_h is the initial condition for the hole (orbital part). In addition to the correlation between spin state and photon polarization, the stationary state $\rho_{\rm ss}^{\uparrow}$ exhibits entanglement between the orbital parts of the hole and photon wavefunctions. In fact, the symmetric and antisymmetric combinations of the localized hole states, $|\xi = +\rangle$ and $|\xi = -\rangle$, result in the generation of the coherent states $|\lambda_+\rangle$ or $|\lambda_-\rangle$, respectively. The complete analytical solution [22] allows also to derive an expression for the minimum measurement duration, and for the corresponding optimum value of κ, as a function of the optical couplings and of the failure probability ϵ. For a Gaussian time profile of the laser switching, these are $T_{\rm min} \approx 3(-\ln\epsilon/\eta)^{1/2}/\tilde{\Omega}_0$ and $\kappa_{\rm opt} \approx \sqrt{\eta/\ln(1/P_0(T))}\tilde{\Omega}_0$. The existence of an optimum value for κ reflects the two-fold role played by the MC: in fact, $\kappa > \kappa_{opt}$ results in a reduced efficiency of the Raman transition, whereas $\kappa < \kappa_{opt}$ implies a slow down of the photon emission, and therefore an increase of T [19]. Experimental developments prompt us to consider as a typical example: $\eta = 0.75$, $\hbar\Delta = 2$meV, $\hbar g_1 = 0.2$meV, $\hbar\kappa = 0.05$meV and $g_1/g_2 = \sqrt{3}$; which lead to $T \approx 200\,$ps for $\epsilon \sim 10^{-2}$, with $\hbar\Omega_1^{(1)} = 0.4\,$meV.

7.6. Conclusions and Outlook

We have presented a proposal for an all-optical implementation of a single-carrier spin. A typical artificial molecule, formed by two vertically-coupled, self-assembled quantum dots was shown to provide a suitable level scheme for such implementation. An additional source of error, besides those considered in the paper, is potentially represented by valence-band mixing. Depending on the microscopic symmetry of the dots, this might produce two kind of effects [28]: reduction of the degree of correlation between hole spin and carrier polarization, or finite probability of optically inducing the hole-spin flip. While the former effect would only increase the number of repetitions required to achieve a given fidelity, the latter one would also spoil the non-destructive character of the measurement.

The generalization of the present scheme to the read-out of electron spins is less straightforward. In particular, a typical level scheme would most likely not be divisible in a "+" and "−" subspaces, due to the interdot coupling experienced by the excess carrier. The large number of repetitions made possible by the long electron-spin relaxation times [2] can however more than compensate such shortcoming. This will be the object of future work.

References

[1] M. A. Nielsen and I. L. Chuang, *Quantum computation and quantum information.* (Cambridge University Press, 1997).

[2] D. D. Awschalom, D. Loss, and N. Samarth, *Semiconductor spintronics and quantum computation.* (Springer-Verlag, 2002).

[3] J. M. Elzerman, R. Hanson, L. H. Willems van Beveren, B. Witkamp, L. M. K. Vandersypen, and L. P. Kouwenhoven, Single-shot read-out of an individual electron spin in a quantum dot, *Nature.* **430**, 431, (2004).

[4] R. Hanson, L. H. Willems van Beveren, I. T. Vink, J. M. Elzerman, W. J. M. Naber, F. H. L. Koppens, L. P. Kouwenhoven, and L. M. K. Vandersypen, Single-shot readout of electron spin states in a quantum dot using spin-dependent tunnel rates, *Phys. Rev. Lett.* **94**, 196802, (2005).

[5] R.-B. Liu, W. Yao, and L. J. Sham, Coherent control of cavity quantum electrodynamics for quantum nondemolition measurements and ultrafast cooling, *Phys. Rev. B.* **72**, 81306, (2005).

[6] C. Piermarocchi, P. Chen, L. J. Sham, and D. G. Steel, Optical RKKY interaction between charged semiconductor quantum dots, *Phys. Rev. Lett.* **89**, 167402, (2002).

[7] F. Troiani, U. Hohenester, and E. Molinari, High-finesse optical quantum gates for electron spins in artificial molecules, *Phys. Rev. Lett.* **90**, 206802, (2003).

[8] S. Cortez, O. Krebs, S. Laurent, M. Senes, P. Marie, X. Voisin, R. Ferreira, J.-M. Bastard, G. Gerard, and T. Amand, Optically driven spin memory in n-doped InAs-GaAs quantum dots, *Phys. Rev. Lett.* **89**, 207401, (2002).

[9] M. Kroutvar, Y. Ducommun, D. Heiss, M. Bichler, D. Schuh, G. Abstreiter, and J. J. Finley, Optically programmable electron spin memory using semiconductor quantum dots, *Nature.* **432**, 81, (2004).

[10] G. S. Solomon, M. Pelton, and Y. Yamamoto, Single-mode spontaneous emission from a single quantum dot in a three-dimensional microcavity, *Phys. Rev. Lett.* **86**, 3903, (2001).

[11] V. B. Braginsky and F. Y. Khalili, *Quantum measurement.* (Cambridge University Press, 1992).

[12] M. O. Scully and M. S. Zubairy, *Quantum optics.* (Cambridge University Press, 1997).

[13] G. Schedelbeck, W. Wegscheider, M. Bichler, and G. Abstreiter, Coupled quantum dots fabricated by cleaved edge overgrowth: from artificial atoms to molecules, *Science.* **278**, 5344, (1997).

[14] W. Sheng and J.-P. Leburton, Spontaneous localization in InAs/GaAs self-assembled quantum-dot molecules, *Appl. Phys. Lett.* **81**, 4449, (2002).

[15] G. Bester, J. Shumway, and A. Zunger, Theory of excitonic spectra and entanglement engineering in dot molecules, *Phys. Rev. Lett.* **93**, 47401, (2004).

[16] U. Hohenester, F. Troiani, E. Molinari, G. Panzarini, and C. Macchiavello, Coherent population transfer in coupled semiconductor quantum dots, *Appl. Phys. Lett.* **77**, 1864, (2000).

[17] J. P. Reithmaier, G. Sek, A. Loffler, C. Hofmann, S. Kuhn, S. Reitzenstein, L. V. Keldysh, V. D. Kulakovskii, T. L. Reinecke, and A. Forchel, Strong coupling in a single quantum dot-semiconductor microcavity system, *Nature.* **432**, 197, (2004).

[18] A. Badolato, K. Hennessy, M. Atature, J. Dreiser, E. Hu, P. M. Petroff, and A. Imamoglu, Deterministic coupling of single quantum dots to single nanocavity modes, *Science.* **308**, 1158, (2005).

[19] C. Maurer, C. Becher, C. Russo, J. Eschner, and R. Blatt, A single-photon source based on a single Ca^+ ion, *New J. of Phys.* **6**, 94, (2004).

[20] J. M. Gerard, B. Sermage, B. Gayral, B. Legrand, E. Costard, and V. Thierry-Mieg, Enhanced spontaneous emission by quantum boxes in a monolithic optical microcavity, *Phys. Rev. Lett.* **81**, 1110, (1998).

[21] F. Troiani, J. I. Perea, and C. Tejedor, Cavity-assisted generation of entangled photon pairs by a quantum-dot cascade decay, *Phys. Rev. B.* **74**, 235310, (2006).

[22] F. Troiani, I. Wilson-Rae, and C. Tejedor, All-optical nondemolition measurement of single hole spin in a quantum-dot molecule, *Appl. Phys. Lett.* **90**, 144103, (2007).

[23] L.-M. Woods, T. L. Reinecke, and R. Kotlyar, Hole spin relaxation in quantum dots, *Phys. Rev. B.* **69**, 125330, (2004).

[24] D. Bulaev and D. Loss, Spin relaxation and decoherence of holes in quantum dots, *Phys. Rev. Lett.* **95**, 76805, (2005).

[25] C. Lu, J. L. Cheng, and M. W. Wu, Hole spin relaxation in semiconductor quantum dots, *Phys. Rev. B.* **71**, 75308, (2005).

[26] T. M. Stace, G. J. Milburn, and C. H. W. Barnes, Entangled two-photon source using biexciton emission of an asymmetric quantum dot in a cavity, *Phys. Rev. B.* **67**, 85317, (2003).

[27] T. C. Ralph, S. D. Bartlett, J. L. O'Brien, G. J. Pryde, and H. M. Wiseman, Spin relaxation and decoherence of holes in quantum dots, *Phys. Rev. A.* **73**, 12113, (2006).

[28] A. V. Koudinov, I. A. Akimov, Y. G. Kusrayev, and F. Henneberger, Optical and magnetic anisotropies of the hole states in Stranski-Krastanov quantum dots, *Phys. Rev. B.* **70**, 241305, (2004).

Chapter 8

Quantum State Transfer from a Photon to an Electron Spin in Quantum Dots and Quantum Dynamics of Electron-Nuclei Coupled System

T. Takagahara and Özgür Ĉakir

Department of Electronics and Information Science, Kyoto Institute of Technology, Matsugasaki, Kyoto 606-8585, Japan

CREST, Japan Science and Technology Agency, 4-1-8 Honcho, Kawaguchi, Saitama 332-0012, Japan

Photons and electrons are the key quantum media for the quantum information processing based on solid state devices. Here we consider the key elements to realize the quantum repeater and reveal the underlying physics of relevant elementary processes. Especially, we discuss the quantum state transfer between a single photon and a single electron and analyze the performance of the operation, clarifying the conditions to achieve the high fidelity of the quantum state transfer. We also investigate quantum dynamics in the electron-nuclei coupled spin system in quantum dots and predict a couple of new phenomena related to the correlations induced by the hyperfine interaction, namely bunching and revival in the electron spin measurements. The underlying mechanism is the squeezing and increase in the purity of the nuclear spin state through the electron spin measurements. These findings will open the way to the long-lived quantum memory based on nuclear spins.

Contents

8.1. Introduction

Coherent manipulation of quantum states is a critical step toward many novel technological applications ranging from manipulation of qubits in quantum logic gates [1–5] to controlling the reaction pathways of molecules. In the field of the quantum state control by optical means, both Rabi oscillation and quantum interference play the central roles. The exciton Rabi splitting was observed in the luminescence spectrum of a single InGaAs quantum dot and the exciton Rabi oscillation was also observed in the spectroscopy of a single GaAs or InGaAs quantum dot [6–12]. The two-qubit CROT gate operation was demonstrated using two orthogonally polarized exciton states and a biexciton state in a GaAs quantum dot [13]. Unfortunately, however, the decoherence/dephasing times of excitons and biexcitons in these quantum dots are limited by the radiative lifetimes ($\sim$ 1 ns) even at low temperatures. Thus a qubit with a longer decoherence time is desirable for the application to the quantum information processing. Electron spins in semiconductor quantum dots(QDs) are considered as one of the most promising candidates of the building blocks for quantum information processing [14, 15] due to their robustness against decoherence effects [16, 17]. In double QD systems, initialization and coherent manipulation of electron spin have been realized, with coherence times extending to 1 μs [18, 19].

Recently, a quantum media converter from a photon qubit to an electron spin qubit was proposed for quantum repeaters [20, 21]. Quantum information can take several different forms and it is preferable to be able to convert among different forms. One form is the photon polarization and another is the electron spin polarization. Photons are the most convenient medium for sharing quantum information between distant locations. However, it is necessary to realize a quantum repeater in order to send the information securely over a very long distance overcoming the photon loss. A quantum repeater requires quantum information storage and the electron spin is a promising candidate of such a quantum memory. A strained InGaAs/InP quantum dot was proposed for such a device based on the g-factor engineering [22]. In the actual operation, the photoexcited holes are to be quickly swept out of the quantum dot to project the photon polarization onto the electron spin polarization, preserving the entanglement. We have analyzed the performance of this operation and clarified the conditions to achieve a high value of the purity of the transferred quantum state or the fidelity of the quantum state transfer.

The electron spin decoherence time reported so far for low temperatures about a few microsecond is not sufficiently long for the secure quantum information processing. Thus the nuclear spin quantum memory will be eventually required and the robust quantum state transfer should be realized between the electron spin and the nuclear spins in order to store and retrieve the quantum memory timely. For that purpose, the fundamental features of dynamics in the electron-nuclei coupled system should be investigated. Recently we discovered a sequence of back-actions between the electron spin and the nuclear spins through the quantum state measurements. These findings will open the way to the quantum state purification of nuclear spins, elongation of the electron spin decoherence time and the nuclear spin quantum memory.

8.2. Quantum State Transfer

In the quantum state transfer between the photon qubit and the electron qubit, the one-to-one correspondence should be established between the photon polarization and the electron spin polarization. For that purpose, the light-hole exciton is preferable than the heavy-hole exciton because of the characteristic optical selection rule and the possible Zeeman splitting in the in-plane magnetic field. In order to have the light-hole states as the ground hole states, the strained quantum well(QW) structure is necessary and can be realized in InGaAs/InP QW structures. In these strained QW structures the light-hole states are written as [23]

$$|\frac{3}{2}\frac{1}{2}\rangle = \sqrt{\frac{2}{3}}|10\rangle\alpha + \sqrt{\frac{1}{3}}|11\rangle\beta\,, \tag{8.1}$$

$$|\frac{3}{2} - \frac{1}{2}\rangle = \sqrt{\frac{1}{3}}|1-1\rangle\alpha + \sqrt{\frac{2}{3}}|10\rangle\beta\,, \tag{8.2}$$

where $|10\rangle\alpha$ represents a p_z-like orbital with the up-spin, for example. Then the relevant Hamiltonian for the light-hole states under an in-plane magnetic field along the x axis is represented as

$$-g_{\ell h}\mu_B B S_x = -\frac{2}{3}g_{\ell h}\mu_B B\begin{pmatrix} 0 & 1 \\ 1 & 0 \end{pmatrix}\,, \tag{8.3}$$

where $g_{\ell h}$ is the g-factor of the light-hole and the eigenstates are given by

$$|\ell h_1\rangle = \frac{1}{\sqrt{2}}\left(|\frac{3}{2}\frac{1}{2}\rangle + |\frac{3}{2} - \frac{1}{2}\rangle\right)\,, \quad |\ell h_2\rangle = \frac{1}{\sqrt{2}}\left(|\frac{3}{2}\frac{1}{2}\rangle - |\frac{3}{2} - \frac{1}{2}\rangle\right)\,. \tag{8.4}$$

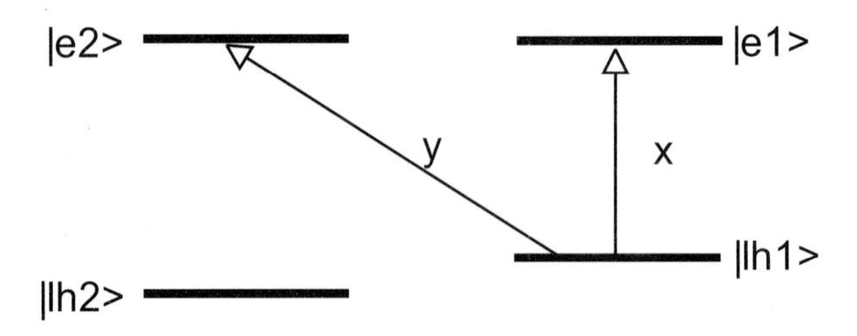

Fig. 8.1. Schematic energy levels in a strained quantum well under an in-plane($\parallel x$) magnetic field.

The selection rules of the optical transition can be derived by taking into account the conservation of the orbital angular momentum and the decomposition of the linear polarization into the circular polarization:

$$\hat{x}\cos\omega t = \frac{1}{2}(\hat{x}+i\hat{y})\cos\omega t + \frac{1}{2}(\hat{x}-i\hat{y})\cos\omega t\ , \tag{8.5}$$

$$\hat{y}\cos\omega t = \frac{1}{2i}(\hat{x}+i\hat{y})\cos\omega t - \frac{1}{2i}(\hat{x}-i\hat{y})\cos\omega t\ , \tag{8.6}$$

where $\hat{x}(\hat{y})$ denotes the unit vector in the $x(y)$-direction. Assuming $|\ell h_1\rangle$ to be the ground hole state (namely, $g_{\ell h} > 0$), we have the optical selection rules:

$$|\ell h\rangle_1 \longrightarrow \frac{1}{\sqrt{2}}(|c\uparrow\rangle - |c\downarrow\rangle) = -\ |e_1\rangle \tag{8.7}$$

for the $x-$polarized light and

$$|\ell h\rangle_1 \longrightarrow \frac{-i}{\sqrt{2}}(|c\uparrow\rangle + |c\downarrow\rangle) = -i\ |e_2\rangle \tag{8.8}$$

for the $y-$polarized light, respectively, where $|e_1\rangle(|e_2\rangle)$ is the conduction band electron state with the spin in the $-x(x)$ direction. The schematic energy levels are plotted in Fig. 8.1.

According to these selection rules, we can excite a linear combination of exciton states which have a common hole state. When a linearly polarized light comes in with polarizaion given by

$$\cos\theta\hat{x} + \sin\theta\hat{y}\ , \tag{8.9}$$

the excited state can be written as

$$-\cos\theta|e_1\rangle|\ell h_1\rangle - i\sin\theta|e_2\rangle|\ell h_1\rangle = -\cos\theta|e_1,h_1\rangle - i\sin\theta|e_2,h_1\rangle\ , \tag{8.10}$$

where the direct product state $|e_1\rangle|\ell h_1\rangle$ is simply denoted by $|e_1,h_1\rangle$ for example and the representation in the form of the electron-hole pair is used instead of the exciton representation. It is beyond the scope of this article to discuss fully the exciton effect in the quantum state transfer. Now we study influences of relaxation processes of the electron and the hole on the quantum state transfer. In the case of a gate-controlled quantum dot, the electron is three-dimensionally confined, whereas the hole is not confined and is absorbed into the negatively biased gate, leaving behind only the photoexcited electron. Thus the important issue is the quantum nature of the electron state after extraction of the hole: How much is the electron spin coherence retained after the process? To answer this question, we will analyze the time evolution of the whole system based on the density matrix formalism. Just after the photoexcitation given by (8.10), the density matrix of the electron-hole system is supposed to be given by

$$\begin{aligned}\rho(t=0) &= \cos^2\theta|e_1,h_1\rangle\langle e_1,h_1| + \sin^2\theta|e_2,h_1\rangle\langle e_2,h_1| \\ &+ i\sin\theta\cos\theta|e_2,h_1\rangle\langle e_1,h_1| - i\sin\theta\cos\theta|e_1,h_1\rangle\langle e_2,h_1|\,. \qquad (8.11)\end{aligned}$$

Then the hole is extracted into the negatively biased gate electrode. In order to simulate this process, we introduce a model consisting of three hole states as depicted in Fig. 8.2; one of them is the hole state $|h_1\rangle$ created in a quantum dot by the photoexcitation, the second one is an intermediate hole state $|h_2\rangle$ representing a delocalized state around the gate electrode and the third one is the swept-out state in the gate. The most important mechanism degrading the electron spin coherence is the electron-hole exchange interaction which induces the spin state mixing. During the hole extraction to the negatively biased electrode the hole spin relaxation occurs and the electron spin states are mixed up, leading to the degradation of quantum state transfer.

8.3. Dynamics of Quantum State Transfer

To set up the equations of motion for the density matrix, we take into account six basis states composed of direct products of two electron states $|e_1\rangle$ and $|e_2\rangle$ and three hole states $|h_1\rangle$, $|h_2\rangle$ and $|h_3\rangle$:

$$|e_1,h_1\rangle, |e_2,h_1\rangle, |e_1,h_2\rangle, |e_2,h_2\rangle, |e_1,h_3\rangle, |e_2,h_3\rangle\,. \qquad (8.12)$$

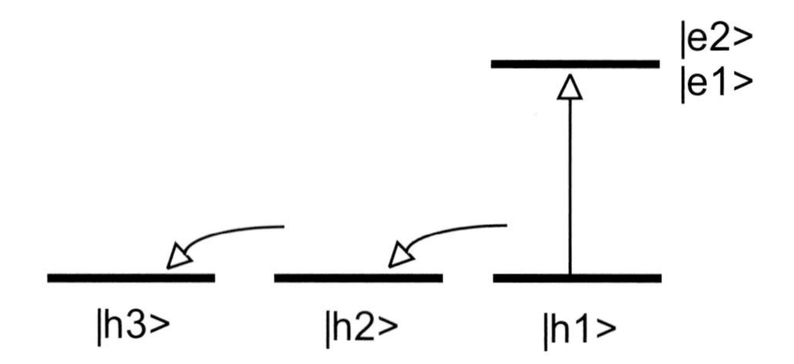

Fig. 8.2. Schematic relaxation paths in the quantum state transfer from a photon to an electron spin. Curved arrows indicate the transfer(tunneling) process of the photoexcited hole. $|h3\rangle$ represents symbolically the final destination of the extracted hole.

In this representation, the equation of motion for the density matrix is given by

$$\dot{\rho} = -\frac{i}{\hbar}[H, \rho] + \Gamma\rho \,, \tag{8.13}$$

$$H = H_0 + H_{\mathrm{exch.}} \,, \tag{8.14}$$

$$\begin{aligned} H_0 &= E_e(|e_1\rangle\langle e_1| + |e_2\rangle\langle e_2|) - E_{h_1}|h_1\rangle\langle h_1| - E_{h_2}|h_2\rangle\langle h_2| \\ &\quad - E_{h_3}|h_3\rangle\langle h_3| \,, \end{aligned} \tag{8.15}$$

$$\begin{aligned} \Gamma\rho &= t_{12}|h_2\rangle\langle h_1|\rho|h_1\rangle\langle h_2| + t_{23}|h_3\rangle\langle h_2|\rho|h_2\rangle\langle h_3| \\ &\quad - t_{12}|h_1\rangle\langle h_1|\rho|h_1\rangle\langle h_1| - t_{23}|h_2\rangle\langle h_2|\rho|h_2\rangle\langle h_2| \\ &\quad - \sum_{i \neq j} \gamma_{ij}^h \{|h_i\rangle\langle h_i|\rho|h_j\rangle\langle h_j| + |h_j\rangle\langle h_j|\rho|h_i\rangle\langle h_i|\} \\ &\quad + \gamma_1^e|e_2\rangle\langle e_1|\rho|e_1\rangle\langle e_2| - \gamma_1^e|e_1\rangle\langle e_1|\rho|e_1\rangle\langle e_1| \\ &\quad - \gamma_{12}^e\{|e_1\rangle\langle e_1|\rho|e_2\rangle\langle e_2| + |e_2\rangle\langle e_2|\rho|e_1\rangle\langle e_1|\} \,, \end{aligned} \tag{8.16}$$

where t_{ij} is the transfer(tunneling) rate from the hole state $|h_i\rangle$ to $|h_j\rangle$, γ_1^e the electron spin relaxation rate from the electron state $|e_1\rangle$ to $|e_2\rangle$, γ_{12}^e the electron spin decoherence rate and γ_{ij}^h is the dephasing rate of the hole state coherence between $|h_i\rangle$ and $|h_j\rangle$. Concerning the electron-hole exchange interaction we employ the following Hamiltonian:

$$H_{\text{exch.}} = W \cdot \begin{pmatrix} & |e_1h_1\rangle & |e_2h_1\rangle & |e_1h_2\rangle & |e_2h_2\rangle & |e_1h_3\rangle & |e_2h_3\rangle \\ |e_1h_1\rangle & 1 & 0.9 & 0.1 & 0.05 & 0.0 & 0.0 \\ |e_2h_1\rangle & 0.9 & 1 & 0.05 & 0.1 & 0.0 & 0.0 \\ |e_1h_2\rangle & 0.1 & 0.05 & 0.2 & 0.18 & 0.0 & 0.0 \\ |e_2h_2\rangle & 0.05 & 0.1 & 0.18 & 0.2 & 0.0 & 0.0 \\ |e_1h_3\rangle & 0.0 & 0.0 & 0.0 & 0.0 & 0.0 & 0.0 \\ |e_2h_3\rangle & 0.0 & 0.0 & 0.0 & 0.0 & 0.0 & 0.0 \end{pmatrix}, \tag{8.17}$$

where the matrix elements associated with the hole state $|h_3\rangle$ is set to be zero because it is the swept-out state into the gate. The short-range electron-hole exchange interaction is small since the spatial overlap between the $|h_3\rangle$ state and other states is negligible and the long-range electron-hole exchange interaction due to the dipole-dipole interaction may also be small. The numerical values in (8.17) are given semi-empirically.

8.4. Purity and Fidelity of Quantum State Transfer

The time evolution of the whole system can be examined by the numerical integration of the equation (8.13). The initial state is supposed to be given by (8.11). After a sequence of tunneling processes, the hole is eventually settled into the state $|h_3\rangle$. At this stage we are concerned with the quantum coherence of the electron spin state. To examine the quantum coherence we calculate the reduced density matrix of the electron by taking the trace of the density matrix for the whole system over the hole states:

$$\rho_{\text{electron}} = \text{Tr}_{\text{hole}}\, \rho\,. \tag{8.18}$$

Then the purity of the electron state is estimated by

$$\mathcal{P} = \text{Tr}\, \rho^2_{\text{electron}}\,. \tag{8.19}$$

We compared the purity for two cases of favorable and unfavorable conditions for the quantum state transfer. In the favorable case, the magnitude of the electron-hole exchange interaction denoted by W is taken to be 3 μeV and the hole transfer(tunneling) time ($1/t_{12} = 1/t_{23}$ =1 ps) is short enough to suppress the spin state mixing by the electron-hole exchange interaction. Results are not sensitive to the polarization angle θ of the excitation light in (8.9) and are exhibited in Fig. 8.3. The purity is high enough (> 0.9999) over a nanosecond to guarantee the secure quantum state transfer. The decay of the purity is determined by the electron spin decoherence/relaxation times which are supposed here to be

$$T_1 = 1/\gamma_1^e = 100\ \mu\text{s}\ , \quad T_2 = 1/\gamma_{12}^e = 1\ \mu\text{s}\ . \tag{8.20}$$

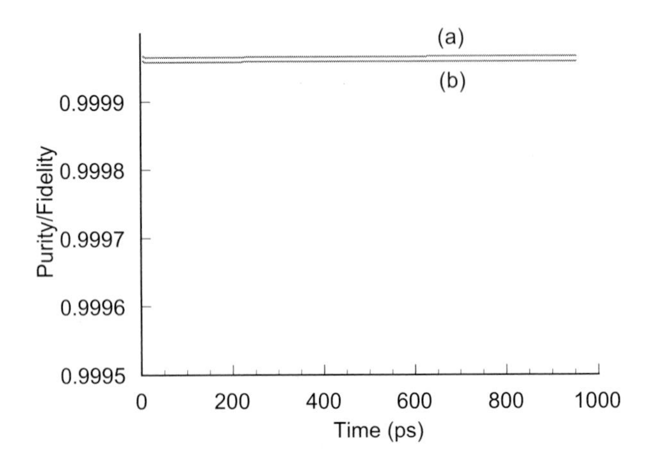

Fig. 8.3. Time evolution of the purity(a) of the electron spin state and the fidelity(b) of the quantum state transfer after photoexcitation for the case of favorable conditions.

In the unfavorable case, the magnitude of the electron-hole exchange interaction ($W = 20\ \mu$eV) is rather large and the hole tunneling time ($1/t_{12} = 1/t_{23} = 10$ ps) is not short enough to suppress the spin state mixing by the electron-hole exchange interaction. In this case also, results are not sensitive to the polarization angle of the excitation light and are exhibited in Fig. 8.4. The purity decreases rapidly within several tens of picoseconds after photoexcitation due to the electron-hole exchange interaction. The characteristic time scale is given by $\hbar/W$ and is $\sim$33 ps. Then the purity is decreasing slowly due to the electron spin decoherence itself ($\sim 1\ \mu$s).

We also examined the fidelity of the quantum state transfer. Ideally we want to prepare the electron spin state corresponding to the photon polarization state in (8.9) as

$$\rho^0_{\text{electron}} = \cos^2\theta|e_1\rangle\langle e_1| + \sin^2\theta|e_2\rangle\langle e_2| \tag{8.21}$$

$$+ i\sin\theta\cos\theta|e_2\rangle\langle e_1| - i\sin\theta\cos\theta|e_1\rangle\langle e_2|\ . \tag{8.22}$$

The fidelity is defined by

$$\mathcal{F}(t) = \text{Tr}\ \rho^0_{\text{electron}}\ \rho_{\text{electron}}(t)\ , \tag{8.23}$$

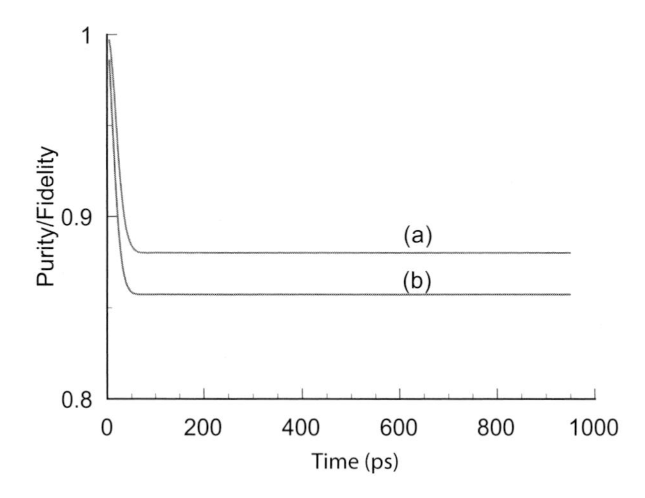

Fig. 8.4. Time evolution of the purity(a) of the electron spin state and the fidelity(b) of the quantum state transfer after photoexcitation for the case of unfavorable conditions.

where $\rho_{\text{electron}}(t)$ is the reduced density matrix of the electron defined in (8.18). This quantity indicates to what extent the actual state is close to the ideally prepared state. The results are exhibited by curves (b) in Figs. 8.3 and 8.4. Qualitative features are the same as in the case of purity.

Based on these results, we can conclude that for the secure quantum state transfer we have to extract the photoexcited hole quickly before the spin state mixing sets in due to the electron-hole exchange interaction. The model used here has an empirical character and a more quantitative analysis would be necessary for the definite design of experiments of quantum state transfer [24, 25].

8.5. Dynamics in Electron-Nuclei Coupled System

As mentioned in the Introduction, the electron spin decoherence time reported so far for low temperatures about a few microsecond is not sufficiently long for the secure quantum information processing. Here the hyperfine(HF) interaction with the host nuclei [26, 27] is considered to be the main decoherence mechanism, dominating over spin-orbit interactions which act on a time scale of tens of milliseconds [28, 29] or even longer. Consequently there have been proposals to reduce the HF induced decoherence by measuring or polarizing the nuclear spins [30–34] and to use nuclear spins as a quantum memory [35].

Extending these studies, we investigate the electron-nuclei spin coupling in quantum dots(QDs) and show that consecutive measurements of the electron spin state following the HF interaction are correlated and lead to purification of the nuclear spin system. More specifically, starting from an unknown initial state of nuclear spins, successive measurements of the electron spin state result in narrowing of the distribution of the nuclear spin field. We predict that the purification of the nuclear spin state would lead to the bunching of results of the electron spin state measurements and also to the reduction in the electron spin decoherence induced by the HF interaction. For the physical realization of the proposals we will in particular discuss a double QD occupied by two electrons, and a single QD occupied by one(two) electron(s). Under sufficiently high magnetic fields compared with the effective HF field(Overhauser field), these systems provide the desired two-level system with a unidirectional HF field.

First of all we consider an electrically gated double QD occupied by two electrons [18, 36]. The excited electronic orbitals of QDs have an energy much greater than the thermal energy and the adiabatic voltage sweeping rates, so that the electrons occupy only the ground state orbitals. Under a high magnetic field where the electron Zeeman splitting is much greater than the HF fields and the exchange energy, dynamics takes place in the spin singlet ground state $|S\rangle$ and triplet state of zero magnetic quantum number $|T\rangle$. For the singlet state each electron can be found in the different or both in the same QD, whereas for the triplet state electrons can only be found in different QDs. Singlet and triplet states are coupled by the HF fields and the system is governed by the Hamiltonian:

$$H_e = JS_z + r\delta h_z S_x \;, \tag{8.24}$$

where $\mathbf{S}$ is the pseudospin operator with $|T\rangle$ and $|S\rangle$ forming the S_z basis. J is the exchange energy and $\delta h_z = h_{1z} - h_{2z}$, where h_{1z} and h_{2z} are the components of nuclear HF field along the external magnetic field in the first and second dot, respectively. $0 \leq r \leq 1$ is the amplitude of the hyperfine coupling. When both electrons are localized in the same dot, $r \to 0$ and $J \gg \delta h_z$, when they are located in different dots HF coupling is maximized $r \to 1$ and $J \to 0$.

8.6. Bunching in Electron Spin Measurements

Now we show that by electron spin measurements in a double QD governed by (8.24), the coherent behavior of nuclear spins can be demonstrated.

Electron spins are initialized in the singlet state and the nuclear spin states are initially in a mixture of δh_z eigenstates:

$$\rho(t=0) = \sum_n p_n \, \rho_n \, |S\rangle\langle S| \,, \tag{8.25}$$

where ρ_n is a nuclear eigenstate with an eigenvalue h_n:

$$\mathrm{Tr}\; \rho_n \, \delta h_z = h_n \,, \quad \mathrm{Tr}\; \rho_n = 1 \,. \tag{8.26}$$

p_n is the probability of the hyperfine field δh_z having the value h_n. In the unbiased regime $r = 1$, the nuclear spins and the electron spins interact for a time span of τ. Then the gate voltage is swept adiabatically, switching off the HF interaction $r \to 0$, in a time scale much shorter than HF interaction time. Next a charge state measurement is performed which detects a singlet or triplet state. Probability to detect the singlet state is

$$\sum_n p_n |\alpha_n|^2 \tag{8.27}$$

and that to detect the triplet state is

$$\sum_n p_n |\beta_n|^2 \,, \tag{8.28}$$

where

$$\alpha_n = \cos \Omega_n \tau/2 + iJ/\Omega_n \sin \Omega_n \tau/2 \;\text{ and }\; \beta_n = -ih_n/\Omega_n \sin \Omega_n \tau/2 \tag{8.29}$$

with

$$\Omega_n = \sqrt{J^2 + h_n^2} \,. \tag{8.30}$$

Subsequently one can again initialize the system in the singlet state of electron spins and turn on the hyperfine interaction for a time span of τ and perform the second measurement. In general over N measurements, the nuclear state conditioned on $k(\leq N)$ times singlet and $N-k$ times triplet detection is

$$\sigma_{N,k} = \binom{N}{k} \sum_n p_n |\alpha_n|^{2k} |\beta_n|^{2(N-k)} \rho_n \,, \tag{8.31}$$

the trace of which yields the probability of k times singlet outcomes:

$$P_{N,k} = \mathrm{Tr}\; \sigma_{N,k} = \binom{N}{k} \langle |\alpha|^{2k} |\beta|^{2(N-k)} \rangle \,, \tag{8.32}$$

where $\langle \ldots \rangle$ is the ensemble averaging over the hyperfine field h_n [27]. Hereafter, this case will be referred to as the *coherent regime*. One can easily contrast this result with that for the *incoherent regime* in which nuclear spins lose their coherence between the successive spin measurements and relax to the equilibrium distribution given by

$$P'_{N,k} = \binom{N}{k} \langle |\alpha|^2 \rangle^k \langle |\beta|^2 \rangle^{(N-k)} . \tag{8.33}$$

If the nuclear spins are coherent over the span of the experiment, then successive electron spin measurements are biased to all singlet(triplet) outcomes. In particular, when the initial nuclear spins are unpolarized and randomly oriented, the distribution of hyperfine field is characterized by a Gaussian distribution with variance σ^2:

$$p[h] = \frac{1}{\sqrt{2\pi\sigma^2}} e^{-\frac{h^2}{2\sigma^2}} \tag{8.34}$$

and the summation is converted to an integration:

$$\sum_n p_n \ldots \rightarrow \int \mathrm{d}h\, p[h] \ldots . \tag{8.35}$$

As the simplest case, let us check the results of two measurements, each following a HF interaction of duration t. The probability for two consecutive singlet detections are given by

$$P_{2,2} = \langle |\alpha|^4 \rangle = \{6 + 2e^{-2t^2} + 8e^{-t^2/2}\}/16 \tag{8.36}$$

which is always greater than

$$P'_{2,2} = \langle |\alpha|^2 \rangle^2 = \{4 + 8e^{-t^2/2} + 4e^{-t^2}\}/16 . \tag{8.37}$$

These results are given particularly for $J = 0$. As J is increased the probabilities approach each other and for $J \gg \sigma$ they become identical [37].

In Fig. 8.5, for $N = 20$ measurements, $P_{N,k}$ is shown for HF interaction times $\sigma\tau = 0.5, 1.5,$ and ∞. For $\tau = 0$, the probability for both (8.32) and (8.33) is peaked at $k = 20$. However, immediately after the HF interaction is introduced, the probability distributions show distinct behavior. The measurement results in the incoherent regime approach a Gaussian distribution. In the coherent case the probabilities bunch at k=0 and 20 for $J = 0$, and when $J/\sigma = 0.5$ those bunch at $k = 20$ only. As J is increased above some critical value, no bunching takes place at $k = 0$ singlet measurement.

We discuss in brief the feasibility to observe the predicted phenomenon. The duration of the cycle involving electron spin initialization and measurement is about 10 μs [18]. The nuclear spin coherence time determined mostly by the nuclear spin diffusion is longer than about several tens of ms [38]. For a HF interaction of duration $\tau = 4\sigma^{-1} \sim 40$ μs [18] in each

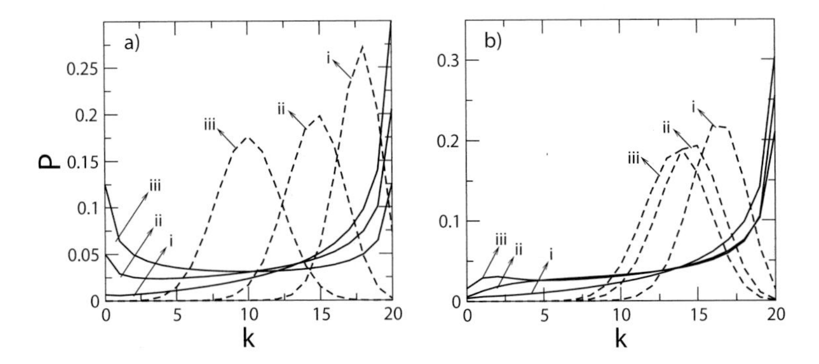

Fig. 8.5. Probability distribution for $N = 20$ measurements as a function of times(k) of singlet detections, for coherent regime(solid lines) and incoherent regime(dashed lines). Two cases of the exchange energy are considered a) $J = 0$, and b) $J/\sigma = 0.5$, for HF interaction times $\sigma\tau =$ i)0.5, ii)1.5, and iii)∞.

step, the bunching can be observed for N successive measurements up to $N \sim 200$.

8.7. Purification of Nuclear Spin State

The nuclear spin state conditioned on the previous electron spin measurements is no longer random even if they are initially random. Accordingly, HF induced electron spin decoherence dynamics is also modified. Depending on the results of previous measurement, one may decrease the singlet-triplet mixing. As a particular example, consider the case: Starting from a random spin configuration, N successive electron spin measurements are performed, each following initialization of electron spins in the spin singlet state and a HF interaction of duration $\tau_i (i = 1 \dots N)$ and all outcomes turn out to be singlet. Here the nuclear spin state is given by $\sigma_{N,N}$. Then again HF interaction is switched on for a time t and the $(N+1)$th measurement is carried out. The conditional probability to recover the initial state, namely to observe again the singlet state, is given by

$$P = \frac{\sum \prod_{i=1}^{N+1} \binom{2}{s_i} e^{-\frac{1}{2}\left[\sum_{j=1}^{N}(s_j-1)\tilde{\tau}_j + (s_{N+1}-1)\tilde{t}\right]^2}}{4\sum \prod_{i=1}^{N} \binom{2}{s_i} e^{-\frac{1}{2}\left[\sum_{j=1}^{N}(s_j-1)\tilde{\tau}_j\right]^2}}, \tag{8.38}$$

where the sums run over $s_i = 0, 1, 2$ and $\tilde{\tau}_i = \sigma\tau_i$. For the particular case $\tau_1 = \tau_2 = \ldots = \tau_N = \tau \gg 1/\sigma$, the initial state is revived at $t = n\tau$, $(n =$

1, 2, ..., N) with a decreasing probability:

$$P \simeq 1/2 + \sum_{s=0}^{N} \binom{2N}{s} e^{\frac{-\sigma^2}{2}(t-(N-s)\tau)^2} / 2 \binom{2N}{N} \quad \text{for } \tau \gg 1/\sigma \,. \tag{8.39}$$

In Fig. 8.6 the conditional probabilities (8.38) are shown for $\sigma\tau = 1.0, 3.0, 6.0$ subject to $N = 0, 1, 2, 5$ times prior singlet measurements in each. Revivals are observable only for $\sigma\tau > 1$, because the modulation period of the nuclear state spectrum characterized by $1/\tau$ should be smaller than the variance σ. In this case the probability for the recovery of the initial state is

$$1/2 + \binom{2N}{s} / 2 \binom{2N}{N} \,, \qquad s = N-1, N-2, \ldots, 0 \tag{8.40}$$

at times $t = \tau, 2\tau, \ldots, N\tau$. The underlying mechanism of revivals is purification of nuclear spins by the electron spin measurements. The purity of a system characterized by the density matrix $\hat{\rho}$ is given by

$$\mathcal{P} = \text{Tr}\, \hat{\rho}^2. \tag{8.41}$$

As an example we are again going to consider the nuclear spin state prepared by N successive electron spin measurements with singlet outcomes, each following a HF interaction of duration $\tau_1 \ldots \tau_N$. The purity of nuclear spins is given by

$$\mathcal{P} = \frac{1}{\mathcal{D}} \frac{\sum_{s_i=0}^{4} \prod_{i=1}^{N} \binom{4}{s_i} e^{-\frac{1}{2}\left[\sum_{j=1}^{N}(s_j-2)\tilde{\tau}_j\right]^2}}{\left[\sum_{s_i=0}^{2} \prod_{i=1}^{N} \binom{2}{s_i} e^{-\frac{1}{2}\left[\sum_{j=1}^{N}(s_j-1)\tilde{\tau}_j\right]^2}\right]^2} \,, \tag{8.42}$$

where $\mathcal{D}$ is the dimension of the Hilbert space for the nuclear spins. For a fixed ratio of $\tau_1 : \tau_2 : \ldots : \tau_N$, the purity (8.42) is a monotonically increasing function of time. For $\tilde{\tau}_i = \sigma\tau_i \gg 1$, one can attain various asymptotic limits for the purity. For instance, for $N = 2$, there are three asymptotic limits:
a) $\tau_1 = 2\tau_2$

$$\mathcal{P} = 11/4\mathcal{D} \,, \tag{8.43}$$

b) $\tau_1 = \tau_2$

$$\mathcal{P} = 35/18\mathcal{D} \,, \tag{8.44}$$

c) otherwise

$$\mathcal{P} = 9/4\mathcal{D} \,. \tag{8.45}$$

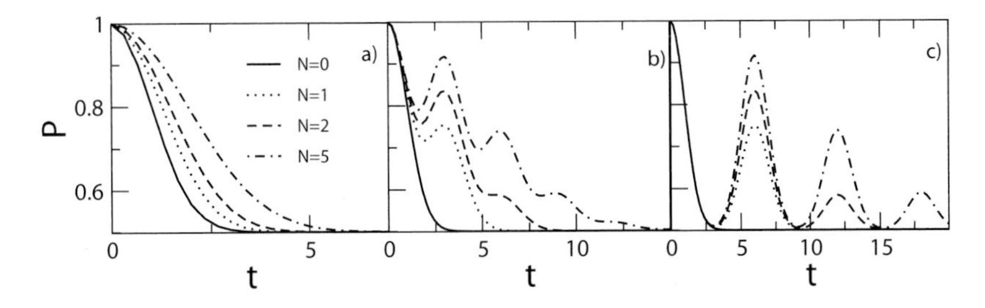

Fig. 8.6. Conditional probability for singlet state detection as a function of HF interaction time σt, subject to $N = 0, 1, 2, 5$ times prior singlet state measurements and for HF interaction times of a) $\sigma\tau = 1.0$, b) $\sigma\tau = 3.0$, and c) $\sigma\tau = 6.0$.

For $N = 2$ with $\tau_1 = 2\tau_2 = 2\tau \gg 1/\sigma$, the conditional probability (8.38) is given as

$$P \simeq 1/2 + \sum_{n=0}^{3}(4-n)\exp[-(\tilde{t} - n\tilde{\tau})^2/2]/8 \,, \tag{8.46}$$

whereas for $\tau_2 = \tau_1 = \tau \gg 1/\sigma$,

$$P \simeq 1/2 + \{e^{-\frac{(\tilde{t}-2\tilde{\tau})^2}{2}} + 4e^{-\frac{(\tilde{t}-\tilde{\tau})^2}{2}} + 6e^{-\frac{\tilde{t}^2}{2}}\}/12 \,. \tag{8.47}$$

It can be seen that as the purity of nuclear spins increases, more revivals are present with an increased amplitude.

Previously various methods have been proposed to measure the HF field in a QD with some precision [32–34]. In the method proposed here, nuclear spin state can be conditionally purified without determining the precise value of the HF field. Although the HF field may be still assuming different values, electron-nuclei correlations lead to revivals at known times. Thus an arbitrary initial state of the electron spin may be recovered at a later time. As an example consider the case when the nuclear spin state is prepared by five HF interaction stages each of which has duration $\tau = 10/\sigma$ and is followed by a singlet detection of the electron spin state. This conditionally prepared nuclear spin state revives any arbitrary electron spin state for $N = 5$ at $t = s\tau$ $(s = 1, 2, \ldots, 5)$ with the probability of $11/12, 31/42, \ldots, 253/504$, respectively. Success probability to prepare such a state is $\sim 1/32$. Here preparation time is $T = N(\tau + \tau_w)$ where τ_w is the time needed for initialization and detection of the electron spin. During the HF interaction time τ, phonon-mediated interations which act on the time scale longer than millisecond [29] or any other electron spin decoherence mechanism except the HF interaction should not take place.

Furthermore the preparation time T should be smaller than the nuclear diffusion time which is of the order of 10 ms [38]. Typically $\tau_w = 10\ \mu$s, and let $\sigma^{-1} = 10\ \mu$s [18]. For the discussed example($\tau = 10/\sigma$, $N = 5$), time needed for the preparation of the desired nuclear spin state is $T = 550\ \mu$s which is shorter than the nuclear spin diffusion time.

8.8. Extension to General Situations

We have discussed the bunching and revival phenomena only for a double QD system. The same predictions can also be made for a single QD occupied by a single electron [39–41]. Consider a single QD occupied by a single electron, under an external magnetic field where the electron Zeeman energy is much greater than the HF energies. Then the system is described by the Hamiltonian:

$$H \simeq g_e \mu_B B S_z + h_z S_z \ , \tag{8.48}$$

where g_e is the electron g-factor, μ_B the Bohr magneton and B is the external field applied in the z direction. Spin flips are suppressed since

$$g_e \mu_B B \gg \sqrt{\langle \mathbf{h}^2 \rangle} \ . \tag{8.49}$$

The states given by

$$|\pm\rangle = (|\uparrow\rangle \pm |\downarrow\rangle)/\sqrt{2} \tag{8.50}$$

are coupled by HF interaction with $|\uparrow (\downarrow)\rangle$ being the eigenstates of S_z. Each time the electron is prepared in the $|+\rangle$ state. Next it is loaded onto the QD, then removed from the QD after some dwelling time τ. The spin measurement is performed in the $|\pm\rangle$ basis. Essentially the same predictions can be made for this system as those for the case of double QD, namely the electron spin bunching and revival. The Hamiltonian (8.24) can also be used to describe a pair of electrons in a single QD [42, 43] and the same predictions as those for a double QD can be made [37].

8.9. Summary

In summary, we have investigated quantum dynamics in coupled systems related to the quantum information processing: one is a model system of the quantum media converter or quantum state transfer between a photon and an electron spin and another one is the electron-nuclei coupled system in QDs. In the former case, we have analyzed the performance of the quantum state transfer and clarified the conditions to achieve a high value of the purity of the transferred quantum state or the fidelity of the operation. In the latter case, we predicted a couple of new phenomena related to the correlations induced by the hyperfine interactions. The underlying mechanism is the squeezing and increase in the purity of the nuclear spin state through the electron spin measurements. This squeezing is expected to lead to the extension of the electron spin coherence time.

We can construct hopefully a secure and robust system of the quantum repeater combining these results, namely the efficient quantum state transfer between a photon and an electron spin and the long-lived quantum memory based on nuclear spins.

Acknowledgments

We would like to thank the financial supports from the CREST project of the Japan Science and Technology Agency and from the Ministry of Education, Culture, Sports, Science and Technology, Japan.

References

[1] Bennett, C. H. and DiVincenzo, D. P. (2000). *Quantum information and computation, Nature* **404**, pp. 247–255.
[2] Barenco, A., Deutsch, D., Ekert, A. and Jozsa, R. (1995). *Conditional Quantum Dynamics and Logic Gates, Phys. Rev. Lett.* **74**, pp. 4083–4086.
[3] Biolatti, E., Iotti, R. C., Zanardi, P. and Rossi, F. (2000). *Quantum Information Processing with Semiconductor Macroatoms, Phys. Rev. Lett.* **85**, pp. 5647–5650.
[4] Troiani, F., Hohenester, U. and Molinari, E. (2000). *Exploiting exciton-exciton interactions in semiconductor quantum dots for quantum-information processing, Phys. Rev. B* **62**, pp. R2263-R2266.
[5] Chen, P., Piermarocchi, C. and Sham, L. J. (2001). *Control of Exciton Dynamics in Nanodots for Quantum Operations, Phys. Rev. Lett.*, **87**, pp. 067401.
[6] Stievater, T. H., Li, Xiaoqin, Steel, D. G., Gammon, D., Katzer, D. S., Park, D., Piermarocchi, C. and Sham, L. J. (2001). *Rabi Oscillations of Excitons in Single Quantum Dots, Phys. Rev. Lett.* **87**, 133603.
[7] Kamada, H., Gotoh, H., Temmyo, J., Takagahara, T., and Ando, H. (2001). *Exciton Rabi Oscillation in a Single Quantum Dot, Phys. Rev. Lett.* **87**, 246401.
[8] Htoon, H., Takagahara, T., Kulik, D., Baklenov, O., Holmes Jr., A. L., and Shih, C. K. (2002). *Interplay of Rabi Oscillations and Quantum Interference in Semiconductor Quantum Dots, Phys. Rev. Lett.* **88**, 087401.

[9] Zrenner, A., Beham, E., Stufler, S., Findeis, F., Bichler, M., and Abstreiter, G. (2002). *Coherent properties of a two-level system based on a quantum-dot photodiode, Nature* **418**, pp. 612 - 614.

[10] Takagahara, T. (ed.) 2003. *Quantum Coherence, Correlation and Decoherence in Semiconductor Nanostructures* (Academic Press, Elsevier, New York).

[11] Wang, Q. Q., Muller, A., Bianucci, P., Rossi, E., Xue, Q. K., Takagahara, T., Piermarocchi, C., MacDonald, A. H., and Shih, C. K. (2005). *Decoherence processes during optical manipulation of excitonic qubits in semiconductor quantum dots, Phys. Rev. B* **72**, 035306.

[12] Wang, Q. Q., Muller, A., Cheng, M. T., Zhou, H. J., Bianucci, P., and Shih, C. K. (2005). *Coherent Control of a V-Type Three-Level System in a Single Quantum Dot, Phys. Rev. Lett.* **95**, 187404.

[13] Li, Xiaoqin, Wu, Y., Steel, D. G., Gammon, D., Stievater, T. H., Katzer, D. S., Park, D., Piermarocchi, C. and Sham, L. J. (2003). *An All-Optical Quantum Gate in a Semiconductor Quantum Dot, Science* **301**, pp. 809–811.

[14] Loss, D. and DiVincenzo, D. P. (1998). *Quantum computation with quantum dots Phys. Rev. A* **57**, pp.120–126.

[15] Imamoglu, A., Awschalom, D. D., Burkard, G., DiVincenzo, D. P., Loss, D., Sherwin, M., and Small, A. (1999). *Quantum Information Processing Using Quantum Dot Spins and Cavity QED, Phys. Rev. Lett.* **83**, pp. 4204–4207.

[16] Golovach, V. N., Khaetskii, A., and Loss, D. (2004). *Phonon-Induced Decay of the Electron Spin in Quantum Dots, Phys. Rev. Lett.* **93**, 016601.

[17] Semenov, Y. G. and Kim, K. W. (2004). *Phonon-Mediated Electron-Spin Phase Diffusion in a Quantum Dot, Phys. Rev. Lett.* **92**, 026601.

[18] Petta, J. R., Johnson, A. C., Taylor, J. M., Laird, E. A., Yacoby, A., Lukin, M. D., Marcus, C. M., Hanson, M. P. and Gossard, A. C. (2005). *Coherent Manipulation of Coupled Electron Spins in Semiconductor Quantum Dots, Science* **309**, pp. 2180–2184.

[19] Koppens, F. H. L., Buizert, C., Tielrooij, K. J., Vink, I. T., Nowack, K. C., Meunier, T., Kouwenhoven, L. P., Vandersypen, L. M. K. (2006). *Driven coherent oscillations of a single electron spin in a quantum dot, Nature* **442**, pp. 766–771.

[20] Yablonovitch, E., Jiang, H. W., Kosaka, H., Robinson, H. D., Rao, D. S. and Szkopek, T. (2003). *Optoelectronic Quantum Telecommunications Based on Spins in Semiconductors, Proc. IEEE* **91**, pp. 761–780 .

[21] Kosaka, H. and Yablonovitch, E. (2003). *Quantum Media Converter from a Photon Qubit to an Electron-Spin Qubit for Quantum Repeaters, Proceedings of the International Symposium on Photonics and Spintronics in Semiconductor Nanostructures (PSSN 2003, Kyoto) ed. by T. Takagahara*, pp. 63–70.

[22] Kosaka, H., Kiselev, A. A., Baron, F. A., Kim, K. W. and Yablonovitch, E. (2001). *Electron g factor engineering in III-V semiconductors for quantum communications, Electron. Lett.* **37**, pp. 464–465.

[23] Similar arguments are given in the following reference: Vrijen, R. and Yablonovitch, E. (2001). *A spin-coherent semiconductor photo-detector for quantum communication, Physica E* **10**, pp. 569-575.

[24] Kosaka, H., Mitsumori, Y., Rikitake, Y. and Imamura, H. (2007). *Polarization transfer from photon to electron spin in g factor engineered quantum wells*, *Appl. Phys. Lett.* **90**, 113511.
[25] Rikitake, Y. and Imamura, H. (2006). *Effect of exchange interaction on the fidelity of quantum state transfer from a photon qubit to an electron-spin qubit*, *Phys. Rev. B* **74**, 081307.
[26] Abragam, A. (1961). *The Principles of Nuclear Magnetism* (Oxford University Press).
[27] Merkulov, I. A., Efros, Al. L., and Rosen, M. (2002). *Electron spin relaxation by nuclei in semiconductor quantum dots*, *Phys. Rev. B* **65**, 205309.
[28] Kroutvar, M., Ducommun, Y., Heiss, D., Bichler, M., Schuh, D., Abstreiter, G., and Finley, J. J. (2004). *Optically programmable electron spin memory using semiconductor quantum dots*, *Nature* **432**, pp. 81–84.
[29] Meunier, T., Vink, I. T., Willems van Beveren, L. H., Tielrooij, K. J., Hanson, R., Koppens, F. H. L., Tranitz, H. P., Wegscheider, W., Kouwenhoven, L. P. and Vandersypen, L. M. K. (2007). *Experimental Signature of Phonon-Mediated Spin Relaxation in a Two-Electron Quantum Dot*, *Phys. Rev. Lett.* **98**, 126601.
[30] Taylor, J. M., Imamoglu, A. and Lukin, M. D. (2003). *Controlling a Mesoscopic Spin Environment by Quantum Bit Manipulation*, *Phys. Rev. Lett.* **91**, 246802.
[31] Imamoglu, A., Knill, E., Tian, L. and Zoller, P. (2003). *Optical Pumping of Quantum-Dot Nuclear Spins*, *Phys. Rev. Lett.* **91**, 017402.
[32] Klauser, D., Coish, W. A. and Loss, D. (2006). *Nuclear spin state narrowing via gate-controlled Rabi oscillations in a double quantum dot*, *Phys. Rev. B* **73**, 205302.
[33] Giedke, G., Taylor, J. M., D'Alessandro, D., Lukin, M. D. and Imamoglu, A. (2006). *Quantum measurement of a mesoscopic spin ensemble*, *Phys. Rev. A* **74**, 032316.
[34] Stepanenko, D., Burkard, G., Giedke, G. and Imamoglu, A. (2006). *Enhancement of Electron Spin Coherence by Optical Preparation of Nuclear Spins*, *Phys. Rev. Lett.* **96**, 136401.
[35] Taylor, J. M., Marcus, C. M. and Lukin, M. D. (2003). *Long-Lived Memory for Mesoscopic Quantum Bits*, *Phys. Rev. Lett.* **90**, 206803.
[36] Coish, W. A. and Loss, D. (2005). *Singlet-triplet decoherence due to nuclear spins in a double quantum dot*, *Phys. Rev. B* **72**, 125337.
[37] Ĉakir, Özgür and Takagahara, T. (2006). *Electron spin dynamics and hyperfine interaction in coupled quantum dots*, *Phys. Stat. Sol. (c)* **3**, pp. 4392–4395; and *Phys. Rev. B* (in press) arXiv:cond-mat/0801.1723.
[38] Paget, D. (1982). *Optical detection of NMR in high-purity GaAs: Direct study of the relaxation of nuclei close to shallow donors*, *Phys. Rev. B* **25**, pp. 4444–4451.
[39] Hanson, R., Witkamp, B., Vandersypen, L. M. K., Willems van Beveren, L. H., Elzerman, J. M. and Kouwenhoven, L. P. (2003). *Zeeman Energy and Spin Relaxation in a One-Electron Quantum Dot*, *Phys. Rev. Lett.* **91**, 196802.

[40] Dutt, M. V. G., Cheng, J., Li, B., Xu, X., Li, Xiaoqin, Berman, P. R., Steel, D. G., Bracker, A. S., Gammon, D., Economou, S. E., Liu, R. B. and Sham, L. J. (2005). *Stimulated and Spontaneous Optical Generation of Electron Spin Coherence in Charged GaAs Quantum Dots*, *Phys. Rev. Lett.* **94**, 227403.

[41] Atatüre, M., Dreiser, J., Badolato, A., Hogele, A., Karrai, K. and Imamoglu, A. (2006). *Quantum-Dot Spin-State Preparation with Near-Unity Fidelity*, *Science* **312**, pp. 551–553.

[42] Fujisawa, T., Austing, D. G., Tokura, Y., Hirayama, Y. and Tarucha, S. (2002). *Allowed and forbidden transitions in artificial hydrogen and helium atoms*, *Nature* **419**, pp. 278–281.

[43] Hanson, R., Willems van Beveren, L. H., Vink, I. T., Elzerman, J. M., Naber, W. J. M., Koppens, F. H. L., Kouwenhoven, L. P. and Vandersypen, L. M. K. (2005). *Single-Shot Readout of Electron Spin States in a Quantum Dot Using Spin-Dependent Tunnel Rates*, *Phys. Rev. Lett.* **94**, 196802.

PART 3
Qubit Decoherence

Chapter 9

Spin Quantum-bits and Decoherence in InAs/GaAs Quantum Dots

Bernhard Urbaszek[1], Thierry Amand[1], Olivier Krebs[2], Pierre Renucci[1], Xavier Marie[1]

[1] *LPCNO, INSA-CNRS, 135 avenue de Rangueil, 31077 Toulouse, France*
[2] *LPN-CNRS, route de Nozay, 91460 Marcoussis, France*

Proposals for using single spins in semiconductor quantum dots as building blocks for future memory or quantum information architectures have been encouraged by experiments that demonstrate long spin relaxation times in these structures. The main interest of this approach relies on the fact that spin states are usually more robust against decoherence than charge states. We review in this paper optical measurements of the electron and exciton spin coherence time in self-organized InAs/GaAs quantum dots. The key role played by the hyperfine interaction with the nuclear spins is investigated in detail.

Contents

9.1. Introduction

In recent years, great attention has been focused on finding the ideal quantum bit (qubit) to perform quantum computation, where it is necessary to create a linear superposition of states with a long coherence time [1].

With respect to the general criteria for quantum computers [2], semiconductor quantum dots (QD) appear as good candidates. In particular, self assembled quantum dots that confine electrons and holes down to 10 nm length scale and can be individually addressed [3], have a strong potential for the realisation of an elementary quantum gate in condensed matter. Two possible routes towards this aim have been investigated up to now:

(i) The elementary quantum bit is carried by the QD ground and first electron-hole excited state [4–6]. The optical polarization relaxation time in such systems is basically limited by the radiative life-time, typically in the nanosecond range [7].

(ii) The spin is taken as the quantum information carrier. This last approach is attractive since a very long coherence time is expected, as a result of the inhibition of classical spin relaxation mechanisms [8]: the discrete energy levels in artificial atoms like semiconductor quantum dots and the corresponding lack of energy dispersion lead to a drastic modification of the spin relaxation dynamics compared to bulk or two-dimensional structures [9, 10]. The absolute lack of energy states between QD energy levels is expected to suppress not only the elastic processes of spin relaxation but also to reduce drastically the inelastic ones such as those involving phonon scattering, i.e. all the spin relaxation processes based on the spin-orbit interaction.

As a consequence, a large variety of schemes were proposed for the implementation of quantum computing using the spin degree of freedom in semiconductors [11]. The earliest suggestions use two electron spins in two semiconductor quantum dots with tunable exchange coupling [12, 13]; the control is achieved with an external magnetic or electric field. The coherent optical control of the exciton spin in a single semiconductor quantum dot has also been proposed to perform basic operations for quantum computing in a two qubit system using two excitons of opposite spins [11, 14]. All optical quantum information processing was also proposed by Imamoglu *et al* using electron spins of doped QDs embedded inside a microcavity [15]. In that case the coupling between the QDs is achieved via virtual photons in the common cavity mode. Finally, the nuclear spins themselves could be used to create long-lived quantum memory for quantum bits. It is indeed well known that nuclear spins possess very long coherence times because of their weak environmental coupling, but single nuclear spins are very difficult to manipulate and measure. Taylor *et al* proposed a very elegant technique for greatly extending the lifetimes of electron-spin qubits in confined structures by coherently mapping an arbitrary electron spin superposition state onto the spins of proximal, polarized nuclei [16].

Recent spectacular experiments demonstrated indeed the possibility of using single spin in quantum dots for quantum information processing. The entanglement between two exciton spin states has for instance been indeed experimentally achieved in GaAs quantum dots [17]. The coherent optical control of the bi-exciton allowed also the realization of an all optical quantum gate in a semiconductor QD [18]. Coherent manipulation of coupled electron spins has also been done successfully at very low temperature ($\sim$100 mK) in gate-defined double quantum dot GaAs devices [19, 20] using electric field sequences.

In this contribution we will focus on the possible implementation of qubits based on *electron* or *exciton spin* in self-organized InAs/GaAs semiconductor quantum dots. We will present detailed investigations of the dephasing mechanisms of the spin states associated to the solid-state environment which constitute probably the main obstacle for the experimental realization of quantum information machines. After a description of the InAs/GaAs quantum dot structures investigated and of the experimental techniques in section 9.2, we review in section 9.3 time-resolved investigations of the *exciton spin* coherence time in non-intentionally doped self-organised InAs/GaAs quantum dots (QDs). We show that, at low temperature, the *exciton spin* is stable, on a time scale more than one order of magnitude larger than the radiative decay time. In section 9.4, we address the topic of optical manipulation and dynamics of *electron spin* states in *p-doped* QDs. We demonstrate in particular that the main electron spin dephasing process is due to the hyperfine interaction with nuclear spins. Finally, in the perspective of future implementation of qubits based on collective nuclear-spin states [16] (which require strong nuclear-spin polarization in the vicinity of confined electrons), we present in section 9.5 a detailed investigation of the nuclear spin polarization in self-organized InAs/GaAs quantum dots.

9.2. Description of Samples and Experimental Techniques

The investigated structures are grown by molecular beam epitaxy on (001) GaAs substrates. The coherent properties of the *exciton spin* will be presented on two samples (Sample I and II). The non-intentionally doped structure (Sample I) consists of 5 InAs lens-shaped QD planes embedded in a GaAs planar cavity, inserted between two GaAs/AlAs Bragg mirrors [21]. The QD layers are localized in the vicinity of the electromagnetic field antinodes in the microcavity which operates in the weak coupling regime. The QD density is $\sim 4 \cdot 10^{10} cm^{-2}$ per layer, and the QD ground state emission is centered around 1.15eV with a Full Width at Half Maximum (FWHM) of about 60meV (at T=10 K) [22].

The second non-intentionally doped structure presented here (Sample II) consists of 40 planes of InAs self-assembled QDs, separated by 15-nm-thick GaAs spacer layers. The average QD density is $\sim 1 \cdot 10^{11} cm^{-2}$ with a typical dot height of 2 nm and a dot diameter of 15 nm. At T=10 K, the ensemble PL transition energy is centered around 1.33 eV, with a FWHM of 60 meV [23].

We have studied the electron spin coherence in p-doped samples. The p-modulation doped structure (Sample III) consists of 10 planes of lens shaped self-assembled InAs/GaAs QDs separated by a 30 nm GaAs layer; a beryllium delta doping layer is located 15 nm below each wetting layer (WL). The nominal acceptor concentration is $N_A = 15 \cdot 10^{10} cm^{-2}$ per layer in this sample. The QD density is about $\sim 4 \cdot 10^{10} cm^{-2}$ per plane. It has been shown that each QD contains on average $\sim$ 1 hole [24].

Finally Sample IV, designed for single dot measurements of the nuclear spin polarization, contains the following layers, starting from the substrate: 200nm of p doped (Be) GaAs / 25nm of GaAs / InGaAs dots with a wetting layer grown in the Stranski-Krastanov mode / 30 nm of GaAs / 100nm $Ga_{0.7}Al_{0.3}As$ / 20nm of GaAs. Placing a doped layer below the dots enables holes to tunnel into the dots [25]. A low dot density between $10^8 cm^{-2}$ and $10^9 cm^{-2}$ adapted to single dot measurements has been obtained by choosing a nominal thickness of 1.7 mono-layers for the InAs layer deposition. The InGaAs quantum dots formed after Gallium and Indium interdiffusion contain typically 45% Indium and 55% Gallium, their typical diameter is around 20nm and the height varies between 4 and 10 nm as determined by TEM measurements.

We have investigated the spin properties in these structures by time-resolved optical orientation experiments [26]. The principle is to transfer the angular momentum of the excitation photons using circularly polarised light to the photogenerated electronic excitations. If the carriers do not loose their spin polarization during their lifetime, the luminescence will also be circularly polarised and will give information on both the symmetry of the carrier wavefunction and the spin relaxation time [8]. These optical orientation experiments are very powerful techniques to (i) measure the decoherence of exciton or electron spin qubits and (ii) perform basic quantum information manipulations [11, 27–29].

The samples are excited by 1.5 ps pulses generated by an Optical Parametric Oscillator or a mode-locked Ti-doped sapphire laser. For sample I and II, the time-resolved photoluminescence (PL) is then recorded by an up-conversion set-up [30, 31]. The time-resolution is limited by the laser pulse-width ($\sim$ 1.5 ps) and the spectral resolution is about 3 meV. For sample III, the time-resolved PL is recorded using a S1 photocathode Hamamatsu Streak Camera with an overall time-resolution of 8 ps. The photoluminescence measurements on individual dots (Sample IV) is carried out with a confocal microscope built around an Attocube nanopositioner placed in the center of a superconducting magnet system at fields between 0 and 4 T. The sample temperature in the variable temperature insert was kept at 1.5 K. The polarization of the excitation as well as the detected signal was controlled with a Glan-Taylor polarizer and a liquid crystal based wave plate. The optical signal was dispersed in a spectrometer with a focal length of 50cm and detected with a Si -CCD camera. The high signals to noise ratio single dot PL spectra were fitted with Lorentzian line shapes that result in a spectral precision of our measurements of +/-2.5 μeV [25]. The linear and the circular polarization degrees of the luminescence are defined as $P_{lin} = (I^X - I^Y)/(I^X + I^Y)$, and $P_c = (I^+ - I^-)/(I^+ + I^-)$ respectively. Here I^X (I^Y) and I^+ (I^-) denote respectively the X (Y) linearly polarized and the right (left) circularly polarized luminescence components, and the X and Y axis are chosen parallel to the (110) and ($1\bar{1}0$) sample directions.

9.3. Exciton Spin Quantum Bits

Several theoretical and experimental works propose to use the neutral *exciton spin* in semiconductors for the implementation of quantum computation schemes [11, 14, 18, 32]. We review in this section the basic properties of the neutral excitons in self-organized InAs/GaAs QDs and present some experimental results on the exciton spin lifetime and coherence time. We show in particular that these characteristic times are much longer than in bulk semiconductors, which is encouraging for future quantum memory applications.

9.3.1. *Exciton fine structure in self organised quantum dots*

In non-intentionally doped QD structures, the electron-hole pair fine structure is strongly affected by the interplay between the exchange interaction and the reduced QD symmetry, usually C_{2v}. In the latter symmetry, the projection on the quantization axis of the total angular momentum is not anymore a good quantum number. Yet, the conduction (valence) single particle states can still be described, to zeroth order, as the $S_c(S_v)$ and $P_c(P_v)$ 2D atomic like orbitals (the growth direction Oz is chosen as the quantization axis for the angular momentum). In the following, we use this convenient labelling for the actual orbitals. Taking into account the spin-orbit interaction, it is convenient to start with the same basis as in (001)-grown type I quantum wells (for which the relevant symmetry is D_{2d}). The lowest conduction band is then S-like, with two spin states $s_{e,z} = \uparrow, \downarrow$; the upper valence band is split into a heavy-hole band with the angular momentum projection $j_{h,z} = \pm 3/2$ and a light-hole band with $j_{h,z} = \pm 1/2$ at the centre of the Brillouin zone. Due to the built in strain in self assembled QDs, the light-hole states lie at much lower energy than the heavy-hole ones, so that we consider only the latter in the following. The heavy-hole exciton states can then be described using the basis set $|J_z\rangle = |j_{h,z}, s_{e,z}\rangle$, i.e. $|J_z = +1\rangle = |3/2, \downarrow\rangle$, $|J_z = -1\rangle = |-3/2, \uparrow\rangle$, $|J_z = +2\rangle = |3/2, \uparrow\rangle$, $|J_z = +2\rangle = |-3/2, \downarrow\rangle$, where the first two states are radiative and the two last ones non-radiative. At this point, it is convenient to describe the heavy-hole states in the presence of spin-orbit interaction by a pseudo spin, the $|\pm 1/2\rangle_h$ states corresponding to the $|\mp 3/2\rangle$ ones respectively [33, 34]. The heavy-hole exciton exchange hamiltonian can then be written, in the C_{2v} symmetry, in the general form [35]:

$$H_{e,h}^{ex} = 2\Delta_0 \hat{j}_z \hat{s}_z + \Delta_1(\hat{j}_x \hat{s}_x - \hat{j}_y \hat{s}_y) + \Delta_2(\hat{j}_x \hat{s}_x + \hat{j}_y \hat{s}_y), \tag{9.1}$$

where $\hat{s}$ and $\hat{j}$ are the electron spin and the heavy-hole pseudo-spin operators (in units of $\hbar$), and Δ_0, Δ_1,Δ_2 are constants equal to the energy distances between centres of the radiative and non-radiative doublets, and between levels within the radiative and non-radiative doublets respectively. Δ_0 and Δ_2 originate from the short range electron-hole coulomb exchange contribution, and one considers usually that $\Delta_2 \ll \Delta_0$ [34]. The anisotropic exchange interaction (AEI) splits the $|\pm 1\rangle$ radiative exciton doublets into two eigenstates labeled $|X\rangle = (|+1\rangle + |-1\rangle)/\sqrt{2}$ and $|Y\rangle = (|+1\rangle - |-1\rangle)/i\sqrt{2}$, linearly polarized along the (110) and $(1\bar{1}0)$ directions respectively [36, 37]. Δ_1, which includes short as well as long range contributions, is comparable to Δ_0 [38]. Cw single dot spectroscopy experiments have clearly evidenced the two corresponding linearly polarized lines in self-organized InGaAs QDs with an exchange splitting Δ_1 of up to $150\mu eV$ [36]. This anisotropic exchange splitting originates from QD elongation and/or interface optical anisotropy [33, 39]. Since the subspace of bright excitons is two dimensional, we can also define a pseudo spin associated to the $|+1\rangle$ and $|-1\rangle$ excitons. The dynamics of these pseudo spins is discussed in the next section.

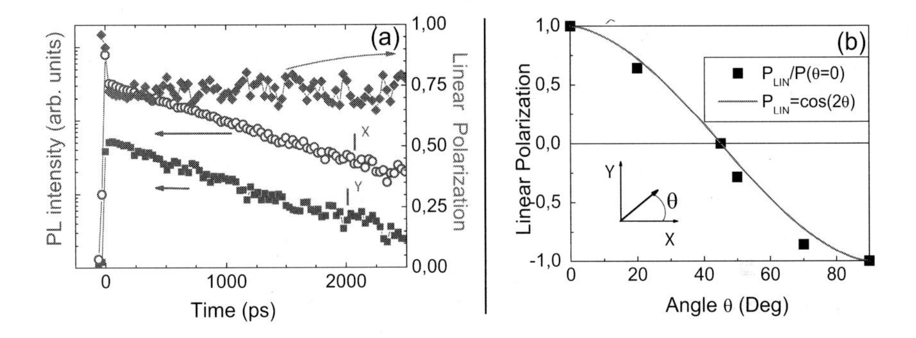

Fig. 9.1. Sample (I). (a) Time resolved PL intensity with polarization parallel I^X (hollow circles) and perpendicular I^Y (solid squares) to the linearly polarized (σ^X) excitation laser (T=10K); the time evolution of the corresponding linear polarization P_{lin} (diamonds) is also displayed. The laser excitation and detection energies are set to 1.137 eV. (b) Dependence of the luminescence linear polarization on the angle of the excitation laser field with respect to the [110] direction (see text).

9.3.2. *Exciton spin lifetime*

Figure 9.1a displays the time-resolved PL intensity with polarization parallel (I^X) and perpendicular (I^Y) to the linearly polarized σ^X excitation laser in sample I (T=10 K). The detection energy is strictly the same as the excitation one; it coincides with the exciton ground state energy. The corresponding linear polarization (P_{lin}) kinetics is also plotted. The experiment is performed here at low excitation power ($\sim 7Wcm^{-2}$) which corresponds to an estimated average density of photo-excited carriers less than one electron-hole pair per QD. The PL intensity decays with a characteristic time $\tau_{rad} \sim 800ps$. After the pulsed excitation, the QD emission exhibits a strong linear polarization ($P_{lin} \sim 0.75$) which remains strictly constant within the experimental accuracy during the exciton emission (i.e. over $\sim$ 2.5 ns). This behaviour differs strongly from the exciton linear polarization dynamics in bulk or type I quantum well structures, characterized by a linear polarization decay time of a few tens of picoseconds [31]. The experimental observation of a QD exciton linear polarization which does not decay with time is the proof that neither the electron, nor the hole spin relax on the exciton lifetime scale. It shows also that the exciton spin eigenstate is maintained during the whole exciton lifetime. From this observation, we can infer that the exciton spin relaxation time is longer than 20 ns, i.e. at least *25 times larger* than the radiative lifetime.

Figure 9.1b displays the dependence of the linear polarization of the luminescence on the angle θ of the excitation laser field with respect to the (110) direction. The linearly polarized luminescence components are still detected along the (110) (X) and $(1\bar{1}0)$ (Y) directions. As expected for exciton eigenstates polarized along the (110) and $(1\bar{1}0)$ directions, the linear polarization follows a simple law as a function of the angle θ given by $P_{lin}(\theta) = cos(2\theta)$.

If a magnetic field is applied along the growth direction ($B \parallel [001]$) and the resulting exciton Zeeman splitting $\hbar\Omega_z = g\mu_B B$ is much larger than the exchange energy Δ_1, the QD exciton eigenstates are no more the $|X\rangle$ and $|Y\rangle$ linearly polarized states but the $|+1\rangle$ and $|-1\rangle$ circular ones (g is the exciton longitudinal Landé factor). As a consequence, the QD emission is circularly polarized after a σ^+ polarized pulsed excitation (not shown here) and again the striking feature is the absence of any polarization decay on the exciton emission time scale [22, 23].

Let us examine now the robustness of the exciton spin when raising the temperature. Figure 9.2 displays the linear polarization of the QD ground state emission as a function of the temperature after a strictly resonant linearly polarised excitation. Up to 30 K, the decay time is longer than 20 ns. At 80 K, the linear decay time drops down to 20 ps with an activation energy $E_a \sim 30meV$. The dependence of the linear polarization on the temperature can be well explained in terms of the second order quasi-elastic interaction between LO phonons and carriers as proposed by Tsitsishvili *et al* [40]. The scattering events occur via the virtual excited states of the exciton built with an S_c and an $P_v^{x(y)}$ state, which are coupled to both $|X\rangle$ and $|Y\rangle$ exciton states via quasi-resonant absorption and emission of LO phonons. This process leads to exciton transitions between the ground $|X\rangle$ and $|Y\rangle$ sublevels with a rate proportionnal to $N_{LO}(1+N_{LO}) \approx e^{-\hbar\omega_{LO}/kT}$, where $\hbar\omega_{LO} = 32meV$ is the InAs LO phonon energy and N_{LO} their occupation factor, and determines the temporal dependence of the linear polarization. The calculated decay times $\tau_{pol} \sim 3.5ns$ (40 K) and $\tau_{pol} \sim 44ps$

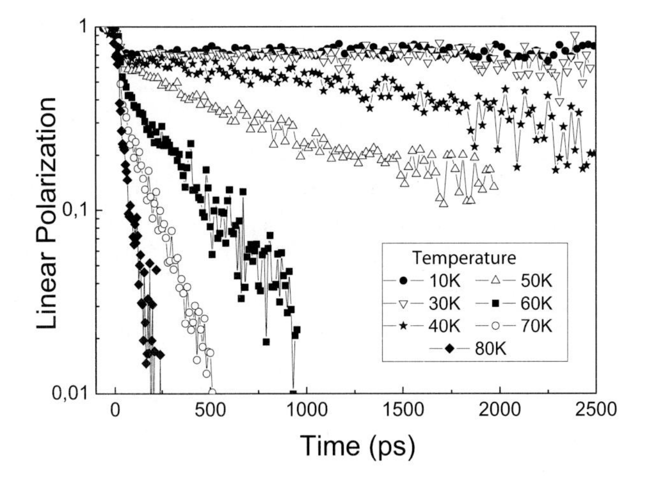

Fig. 9.2. (Sample I) Temperature dependence of the linear polarization dynamics of neutral excitons.

(80 K) closely agree with the experimental results. The future implementation of quantum bits based on *exciton spin* in InAs/GaAs QDs would thus require low temperature operation ($\leq 40K$).

9.3.3. *Exciton spin coherence*

A linearly polarised laser excitation along the (100) or (010) directions (45 degree tilt with respect to the (110) and $(1\bar{1}0)$ directions) should lead to the observation of beats of P_{lin} at the pulsation corresponding to the exchange splitting Δ_1. This has been clearly observed in a single CdSe dot experiment performed by Flissikowski *et al* showing that the exciton spin coherence time T_2 is longer than the radiative lifetime [41].

Similarly a circularly polarized (σ^+) excitation should lead to the observation of circular polarization P_c quantum beats at the angular frequency $\Delta_1/\hbar$ [42]. We have performed this experiment in an ensemble of self-organized InAs/GaAs QDs [23, 43]. Following σ^+ excitation, a coherent superposition of the two linearly polarized eigenstates $|X\rangle$ and $|Y\rangle$ is generated as the laser linewidth is larger than the AEI splitting Δ_1, resulting in an oscillating PL signal during the first few hundred picoseconds ($t < 300$ ps) shown in figure 9.3a (sample II), with a period corresponding to the energy splitting between $|X\rangle$ and $|Y\rangle$.

We obtain an oscillation period of 135 ± 10 ps, corresponding to an average energy splitting $\Delta_1 = 30 \pm 3\mu eV$, and an inhomogeneous decay

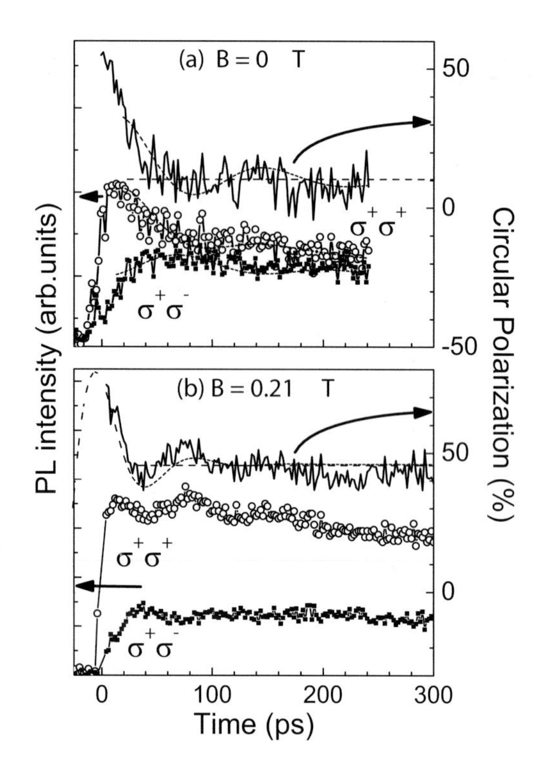

Fig. 9.3. (Sample II). Time resolved photoluminescence, co- (open circles) and counter-polarised (solid squares) after quasi-resonant σ^+ excitation. The dashed line is a fit of the oscillations using $P_c = P_c(0)e^{-t/T_2^*} cos(2\pi t/T)$ [23]. (a) For B=0 oscillations in the *circular* polarization degree P_C with a period of T=135ps are observed. (b) For B=0.21T along Oz the period shortens to T=90ps.

time of $T_2^* = 30ps$, implying a dispersion of the AEI splitting from dot to dot of the order of 40 μeV. The measured Δ_1 is in agreement with values reported by other groups on similar samples, from single-dot PL and transmission [36], differential transmission experiments [42] and transient Four-Wave Mixing experiments on ensembles [44].

As explained above, a magnetic field applied along the growth axis of the sample (Faraday configuration) can be used to control the exciton spin eigenstates. At zero magnetic field the optically active eigenstates for neutral dots are $|X\rangle$ and $|Y\rangle$. By increasing the magnetic field, the Zeeman splitting between the $|+1\rangle$ and $|-1\rangle$ states starts to dominate the AEI splitting, and $|+1\rangle$ and $|-1\rangle$ become the optically active eigenstates [36, 45].

In analogy to the quantum beats observed at B=0 for $t < 300$ ps (figure 9.3a), we expect to measure quantum beats in magnetic fields (with different oscillation periods). Indeed for an intermediate field of B=0.2 T we observe a beating of the *circular* polarization following *circularly* polarized excitation, as at B=0, but with a shorter period of 90 ps (see figure 9.3b, sample II). This coherent oscillation period Ω is controlled by both the AEI exchange energy Δ_1 and the exciton g factor through :

$$\Omega = \sqrt{(\Delta_1/\hbar)^2 + (g\mu_B B/\hbar)^2}, \tag{9.2}$$

We emphasize that the fast damping of the oscillations observed in figure 9.3 is not due to an intrinsic decoherence process. It simply comes from the inhomogeneous character of the experiment performed on an ensemble of dots. Similarly to CdSe dots [41], the exciton spin coherence time T_2 is of the order of the radiative lifetime (a few hundreds of picoseconds). This results is encouraging for the quantum computation proposals based on *exciton spin* in quantum dots.

9.3.4. *Exciton/Biexciton spin coherence*

All the results presented up to now have been obtained at low excitation intensity (i.e. when the number of photo-generated electron-hole pairs is small compared to the number of QD). Figure 9.4 displays the linear polarization dynamics under resonant linear polarization excitation for three excitation intensities (Sample I). We note that the linear polarization never decays with time but the amplitude of this polarization decreases when the excitation intensity increases. This effect is due to the photo-generation of biexciton states. Under linearly polarized pulsed excitation, a quantum dot in which the ground state is already occupied by an exciton $|X\rangle = (|+1\rangle + |-1\rangle)/\sqrt{2}$ can absorb another photon, which leads to the photogeneration of a biexciton state labeled $(|XX\rangle + |YY\rangle)/\sqrt{2}$. This biexciton state will emit σ^X or σ^Y photons with nearly the same probabilities, whereas the QDs occupied only by one exciton after the laser excitation will still emit fully polarized σ^X luminescence. As a consequence, the amplitude of the measured linear polarization of the spectrally integrated signal decays when the number of photogenerated biexciton states increases. We emphasize that the spectral resolution of our setup (~3 meV) does not allow to resolve the exciton and the biexciton emission lines, which are often separated by less than 1meV.

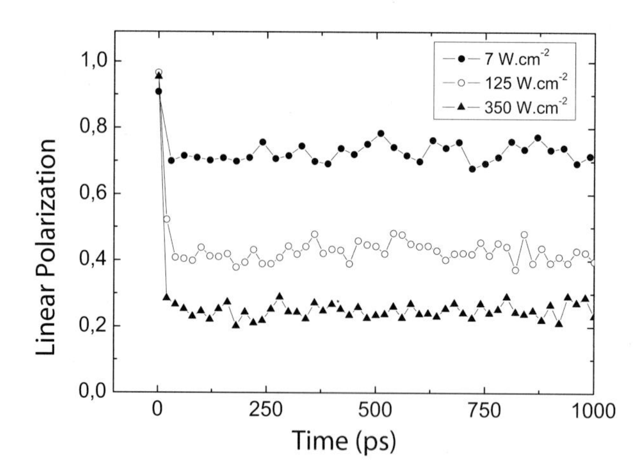

Fig. 9.4. (Sample I). Linear polarization dynamics for three different laser excitation powers

The biexciton states corresponds to an entangled two-exciton state, and many approaches in the perspective of building quantum logic gates rely on the use of exciton and biexciton states in quantum dots [11, 18, 46–48]. Due to the symmetry of the biexciton state, the absorption of a σ^X photon resonant with the exciton-biexciton transition is only possible if the initially photocreated exciton is in the $|X\rangle$ state. One may thus consider this system as a building block for a conditional-not quantum logic gate. Note that two-photon coherent control in the time domain of the biexciton state in a single CdSe quantum dot has also been experimentally achieved, which is a first step towards the realization of this kind of quantum logic gate [14, 29].

9.4. Electron Spin Quantum Bits - Dephasing Due to the Hyperfine Interaction

In many theoretical proposals for the implementation of quantum computation, the quantum bit is based on the spin of an excess electron in a single dot. Indeed it has been experimentally shown that the electron spin relaxation time in semiconductor quantum dots can be very long (up to the ms range) [49–51]. However quantum bit applications require not only long spin lifetimes but also long spin coherence times, which is usually more difficult to achieve in semiconductors. In this section, we present a detailed investigation of the decoherence processes of electron spins in self-organized InAs/GaAs QDs.

Several theoretical studies have predicted that the dominant mechanism of electron spin relaxation in QDs at low temperature is due to the hyperfine interaction with nuclear spins [52–54]. An electron spin in a quantum dot interacts with a large but finite number of nuclei $N \sim 10^3$ to 10^5 [28]. In the frozen fluctuation model, the sum over the interacting nuclear spins gives rise to a local effective hyperfine field B_N [52]. The electron spin can thus coherently precess around B_N. However, the amplitude and direction of the effective nuclear field vary strongly from dot to dot. The average electron spin $\langle S(t) \rangle$ in an ensemble of dots will thus decay as a consequence of the random distribution of the local nuclear effective field. For the sake of simplicity this spin dephasing mechanism of the QD ensemble is often termed spin relaxation. Note that for repeated measurements on a single QD the hyperfine interaction has the same effect as for an ensemble of dots [52, 54].

The spin dynamics of carriers in III-V or II-VI semiconductor QDs have been studied experimentally by different groups in recent years [22, 40, 42, 49, 51, 55–61]. In all these experiments no manifestation of the electron spin relaxation due to the interaction with nuclei has been observed for the following reasons:

- In undoped QDs the photogenerated electron feels a strong effective magnetic field due to the exchange interaction with the hole [36]. This exchange field is much stronger than the effective hyperfine field of the nuclei, which thus plays a negligible role [62].

- In the experiments performed on n-doped QDs the ground state luminescence corresponds to the radiative recombination of the negatively charged exciton X^- formed by one hole, and a pair of electrons with opposite spins in a singlet state [38]. In this case, no effect of the hyperfine interaction with nuclei is expected during the radiative lifetime since the total electron spin in the charged exciton is zero and the hole spin is only weakly coupled to the nuclear spins due to the p-symmetry of the hole Bloch function [63, 64].

The positively charged excitons X^+ (consisting of one electron and two holes forming a spin singlet) is the ideal configuration to probe the electron spin relaxation mediated by nuclei in QDs with optical experiments. The exchange interaction between the electron and the two holes cancels in the X^+ ground state as in the case of X^-. The analysis of the circular polarization of the X^+ luminescence in p-doped QDs following a circularly polarized laser excitation will thus probe directly the spin polarization of the electron. We have performed this experiment in self-organized InAs/GaAs

QDs [24].

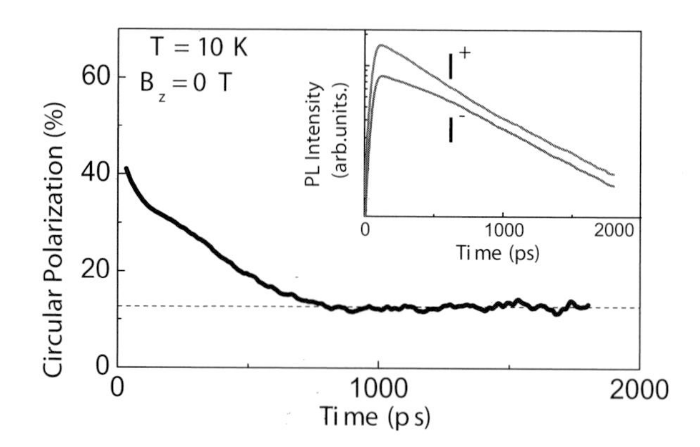

Fig. 9.5. (Sample III). Circular polarization dynamics of the QD luminescence after a circularly polarized σ^+ laser excitation. Inset: photoluminescence intensity co-polarized I^+ and counter-polarized I^- with the laser (semi-logarithmic scale). The detection energy is centred at $E_{det} = 1.11eV$.

Figure 9.5 displays the circular polarization dynamics of the QD ground state luminescence in a p-modulation doped InAs/GaAs QD structure (Sample III). The inset presents the time evolution of the luminescence intensity components I^+ and I^-. We consider for the interpretation that (i) the dots contain a single resident hole and (ii) a single electron-hole pair is optically injected into the dot. Following excitation into the wetting layer (WL) with $E_{exc.} = 1.44$eV, it is well known that the electron spin does not relax significantly during the capture and energy relaxation process in the QD whereas the initial hole spin orientation is lost due to efficient spin relaxation processes in the WL [49, 65, 66]. The recorded PL in this p-doped QD sample corresponds essentially to the radiative recombination of positively charged exciton X^+ formed with a spin polarized electron and two holes with opposite spin : $|X^+\rangle = (|\frac{3}{2}, -\frac{3}{2}, \downarrow\rangle - |-\frac{3}{2}, \frac{3}{2}, \downarrow\rangle)/\sqrt{2}$.

The circular polarization dynamics in figure 9.5 presents two regimes. The polarization decays within the first 800ps down to a value of about 12%; then it remains stable with no measurable decay on the radiative lifetime scale. We can infer that the spin relaxation in this second regime is longer than 10 ns. This specific circular polarization dynamics has been observed for any detection energy in the PL spectrum of the QD ground state ensemble. Moreover, we have measured similar kinetics in all the p-doped samples with similar doping levels that we have studied. All these results are in very good agreement with the predicted electron spin relaxation due to the hyperfine interaction with the nuclei [52–54]. The theoretical time dependence of the average electron spin due to this process can be written as [24, 52]:

$$\langle \mathbf{S}(t) \rangle = \frac{\mathbf{S}_0}{3} \left\{ 1 + 2 \left[1 - 2 \left(\frac{t}{2T_\Delta} \right)^2 \right] exp \left[- \left(\frac{t}{2T_\Delta} \right)^2 \right] \right\} \tag{9.3}$$

where $\mathbf{S}_0$ is the initial spin, $T_\Delta = \hbar/(g_e \mu_B \Delta_B)$ is the dephasing time due to the random electron precession frequencies in the randomly distributed frozen fluctuation of the nuclear hyperfine field, μ_B is the Bohr magneton, g_e is the effective electron Landé factor. The dispersion of the nuclear hyperfine field B_N is described here by a Gaussian distribution characterized by its width Δ_B : $W(\mathbf{B}_N) \approx exp(-(\mathbf{B}_N)^2/\Delta_B^2)$ where $\Delta_B^2 = 2\langle \mathbf{B}_N^2 \rangle /3$ [52]. Note that it is assumed here that the electron spin does not polarize the nuclear spins. It is clear from equation (9.3) that the time dependence of the average electron spin polarization exhibits two regimes. After a strong initial decay with a characteristic time T_Δ the average electron spin polarization is expected to reach a constant value of 1/3 of the initial polarization. The initial value of the average electron spin polarization in figure 9.5 is about 40%; it then drops down to about 1/3 ($\approx$ 12%) of its initial value in agreement with the predictions of equation (9.3). In our experiments, we find $T_\Delta \approx 500ps$ and a value for Δ_B of about 45mT [24]. After the initial drop the average electron spin polarization remains stable on the radiative lifetime scale. Merkulov *et al* calculated that the subsequent electron spin dephasing, which is the result of the nuclear spin precession around the average electron spin orientation in each dot, occurs on a time scale of the order of $\sqrt{N} \cdot T_\Delta$, i.e. typically 100 times longer than T_Δ for $N = 10^4 - 10^5$ nuclei per dot [52]. It can thus not be observed on the radiative lifetime scale in figure 9.5.

A key argument for the hyperfine interaction being responsible for the initial polarization decay comes from magnetic field dependent measurements. We have recorded the circular polarization dynamics of the QD ground state luminescence with a magnetic field applied along the Oz growth axis (Faraday configuration). Merkulov *et al* and Semenov *et al* predict that the electron spin dephasing induced by hyperfine interaction can be strongly suppressed in an external magnetic field [52, 54]. The required magnetic field must be larger than B_N, which is of the order of 10mT [53], to ensure that the Zeeman interaction of the electron spin with the magnetic field is stronger than the interaction with the nuclei. Figure 9.6 displays the circular polarization dynamics of the QD ground state luminescence with magnetic fields $B_z = 100mT$ and $B_z = 400mT$; the dynamics for $B_z = 0$ is also presented for comparison. By applying a field of $B_z = 100mT$ we drastically increase the initial decay time to ~4000ps as compared to ~500ps at $B_z = 0$. Note that the Zeeman splitting energy of the electron in this weak magnetic field is at least 50 times smaller than $k_B T$ at T=10K [62]. This pronounced effect of the small external magnetic field observed in figure 9.6 agrees very well with the expected influence of the external magnetic field on the QD electron spin relaxation by nuclei [52, 54]. The effect observed here is similar to the suppression of the effect of nuclear hyperfine interaction measured for localized electrons in lightly doped bulk n-GaAs [67, 68].

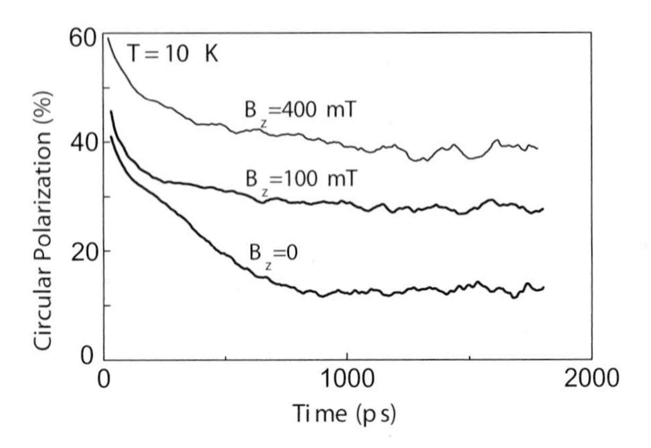

Fig. 9.6. (Sample III). Circular polarization dynamics of the QD ground state luminescence (semi-log. scale) for $B_z = 0$, $B_z = 100mT$ and $B_z = 400mT$.

These results show that the dominant electron spin dephasing mechanism at low temperature in QDs is the hyperfine interaction with nuclei. The effect of this hyperfine interaction on the electron spin coherence can also be directly measured by recording the electron spin dynamics in a transverse magnetic field (perpendicular to the growth direction, Voigt configuration). This can be done by recording the PL circular polarization dynamics of positively charged excitons X^+ [69]. The polarization oscillations as a function of time reflect simply the precession of the electron spin around the applied magnetic field. Fitting the oscillation damping with an exponential decay law yields an estimate of the dephasing time T_2^*, which, in a classical view, is the decoherence time of the spin ensemble during the precession around the applied magnetic field.

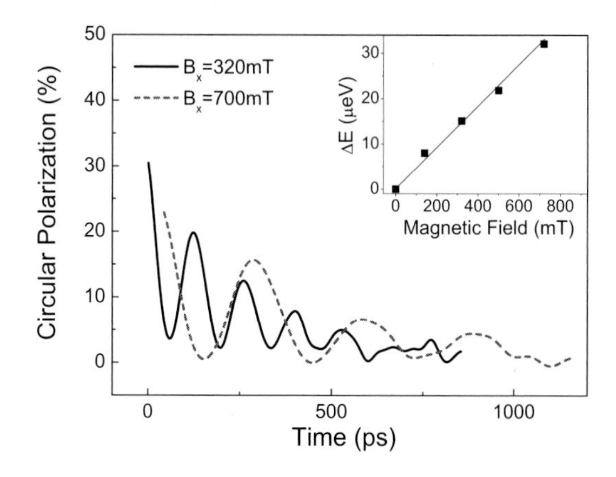

Fig. 9.7. (Sample III). Time-resolved circular polarization decay of the X^+ trion luminsecence under σ^+ excitation for two different magnetic fields B_x. Inset: Zeeman splitting as a function of B_x.

The oscillation period observed, corresponding to the X^+ ground state Zeeman splitting Δ_z, decreases when the magnetic field increases. The inset in figure 9.7 shows this linear dependence: $\Delta_z = g_e \mu_B B_x$. We can thus find the average electron transverse g_e-factor $|g_e| \approx 0.8$. From the curve $P_c(t)$ in figure 9.7 we also see that despite the relatively high magnetic field value, the electron spin coherence decays with a typical time T_2^* of about 300 ps, one order of magnitude shorter than previously found in Faraday configuration (see figure 9.6). As demonstrated below, this fast damping is mainly due to the hyperfine interaction with nuclei [69].

Increasing the magnetic field also leads to an increase of the damping of the polarization oscillations. This magnetic field dependent damping arises from variations of electron g_e factor over the QD ensemble, leading to a spreading of the angular Larmor frequency with increasing B_x. The same behaviour has been observed in reference [70] when the authors recorded

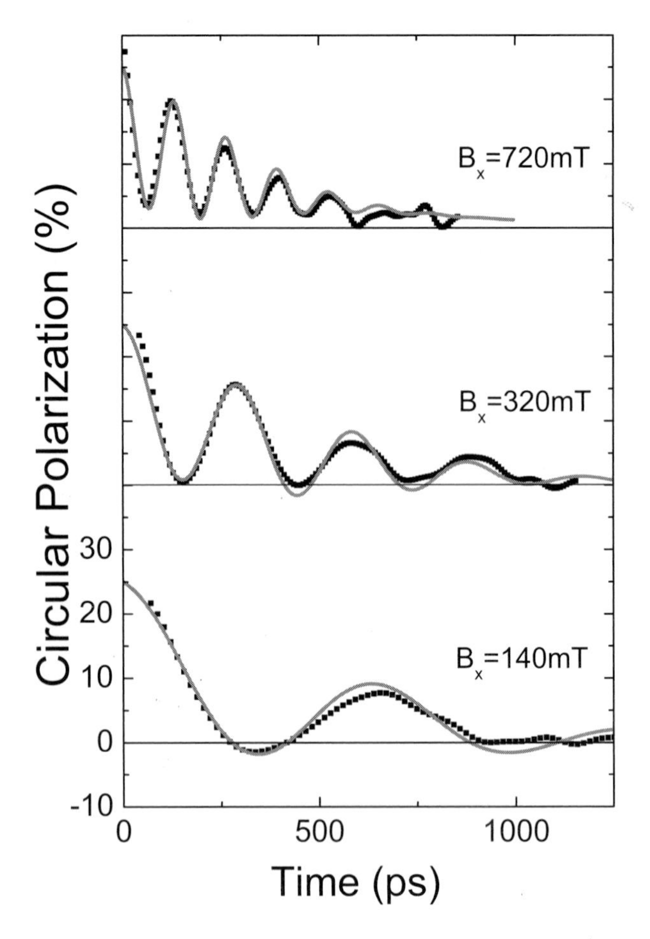

Fig. 9.8. (Sample III). Time-resolved circular polarization decay of X^+ trion in transverse magnetic field B_x under σ^+ excitation. Black line: experimental data. Grey line: fit using a model [69] including both an electron g factor dispersion of 7% and nuclear field fluctuations of Δ_B= 28mT (T_Δ = 400ps).

the time evolution of optically generated electron spin coherence in negatively charged InAs/GaAs QDs using time resolved Kerr pump-probe spectroscopy. This contribution is the second source of spin dephasing for an ensemble of quantum dots. Figure 9.8 represents the experimental curves of the photoluminescence circular polarization $P_c(t)$ (bold line) in comparison with $P_c(t)$ from the theoretical model including (i) the inhomogeneous variation of the electron g_e factor and (ii) the random fluctuation of the nuclear hyperfine field (grey line) [69].

The experiments presented in this section demonstrate the key role played by the nuclear field fluctuations on the electron spin coherence dynamics in QDs; we found a decoherence time of about 1000 ps for this contribution alone in InAs/GaAs QDs. In spite of this efficient inhomogeneous dephasing mechanism, Greilich *et al.* demonstrated possible applications based on robust electron spin quantum coherence within an ensemble of InAs/GaAs dots for time scales up to $1\mu s$ [71]. Note that the electron spin dephasing due to the hyperfine interaction was also clearly evidenced by electrical measurements in gate-defined GaAs/AlGaAs quantum dots [19, 20].

9.5. Quantum Bits Based on Collective Nuclear Spin States

As we just described above, the hyperfine interaction with randomly oriented nuclear spins is a fundamental decoherence mechanism for electron spin in a semiconductor quantum dot. However, the nuclear spins themselves could be used to create long-lived quantum memory for quantum bits. Nuclear spins can possess very long coherence times because of their weak environmental coupling, but single nuclear spins are very difficult to manipulate and measure [8]. A recent proposal by Taylor *et al* is motivated by the possibility to use the nuclear spins as a quantum memory and aims to combine the strengths of electron-spin (or charge) manipulation with the long-term memory provided by the nuclear spin system [16]. The target is to achieve a nuclear polarisation degree that is high enough to transfer a coherent spin state from a single electron to the nuclear spin system. This is achieved through the effective control of the spin-exchange part of hyperfine contact interaction (via an external magnetic field pulse for instance). After the transfer is completed, the resulting superpositions could be stored for very long times and mapped back into the electron spin degrees of freedom on demand. It would thus be possible to reliably map the quantum state of a spin qubit onto long-lived collective nuclear-spin states. This proposal is quite similar to the use of atomic ensembles as quantum information carriers [72]. The experimental implementation of this technique will require on the one hand, a strong strong nuclear-spin polarization in the vicinity of confined electrons, and on the other hand long carrier spin life and coherence times, see section 9.4. We have thus performed a detailed investigation of the nuclear spin polarization in self-organized InAs/GaAs quantum dots. Achieving a very large nuclear spin polarization would (i) limit the fundamental decoherence mechanism for electron spin in a quantum dot [28, 73] (see section 9.4) and (ii) opens the route towards the implementation of quantum memories based on collective nuclear spin states.

Let us recall that following injection of electrons with a preferred spin direction, the electron spin can be imprinted on the nuclei in the dot via the hyperfine interaction. Dynamical nuclear polarization through optical pumping leads to the construction of an effective nuclear field B_n. In an applied magnetic field the Zeeman splitting of an electron is given by the contributions of the external field and B_n. The contribution due to B_n is the Overhauser shift (OHS) δ_n and can be measured by single dot photoluminescence (PL) spectroscopy in the Faraday geometry. Previous work by Gammon *et al* on GaAs interface fluctuation quantum dots on neutral and charged excitons show that dynamical polarization is very effective in their samples with 60% of the nuclei in the quantum dot polarized through optical pumping [55, 65]. To evaluate the nuclear polarization these authors compare the measured OHS with the maximum value theoretically obtainable for a nuclear polarization of 100%. The small confinement potentiel in these samples compared to InAs quantum dots could be a limitation for future applications above 4K. The first experiments on neutral (charged) excitons in self-assembled InAlAs (InGaAs) quantum dots show a nuclear polarization of only 6% (10%) [74–76]. For self-assembled InP dots effective nuclear fields B_n of only a few mT have been observed [58, 77] compared to effective fields of several Teslas in GaAs dots. This is very surprising as for In containing compounds, in principle large nuclear effects are expected due to the nuclear spin of 9/2 as compared to only 3/2 for both Ga and As. We showed recently that in order to achieve a substantial nuclear polarization through optical pumping in self-assembled InGaAs dots a careful analysis of the interdependence of applied magnetic field strength, optical pumping power, and electron spin polarization effects is necessary [25]. We managed to polarize up to 47% of the nuclei in individual QDs corresponding to $\delta_n = 110\mu eV$ [78]. This is a first step towards the implementation of quantum bits based on the proposal by Taylor *et al* [16].

As in the investigated temperature range (1.5-10K) the thermal energy is much larger than the nuclear Zeeman splitting, the presence of spin polarised electrons is essential for building up a nuclear field. This is the case during the radiative lifetime of the positively charged exciton X^+, where the holes form a spin singlet and the single electron interacts with the nuclei. The radiative lifetime of these pseudo-particles is about 1 ns. As shown in section 9.4, the analysis of the circular polarization of the X^+ luminescence in QDs following circularly polarized laser excitation probes *directly* the spin polarization of the electron as $\langle \hat{S}_z^e \rangle = -P_c/2$. The hyperfine interaction between an electron of spin $\hat{\mathbf{S}}^e = \frac{1}{2}\hat{\sigma}^e$ confined to a quantum dot and N nuclei is described by the Fermi contact Hamiltonian.

$$\hat{H}_{hf} = \frac{\nu_0}{2} \sum_j A^j |\psi(\bar{r}_j)|^2 \left(2\hat{I}_z^j \hat{S}_z^e + [\hat{I}_+^j \hat{S}_-^e + \hat{I}_-^j \hat{S}_+^e] \right) \tag{9.4}$$

where ν_0 is the two atom unit cell volume, $\bar{r}_j$ is the position of the nuclei j with spin $\hat{I}^j$, the nuclear species are In, As and Ga. A^j is the constant of the hyperfine interaction with the electron and $\psi(\bar{r})$ is the electron envelope function.

We present here the dynamical polarization of nuclei in InGaAs quantum dots in an external magnetic field B_z parallel to the sample growth direction, that is larger than both the local magnetic field B_L (characterising the strength of the nuclear dipole-dipole interaction) and the Knight field B_e (the effective magnetic field seen by the nuclei due to the presence of a spin polarised electron), which are in the order of mT. For recent discussions in quantum dots see Refs [76, 79, 80].

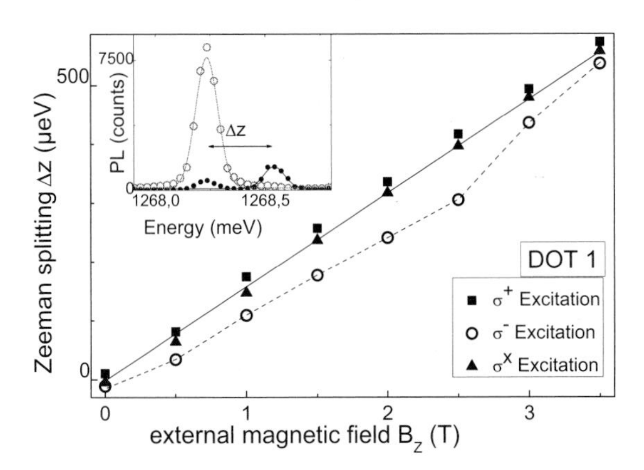

Fig. 9.9. (Sample IV). The Zeeman splitting ΔZ is measured for an individual quantum dot for three different excitation laser polarizations: σ^+ (solid squares), σ^- (hollow circles) and linear (solid triangles) for an excitation power of P=8.7μW. The dotted line is a guide to the eye. Inset: Single dot PL at 4T showing a clear Zeeman splitting between σ^+ PL (solid circles) and σ^- PL (hollow circles) following σ^- excitation.

Figure 9.9 shows the Zeeman splitting for a single quantum dot following linearly and circularly polarised excitation. We have verified experimentally that changing the direction of $\langle S_z^e \rangle$ has the same effect as changing the field direction from $+B_z$ to $-B_z$. In this work we only show measurements for the same direction of B_z for changing $\langle S_z^e \rangle$. The Zeeman splitting following linear excitation is due to the external applied fields and grows, as expected, linearly with B_z. In contrast, when exciting the sample with $\sigma^+(\sigma^-)$ polarised light, the Zeeman splitting increases (decreases). This has been observed in GaAs interface fluctuation dots [81]. Figure 9.10 shows the values of the OHS for fields up to 2 Tesla in greater detail for another dot. We note:

(i) δ_n does not reach the same absolute values for σ^+ and σ^- excitation. The total magnetic field seen by the electron is smaller in the case of σ^- excitation than in the case of σ^+ excitation, as in the case of σ^- excitation the external magnetic field and the nuclear field are anti-parallel. This can be seen directly from the smaller Zeeman splitting in figure 9.9. As the spin flip of the electron means going from one Zeeman level to another, nuclear polarization due to the simultaneous flip of an electron spin and a nuclear spin (flip-flop) is more efficient when the total magnetic field is small, here in the case of σ^- excitation. This would explain the larger absolute value of the OHS measured for σ^- excitation.

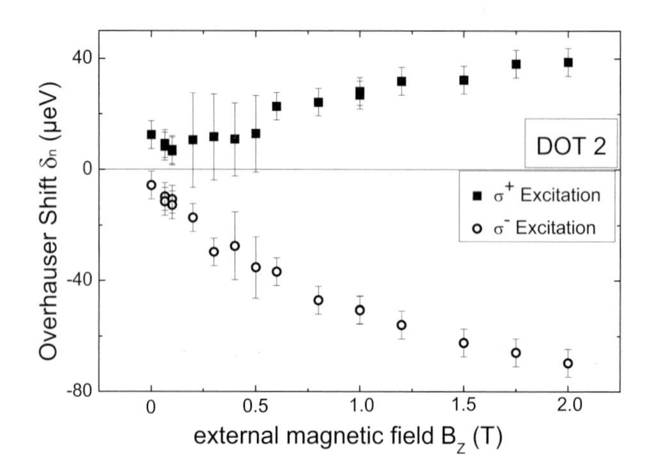

Fig. 9.10. (Sample IV). The Overhauser shift δ_n is measured for magnetic fields up to 2 Tesla following σ^+ (solid squares) and σ^- (hollow circles) excitation.

(ii) The OHS does increase with increasing magnetic field up to about 2T. For small fields in the order of mT this can be understood in terms of the suppression of the dipole-dipole interaction between nuclei, that leads to nuclear spin relaxation [8]. Once this mechanism is supressed it is surprising to find a further increase of the OHS with the applied magnetic field. In reference [81] for pure GaAs dots the OHS does not change measurably between 0.2T to 3.5T. We interpret the increase in OHS observed in our experiment as a function of the magnetic field as a gradual suppression of one or several nuclear depolarization mechanisms, induced, for instance, by nuclear quadrupole coupling. Self-assembled, nominally pure InAs QDs are in reality InGaAs QDs due to In and Ga interdiffusion. This will induce local lattice distortion and strain [82], giving rise to electric field gradients that lead to quadrupole coupling between the nuclear spin states [77, 83]. It has already been suggested that nuclear quadrupole coupling is responsible for the small nuclear fields in self assembled InP dots [58]. This coupling, which exists for all nuclei with a spin $I \geq 1$ in a lattice with a symmetry that is lower than cubic, does not induce any nuclear spin relaxation itself, but it mixes the different nuclear spin states. Transitions are then possible due to fluctuations of the occupation of the dot by an electron. The degree of the mixing depends on the energy separation i.e. the Zeeman splitting between the different nuclear spin states. Increasing the external field will increase the nuclear Zeeman splitting and hence decouple the nuclear spin states, as a simple perturbation treatment shows [83].This would explain why the measurements in figure 9.10 show a steady increase in the OHS with the applied external field. As strain gradients and alloy composition vary from dot to dot on a microscopic scale, the maximum OHS will not reach the same value for every dot. The quadrupole coupling in GaAs interface fluctuation dots is certainly much weaker as the dots consist of a binary compound and are not strained, which could explain that the OHS does not measurably depend on the applied magnetic field for these dots.

As explained above, the quantum bits based on collective nuclear spin states proposed by Taylor *et al* require strong nuclear polarization in the vicinity of the QD electron [16]. To obtain a strong nuclear polarization in InAs/GaAs QDs, we have shown that the applied magnetic field strength, the optical pump power and the optically created electron spin polarization have to be optimized. We have demonstrated that on the one hand, B_z should not cause a too large electron Zeeman splitting, that makes spin flip-flop processes necessary to build up a nuclear polarization too costly in energy. On the other hand, we believe that B_z has to be strong enough to decouple the nuclear spin states that are mixed by the quadrupolar interaction. We find experimentally that fields between 1.5 and 2.5 T fulfil both criteria.

9.6. Conclusions

We have shown that optical orientation experiments are efficient tools to investigate the *electron* or *exciton spin* coherences in self-organized quantum dots. Though the hyperfine interaction with nuclei might raise difficulties for large scale implementation of quantum information processing based on electron spin alone, the recent experimental results on dynamical polarization of nuclei in single dots demonstrate the great potential of quantum bits based on collective nuclear spin states.

Acknowledgments

We thank J.M. Gérard, V. Ustinov and A. Lemaître for the sample growth. We are grateful to M. Paillard, M. Sénès, P.F. Braun, L. Lombez, D. Lagarde, S. Laurent, B. Eble, H. Carrère, P. Voisin, K.V. Kavokin, V.K. Kalevich for their contributions to this work.

References

[1] D.P.DiVicenzo, Science **270**, 255 (1995).
[2] D. Loss, D.D. Awschalom and N. Samarth (Editors) *Semiconductor Spintronics and Quantum Computation* (Springer, Berlin 2002).
[3] J.Y. Marzin *et al.*, Phys. Rev. Lett **73**, 716 (1994).
[4] T. H. Stievater, Xiaoqin Li, D. G. Steel, D. Gammon, D. S. Katzer, D. Park, C. Piermarocchi, and L. J. Sham, Phys. Rev. Lett **87**, 133603 (2001).
[5] H. Kamada, H. Gotoh, J. Temmyo, T. Takagahara, and H. Ando, Phys. Rev. Lett **87**, 246401 (2001).
[6] H. Htoon, T. Takagahara, D. Kulik, O. Baklenov, A. L. Holmes, Jr., and C. K. Shih, Phys. Rev. Lett. **87**, 087401 (2002).
[7] P. Borri *et al.* , Phys. Rev. Lett **87**, 157401(2001).
[8] F. Meier and B. Zakharchenya, *Optical Orientation, Modern Problem in Condensed Matter Sciences* vol. 8 (North-Holland, Amsterdam, 1984).
[9] M. Chamaro, C. Gourdon, P. Lavallard, O. Lublinskaya, A.I. Eimov, Phys. Rev. B **53**, 1336 (1996).
[10] A.V. Khaetskii, Y. Nazarov, Phys. Rev. B **61**, 12639-12642 (2000).
[11] P. Chen, C. Piermarocchi, and L. J. Sham, Phys. Rev. Lett. **87**, 067401 (2001).
[12] D. Loss *et al.*, Phys. Rev. A **57**, 120 (1998).
[13] G. Burkard *et al.*, Phys. Rev. B **62**, 2581 (2000).
[14] T. Amand, V. Blanchet, B. Girard, X. Marie, *Coherent control in atoms, molecules and solids* in Femtosecond Laser Pulses, Springer, 2nd edition (2005), Ed. C. Rulliere.
[15] A. Imamoglu *et al.*, Phys. Rev. Lett. **83**, 4204 (1999).

[16] J.M. Taylor *et al.*, Phys. Rev. Lett. **90**, 206803 (2003).
[17] G. Chen *et al.*, Science **289**, 1906(2000).
[18] X. Li *et al.*, Science **301**, 809(2003).
[19] J.R. Petta *et al.*, Science **309**, 2180(2005).
[20] H.L. Koppens *et al.*, Nature **442**, 766(2006).
[21] J.M. Gérard, B. Sermage, B. Gayral, B. Legrand, E. Costard, V. Thierry-Mieg, Phys. Rev. Lett. **81**, 1110 (1998).
[22] M. Paillard, X. Marie, P. Renucci, T. Amand, A. Jbeli, J.M. Gérard, Phys. Rev. Lett. **86**, 1634 (2001).
[23] M.Senes, B. Urbaszek, X. Marie, T. Amand, J. Tribollet, F. Bernardot, C. Testelin, M. Chamarro, and J.-M. Gérard, Phys. Rev. B **71**, 115334 (2005).
[24] P.-F. Braun, X. Marie, L. Lombez, B. Urbaszek, T. Amand, P. Renucci, V. K. Kalevich, K. V. Kavokin, O. Krebs, P. Voisin, et al., Phys. Rev. Lett. **94**, 116601 (2005).
[25] P.-F. Braun, B. Urbaszek, T. Amand, X. Marie, O. Krebs, B. Eble, A. Lemaitre, and P. Voisin, Phys. Rev. B **74**,245306 (2006).
[26] B. Dareys *et al.*, Superlattices and Microstr. **13**, 353 (1993).
[27] T. Amand, X. Marie, P. Le Jeune, M. Brousseau, D. Robart, J. Barrau, R. Planel, Phys. Rev. Lett. **78**, 1355 (1997).
[28] A. Imamoglu, E. Knill, L. Tian, and P. Zoller, Phys. Rev. Lett. **91**, 017402 (2003).
[29] T. Flissikowski, A. Betke, I. Akimov, F. Henneberger, Phys. Rev. Lett. **92**, 227401 (2004).
[30] M. Paillard, X. Marie, E. Vanelle, T. Amand, V.K. Kalevich, A.R. Kovsh, A.E. Zhukov, V.M. Ustinov, Appl. Phys. Lett. **76**, 76 (2000).
[31] X. Marie, P. LeJeune, T. Amand, M. Brousseau, J. Barrau, M. Paillard, R. Planel, Phys. Rev. Lett. **79**, 3222 (1997).
[32] M. Oestreich *et al.*, Semicond. Sci. Technol. **17**, 285 (2002).
[33] R.I. Dzhioev *et al.*, Phys. of Solid State **40**, 790 (1998).
[34] E.L. Ivchenko, Springer series in Solid-State Science, vol. 110 (Springer-Verlag, Berlin, 1997).
[35] I. E. Kozin *et al.*, Phys. Rev. B **65**, 241312 (2002).
[36] M. Bayer *et al.*, Phys. Rev. B **65**, 195315 (2002), and references therein.
[37] E. L. Ivchenko, A. Y. Kaminski, and I. L. Aleiner, JETP 77, 609 (1993).
[38] B. Urbaszek, R. J.Warburton, K. Karrai, B. D. Gerardot, P.M. Petroff, and J.M. Garcia, Phys. Rev. Lett. **90**, 247403(2003).
[39] D. Gammon *et al*, Phys. Rev. Lett. **76**, 3005(1996).
[40] E.Tsitsishvili *et al*, Phys. Rev. B 66, 161405 (2002).
[41] T. Flissikowski, A. Hundt, M. Lowisch, M. Rabe, and F. Henneberger, Phys. Rev. Lett. **86**, 3172 (2001).
[42] A. S. Lenihan *et al*, Phys. Rev. Lett. **88**, 223601 (2002).
[43] F. Bernardot, E. Aubry, J. Tribollet, C. Testelin, M. Chamarro, L. Lombez, P.-F. Braun, X. Marie, T. Amand and J.-M. Gérard Phys. Rev. B, **73**, 085301 (2006).
[44] W. Langbein, P. Borri, U. Woggon, V. Stavarache, D. Reuter, and A. D. Wieck, Phys. Rev. B. **69**, R161301 (2004).

[45] R.I. Dzhioev *et al*, Phys. Rev. B. **56**, 13405 (1997).
[46] E. Biolatti, R. C. Iotti, P. Zanardi, and F. Rossi, Phys. Rev. Lett. **85**, 5647 (2000).
[47] F. Troiani, U. Hohenester, and E. Molinari, Phys. Rev. B **62**, R2263 (2000).
[48] G. Chen, T. H. Stievater, E. T. Batteh, Xiaoqin Li, D. G. Steel, D. Gammon, D. S. Katzer, D. Park, and L. J. Sham, Phys. Rev. Lett. **88**, 117901 (2002).
[49] S. Cortez, O. Krebs, S. Laurent, M. Senes, X. Marie, P. Voisin, R. Ferreira, G. Bastard J.-M. Gérard, T. Amand, Phys. Rev. Lett. **89**, 207 (2002).
[50] R. Oulton *et al*, Phys. Stat. Sol. (b) **243**, 3922 (2006).
[51] M. Kroutvar, Y. Ducommun, D. Heiss, M. Bichler, D. Schuh, G. Abstreiter, and J. J. Finley, Nature **432**, 81 (2004).
[52] I. A. Merkulov *et al.*, Phys. Rev. B **65**, 205309 (2002).
[53] A. Khaetskii *et al.*, Phys. Rev. Lett. **88**, 186802 (2002).
[54] Y. Semenov *et al.*, Phys. Rev. B **67**, 73301 (2003)
[55] D. Gammon *et al.*, Science **277**, (1997).
[56] J. A. Gupta *et al.*, Phys. Rev. B **59**, 10421 (1999).
[57] I. Ignatiev *et al.*, Physica E **17**, 361 (2003).
[58] R.I. Dzhioev *et al.*, Phys. Sol. State **41**, 2014 (1999).
[59] T. Amand *et al.*, Superlattices and Microstructures **32**, 157 (2002).
[60] T. Flissikowski *et al.*, Phys. Rev. B **68**, 161309 (2003).
[61] K. Gundogdu *et al.*, Appl. Phys Lett. **84**, 2793 (2004).
[62] S.I. Erlingson *et al.*, Phys. Rev. B **64**, 195306 (2001).
[63] A. Abragam, *Principles of Nuclear Magnetism*, (Oxford University Press, 1961).
[64] E. I. Gryncharova and V. I. Perel, Sov. Phys. Semicond. **11**, 997 (1977).
[65] A. Bracker *et al.*, Phys. Rev. Lett. **94**, 047402 (2005).
[66] T.C. Damen *et al.*, Phys. Rev. Lett. **67**, 3432 (1991).
[67] R. I. Dzhioev *et al.*, Phys. Rev. B **66**, 245204 (2002).
[68] J. S. Colton *et al.*, Phys. Rev. B **69**, 121307 (2004).
[69] L. Lombez, P.-F. Braun, X. Marie, P. Renucci, B. Urbaszek, T. Amand, O. Krebs and P. Voisin, Phys. Rev. B **75**, 195314 (2007).
[70] A. Greilich *et al.*, Phys. Rev. Lett. **96**, 227401 (2006).
[71] A. Greilich *et al.*, Science **313**, 341 (2006).
[72] L.-M. Duan *et al.*, Nature **414**, 413(2001).
[73] G. Giedke *et al.*, Phys. Rev. A **74**, 32316 (2006).
[74] T. Yokoi, S. Adachi, H. Sasakura, S. Muto, H. Z. Song, T. Usuki, and S. Hirose, Phys. Rev. B **71**, R041307, (2005).
[75] B. Eble, O. Krebs, A. Lemaitre, K. Kowalik, A. Kudelski, P. Voisin, B. Urbaszek, X. Marie, and T. Amand, Phys. Rev. B **74**, R081306 (2006).
[76] C. W. Lai, P. Maletinsky, A. Badolato, and A. Imamoglu, Phys. Rev. Lett. **96**, 167403 (2006).
[77] I. V. Ignatiev, I. Ya. Gerlovin, S. Yu. Verbin, W. Maruyama, and Y. Masumoto, *13th International Symposium on Nanostructures: Physics and Technology* (2005), p. 47.
[78] B. Urbaszek, P.-F. Braun, X. Marie, O. Krebs, A. Lemaitre, P. Voisin and T. Amand, arXiv:0707.0370 (2007).

[79] I. Akimov, D. H. Feng, and F. Henneberger, Phys. Rev. Lett. **97**, 056602 (2006).
[80] A. Tartakovskii *et al.*, Phys. Rev. Lett. **98**, 026806 (2007).
[81] D. Gammon, A. L. Efros, T. A. Kennedy, M. Rosen, D. S. Katzer, D. Park, S. W. Brown, V. L. Korenev, and I. A. Merkulov, Phys. Rev. Lett. **86**, 5176 (2001).
[82] L. He, G. Bester, and A. Zunger, Phys. Rev. B **70**, 235316 (2004).
[83] C. Deng and X. Hu, Phys. Rev. B **71**, 033307 (2005).

Chapter 10

Electron and Hole Spin Dynamics and Decoherence in Quantum Dots

D. Klauser, D. V. Bulaev, W. A. Coish, and Daniel Loss

Department of Physics and Astronomy, University of Basel, Klingelbergstrasse 82, CH-4056 Basel, Switzerland

In this article we review our work on the dynamics and decoherence of electron and hole spins in single and double quantum dots. The first part, on electron spins, focuses on decoherence induced via the hyperfine interaction while the second part covers decoherence and relaxation of heavy-hole spins due to spin-orbit interaction as well as the manipulation of heavy-hole spin using electric dipole spin resonance.

Contents

10.1. Introduction

The Loss-DiVincenzo proposal [1] to use the spin of a single electron confined in a quantum dot as a qubit for quantum computation has triggered significant interest in the dynamics and control of single spins in quantum dots. This has led to numerous exciting experimental achievements, among them the realization of single electrons in single dots [2, 3] as well as double dots [4–6], the implementation of single-spin read out [7, 8], the demonstration of the $\sqrt{\mathrm{SWAP}}$ operation via pulsed exchange interaction [9] and the measurement of single-spin ESR [10]. For a detailed account of the progress in implementing the Loss-DiVincenzo proposal, see the extensive reviews in Refs. 11 and 12.

On the theoretical side, one focus was, and still is, the investigation of the decoherence induced by the nuclei in the host material via the hyperfine interaction. The first part of this review article is devoted to the discussion of the rich spin dynamics that results from the hyperfine interaction. We first give an introduction to hyperfine interaction in quantum dots (Sec. 10.2). Subsequently, we discuss dynamics under the influence of hyperfine interaction for the case of a single spin in a single dot (Sec. 10.3) and for a double dot with one electron in each dot (Sec. 10.4). To conclude the part about hyperfine interaction, we discuss the idea of narrowing the nuclear spin state in order to increase the spin coherence time (Sec. 10.5).

The second part of the article is devoted to the dynamics and the manipulation of heavy-hole spins in quantum dots. The motivation to study hole spins comes from the fact that the valence band has p-symmetry and thus the hyperfine interaction with lattice nuclei for holes is suppressed in comparison to that of the conduction band electrons. As a consequence, the main interest for hole spin dynamics is the relaxation and decoherence due to spin-orbit interaction and we discuss this in Sec. 10.6. The next task towards using hole spins as qubits for quantum computation is of course the coherent manipulation of single hole spins. A potentially powerful method to achieve coherent manipulation of spins is electric dipole spin resonance (EDSR). An analysis of EDSR for heavy holes in quantum dots will be presented in Sec. 10.7.

10.2. Hyperfine Interaction for Electrons in Quantum Dots

In this part of the article concerning electron spin decoherence we assume that the orbital level spacing is much larger than the typical energy scale of the hyperfine interaction. This is the case in typical lateral quantum dots containing single electrons and allows one to write an effective hyperfine Hamiltonian H_{hf} for a single electron confined to the quantum-dot orbital ground state ψ_0

$$H_{\mathrm{hf}} = \mathbf{h} \cdot \mathbf{S}, \;\; \mathbf{h} = A\nu \sum_k |\psi_0(\mathbf{r}_k)|^2 \mathbf{I}_k, \tag{10.1}$$

where $\mathbf{S}$ is the spin-1/2 operator for a single electron and $\mathbf{I}_k$ is the spin operator for the nuclear spin at lattice site k, while ν is the volume of the crystal unit cell and A is the hyperfine coupling strength. For GaAs, which is mostly used for the fabrication of lateral dots, the average hyperfine coupling strength weighted by the natural abundance of each isotope is $A \approx 90\mu eV$ [13]. In Fig. 10.1 the hyperfine coupling of the electron spin in a lateral double quantum dot is illustrated. The electron spin dynamics

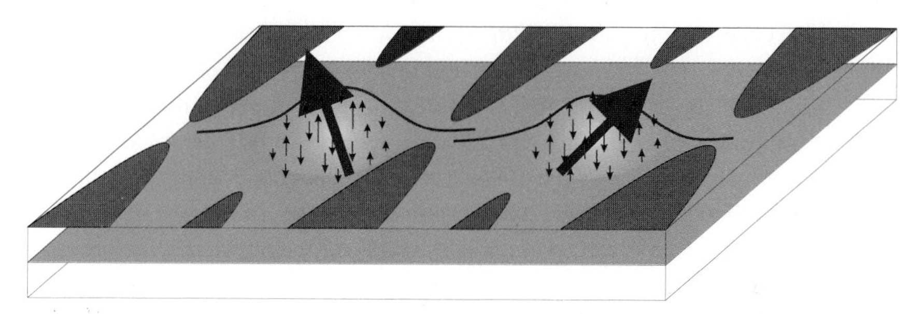

Fig. 10.1. A double quantum dot. Top-gates are set to a voltage configuration that confines the electrons in the two-dimensional electron gas (green) to quantum dots (yellow). The blue line indicates the envelope wave function of the electron (blue arrow). The hyperfine interaction with a particular nuclear spins (red arrows) is proportional to the envelope wave function squared at the position of the nuclei. Thus the nuclear spins in the center are drawn bigger since they couple stronger to the electron spin.

under $H_{\rm hf}$ have been studied under various approximations and in different parameter regimes. For an extensive overview, see reviews in Refs. 11, 14 and 15. Here, we briefly mention parts of this study before we focus on a few cases of special interest. The first analysis of electron spin dynamics under $H_{\rm hf}$ in this context showed that the long-time longitudinal spin-flip probability is $\sim 1/p^2N$ [16], i.e., this probability is suppressed in the limit of large nuclear spin polarization p and large number N of nuclear spins in the dot. An exact solution for the case of full polarization ($p = 1$) gives, for both transverse and longitudinal electron spin components, a long-time power law decay $\sim 1/t^{3/2}$ by a fraction $\sim 1/N$ on a timescale of $\tau \sim \hbar N/A \sim 1\mu s$ (for a GaAs dot with $N \sim 10^5$) [17]. The fact that this exact solution shows a non-exponential decay demonstrates the non-Markovian behavior of the nuclear spin bath. For non-fully polarized systems $p < 1$ and in the limit of large magnetic fields (or high polarization $p \gg 1/\sqrt{N}$), the transverse electron spin undergoes a Gaussian decay [17, 18] on a timescale $\tau \sim \hbar\sqrt{N}/A\sqrt{1-p^2}$ ($\tau \sim 5ns$ for GaAs with $p \ll 1$ and $N \sim 10^5$) [19]. This Gaussian decay can be reversed using a spin-echo sequence or by preparing the nuclear spin system in an eigenstate of h_z [19].

Several methods to prepare the nuclear spin system have recently been suggested [20–22] and we discuss one of these methods [21] in Sec. 10.5. Once the nuclear spin system is prepared in an eigenstate, the electron spin coherence is on one hand limited by dynamics in the nuclear spin system driven by the dipole-dipole interaction for which a worst case estimate [21] gives $\sim 100\mu s$ and on the other hand, even for an eigenstate of h_z, there is decoherence due to the flip-flop dynamics which can be important at times $\sim \hbar N/A \sim 1\mu s$ (or less, depending on the size of the electron Zeeman splitting). For the decay of nuclear spin polarization experiments suggest timescales up to tens of seconds [23–26] and hysteretic behavior of the nuclear spin polarization with respect to the external magnetic field has been observed [27]. Further, measurements of the transport current in the so-called spin-blockade regime [28] revealed hysteretic behavior with respect to the magnetic field [23, 29] and bistable behavior in time [23], which is attributed to a bistability in the nuclear spin polarization. Further, very recent experiments suggest a strong dependence of the nuclear field correlation time depending on whether an electron is present in the dot or not and thus hyperfine mediated nuclear spin flips are a possible mechanism for nuclear spin diffusion [24]. This last mechanism has been estimated to lead to fluctuations of nuclear spin polarization on a timescale of $\sim 100\mu s$ [21].

10.3. Single-Electron Spin Decoherence

In this section we look in more detail at hyperfine-induced decoherence for a single spin in a quantum dot in the regime of large Zeeman splitting $\epsilon_z = g\mu_B b_z$ (due to an externally applied magnetic field b_z). If ϵ_z is much larger than $\sigma = [\mathrm{Tr}\{\rho_I(h_z - h_0)^2\}]^{1/2}$, with $h_0 = \mathrm{Tr}\{\rho_I h_z\}$ (where ρ_I is the density matrix of the nuclear spin system), we may neglect the transverse term $S_\perp \cdot h_\perp$ and find that the Hamiltonian is simply

$$H_0 = (\epsilon_z + h_z)S_z. \tag{10.2}$$

This Hamiltonian just induces precession around the z-axis with a frequency that is determined by the eigenvalue h_z^n of h_z, where $h_z|n\rangle = h_z^n|n\rangle$. For a large but finite number of nuclear spins ($N \sim 10^5$ for lateral GaAs dots) the eigenvalues h_z^n are Gaussian distributed (due to the central limit theorem) with mean h_0 and variance $\sigma \approx \hbar A/\sqrt{N}$ [19]. Calculating the dynamics under H_0 (which is valid up to a timescale of $\sim \epsilon_z/\sigma^2 \sim 1\mu s$, where the transverse terms become relevant) leads to a Gaussian decay of the transverse electron spin state $|+\rangle = (|\uparrow\rangle + |\downarrow\rangle)/\sqrt{2}$ [19]:

$$\begin{aligned} C^0_{++}(t) &= \frac{1}{\sqrt{2\pi}\sigma}\int_{-\infty}^{\infty} dh_z^n e^{\left(-\frac{(h_z^n - h_0)^2}{2\sigma^2}\right)} |\langle n| \otimes \langle +|e^{(-iH_0 t)}|+\rangle \otimes |n\rangle|^2 \\ &= \frac{1}{2} + \frac{1}{2}e^{\left(-\frac{t^2}{2\tau^2}\right)} \cos\left[(\epsilon_z + h_0)t\right]; \quad \tau = \frac{1}{\sigma} = \sqrt{\frac{N}{1-p^2}\frac{2\hbar}{A}}. \end{aligned} \tag{10.3}$$

Here again, p denotes the polarization and for an unpolarized GaAs quantum dot with $N \sim 10^5$ we find $\tau \sim 5ns$. Applying an additional ac driving field with amplitude b along the x-direction leads to single-spin ESR. Assuming again that $\epsilon_z \gg \sigma$, we have the Hamiltonian

$$H_{\mathrm{ESR}} = H_0 + b\cos(\omega t)S_x. \tag{10.4}$$

In a rotating-wave approximation (which is valid for $(b/\epsilon_z)^2 \ll 1$) the decay of the driven Rabi oscillations is given by [30]

$$C^{\mathrm{ESR}}_{\uparrow\uparrow}(t) \sim 1 - C + \sqrt{\frac{b}{8\sigma^2 t}}\cos\left(\frac{b}{2}t + \frac{\pi}{4}\right) + \mathcal{O}\left(\frac{1}{t^{3/2}}\right), \tag{10.5}$$

for $t \gtrsim \max\left(1/\sigma, 1/b, b/2\sigma^2\right)$ and $\epsilon_z + h_0 - \omega = 0$. Here, $C^{\mathrm{ESR}}_{\uparrow\uparrow}(t)$ is defined in the same way as $C^0_{++}(t)$ in Eq. (10.3). The time-independent constant is given by $C = \exp(b^2/8\sigma^2)\mathrm{erfc}(b/\sqrt{8}\sigma)\sqrt{2\pi}b/8\sigma$. The two interesting features of the decay are the slow ($\sim 1/\sqrt{t}$) power law and the universal phase shift of $\pi/4$. The fact that the power law already becomes valid after a short time $\tau \sim 15ns$ (for $b \approx \sigma$) preserves the coherence over a long time, which makes the Rabi oscillations visible even when the Rabi period is much longer than the timescale $\tau \sim 15ns$ for transverse spin decay. Both the universal phase shift and the non-exponential decay have recently been observed in experiment [30]. In order to take corrections due to the transverse terms $S_\perp \cdot h_\perp$ into account, a more elaborate calculation is required. The Hamiltonian with flip-flop terms (but without a driving field) takes the form

$$H_{\mathrm{ff}} = H_0 + \frac{1}{2}(S_+h_- + S_-h_+). \tag{10.6}$$

In Ref. 19 a systematic calculation taking into account these so-called flip-flop terms was performed using a generalized master equation, valid in the limit of large magnetic field or large polarization. This calculation shows that even for an eigenstate of h_z, for which the Gaussian decay in Eq. (10.3) vanishes, the electron spin undergoes nontrivial non-Markovian decay on a timescale $\hbar N/A \sim 10\mu s$.

Other calculations [31–33] give microsecond timescales for the electron spin decoherence due to electron-nuclear spin flip-flops processes. The results in Ref. 31 suggest that also the decoherence due to dynamics in the nuclear spin system via electron mediated nuclear dipole-dipole interaction is suppressed by a spin echo and thus that the spin-echo decay time may be considerably different from the (not ensemble averaged) free-induction decay.

10.4. Singlet-Triplet Decoherence in a Double Quantum Dot

We now move on to discuss hyperfine induced decoherence in a double quantum dot. The effective Hamiltonian in the subspace of one electron on each dot is best written in terms of the sum and difference of electron spin and collective nuclear spin operators: $\mathbf{S} = \mathbf{S}_1 + \mathbf{S}_2, \delta\mathbf{S} = \mathbf{S}_1 - \mathbf{S}_2$ and $\mathbf{h} = \frac{1}{2}(\mathbf{h}_1 + \mathbf{h}_2), \delta\mathbf{h} = \frac{1}{2}(\mathbf{h}_1 - \mathbf{h}_2)$:

$$H_{\mathrm{dd}}(t) = \epsilon_z S_z + \mathbf{h}\cdot\mathbf{S} + \delta\mathbf{h}\cdot\delta\mathbf{S} + \frac{J}{2}\mathbf{S}\cdot\mathbf{S} - J. \tag{10.7}$$

Here, J is the Heisenberg exchange coupling between the two electron spins. Similar to the single-dot case, we assume that the Zeeman splitting is much larger than $\langle\delta\mathbf{h}\rangle_{\mathrm{rms}}$ and $\langle\mathbf{h}_i\rangle_{\mathrm{rms}}$, where $\langle\mathcal{O}\rangle_{\mathrm{rms}} = [\mathrm{Tr}\{\rho_I(\mathcal{O} - \langle\mathcal{O}\rangle)^2\}]^{1/2}$ is the root-mean-square expectation value of the operator $\mathcal{O}$ with respect to the nuclear spin state ρ_I. Under these conditions the relevant spin Hamiltonian becomes block diagonal with blocks labeled by the total electron spin projection along the magnetic field S_z. In the subspace of $S_z = 0$ (singlet $|S\rangle$, and triplet $|T_0\rangle$) the Hamiltonian can be written as [21, 34]

$$H_{\mathrm{sz0}}(t) = \frac{J}{2}\mathbf{S}\cdot\mathbf{S} + (\delta h_z + \delta b_z)\delta S_z \tag{10.8}$$

Here, δb_z is the inhomogeneity of the externally applied classical static magnetic field with $\delta b_z \ll \epsilon_z$, while the nuclear difference field δh_z is Gaussian distributed, as was h_z in the single dot case. A full account of the rich pseudo-spin dynamics under $H_{\mathrm{sz0}}(t)$ can be found in Refs. 21 and 34. Here we only discuss the most prominent features for $C_{SS}^{\mathrm{sz0}}(t)$, which gives the probability to find the singlet $|S\rangle$, if the system was initialized to $|S\rangle$ at $t = 0$. The parameters that determine the dynamics are the exchange coupling J, the expectation value of the total difference field $x_0 = \delta b_z + \delta h_0$ and the width of the difference field σ_δ (with $\delta h_0 = \langle\psi_{\mathrm{I}}|\delta h_z|\psi_I\rangle$ and $\sigma_\delta = \langle\psi_{\mathrm{I}}|(\delta h_z - \delta h_0)^2|\psi_I\rangle^{1/2}$). For the asymptotics one finds that the singlet probability does not decay to zero, but goes to a finite, parameter-dependent value [34]. In the case of strong exchange coupling $|J| \gg \max(|x_0|, \sigma_\delta)$ the singlet only decays by a small fraction quadratic in σ_δ/J or x_0/J:

$$C_{SS}^{\mathrm{sz}0}(t \to \infty) \sim \begin{cases} 1 - 2\left(\frac{\sigma_\delta}{J}\right)^2, & |J| \gg \sigma_\delta \gg |x_0|, \\ 1 - 2\left(\frac{x_0}{J}\right)^2, & |J| \gg |x_0| \gg \sigma_\delta. \end{cases} \quad (10.9)$$

At short times $C_{SS}^{\mathrm{sz}0}(t)$ undergoes a Gaussian decay on a timescale $\sqrt{J^2 + 4x_0^2}/4|x_0|\sigma_\delta$ while at long times $t \gg |J|/4\sigma_\delta^2$ we have a power law decay

$$C_{SS}^{\mathrm{sz}0}(t) \sim C_{SS}^{\mathrm{sz}0}(t \to \infty) + e^{-\frac{x_0^2}{2\sigma_\delta^2}} \frac{\cos(|J|t + \frac{3\pi}{4})}{4\sigma_\delta\sqrt{|J|}\, t^{\frac{3}{2}}}. \quad (10.10)$$

As in the case of single-spin ESR, we again have a power-law decay, now with $1/t^{3/2}$ and a universal phase shift, in this case: $3\pi/4$. Measurements [35] of the correlator $C_{SS}^{\mathrm{sz}0}(t)$ confirmed the parameter dependence of the saturation value and were consistent with the theoretical predictions concerning the decay. Using the same methods, one may also look at transverse correlators in the $S_z = 0$ subspace and find again power-law decays and a universal phase shift, albeit, with different decay power and different value of the universal phase shift [21]. Looking at the short-time behavior of the transverse correlators also allows one to analyze the fidelity of the $\sqrt{\mathrm{SWAP}}$ gate [21].

10.5. Nuclear Spin State Narrowing

The idea to prepare the nuclear spin system in order to prolong the electron spin coherence was put forward in Ref. 34. Specific methods for nuclear spin state narrowing have been described in Ref. 21 in the context of a double dot with oscillating exchange interaction, in Ref. 22 for phase-estimation of a single (undriven) spin in a single dot and in an optical setup in Ref. 20. Here, we discuss narrowing for the case of a driven single spin in a single dot, for which the details are very similar to the treatment in Ref. 21. The general idea behind state narrowing is that the evolution of the electron spin system depends on the value of the nuclear field since the effective Zeeman splitting is given by $\epsilon_z + h_z^n$. This leads to a nuclear field dependent resonance condition $\epsilon_z + h_z^n - \omega = 0$ for ESR and thus measuring the evolution of the electron spin system determines h_n^z and thus the nuclear spin state.

We start from the Hamiltonian for single-spin ESR as given in Eq. (10.4). The electron spin is initialized to the $|\uparrow\rangle$ state at time $t = 0$ and evolves under H_{esr} up to a measurement performed at time t_m. The probability to find $|\downarrow\rangle$ for a given eigenvalue h_z^n of the nuclear field operator ($h_z|n\rangle = h_z^n|n\rangle$) is then given by

$$P_\downarrow^n(t) = \frac{1}{2}\frac{b^2}{b^2+4\delta_n^2}\left[1-\cos\left(\frac{t}{2}\sqrt{b^2+4\delta_n^2}\right)\right] \tag{10.11}$$

where $\delta_n = \epsilon_z + h_z^n - \omega$ and b is the amplitude of the driving field. As mentioned above, in equilibrium we have a Gaussian distribution for the eigenvalues h_z^n, i.e., for the diagonal elements of the nuclear spin density matrix $\rho_I(h_z^n, 0) = \langle n|\rho_I|n\rangle = \exp\left(-(h_z^n - h_0)^2/2\sigma^2\right)/\sqrt{2\pi}\sigma$. Thus, averaged over the nuclear distribution we have the probability $P_\downarrow(t)$ to find the state $|\downarrow\rangle$, i.e., $P_\downarrow(t) = \int dh_z^n \rho_I(h_z^n, 0)P_\downarrow^n(t)$. After one measurement with outcome $|\downarrow\rangle$, we thus find for the diagonal of the nuclear spin density matrix [36]

$$\rho_I(h_z^n, 0) \xrightarrow{|\downarrow\rangle} \rho_I^{(1,\downarrow)}(h_z^n, t_m) = \rho_I(h_z^n, 0)\frac{P_\downarrow^n(t_m)}{P_\downarrow(t_m)}. \tag{10.12}$$

Assuming now that the measurement is performed in such a way that it gives the time averaged value (i.e., with a time resolution less than $1/b$) we have for the probability $P_\downarrow^n$ of measurement result $|\downarrow\rangle$ as a function of the nuclear field eigenvalue $P_\downarrow^n = \frac{1}{2}\frac{b^2}{b^2+4\delta_n^2}$. Thus, by performing a measurement on the electron spin (with outcome $|\downarrow\rangle$), the nuclear-spin density matrix is multiplied by a Lorentzian with width b centered around the h_z^n that satisfies the resonance condition $\epsilon_z + h_z^n - \omega = 0$. This results in a narrowed nuclear spin distribution, and thus an extension of the electron spin coherence, if $b < \sigma$. In the case of measurement outcome $|\uparrow\rangle$ we find

$$\rho_I(h_z^n, 0) \xrightarrow{|\uparrow\rangle} \rho_I^{(1,\uparrow)}(h_z^n, t_m) = \rho_I(h_z^n, 0)\frac{1-P_\downarrow^n(t_m)}{1-P_\downarrow(t_m)}, \tag{10.13}$$

i.e., the Gaussian nuclear spin distribution is multiplied by one minus a Lorentzian, thus reducing the probability for the nuclear field to have a value matching the resonance condition $\epsilon_z + h_z^n - \omega = 0$. Due to the slow dynamics of the nuclear spin system (see discussion at the end of Sec. 10.2), many such measurements of the electron spin are possible (with re-initialization of the electron spin between measurements). Under the assumption of a static nuclear field during M such initialization and measurement cycles we find

$$\rho_I(h_z^n, 0) \longrightarrow \rho^{(M,\alpha_\downarrow)}(h_z^n) = \frac{1}{N}\rho_I(h_z^n, 0)(P_\downarrow^n)^{\alpha_\downarrow}(1 - P_\downarrow^n)^{M-\alpha_\downarrow}, \quad (10.14)$$

where $\alpha_\downarrow$ is the number of times the measurement outcome was $|\downarrow\rangle$. The simplest way to narrow is to perform single measurements with $b \ll \sigma$. If the outcome is $|\downarrow\rangle$, narrowing has been achieved. Otherwise, the nuclear system should be allowed to re-equilibrate before the next measurement [37]. In order to achieve a systematic narrowing, one can envision adapting the driving frequency (and thus the resonance condition) depending on the outcome of the previous measurements. Such an adaptive scheme is described in detail in Refs. 20 and 21. With this we conclude the part on hyperfine-induced decoherence of electron spins in quantum dots and move on to the heavy holes.

10.6. Spin Decoherence and Relaxation for Heavy Holes

Now we consider the spin coherence of heavy holes in quantum dots. The contact hyperfine interaction between lattice nuclei and heavy-hole spin is much weaker than that for electrons, since the valence band has p symmetry. Thus (neglecting *sp* hybridization) only the weaker anisotropic hyperfine interaction is present. Therefore, the decoherence due to hyperfine interaction is suppressed for heavy holes and in this section we focus only on the spin decoherence due to spin-orbit interaction induced by heavy-hole - phonon coupling.

From the two-band Kane model, the Hamiltonian for the valence band of III–V semiconductors is given by

$$H_{\text{bulk}} = H_{\text{LK}} + \eta \mathbf{J} \cdot \mathbf{\Omega} + H_{\text{Z}}, \quad (10.15)$$

where H_{LK} is the Luttinger-Kohn Hamiltonian [38]. The second term is the Dresselhaus spin-orbit coupling (due to bulk inversion asymmetry) for the valence band [39, 40], $\mathbf{J} = (J_x, J_y, J_z)$ are 4×4 matrices corresponding to spin 3/2, $\Omega_z = P_z(P_x^2 - P_y^2)$, and Ω_x, Ω_y are given by cyclic permutations. The last term in Eq. (10.15) $H_{\text{Z}} = -2\kappa\mu_B\mathbf{B}\cdot\mathbf{J} - 2q\mu_B\mathbf{B}\cdot\mathcal{J}$ is the Zeeman term for the valence band [41] (κ and q are the Luttinger parameters [41] and $\mathcal{J} = (J_x^3, J_y^3, J_z^3)$).

We consider a [001]-grown two-dimensional system. In the case of an asymmetric quantum well, due to structure inversion asymmetry along the growth direction, there is an additional spin-orbit term, the Bychkov-Rashba spin-orbit term, which, in the two-band model is given by [42]

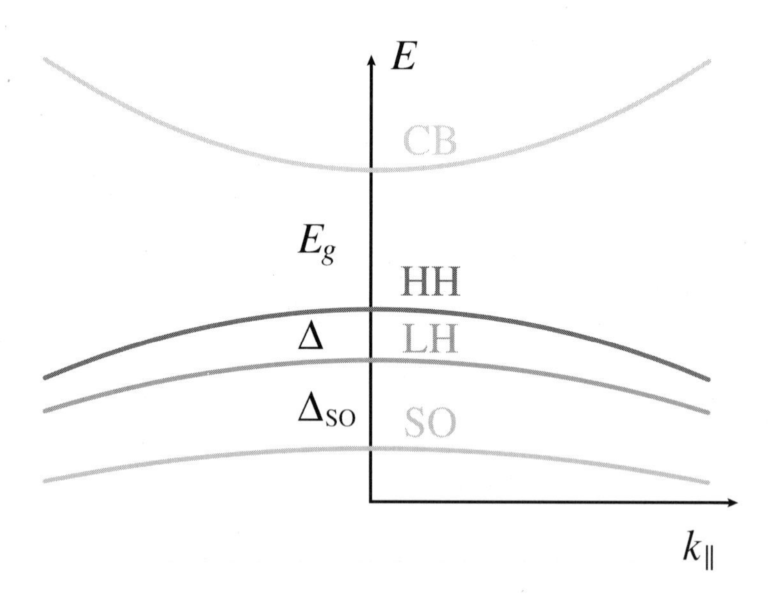

Fig. 10.2. Band structure of a III-V semiconductor quantum well with the [001]-growth direction, where E_g is the band gap, Δ is the splitting between the light- and heavy-hole subbands due to quantum-well confinement, and Δ_{SO} is the splitting of the valence band due to spin-orbit interaction.

$\alpha_R \mathbf{P} \times \mathbf{E} \cdot \mathbf{J}$, where α_R is the Bychkov-Rashba spin-orbit coupling constant and $\mathbf{E}$ is an effective electric field along the growth direction. Due to confinement along the growth direction, the valence band splits into a heavy-hole subband with $J_z = \pm 3/2$ and a light-hole subband with $J_z = \pm 1/2$ (see Fig. 10.2 and Ref. 40). If the splitting Δ of heavy-hole and light-hole subbands is large, we describe the properties of heavy-holes and light-holes separately, using only the 2×2 submatrices for the $J_z = \pm 3/2$ and $J_z = \pm 1/2$ states, respectively. The heavy-hole submatrices have the property that $\tilde{J}_x = \tilde{J}_y = 0$ and $\tilde{J}_z = \frac{3}{2}\sigma_z$. For such a system and low temperatures, only the lowest heavy-hole subband is significantly occupied. In this case, we consider heavy holes only. In the framework of perturbation theory [39], using Eq. (10.15) and taking into account the Zeeman energy and the Bychkov-Rashba spin-orbit coupling term, the effective Hamiltonian for heavy holes of a quantum dot with lateral confinement potential $U(x, y)$ is given by

$$H = \frac{1}{2m}(P_x^2 + P_y^2) + U(x, y) + H_{\text{SO}}^{\text{HH}} - \frac{1}{2} g_\perp \mu_B B_\perp \sigma_z, \tag{10.16}$$

where m is the effective heavy-hole mass, $\mathbf{P} = \mathbf{p} + |e|\mathbf{A}(\mathbf{r})/c$, $\mathbf{A}(\mathbf{r}) = (-yB_\perp/2, xB_\perp/2, yB_x - xB_y)$, $g_\perp$ is the component of the g-factor tensor along the growth direction, and

$$H_{\rm SO}^{\rm HH} = i\alpha P_-^3\sigma_+ + \beta P_- P_+ P_- \sigma_+ + \gamma B_- P_-^2 \sigma_+ + {\rm H.c.} \qquad (10.17)$$

is the spin-orbit coupling of heavy holes consisting of three contributions: the Dresselhaus term (β) [40], the Rashba term (α) [43], and the last term (γ) combines two effects: orbital coupling via non-diagonal elements in the Luttinger-Kohn Hamiltonian ($\propto P_\pm^2$) and magnetic coupling via non-diagonal elements in the Zeeman term ($\propto B_\pm$). This latter term represents a new type of spin-orbit interaction which is unique for heavy holes [44]. Here, $\alpha = 3\gamma_0\alpha_R\langle E_z\rangle/2m_0\Delta$, $\beta = -3\gamma_0\eta\langle P_z^2\rangle/2m_0\Delta$, $\gamma = 3\gamma_0\kappa\mu_B/m_0\Delta$, $\sigma_\pm = (\sigma_x \pm i\sigma_y)/2$, $P_\pm = P_x \pm iP_y$, $B_\pm = B_x \pm iB_y$, m_0 is the free electron mass, γ_0 is the Luttinger parameter [41], $\langle E_z\rangle$ is the averaged effective electric field along the growth direction of a quantum dot, and Δ is the splitting of light-hole and heavy-hole subbands. The splitting between heavy-hole and light-hole subbands $\Delta \sim h^{-2}$, where h is the quantum-dot height.

The spectrum of (10.16) for parabolic lateral confinement $[U(x,y) = m\omega_0^2(x^2+y^2)/2]$ and for vanishing spin-orbit interaction ($H_{\rm SO}^{\rm HH} = 0$) is the Fock-Darwin spectrum split by the Zeeman term [45, 46]. From Eq. (10.17), it can be seen that $H_{\rm SO}^{\rm HH}$ leads to coupling of the two lowest states $|0, \pm 3/2\rangle$ to the states with the opposite spin orientations and different orbital momenta $|l, \mp 3/2\rangle$. Note that the three spin-orbit terms in Eq. (10.17) differ by symmetry in momentum space and hence mix different states resulting in avoided crossings of the energy levels (see inset of Fig. 10.3). Due to this spin-orbit mixing of the heavy-hole states, the transitions between the states $|0, \pm 3/2\rangle$ with emission or absorption of an acoustic phonon become possible and this is the main source of spin relaxation and decoherence for heavy-holes [40].

We consider a single-particle quantum dot, in which a heavy hole can occupy one of the low-lying levels. In the following, we study the relaxation of an n-level system, the first $n-1$ levels have the same spin and the n-th level has the opposite spin orientation. In the framework of Bloch-Redfield theory [47], the Bloch equations for heavy-hole spin motion for such a system in the interaction representation are given by

$$\langle \dot{S}_z\rangle = \left(S_T - \langle S_z\rangle\right)/T_1 - R(t), \qquad (10.18)$$

$$\langle \dot{S}_x\rangle = -\langle S_x\rangle/T_2,\ \langle \dot{S}_y\rangle = -\langle S_y\rangle/T_2, \qquad (10.19)$$

where $R(t) = W_{n1}\rho_{nn}(t) + \sum_{i=1}^{n-1} W_{ni}\rho_{ii}(t)$, $\rho(t)$ is the density matrix, W_{ij} is the transition rate from state j to state i, S_T is a constant (which has the value of $\langle S_z \rangle$ in thermodynamic equilibrium if $R(t) = 0$),

$$\frac{1}{T_1} = W_{n1} + \sum_{i=1}^{n-1} W_{in}, \quad \frac{1}{T_2} = \frac{1}{2T_1} + \frac{1}{2}\sum_{i=2}^{n-1} W_{i1}, \tag{10.20}$$

where pure dephasing (due to fluctuations along z direction) is absent in the spin decoherence time T_2 since the spectral function is superohmic. As can be seen from Eq. (10.18), the spin motion has a complex dependence on the density matrix and, in the general case, there are $n-1$ spin relaxation rates. However, in the case of low temperatures ($\hbar q s_\alpha \gg k_B T$), when phonon absorption becomes strongly suppressed, solving the master equation, we find that $R(t) \approx 0$. Therefore, there is only one spin relaxation time T_1. In this limit, the last sum in Eq. (10.20) is negligible and the spin decoherence time saturates, i.e., $T_2 = 2T_1$.

Note that in contrast to electrons [48] there are no interference effects between different spin-orbit coupling terms, thus the total spin relaxation rate $1/T_1$ is the sum of rates $1/T_1 = 1/T_1^{\mathrm{D}} + 1/T_1^{\mathrm{BR}} + 1/T_1^{\parallel}$ [40, 44]:

$$\begin{aligned}
\frac{1}{T_1^{\mathrm{BR}}} &\propto \alpha^2 \omega_Z^7 \left(\frac{\omega_+^3}{3\omega_+ + \omega_Z} - \frac{\omega_-^3}{3\omega_- - \omega_Z} \right)^2, \\
\frac{1}{T_1^{\mathrm{D}}} &\propto \beta^2 \omega_Z^3 \left(\frac{\omega_+}{\omega_+ + \omega_Z} - \frac{\omega_-}{\omega_- - \omega_Z} \right)^2, \\
\frac{1}{T_1^{\parallel}} &\propto \gamma^2 B_\parallel^2 \omega_Z^5 \left(\frac{\omega_+^2}{2\omega_+ + \omega_Z} + \frac{\omega_-^2}{2\omega_- - \omega_Z} \right)^2,
\end{aligned} \tag{10.21}$$

where $\omega_\pm = \sqrt{\omega_0^2 + \omega_c^2/4} \pm \omega_c/2$, $\omega_Z = g_\perp \mu_B B_\perp/\hbar$, $B_\parallel = \sqrt{B_x^2 + B_y^2}$. In Fig. 10.3 the total spin relaxation rate $1/T_1$ is plotted as a function of perpendicular magnetic field $B_\perp$. There are three peaks in the relaxation rate curve at $\omega_Z = \omega_-$, $2\omega_-$, and $3\omega_-$, which are caused by strong spin mixing at the anticrossing points. In the inset, the first (third) avoided crossing resulting from Dresselhaus (Rashba) spin-orbit coupling corresponds to the first (third) peak of the spin relaxation curve in Fig. 10.3. At non-zero in-plane magnetic fields ($B_\parallel$), there is an additional peak which is due to an anticrossing between the energy levels $E_{0,+3/2}$ and $E_{2,-3/2}$ (see the second avoided crossing in the inset). Note that the spin relaxation rate for heavy holes is comparable to that for electrons [40, 50] due to the fact that

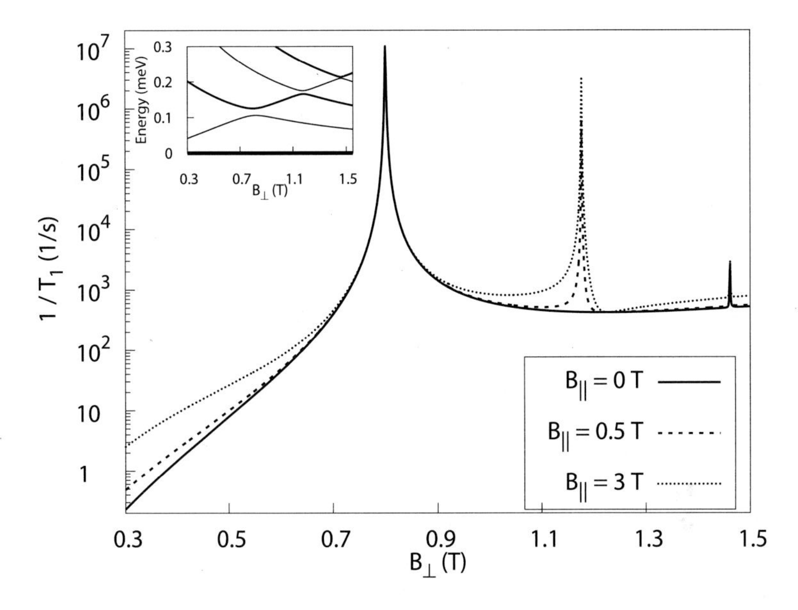

Fig. 10.3. Heavy hole spin relaxation rate $1/T_1$ in a GaAs quantum dot versus an applied perpendicular magnetic field $B_\perp$ (the height of the quantum dot is chosen to be $h = 5$ nm, the lateral size $l_0 = \sqrt{\hbar/m\omega_0} = 40$ nm, $\kappa = 1.2$, $\gamma_0 = 2.5$, $g_\perp = 2.5$ [49], and the other parameters are given in Ref. [40]). Inset: Energy differences of lowest excited levels with respect to the ground state $E_{0,+3/2}$. The second avoided crossing comes from the spin-orbit interaction and the in-plane magnetic field $B_\parallel$ (3rd term in Eq. 10.17). The anticrossing gap is proportional to $B_\parallel$, implying that the coupling between corresponding states can be controlled externally.

spin-orbit coupling of heavy holes is strongly suppressed for flat quantum dots (see Eq. (10.17)), as confirmed by a recent experiment [51].

10.7. Electric Dipole Spin Resonance for Heavy Holes

Let us now consider methods for the manipulation and detection of the heavy-hole spin in quantum dots. For electrons in two-dimensional structures, an applied oscillating in-plane magnetic field couples spin-up and spin-down states via magnetic-dipole transitions and is commonly used in electron spin resonance, Rabi oscillation, and spin echo experiments [10]. It can be shown that magnetic-dipole transitions ($\Delta n = 0$, $\Delta\ell = 0$, and $\Delta s = \pm 1$) are forbidden and, due to spin-orbit mixing of the states $|0, \pm 3/2\rangle$ with $i\beta_1^{\pm}|1, \mp 3/2\rangle$, electric-dipole transitions ($\Delta n = \pm 1$, $\Delta\ell = \pm 1$, and $\Delta s = 0$) are most likely to occur. Therefore, the heavy holes are affected by the oscillating electric field component and not by the magnetic one.

We consider a circularly polarized electric field rotating in the XY-plane with frequency ω: $\mathbf{E}(t) = E(\sin\omega t, -\cos\omega t, 0)$. Therefore, the interaction of heavy holes with the electric field is described by the Hamiltonian $H^E(t) = (|e|E/m\omega)(\cos\omega t P_x + \sin\omega t P_y)$. The coupling between the states $|\pm\rangle$ is given by $\langle +|H^E(t)|-\rangle = H^E_{+-} = \left(H^E_{-+}\right)^* = d_{\rm SO}Ee^{-i\omega t}$, where

$$d_{\rm SO} = (|e|l/2\omega)(\beta_1^-\omega_- + \beta_1^+\omega_+) \qquad (10.22)$$

is an effective dipole moment of a heavy hole depending on Dresselhaus spin-orbit coupling constants, perpendicular magnetic field $B_\perp$, lateral size of a quantum dot, and frequency ω of an rf electric field.

In the framework of the Bloch-Redfield theory [47] (taking into account also off-diagonal matrix elements), the effective master equation for the density matrix ρ_{nm} assumes the form of Bloch equations [47], with the detuning of the rf field given by $\delta_{\rm rf} = \omega_{\rm Z} - \omega$. $2d_{\rm SO}E/\hbar$ is the Larmor frequency, $T_1 = 1/(W_{+-} + W_{-+})$ the spin relaxation time (W_{nm} is the transition rate from state m to state n), $T_2 = 2T_1$ [40] the spin decoherence time, and $\rho_z^T = (W_{+-} - W_{-+})T_1$ the equilibrium value of ρ_z without rf field.

The coupling energy between a heavy hole and an oscillating electric field is given by

$$\langle H^E(t)\rangle = {\rm Tr}(\rho H^E(t)) = -\mathbf{d}_{\rm SO}\cdot\mathbf{E}(t), \qquad (10.23)$$

where $\mathbf{d}_{\rm SO} = d_{\rm SO}(i\rho_{-+} - i\rho_{+-}, \rho_{+-} + \rho_{-+}, 0)$ is the dipole moment of a heavy hole. Therefore, the rf power $P = -d\langle H^E(t)\rangle/dt = -\omega d_{\rm SO}E\rho_-$ absorbed by a heavy-hole spin system in the stationary state is given by [52]

$$P = \frac{2\omega(d_{\rm SO}E)^2T_2\rho_z^T/\hbar}{1 + \delta_{\rm rf}^2T_2^2 + (2d_{\rm SO}E/\hbar)^2T_1T_2}. \qquad (10.24)$$

In Fig. 10.4, the dependence of P on a perpendicular magnetic field $B_\perp$ and frequency ω of the oscillating electric field is plotted. The rf power P absorbed by the system has three resonances and one resonant dip. The first resonance appears when the energy of rf radiation equals the Zeeman energy of heavy holes: $B_\perp^{\rm r,1} = \hbar\omega/g_\perp\mu_{\rm B}$. The shape of this resonance (at certain ω) is given by $P \approx \hbar\omega\rho_z^T/2\hbar[1 + \hbar^2\delta_{\rm rf}^2T_2/(2d_{\rm SO}E)^2T_1]$.

If the first and second resonances are well separated ($\omega \ll \omega_-$), then the absorbed power can be estimated as

$$P \approx 2\omega(d_{\rm SO}E)^2\rho_z^T/\hbar\delta_{\rm rf}^2T_2 \qquad (10.25)$$

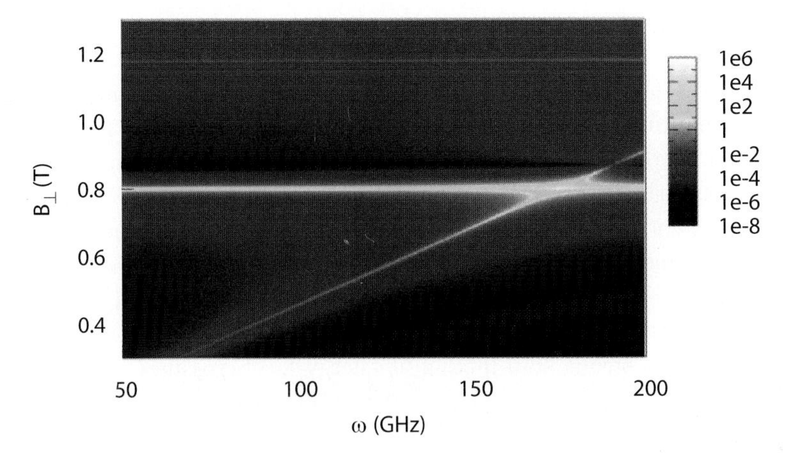

Fig. 10.4. Absorbed power P (meV/s) as a function of perpendicular magnetic field $B_\perp$ and rf frequency ω ($T_2 = 2T_1$, $E = 2.5$ V/cm, $B_\| = 1$ T, and the other parameters are the same as those in Fig. 10.3).

in the region of the second and third resonances and the resonant dip. The second resonance corresponds to an anticrossing of the levels $E_{0,-3/2}$ and $E_{1,3/2}$ (see the first avoided crossing in the inset of Fig. 10.3) at $\omega_- = \omega_Z$ [40] (at $B = B_\perp^{r,2}$). At the anticrossing point, there is strong mixing of the spin-up and spin-down states and the dipole moment of a heavy-hole spin system is maximal $d_{SO}^{max} = |e|l\omega_Z/2\omega$ and equals half of the lowest electric dipole moment of a quantum dot ($|e|l\omega_-/\omega$). Therefore, the height of the second resonance is given by $(el\omega_Z E)^2/2\hbar\omega\delta_{rf}^2 T_2$. The resonant dip appears at $B_\perp^d = (\hbar\omega_0/2g_\perp\mu_B)\sqrt{2m_0/g_\perp m}$, which corresponds to $\beta_1^-\omega_- + \beta_1^+\omega_+ = 0$ and to zero dipole moment (see Eq. (10.22)). The third resonance reflects the peak in the spin decoherence rate T_2^{-1} due to an applied in-plane magnetic field (see Fig. 10.3) at the second anticrossing point (the second avoided crossing in inset of Fig. 10.3) at $2\omega_- = \omega_Z$ ($B_\perp^{r,3} = 4\hbar\omega_0/g_\perp\mu_B\sqrt{1 + 4m_0/g_\perp m}$). From the positions of the resonances we can determine $g_\perp$, m, and ω_0, from the shape and the height of those we can extract information about the spin-orbit interaction constants α, β, and spin-orbit interaction strength due to in-plane magnetic field (which is proportional to $\gamma_0\kappa/\Delta$). Moreover, we can determine the dependence of the spin relaxation and decoherence times on $B_\perp$.

10.8. Conclusions

We have discussed the rich dynamics of single electron spins in single and double quantum dots due to hyperfine interaction with the nuclei. Key features are non-exponential decays of various kinds and a remarkable universal phase-shift. Further, we have studied spin decoherence and relaxation of heavy holes in quantum dots due to spin-orbit coupling. The spin relaxation time T_1 for heavy holes in flat quantum dots can be comparable to that for electrons [40, 50] as confirmed by experiment [51]. The spin decoherence time for heavy holes is given by $T_2 = 2T_1$ at low temperatures. There is strong spin mixing at energy-level crossings resulting in a non-monotonic dependence $T_1(B)$. We have proposed a new method for manipulation of a heavy-hole spin in a quantum dot via rf electric fields. This method can be used for detection of heavy-hole spin resonance signals, for spin manipulation, and for determining important parameters of heavy holes [44].

Aknowledgments

We aknowledge financial support from the Swiss NSF, the NCCR Nanoscience and JST ICORP.

References

[1] D. Loss and D. P. DiVincenzo, Quantum computation with quantum dots, *Phys. Rev. A.* **57**, 120–126, (1998).
[2] S. Tarucha, D. G. Austing, T. Honda, R. J. van der Hage, and L. P. Kouwenhoven, Shell filling and spin effects in a few electron quantum dot, *Phys. Rev. Lett.* **77**(17), 3613–3616, (1996).
[3] M. Ciorga, A. S. Sachrajda, P. Hawrylak, C. Gould, P. Zawadzki, S. Jullian, Y. Feng, and Z. Wasilewski, Addition spectrum of a lateral dot from coulomb and spin-blockade spectroscopy, *Phys. Rev. B.* **61**(24), R16315–R16318, (2000).
[4] J. M. Elzerman, R. Hanson, J. S. Greidanus, L. H. Willems van Beveren, S. De Franceschi, L. M. K. Vandersypen, S. Tarucha, and L. P. Kouwenhoven, Few-electron quantum dot circuit with integrated charge read out, *Phys. Rev. B.* **67**(16), 161308, (2003).
[5] T. Hayashi, T. Fujisawa, H. D. Cheong, Y. H. Jeong, and Y. Hirayama, Coherent manipulation of electronic states in a double quantum dot, *Phys. Rev. Lett.* **91**(22), 226804, (2003).
[6] J. R. Petta, A. C. Johnson, C. M. Marcus, M. P. Hanson, and A. C. Gossard,

Manipulation of a Single Charge in a Double Quantum Dot, *Phys. Rev. Lett.* **93**(18), 186802, (2004).

[7] J. M. Elzerman, R. Hanson, L. H. Willems van Beveren, B. Witkamp, L. M. K. Vandersypen, and L. P. Kouwenhoven, Single-shot read-out of an individual electron spin in a quantum dot, *Nature.* **430**, 431–435, (2004).

[8] R. Hanson, L. H. van Beveren, I. T. Vink, J. M. Elzerman, W. J. Naber, F. H. Koppens, L. P. Kouwenhoven, and L. M. Vandersypen, Single-Shot Readout of Electron Spin States in a Quantum Dot Using Spin-Dependent Tunnel Rates, *Phys. Rev. Lett.* **94**(19), 196802, (2005).

[9] J. R. Petta, A. C. Johnson, J. M. Taylor, E. A. Laird, A. Yacoby, M. D. Lukin, C. M. Marcus, M. P. Hanson, and A. C. Gossard, Coherent Manipulation of Coupled Electron Spins in Semiconductor Quantum Dots, *Science.* **309**(5744), 2180–2184, (2005).

[10] F. H. L. Koppens, C. Buizert, K. J. Tielrooij, I. T. Vink, K. C. Nowack, T. Meunier, L. P. Kouwenhoven, and L. M. K. Vandersypen, Driven coherent oscillations of a single electron spin in a quantum dot, *Nature.* **442**, 766–771, (2006).

[11] V. Cerletti, W. A. Coish, O. Gywat, and D. Loss, TUTORIAL: Recipes for spin-based quantum computing, *Nanotechnology.* **16**, 27, (2005).

[12] W. A. Coish and D. Loss, Quantum computing with spins in solids, *http://arXiv.org/cond-mat/0606550.* (2006).

[13] D. Paget, G. Lampel, B. Sapoval, and V. I. Safarov, Low field electron-nuclear spin coupling in gallium arsenide under optical pumping conditions, *Phys. Rev. B.* **15**, 5780–5796, (1977).

[14] J. Schliemann, A. Khaetskii, and D. Loss, TOPICAL REVIEW: Electron spin dynamics in quantum dots and related nanostructures due to hyperfine interaction with nuclei, *J. Phys.: Condens. Matter.* **15**, 1809 (Dec., 2003).

[15] J. M. Taylor, J. R. Petta, A. C. Johnson, A. Yacoby, C. M. Marcus, and M. D. Lukin, Relaxation, dephasing, and quantum control of electron spins in double quantum dots, *Phys. Rev. B.* **76**, 035315, (2007).

[16] G. Burkard, D. Loss, and D. P. DiVincenzo, Coupled quantum dots as quantum gates, *Phys. Rev. B.* **59**, 2070–2078, (1999).

[17] A. V. Khaetskii, D. Loss, and L. Glazman, Electron Spin Decoherence in Quantum Dots due to Interaction with Nuclei, *Phys. Rev. Lett.* **88**(18), 186802, (2002).

[18] I. A. Merkulov, A. L. Efros, and M. Rosen, Electron spin relaxation by nuclei in semiconductor quantum dots, *Phys. Rev. B.* **65**(20), 205309, (2002).

[19] W. A. Coish and D. Loss, Hyperfine interaction in a quantum dot: Non-Markovian electron spin dynamics, *Phys. Rev. B.* **70**(19), 195340, (2004).

[20] D. Stepanenko, G. Burkard, G. Giedke, and A. Imamoğlu, Enhancement of electron spin coherence by optical preparation of nuclear spins, *Phys. Rev. Lett.* **96**(13), 136401, (2006).

[21] D. Klauser, W. A. Coish, and D. Loss, Nuclear spin state narrowing via gate–controlled Rabi oscillations in a double quantum dot, *Phys. Rev. B.* **73** (20), 205302, (2006).

[22] G. Giedke, J. M. Taylor, D. D'Alessandro, M. D. Lukin, and A. Imamoğlu,

Quantum measurement of a mesoscopic spin ensemble, *Phys. Rev. A.* **74**(3), 032316, (2006).

[23] F. H. L. Koppens, J. A. Folk, J. M. Elzerman, R. Hanson, L. H. W. van Beveren, I. T. Vink, H. P. Tranitz, W. Wegscheider, L. P. Kouwenhoven, and L. M. K. Vandersypen, Control and Detection of Singlet-Triplet Mixing in a Random Nuclear Field, *Science.* **309**, 1346–1350, (2005).

[24] P. Maletinsky, A. Badolato, and A. Imamoğlu, Dynamics of quantum dot nuclear spin polarization controlled by a single electron, *Phys. Rev. Lett.* **99**, 056804, (2007).

[25] J. Baugh, Y. Kitamura, K. Ono, and S. Tarucha, Large nuclear overhauser fields detected in vertically-coupled double quantum dots, *Phys. Rev. Lett.* **99**, 096804, (2007).

[26] D. Reilly, J. M. Taylor, E. A. Laird, J. R. Petta, C. M. Macrus, M. P. Hanson, and A. C. Gossard, The dynamic nuclear environment in a double quantum dot, http://arxiv.org/abs/0712.4033.

[27] P. Maletinsky, C. W. Lai, A. Badolato, and A. Imamoğlu, Nonlinear dynamics of quantum dot nuclear spins, *Phys. Rev. B.* **75**, 035409, (2007).

[28] K. Ono, D. G. Austing, Y. Tokura, and S. Tarucha, Current Rectification by Pauli Exclusion in a Weakly Coupled Double Quantum Dot System, *Science.* **297**, 1313–1317, (2002).

[29] K. Ono and S. Tarucha, Nuclear-Spin-Induced Oscillatory Current in Spin-Blockaded Quantum Dots, *Phys. Rev. Lett.* **92**(25), 256803, (2004).

[30] F. H. L. Koppens, D. Klauser, W. A. Coish, K. C. Nowack, L. P. Kouwenhoven, D. Loss, and L. M. K. Vandersypen, Universal phase shift and non-exponential decay of driven single-spin oscillations, *Phys. Rev. Lett.* **99**, 106803, (2007).

[31] W. Yao, R.-B. Liu, and L. J. Sham, Theory of electron spin decoherence by interacting nuclear spins in a quantum dot, *Phys. Rev. B.* **74**(19), 195301, (2006).

[32] N. Shenvi, R. de Sousa, and K. B. Whaley, Nonperturbative bounds on electron spin coherence times induced by hyperfine interactions, *Phys. Rev. B.* **71**(14), 144419, (2005).

[33] C. Deng and X. Hu, Analytical solution of electron spin decoherence through hyperfine interaction in a quantum dot, *Phys. Rev. B.* **73**(24), 241303, (2006). Erratum: *Phys. Rev. B* **74**, 129902.

[34] W. A. Coish and D. Loss, Singlet-triplet decoherence due to nuclear spins in a double quantum dot, *Phys. Rev. B.* **72**(12), 125337, (2005).

[35] E. A. Laird, J. R. Petta, A. C. Johnson, C. M. Marcus, A. Yacoby, M. P. Hanson, and A. C. Gossard, Effect of exchange interaction on spin dephasing in a double quantum dot, *Phys. Rev. Lett.* **97**(5), 056801, (2006).

[36] A. Peres, *Quantum Theory: Concepts and Methods.* (Kluwer Academic Publishers, 1993).

[37] D. Klauser, W. A. Coish, and D. Loss, Quantum-dot spin qubit and hyperfine interaction, *Adv. Solid State Phys.* **47**, 17–29, (2007).

[38] J. M. Luttinger and W. Kohn, Motion of electrons and holes in perturbed periodic fields, *Phys. Rev.* **97**(4), 869–883, (1955).

[39] B. P. Zakharchenya and F. Mier, Eds., *Optical orientation.* (North-Holland, Amsterdam, 1984).
[40] D. V. Bulaev and D. Loss, Spin relaxation and decoherence of holes in quantum dots, *Phys. Rev. Lett.* **95**, 076805, (2005).
[41] J. M. Luttinger, Quantum Theory of Cyclotron Resonance in Semiconductors: General Theory, *Phys. Rev.* **102**, 1030, (1956).
[42] R. Winkler, Rashba spin splitting in two-dimensional electron and hole systems, *Phys. Rev. B.* **62**, 4245, (2000).
[43] R. Winkler, H. Noh, E. Tutuc, and M. Shayegan, Anomalous Rashba spin splitting in two-dimensional hole systems, *Phys. Rev. B.* **65**, 155303, (2002).
[44] D. V. Bulaev and D. Loss, Electric dipole spin resonance for heavy holes in quantum dots, *Phys. Rev. Lett.* **98**, 097202, (2007).
[45] V. Fock, Bemerkung zur Quantelung des harmonischen Oszillators im Magnetfeld, *Z. Phys.* **47**, 446–448, (1928).
[46] C. G. Darwin, The Diamagnetism of the Free Electron, *Proc. Cambridge Philos. Soc.* **27**, 86–90, (1930).
[47] K. Blum, *Density Matrix Theory and Applications.* (Plenum, New York, 1996).
[48] V. N. Golovach, A. Khaetskii, and D. Loss, Phonon-Induced Decay of the Electron Spin in Quantum Dots, *Phys. Rev. Lett.* **93**, 016601, (2004).
[49] H. W. van Kestern, E. C. Cosman, W. A. J. A. van der Poel, and C. T. Foxton, Fine structure of excitons in type-II GaAs/AlAs quantum wells, *Phys. Rev. B.* **41**, 5283, (1990).
[50] D. V. Bulaev and D. Loss, Spin relaxation and anticrossing in quantum dots: Rashba versus dresselhaus spin-orbit coupling, *Phys. Rev. B.* **71**, 205324, (2005).
[51] D. Heiss, S. Schaeck, H. Huebl, M. Bichler, G. Abstreiter, J. J. Finley, D. V. Bulaev, and D. Loss, Observation of extremely slow hole spin relaxation in self-assembled quantum dots, *Phys. Rev. B* **76**, 241306, (2007).
[52] A. Abragam, *The Principles of Nuclear Magnetism.* (Oxford University Press, London, 1961).

Chapter 11

Microscopic Theory of Energy Dissipation and Decoherence in Semiconductor Nanodevices

Fausto Rossi

Dipartimento di Fisica, Politecnico di Torino, Corso Duca degli Abruzzi 24, 10129 Torino, Italy

A general density-matrix formulation of electronic quantum phenomena in semiconductor nanodevices is presented. More specifically, contrary to the conventional single-particle correlation expansion, we shall investigate the effect of the adiabatic or Markov limit, without considering/performing any reduction procedure. Our fully operatorial approach allows us to better identify the general properties of the scattering superoperators entering our effective quantum-transport theory at various description levels. In particular, we shall revisit the conventional adiabatic or Markov approximation, showing its intrinsic failure in describing the proper quantum-mechanical evolution of a generic subsystem interacting with its environment. In particular, we shall show that — contrary to the semiclassical case— the Markov limit does not preserve the positive-definite character of the corresponding density matrix, thus leading to highly non-physical results. To overcome this problem, we shall propose an alternative adiabatic procedure which (i) in the semiclassical limit reduces to the standard Fermi's golden rule, and (ii) describes a genuine Lindblad evolution, thus providing a reliable/robust treatment of energy-dissipation and dephasing processes in electronic quantum devices. Compared to standard master-equation formulations, the proposed approach does not involve/require any reduction or average procedure, exactly as for the derivation of the well known Fermi's golden rule.

Contents

11.1. Introduction

Present-day technology pushes device dimensions toward limits where the traditional semiclassical or Boltzmann theory [1] can no longer be applied, and more rigorous quantum-kinetic approaches are imperative [2]. However, in spite of the quantum-mechanical nature of electron and photon dynamics in the core region of typical solid-state nanodevices —e.g., superlattices [3] and quantum-dot structures [4]— the overall behavior of such quantum systems is often governed by a complex interplay between phase coherence and energy relaxation/dephasing [5], the latter being also due to the presence of spatial boundaries [6]. Therefore, a proper treatment of such novel nanoscale devices requires a theoretical modeling able to properly account for both coherent and incoherent —i.e., phase-breaking— processes on the same footing.

The wide family of so-called solid-state quantum devices can be schematically divided into two main classes: (i) a first one which comprises low-dimensional nanostructures whose electro-optical response may be safely treated within the semiclassical picture [7] (e.g., quantum-cascade lasers [8]), and (ii) a second one grouping solid-state devices characterized by a genuine quantum-mechanical behavior of their electronic subsystem (e.g., solid-state quantum logic gates [9]) whose quantum evolution is only weakly disturbed by decoherence processes.

For purely atomic and/or photonic quantum logic gates, decoherence phenomena are successfully described via adiabatic-decoupling procedures [10] in terms of extremely simplified models via phenomenological parameters; within such effective treatments, the main goal/requirement is to identify a suitable form of the Liouville superoperator, able to ensure/maintain the positive-definite character of the corresponding density-matrix operator [11]. This is usually accomplished by identifying proper Lindblad-like decoherence superoperators [11, 12], expressed in terms of a few crucial system-environment coupling parameters [13].

In contrast, solid-state devices are often characterized by a complex many-electron quantum evolution, resulting in a non-trivial interplay between coherent dynamics and energy-relaxation/decoherence processes; it follows that for a quantitative description of such coherence/dissipation coupling the latter need to be treated via fully microscopic models.

To this aim, motivated by the power and flexibility of the semiclassical kinetic theory [1] in describing a large variety of interaction mechanisms, a quantum generalization of the standard Boltzmann collision operator has been proposed [5]; the latter, obtained via the conventional Markov limit, describes the evolution of the reduced density matrix in terms of in- and out-scattering superoperators. However, contrary to the semiclassical case, such collision superoperator does not preserve the positive-definite character of the density-matrix operator.

To overcome this serious limitation, in this chapter we shall present an alternative adiabatic procedure which (i) in the semiclassical limit reduces to the well known Fermi's golden rule [14], and (ii) describes a genuine Lindblad evolution, thus providing a reliable/robust treatment of energy-dissipation and dephasing processes in semiconductor quantum devices. We stress that, contrary to standard master-equation formulations, the proposed adiabatic-decoupling approach does not involve/require any reduction or average procedure (see below) exactly as in the standard derivation of the conventional Fermi's golden rule.

The paper is organized as follows. In Sect. 11.2 we shall review the fundamentals of the density-matrix formalism applied to the analysis of solid-state quantum devices; Section 11.3 will provide a fully-operatorial derivation of the conventional adiabatic or Markov approximation; In Sect. 11.4 we shall propose an alternative formulation of the Markov limit, leading to the derivation of a Lindblad-like scattering superoperator in terms of a generalized or quantum Fermi's golden rule. Section 11.5 will address the derivation of effective scattering superoperators describing the electronic subsystem interacting with various quasiparticles, e.g., phonons, photons, plasmons, etc. Finally, in Sect. 11.6 we shall summarize and draw some conclusions.

11.2. Fundamentals of Density-matrix Theory Applied to Solid-state Nanodevices

The idealized behavior of a so-called "quantum device" [7] is usually described via the elementary physical picture of the square-well potential and/or in terms of a simple quantum-mechanical n-level system. For a quantitative investigation of state-of-the-art quantum optoelectronic devices, however, two features strongly influence and modify such simplified scenario: (i) the intrinsic many-body nature of the carrier system under investigation, and (ii) the potential coupling of the electronic subsystem of interest with a variety of interaction mechanisms, including the presence of spatial boundaries [6]. These systems are characterized by a strong interplay between coherent dynamics and energy-relaxation/dephasing processes; it follows that for a quantitative description of such non-trivial coherence/dissipation coupling the latter need to be treated via fully microscopic models.

In this chapter we shall provide a comprehensive microscopic theory of charge quantum transport in semiconductor nanostructures based on the well-known density-matrix approach. It is worth mentioning that an alternative approach, equivalent to the density-matrix formalism employed in our work, is given by the nonequilibrium Green's function technique; the latter can be regarded as an extension of the well-known equilibrium or zero-temperature Green's function theory to nonequilibrium regimes, introduced in the 1960s by Kadanoff and Baym [15] and Keldysh [16]. An introduction to the theory of nonequilibrium Green's functions with applications to many problems in transport and optics of semiconductors can be found in the book by Haug and Jauho [17]. By employing —and further developing and extending— such nonequilibrium Green's function formalism, a number of groups have proposed efficient quantum-transport treatments for the study of various semiconductor nanostructures as well as of modern micro/optoelectronic devices [18].

Within the general density-matrix formalism two different strategies are commonly employed: (i) the quantum-kinetic treatment [5], and (ii) the description based on the Liouville-von Neumann equation [19].

The primary goal of a quantum-kinetic theory is to evaluate the temporal evolution of a reduced set of single- or few-particle quantities directly related to the electro-optical phenomenon under investigation, the so-called kinetic variables of the system. However, due to the many-body nature of the problem, an exact solution in general is not possible; it follows that for a detailed understanding realistic semiconductor models have to be considered, which then can only be treated approximately. Within the kinetic-theory approach one starts directly with the equations of motion for the single-particle density matrix. Due to the many-body nature of the problem, the resulting set of equations of motion is not closed; instead, it constitutes the starting point of an infinite hierarchy of higher-order density matrices. Besides differences related to the quantum statistics of the quasi-particles involved, this is equivalent to the BBGKY hierarchy in classical gas dynamics [20]. The central approximation in this formalism is the truncation of the hierarchy. This can be based on different physical pictures. A common approach is to use the argument that correlations involving an increasing number of particles will become less and less important [5]. An alternative quantum-kinetic scheme —based on an expansion in powers of the exciting laser field— has been introduced by Axt and Stahl, the so-called "dynamics controlled truncation" (DCT) [21].

Within the treatment based on the Liouville-von Neumann equation, the starting point is the equation of motion for the global density-matrix operator, describing many electron plus various quasiparticle excitations. The physical quantities of interest for the electronic subsystem are then typically derived via a suitable "reduction procedure", aimed at tracing out non-relevant degrees of freedom. Contrary to the kinetic theory, this approach has allowed for a fully quantum-mechanical treatment of high-field transport in semiconductors [22], thus overcoming some of the basic limitations of conventional kinetic treatments, e.g., the completed-collision limit and the Markov approximation.

As anticipated, primary goal of the present chapter is to discuss in very general terms the physical properties and validity limits of the so-called "adiabatic" or Markov approximation. Within the traditional semiclassical or Boltzmann theory, this approximation is typically introduced together with the so-called diagonal approximation, i.e., the neglect of non-diagonal density-matrix elements. However, as described in Ref. 23, the Markov limit can be also performed within a fully non-diagonal density-matrix treatment of the problem; this leads to the introduction of generalized in- and out-scattering superoperators, whose general properties and physical interpretation are not straightforward. In particular, it is imperative to understand if —and under which conditions— the adiabatic or Markov approximation preserves the positive-definite character of our reduced density matrix; indeed, this distinguished property is generally lost within the quantum-kinetic approaches previously mentioned [5, 23]. To this end, starting from the Liouville-von Neumann-equation approach, we shall propose a very general derivation of the Markov approximation.

11.3. Fully Operatorial Derivation of the Adiabatic or Markov Limit

In order to discuss the main features and intrinsic limitations of the conventional adiabatic or Markov limit, let us recall its general derivation following the fully operatorial approach proposed in Ref. 24. Given a generic physical quantity A —described by the operator $\hat{A}$— its quantum plus statistical average value is given by

$$A = \mathrm{tr}\left\{\hat{A}\hat{\rho}\right\} , \tag{11.1}$$

where $\hat{\rho}$ is the so-called density-matrix operator. Its time evolution is dictated by the total (system plus environment) Hamiltonian. Within the usual interaction picture, the latter can be regarded as the sum of a noninteracting (system plus environment) contribution plus a system-environment coupling term:

$$\hat{H} = \hat{H}_\circ + \hat{H}' . \tag{11.2}$$

The corresponding equation of motion for the density-matrix operator —also known as Liouville-von Neumann equation— in the interaction picture is given by:

$$\frac{d\hat{\rho}^i}{dt} = -i\left[\hat{\mathcal{H}}^i, \hat{\rho}^i\right] , \tag{11.3}$$

where $\hat{\mathcal{H}}$ denotes the interaction Hamiltonian $\hat{H}'$ written in units of $\hbar$.

The key idea beyond any perturbation approach is that the effect of the interaction Hamiltonian $\hat{H}'$ is "small" compared to the free evolution dictated by the noninteracting Hamiltonian $\hat{H}_\circ$. Following this spirit, by formally integrating Eq. (11.3) from $-\infty$ to the current time t, and inserting such formal solution for $\hat{\rho}^i(t)$ on the right-hand side of Eq. (11.3), we obtain an integro-differential equation of the form:

$$\frac{d}{dt}\hat{\rho}^i(t) = -i\left[\hat{\mathcal{H}}^i(t), \hat{\rho}^i(-\infty)\right] - \int_{-\infty}^{t} dt' \left[\hat{\mathcal{H}}^i(t), \left[\hat{\mathcal{H}}^i(t'), \hat{\rho}^i(t')\right]\right] . \tag{11.4}$$

We stress that so far no approximation has been introduced: Equations (11.3) and (11.4) are fully equivalent, we have just isolated the first-order contribution from the full time evolution in Eq. (11.3).

Let us now focus on the time integral in Eq. (11.4). Here, the two quantities to be integrated over t' are the interaction Hamiltonian $\hat{\mathcal{H}}^i$ and the density-matrix operator $\hat{\rho}^i$. In the spirit of the perturbation approach previously recalled, the time variation of $\hat{\rho}^i$ can be considered adiabatically slow compared to that of the Hamiltonian $\hat{\mathcal{H}}$ written in the interaction picture, i.e.,

$$\hat{\mathcal{H}}^i(t') = \hat{U}_\circ^\dagger(t')\hat{\mathcal{H}}\hat{U}_\circ(t') ; \tag{11.5}$$

indeed, the latter exhibits rapid oscillations due to the noninteracting evolution operator

$$\hat{U}_\circ(t) = e^{-\frac{i\hat{H}_\circ t}{\hbar}} . \tag{11.6}$$

As a result, the density-matrix operator $\hat{\rho}^i$ can be taken out of the time integral and evaluated at the current time t.

Following such prescription, the second-order contribution to the system dynamics written in the Schrödinger picture for the case of a time-independent interaction Hamiltonian $\hat{\mathcal{H}}$ comes out to be:

$$\frac{d\hat{\rho}}{dt} = -\frac{1}{2}\left[\hat{\mathcal{H}}, \left[\hat{\mathcal{K}}, \hat{\rho}\right]\right] \tag{11.7}$$

with

$$\hat{\mathcal{K}} = 2\int_{-\infty}^{0} dt' \hat{\mathcal{H}}^i(t') = 2\int_{-\infty}^{0} dt' \hat{U}_\circ^\dagger(t')\hat{\mathcal{H}}\hat{U}_\circ(t') . \tag{11.8}$$

As discussed extensively in [24], the operator $\hat{\mathcal{K}}$ describes energy-conserving scattering events (real processes) as well as energy-renormalization contributions (virtual processes); the latter are known to play a minor role in semiconducting materials, and in general may be safely neglected; such approximation amounts to impose the following time-reversal symmetry on the system dynamics:

$$\hat{\mathcal{H}}^i(t') = \hat{\mathcal{H}}^i(-t') . \tag{11.9}$$

This corresponds to neglect higher-order effects of the interaction Hamiltonian $\hat{\mathcal{H}}$ on the free system evolution $\hat{U}_\circ$. This is accomplished by neglecting contributions related to the commutator between the "dressed" evolution and the free one, i.e.,

$$\left[\left[\hat{U}_\circ, \hat{\mathcal{H}}\right], \hat{U}_\circ^\dagger\right] = 0 . \tag{11.10}$$

Indeed, this is the condition needed in order to get the desired time-reversal symmetry $\hat{\mathcal{H}}^i(t) = \hat{\mathcal{H}}^i(-t)$.

Within such approximation scheme, the operator $\hat{\mathcal{K}}$ in (11.8) may be rewritten extending the time integration from $-\infty$ to $+\infty$:

$$\hat{\mathcal{K}} = \int_{-\infty}^{+\infty} dt' \hat{U}_\circ^\dagger(t')\hat{\mathcal{H}}\hat{U}_\circ(t') . \tag{11.11}$$

The effective equation in (11.7) has still the double-commutator structure in (11.4) but it is now local in time. The Markov limit recalled so far leads to significant modifications to the system dynamics: while the exact quantum-mechanical evolution in (11.3) corresponds to a fully reversible and isoentropic unitary transformation, the instantaneous double-commutator structure in (11.7) describes, in general, a non-reversible (i.e., non unitary) dynamics characterized by energy dissipation and dephasing. However, since any effective Liouville superoperator should describe correctly the time evolution of $\hat{\rho}$ and since the latter, by definition, needs to be trace-invariant and positive-definite at any time, it is imperative to determine if the Markov superoperator in (11.7) fulfills this two basic requirements. As far as the first issue is concerned, in view of its commutator structure, it is easy to show that this effective superoperator is indeed trace-preserving. In contrast, as discussed extensively in [24], the latter does not ensure that for any initial condition the density-matrix operator will be positive-definite at any time. This is by far the most severe limitation of the conventional Markov approximation.

By denoting with $\{|\lambda\rangle\}$ the eigenstates of the noninteracting Hamiltonian $\hat{H}_\circ$, the effective equation (11.7) written in this basis is of the form:

$$\frac{d\rho_{\lambda_1\lambda_2}}{dt} = \frac{1}{2}\sum_{\lambda'_1\lambda'_2}\left[\mathcal{P}_{\lambda_1\lambda_2,\lambda'_1\lambda'_2}\rho_{\lambda'_1\lambda'_2} - \mathcal{P}_{\lambda_1\lambda'_2,\lambda'_1\lambda'_1}\rho_{\lambda'_2\lambda_2}\right] + \text{H.c.} \qquad (11.12)$$

with generalized scattering rates given by:

$$\mathcal{P}_{\lambda_1\lambda_2,\lambda'_1\lambda'_2} = \frac{2\pi}{\hbar}H'_{\lambda_1\lambda'_1}H'^{*}_{\lambda_2\lambda'_2}\delta(\epsilon_{\lambda_2} - \epsilon_{\lambda'_2}) , \qquad (11.13)$$

ϵ_λ denoting the energy corresponding to the noninteracting state $|\lambda\rangle$.

The well-known semiclassical or Boltzmann theory [1] can be easily derived from the quantum formulation presented so far, by introducing the so-called diagonal or semiclassical approximation. The latter corresponds to neglecting all non-diagonal density-matrix elements (and therefore any quantum-mechanical phase coherence between the generic states λ_1 and λ_2), i.e.,

$$\rho_{\lambda_1\lambda_2} = f_{\lambda_1}\delta_{\lambda_1\lambda_2} , \qquad (11.14)$$

where the diagonal elements f_λ describe the semiclassical distribution function over our noninteracting basis states.

Within such approximation scheme, the effective quantum equation (11.12) reduces to the well-known Boltzmann equation:

$$\frac{df_\lambda}{dt} = \sum_{\lambda'}\left(P_{\lambda\lambda'}f_{\lambda'} - P_{\lambda'\lambda}f_\lambda\right) , \qquad (11.15)$$

where

$$P_{\lambda\lambda'} = \mathcal{P}_{\lambda\lambda,\lambda'\lambda'} = \frac{2\pi}{\hbar}|H'_{\lambda\lambda'}|^2\delta\left(\epsilon_\lambda - \epsilon_{\lambda'}\right) \tag{11.16}$$

are the conventional semiclassical scattering rates given by the well-known Fermi's golden rule [14].

At this point it is crucial to stress that, contrary to the non-diagonal density-matrix description previously introduced, the Markov limit combined with the semiclassical or diagonal approximation ensures that at any time t our semiclassical distribution function f_λ is always positive-definite. This explains the "robustness" of the Boltzmann transport equation (11.15), and its extensive application in solid-state-device modeling as well as in many other areas, where quantum effects play a very minor role. In contrast, in order to investigate genuine quantum-mechanical phenomena, the conventional Markov superoperator in (11.7) cannot be employed, since it does not preserve the positive-definite character of the density matrix $\rho_{\lambda_1\lambda_2}$.

11.4. Alternative Formulation of the Markov Limit: Derivation of a "Quantum Fermi's Golden Rule"

As anticipated, aim of the present chapter is to propose an alternative formulation of the standard Markov limit, able to provide a Lindblad-like scattering superoperator, thus preserving the positive-definite character of our density matrix. To this end, let us go back to the integro-differential equation (11.4). As previously discussed, the crucial step in the standard derivation is to replace $\hat{\rho}^i(t')$ with $\hat{\rho}^i(t)$. Indeed, since in the adiabatic limit the time variation of the density matrix within the interaction picture is negligible, the latter can be evaluated at any time. Based on this remark, what we propose is the following time symmetrization: given the two times t' and t, we shall introduce the "average" or "macroscopic" time $T = \frac{t+t'}{2}$ and the "relative" time $\tau = t - t'$. The basic idea is that the relevant time characterizing/describing our effective system evolution is the macroscopic time T. Following this spirit, it is easy to rewrite the second-order contribution in Eq. (11.4) in terms of the new time variables T and τ:

$$\frac{d}{dT}\hat{\rho}^i(T) = -\int_0^\infty d\tau \left[\hat{\mathcal{H}}^i\left(T + \frac{\tau}{2}\right), \left[\hat{\mathcal{H}}^i\left(T - \frac{\tau}{2}\right), \hat{\rho}^i\left(T - \frac{\tau}{2}\right)\right]\right] . \tag{11.17}$$

Following again the spirit of the adiabatic decoupling, we shall now replace $\hat{\rho}^i\left(T - \frac{\tau}{2}\right)$ with $\hat{\rho}^i(T)$; the resulting effective equation rewritten in the original Schrödinger picture comes out to be:

$$\frac{d\hat{\rho}}{dT} = -\int_0^\infty d\tau \left[\hat{\mathcal{H}}^i\left(\frac{\tau}{2}\right), \left[\hat{\mathcal{H}}^i\left(-\frac{\tau}{2}\right), \hat{\rho}\right]\right] . \tag{11.18}$$

Neglecting again renormalization contributions [see Eq. (11.9)], Eq. (11.18) may be rewritten by extending the time integration over τ from $-\infty$ to $+\infty$:

$$\frac{d\hat{\rho}}{dT} = -\frac{1}{2}\int_{-\infty}^{+\infty} d\tau \left[\hat{\mathcal{H}}^i\left(\frac{\tau}{2}\right), \left[\hat{\mathcal{H}}^i\left(-\frac{\tau}{2}\right), \hat{\rho}\right]\right] . \tag{11.19}$$

By Fourier expanding the above symmetric convolution integral we finally get the desired Lindblad-like scattering superoperator:

$$\frac{d\hat{\rho}}{dT} = -\frac{1}{2}\int d\omega \left[\hat{\mathcal{L}}(\omega), \left[\hat{\mathcal{L}}(\omega), \hat{\rho}\right]\right] \tag{11.20}$$

with

$$\hat{\mathcal{L}}(\omega) = \frac{1}{\sqrt{2\pi}}\int_{-\infty}^{\infty} d\tau \hat{U}_\circ^\dagger\left(\frac{\tau}{2}\right)\hat{\mathcal{H}}\hat{U}_\circ\left(\frac{\tau}{2}\right)e^{i\omega\tau} . \tag{11.21}$$

We stress how the proposed time symmetrization gives rise to a fully symmetric Lindblad-like superoperator (expressed in terms of the operator $\hat{\mathcal{L}}$ only), compared to the strongly asymmetric Markov superoperator in (11.7).

If we now rewrite the new Markov superoperator in (11.19) in our non-interacting basis λ, we obtain again the effective equation of motion in (11.12), but now the generalized scattering rates in (11.13) are replaced by the following symmetrized quantum scattering rates:

$$\tilde{\mathcal{P}}_{\lambda_1\lambda_2,\lambda_1'\lambda_2'} = \frac{2\pi}{\hbar}H'_{\lambda_1\lambda_1'}H'^*_{\lambda_2\lambda_2'}\delta\left(\frac{\epsilon_{\lambda_1}+\epsilon_{\lambda_2}}{2} - \frac{\epsilon_{\lambda_1'}+\epsilon_{\lambda_2'}}{2}\right) . \tag{11.22}$$

The above scattering superoperator can be regarded as the quantum-mechanical generalization of the conventional Fermi's golden rule; indeed, in the semiclassical limit ($\lambda_1 = \lambda_2, \lambda_1' = \lambda_2'$) the standard formula in (11.16) is readily recovered.

11.5. Derivation of Effective Scattering Superoperators for the Electronic Subsystem

Let us now specialize our general result in (11.20) to the case of a semiconductor quantum device, for which the unperturbed Hamiltonian $\hat{H}_\circ$ in (11.2) can be written as

$$\hat{H}_\circ = \hat{H}_e + \hat{H}_{qp} . \tag{11.23}$$

This is the sum of the electronic (e) Hamiltonian and of the free-quasiparticle (qp) Hamiltonian

$$\hat{H}_{qp} = \sum_{\mathbf{q}} \epsilon_{\mathbf{q}} \hat{b}^{\dagger}_{\mathbf{q}} \hat{b}_{\mathbf{q}} , \tag{11.24}$$

where the Bosonic operators $\hat{b}^{\dagger}_{\mathbf{q}}$ ($\hat{b}_{\mathbf{q}}$) denote creation (destruction) of a generic quasiparticle excitation with wavevector $\mathbf{q}$ and energy $\epsilon_{\mathbf{q}}$, i.e., phonons, photons, plasmons, etc.

In this case, the noninteracting (carrier-plus-quasiparticle) basis states

$$|\lambda\rangle = |\alpha\rangle \otimes |\{n_{\mathbf{q}}\}\rangle \tag{11.25}$$

are given by the tensor product of electronic states $|\alpha\rangle$ and quasiparticle states $|\{n_{\mathbf{q}}\}\rangle$ corresponding to the set of occupation numbers $\{n_{\mathbf{q}}\}$; the noninteracting energy spectrum

$$\epsilon_{\lambda} = \epsilon_{\alpha} + \sum_{\mathbf{q}} \epsilon_{\mathbf{q}} n_{\mathbf{q}} \tag{11.26}$$

is then the sum of the electronic and total quasiparticle energies.

For all relevant carrier-quasiparticle interaction mechanisms in semiconductor nanostructures —e.g., carrier-phonon, carrier-photon, carrier-plasmon, etc.— the perturbation Hamiltonian $\hat{H}'$ in (11.2) can be written as:

$$\hat{H}' = \sum_{\mathbf{q}} \left(\hat{H}_{\mathbf{q}} \hat{b}_{\mathbf{q}} + \hat{H}^{\dagger}_{\mathbf{q}} \hat{b}^{\dagger}_{\mathbf{q}} \right) = \hat{H}^{ab} + \hat{H}^{em} . \tag{11.27}$$

Here $\hat{H}_{\mathbf{q}} = \hat{H}^{\dagger}_{-\mathbf{q}}$ are electronic operators (parameterized by the quasiparticle wavevector $\mathbf{q}$) acting on the α subsystem only. The two terms in (11.27) —corresponding to quasiparticle destruction and creation— describe electronic absorption and emission processes.

As anticipated, the average value of any given physical quantity A can be easily expressed in terms of the density-matrix operator $\hat{\rho}$ according to Eq. (11.1). In the study of electronic quantum phenomena in semiconductor nanostructures, most of the physical quantities of interest depend on the electronic-subsystem coordinates α only (carrier energy loss, electronic coherence length, carrier-carrier correlation functions, etc.), i.e.,

$$A_{\lambda\lambda'} = A_{\alpha,\{n_{\mathbf{q}}\};\alpha',\{n'_{\mathbf{q}}\}} = A_{\alpha\alpha'} \delta_{\{n_{\mathbf{q}}\}\{n'_{\mathbf{q}}\}} . \tag{11.28}$$

In this case it is convenient to write Eq. (11.1) as

$$A = \mathrm{tr}\left\{ \hat{A}\hat{\rho} \right\} = \sum_{\lambda\lambda'} A_{\lambda\lambda'} \rho_{\lambda'\lambda} = \sum_{\alpha,\alpha'} A_{\alpha\alpha'} \rho^{e}_{\alpha'\alpha} , \tag{11.29}$$

where

$$\rho^e_{\alpha\alpha'} = \sum_{\{n_{\mathbf{q}}\}} \rho_{\alpha,\{n_{\mathbf{q}}\};\alpha',\{n_{\mathbf{q}}\}} \tag{11.30}$$

is the so-called reduced or electronic density matrix. Equation (11.30) can be also written in an operatorial form as:

$$\hat{\rho}^e = \mathrm{tr}\left\{\hat{\rho}\right\}_{qp} , \tag{11.31}$$

which shows that the electronic density-matrix operator $\hat{\rho}^e$ is obtained by performing a trace operation over the quasi-particle variables $\{n_{\mathbf{q}}\}$. Since $\hat{\rho}^e$ is the only quantity entering the evaluation of the average value in (11.29), it is desirable to derive a corresponding equation of motion for the reduced density-matrix operator. However, such trace over the quasiparticle coordinates does not commute with the scattering superoperator in (11.20), which does not allow to obtain a closed equation of motion for the reduced density-matrix operator $\hat{\rho}^e$. In order to overcome this problem, the typical assumption is to consider the quasiparticle subsystem as characterized by a huge number of degrees of freedom (compared to the subsystem α). In other words this amounts to say that the quasiparticle subsystem has an infinitely high heat capacity, i.e., it behaves as a thermal bath; this allows to consider this subsystem always in thermal equilibrium, i.e., not significantly perturbed by the electronic subsystem. Within such approximation scheme, the global (e plus qp) density-matrix operator $\hat{\rho}$ can be written as the product of the equilibrium density-matrix operator for the quasiparticle subsystem $\hat{\rho}^{qp}$ and the reduced density-matrix operator $\hat{\rho}^e$:

$$\hat{\rho} = \hat{\rho}^e \hat{\rho}^{qp} , \qquad \hat{\rho}^{qp} = \frac{e^{-\frac{\hat{H}_{qp}}{k_B T}}}{\mathrm{tr}\left\{e^{-\frac{\hat{H}_{qp}}{k_B T}}\right\}} . \tag{11.32}$$

More specifically, let us start by applying the trace over the quasiparticle coordinates to Eq. (11.19); by expanding the double commutator and employing the factorization scheme in (11.32) we get:

$$\begin{aligned}
\frac{d\hat{\rho}^e}{dT} &= -\frac{1}{2}\int_{-\infty}^{+\infty} d\tau \mathrm{tr}\left\{\left[\hat{\mathcal{H}}^i\left(\frac{\tau}{2}\right), \left[\hat{\mathcal{H}}^i\left(-\frac{\tau}{2}\right), \hat{\rho}^e\hat{\rho}^{qp}\right]\right]\right\}_{qp} \\
&= -\frac{1}{2}\int_{-\infty}^{+\infty} d\tau \mathrm{tr}\left\{\hat{\mathcal{H}}^i\left(\frac{\tau}{2}\right)\hat{\mathcal{H}}^i\left(-\frac{\tau}{2}\right)\hat{\rho}^e\hat{\rho}^{qp}\right\}_{qp} \\
&\quad +\frac{1}{2}\int_{-\infty}^{+\infty} d\tau \mathrm{tr}\left\{\hat{\mathcal{H}}^i\left(\frac{\tau}{2}\right)\hat{\rho}^e\hat{\rho}^{qp}\hat{\mathcal{H}}^i\left(-\frac{\tau}{2}\right)\right\}_{qp} \\
&\quad +\frac{1}{2}\int_{-\infty}^{+\infty} d\tau \mathrm{tr}\left\{\hat{\mathcal{H}}^i\left(-\frac{\tau}{2}\right)\hat{\rho}^e\hat{\rho}^{qp}\hat{\mathcal{H}}^i\left(\frac{\tau}{2}\right)\right\}_{qp} \\
&\quad -\frac{1}{2}\int_{-\infty}^{+\infty} d\tau \mathrm{tr}\left\{\hat{\rho}^e\hat{\rho}^{qp}\hat{\mathcal{H}}^i\left(-\frac{\tau}{2}\right)\hat{\mathcal{H}}^i\left(\frac{\tau}{2}\right)\right\}_{qp} .
\end{aligned} \tag{11.33}$$

To this end, it is imperative to evaluate the explicit form of the interaction Hamiltonian $\hat{\mathcal{H}}^i(t')$ in (11.5). Recalling the explicit form of the noninteracting Hamiltonian $\hat{H}_\circ$ in (11.23), we have:

$$\hat{U}_\circ(t') = e^{-\frac{i\hat{H}_\circ t'}{\hbar}} = e^{-\frac{i\hat{H}_e t'}{\hbar}} e^{-\frac{i\hat{H}_{qp} t'}{\hbar}} = \hat{U}_e(t')\hat{U}_{qp}(t') . \tag{11.34}$$

Moreover, employing the bosonic commutation relations for the creation and destruction operators, we get:

$$\begin{aligned}
\hat{b}^i_{\mathbf{q}}(t') &= \hat{U}^\dagger_{qp}(t')\hat{b}_{\mathbf{q}}\hat{U}_{qp}(t') = e^{-\frac{i\epsilon_{\mathbf{q}} t'}{\hbar}}\hat{b}_{\mathbf{q}} \\
\hat{b}^{i\dagger}_{\mathbf{q}}(t') &= \hat{U}^\dagger_{qp}(t')\hat{b}^\dagger_{\mathbf{q}}\hat{U}_{qp}(t') = e^{+\frac{i\epsilon_{\mathbf{q}} t'}{\hbar}}\hat{b}^\dagger_{\mathbf{q}} .
\end{aligned} \tag{11.35}$$

In view of these two properties, the explicit form of the electron-quasiparticle coupling Hamiltonian (11.27) written in the interaction picture comes out to be:

$$\hat{\mathcal{H}}^i(t') = \sum_{\mathbf{q}} \left(\hat{\mathcal{H}}^-_{\mathbf{q}}(t')\hat{b}_{\mathbf{q}} + \hat{\mathcal{H}}^+_{\mathbf{q}}(t')\hat{b}^\dagger_{\mathbf{q}}\right) \tag{11.36}$$

with

$$\hat{\mathcal{H}}^-_{\mathbf{q}}(t') = \frac{\hat{H}^i_{\mathbf{q}}(t')}{\hbar} e^{-\frac{i\epsilon_{\mathbf{q}} t'}{\hbar}} , \qquad \hat{\mathcal{H}}^+_{\mathbf{q}}(t') = \frac{\hat{H}^{i\dagger}_{\mathbf{q}}(t')}{\hbar} e^{+\frac{i\epsilon_{\mathbf{q}} t'}{\hbar}} . \tag{11.37}$$

By inserting the above result into Eq. (11.33) and recalling that

$$\mathrm{tr}\left\{\hat{b}_{\mathbf{q}}\hat{b}_{\mathbf{q}'}\hat{\rho}^{qp}\right\} = \mathrm{tr}\left\{\hat{b}^\dagger_{\mathbf{q}}\hat{b}^\dagger_{\mathbf{q}'}\hat{\rho}^{qp}\right\} = 0 \tag{11.38}$$

as well as $\hat{\mathcal{H}}^{\pm}_{\mathbf{q}}(t') = \hat{\mathcal{H}}^{\mp\dagger}_{\mathbf{q}}(t')$, we obtain:

$$\begin{aligned}
\frac{d\hat{\rho}^e}{dT} = &-\frac{1}{2}\sum_{\mathbf{q}\pm}\left(N_{\mathbf{q}} + \frac{1}{2} \mp \frac{1}{2}\right)\int_{-\infty}^{+\infty} d\tau \hat{\mathcal{H}}^{\pm}_{\mathbf{q}}\left(\frac{\tau}{2}\right)\hat{\mathcal{H}}^{\pm\dagger}_{\mathbf{q}}\left(-\frac{\tau}{2}\right)\hat{\rho}^e \\
&+\frac{1}{2}\sum_{\mathbf{q}\pm}\left(N_{\mathbf{q}} + \frac{1}{2} \pm \frac{1}{2}\right)\int_{-\infty}^{+\infty} d\tau \hat{\mathcal{H}}^{\pm}_{\mathbf{q}}\left(\frac{\tau}{2}\right)\hat{\rho}^e\hat{\mathcal{H}}^{\pm\dagger}_{\mathbf{q}}\left(-\frac{\tau}{2}\right) \\
&+\frac{1}{2}\sum_{\mathbf{q}\pm}\left(N_{\mathbf{q}} + \frac{1}{2} \mp \frac{1}{2}\right)\int_{-\infty}^{+\infty} d\tau \hat{\mathcal{H}}^{\pm\dagger}_{\mathbf{q}}\left(-\frac{\tau}{2}\right)\hat{\rho}^e\hat{\mathcal{H}}^{\pm}_{\mathbf{q}}\left(\frac{\tau}{2}\right) \\
&-\frac{1}{2}\sum_{\mathbf{q}\pm}\left(N_{\mathbf{q}} + \frac{1}{2} \pm \frac{1}{2}\right)\int_{-\infty}^{+\infty} d\tau \hat{\rho}^e\hat{\mathcal{H}}^{\pm\dagger}_{\mathbf{q}}\left(-\frac{\tau}{2}\right)\hat{\mathcal{H}}^{\pm}_{\mathbf{q}}\left(\frac{\tau}{2}\right) \quad (11.39)
\end{aligned}$$

Here

$$N_{\mathbf{q}} = \mathrm{tr}\left\{\hat{b}^{\dagger}_{\mathbf{q}}\hat{b}_{\mathbf{q}}\hat{\rho}^{qp}\right\} \quad (11.40)$$

denotes the average occupation number for the quasiparticle state $\mathbf{q}$. As we can see, for each quasiparticle state/mode $\mathbf{q}$ we have always two contributions ($\pm$) describing quasiparticle emission and absorption.

By regrouping/rearranging the various terms entering the above four contributions, Eq. (11.39) can be also written as:

$$\begin{aligned}
\frac{d\hat{\rho}^e}{dT} = &-\frac{1}{2}\sum_{\mathbf{q}}\left(N_{\mathbf{q}} + \frac{1}{2}\right)\sum_{\pm}\int_{-\infty}^{+\infty} d\tau \left[\hat{\mathcal{H}}^{\pm}_{\mathbf{q}}\left(\frac{\tau}{2}\right), \left[\hat{\mathcal{H}}^{\pm\dagger}_{\mathbf{q}}\left(-\frac{\tau}{2}\right), \hat{\rho}^e\right]\right] \\
&+\frac{1}{2}\sum_{\mathbf{q}\pm} \pm \int_{-\infty}^{+\infty} d\tau \left[\hat{\mathcal{H}}^{\pm}_{\mathbf{q}}\left(\frac{\tau}{2}\right), \left\{\hat{\mathcal{H}}^{\pm\dagger}_{\mathbf{q}}\left(-\frac{\tau}{2}\right), \hat{\rho}^e\right\}\right], \quad (11.41)
\end{aligned}$$

where $\{\hat{A}, \hat{B}\} = \hat{A}\hat{B} + \hat{B}\hat{A}$ denotes the anticommutation parenthesis.

Let us now discuss the two terms in (11.41). To this end, it is crucial to recall the physical origin of the quantity $\left(N_{\mathbf{q}} + \frac{1}{2} \pm \frac{1}{2}\right)$ in (11.39). Since here the $\pm$ sign refers, respectively, to emission and absorption processes [see also Eq. (11.36)] it follows that all absorption contributions are proportional to $N_{\mathbf{q}}$, while the emission ones are proportional to $N_{\mathbf{q}} + 1$, the latter being usually interpreted as the sum of a "stimulated" plus a "spontaneous" contribution. Indeed, here the quantum nature of the quasiparticle subsystem manifest itself via the $+1$ term (i.e., the difference between total emission and absorption interaction processes), which follows directly from the Bosonic commutation relations for our creation and destruction quasiparticle operators. In view of the above, it is easy to conclude that the first term in (11.41) describes the interaction of the quantum electronic subsystem with a classical quasiparticle field, while the second term describes the quantum nature of the quasiparticle subsystem via spontaneous-emission contributions only. Moreover, while the first term exhibits again the same double-commutator structure of the global scattering superoperator in (11.19), the second one contains anticommutator structures as well. We are then forced to conclude that the total scattering superoperator (11.41) is no more Lindblad-like; this is ascribed to the presence of spontaneous-emission contributions [see last term in (11.41)] previously discussed.

In the so-called classical or macroscopic limit (i.e., $N_{\mathbf{q}} \gg 1$), the last term in (11.41) vanishes. More specifically, considering that in this limit $N_{\mathbf{q}} + \frac{1}{2} \simeq N_{\mathbf{q}}$, we obtain:

$$\frac{d\hat{\rho}^e}{dT} = -\frac{1}{2}\sum_{\mathbf{q}} N_{\mathbf{q}} \sum_{\pm} \int_{-\infty}^{+\infty} d\tau \left[\hat{\mathcal{H}}_{\mathbf{q}}^{\pm}\left(\frac{\tau}{2}\right), \left[\hat{\mathcal{H}}_{\mathbf{q}}^{\pm\dagger}\left(-\frac{\tau}{2}\right), \hat{\rho}^e\right]\right] . \quad (11.42)$$

As anticipated, also for the present reduced (electronic) description we get again the double-commutator structure of Eq. (11.19). Indeed, by Fourier expanding the above symmetric convolution time integral we finally get the desired electronic scattering superoperator written again in a Lindblad form:

$$\frac{d\hat{\rho}^e}{dT} = -\frac{1}{2}\sum_{\mathbf{q}} N_{\mathbf{q}} \int d\omega \sum_{\pm} \left[\hat{\mathcal{L}}_{\mathbf{q}}^{\pm}(\omega), \left[\hat{\mathcal{L}}_{\mathbf{q}}^{\pm\dagger}(\omega), \hat{\rho}^e\right]\right] \quad (11.43)$$

with

$$\hat{\mathcal{L}}_{\mathbf{q}}^{\pm}(\omega) = \frac{1}{\sqrt{2\pi}} \int_{-\infty}^{\infty} d\tau \hat{\mathcal{H}}_{\mathbf{q}}^{\pm}\left(\frac{\tau}{2}\right) e^{i\omega\tau} . \quad (11.44)$$

The above result is extremely important: contrary to conventional quantum-kinetic approaches, we have shown that for the case of a macroscopic or classical quasiparticle subsystem, e.g., an external coherent laser field, the reduction procedure applied to our general formulation based on the proposed quantum Fermi's golden rule preserves the Lindblad-like character of the electronic quantum dynamics. However, in the presence of significant spontaneous-emission contributions, e.g., electron-phonon interaction, the reduction procedure previously discussed does not preserve the Lindblad character of the original scattering superoperator in (11.20).

Let us finally rewrite the scattering superoperator (11.42) in our electronic basis $\{|\alpha\rangle\}$. By denoting with

$$H^{\mathbf{q}}_{\alpha_1\alpha_2} = \langle\alpha_1|\hat{H}_{\mathbf{q}}|\alpha_2\rangle \quad (11.45)$$

the matrix elements of the electronic interaction operator $\hat{H}_{\mathbf{q}}$ we get:

$$\frac{d\rho^e_{\alpha_1\alpha_2}}{dT} = \frac{1}{2}\sum_{\alpha'_1\alpha'_2}\left[\tilde{\mathcal{P}}^e_{\alpha_1\alpha_2,\alpha'_1\alpha'_2}\rho^e_{\alpha'_1\alpha'_2} - \tilde{\mathcal{P}}^e_{\alpha_1\alpha'_2,\alpha'_1\alpha'_1}\rho^e_{\alpha'_2\alpha_2}\right] + \text{H.c.} \tag{11.46}$$

with generalized scattering rates

$$\tilde{\mathcal{P}}^e_{\alpha_1\alpha_2,\alpha'_1\alpha'_2} = \frac{2\pi}{\hbar}\sum_{\mathbf{q}} N_{\mathbf{q}} \sum_{\pm} H^{\mathbf{q}}_{\alpha_1\alpha'_1} H^{\mathbf{q}*}_{\alpha_2\alpha'_2}\,\delta\left(\frac{\epsilon_{\alpha_1}+\epsilon_{\alpha_2}}{2} - \frac{\epsilon_{\alpha'_1}+\epsilon_{\alpha'_2}}{2} \pm \epsilon_{\mathbf{q}}\right). \tag{11.47}$$

Again, the above scattering superoperator can be regarded as the quantum-mechanical generalization of the conventional Fermi's golden rule describing our electronic quantum subsystem interacting with a classical quasi-particle field with amplitude $\sqrt{N_{\mathbf{q}}}$. Indeed, in the semiclassical limit ($\rho^e_{\alpha\alpha'} = f^e_\alpha \delta_{\alpha\alpha'}$), the effective quantum equation (11.46) reduces to the following Boltzmann equation for the electronic subsystem:

$$\frac{df_\alpha}{dT} = \sum_{\alpha'}\left(P^e_{\alpha\alpha'} f^e_{\alpha'} - P^e_{\alpha'\alpha} f^e_\alpha\right), \tag{11.48}$$

with

$$P^e_{\alpha\alpha'} = \tilde{\mathcal{P}}^e_{\alpha\alpha,\alpha'\alpha'} = \frac{2\pi}{\hbar}\sum_{\mathbf{q}} N_{\mathbf{q}} \sum_{\pm} |H^{\mathbf{q}}_{\alpha\alpha'}|^2 \delta\left(\epsilon_\alpha - \epsilon_{\alpha'} \pm \epsilon_{\mathbf{q}}\right). \tag{11.49}$$

We finally stress that, as for the general result in (11.22), our Lindblad-like scattering superoperator corresponds again to fully symmetric scattering rates; this is a distinguished advantage of the proposed adiabatic-decoupling procedure, compared to the conventional (i.e., non-Lindblad) quantum scattering rates discussed in Ref. 24:

$$\mathcal{P}^e_{\alpha_1\alpha_2,\alpha'_1\alpha'_2} = \frac{2\pi}{\hbar}\sum_{\mathbf{q}} N_{\mathbf{q}} \sum_{\pm} H^{\mathbf{q}}_{\alpha_1\alpha'_1} H^{\mathbf{q}*}_{\alpha_2\alpha'_2}\,\delta\left(\epsilon_{\alpha_2} - \epsilon_{\alpha'_2} \pm \epsilon_{\mathbf{q}}\right). \tag{11.50}$$

11.6. Summary and Conclusions

To summarize, we have critically reviewed the standard adiabatic or Markov procedure, showing its intrinsic failure in describing the proper quantum-mechanical evolution of a generic subsystem interacting with its environment. More specifically, we have shown that within the Markov approximation the density-matrix operator is not necessarily positive-definite, thus leading to highly non-physical results. To overcome this serious limitation, we have identified an alternative adiabatic procedure which (i) in the semiclassical limit reduces to the standard Fermi's golden rule, and (ii) describes a genuine Lindblad evolution, thus providing a reliable/robust treatment of energy-dissipation and dephasing in state-of-the-art quantum devices.

We stress again that, contrary to standard master-equation formulations, the proposed adiabatic-decoupling approach does not involve/require any reduction or average procedure, exactly as for the standard derivation of the conventional Fermi's golden rule. Moreover, the quality of the proposed adiabatic decoupling is confirmed by our reduced-description analysis: the resulting electronic scattering superoperators, describing the interaction of our electronic quantum system with a classical quasiparticle field, are still Lindblad-like, thus confirming the good property of such effective electronic description. However, the quantum nature of the quasiparticle subsystem —resulting in additional spontaneous-emission contributions— cannot be described via a Lindblad-like superoperator; to overcome this limitation, effective/approximated thermal models, not discussed here, may be introduced.

At this point a few final comments are in order. As discussed extensively in Ref. 24, also for the simplest case of a standard two-level system —i.e., a generic quantum bit— the standard Markov superoperator predicts a non-trivial coupling between level population and polarization described by the so-called T_3 contributions. In contrast, for a two-level system coupled to its environment, the proposed quantum Fermi's golden rule does not predict any T_3 coupling term, thus providing a rigorous derivation of the well-known and successfully employed T_1T_2 dephasing model [3]. Moreover, it is imperative to stress that in the presence of a strong system-environment interaction the adiabatic decoupling investigated so far needs to be replaced by more realistic treatments, expressed via non-Markovian integro-differential equations of motion (i.e., with "memory effects") [5]. Again, while for purely atomic and/or photonic systems it is possible to identify effective non-Markovian evolution operators [25], for solid-state quantum devices this is still an open problem.

Acknowledgments

We are grateful to David Taj and Paolo Zanardi for stimulating and fruitful discussions.

References

[1] See, e.g., C. Jacoboni and P. Lugli, *The Monte Carlo Method for Semiconductor Device Simulations* (Springer, Wien 1989).
[2] See, e.g., *Hot Carriers in Semiconductor Nanostructures: Physics and Applications*, edited by J. Shah (Academic Press inc., Boston, 1992); *Theory of Transport Properties of Semiconductor Nanostructures*, edited by E. Schöll (Chapman and Hall, London, 1998).
[3] See, e.g., J. Shah, *Ultrafast Spectroscopy of Semiconductors and Semiconductor Nanostructures* (Springer, Berlin, 1996).
[4] See, e.g., D. Bimberg *et al.*, *Quantum Dot Heterostructures* (Wiley, Chichester, 1998); L. Jacak, P. Hawrylak, and A. Wojs, *Quantum Dots* (Springer, Berlin, 1998); S.M. Reimann and M. Manninen, Rev. Mod. Phys. **74**, 1283 (2002).
[5] See, e.g., F. Rossi and T. Kuhn, Rev. Mod. Phys. **74**, 895 (2002).
[6] See, e.g., W. Frensley, Rev. Mod. Phys. **62**, 3 (1990); F. Rossi, A. Di Carlo, and P. Lugli, Phys. Rev. Lett. **80**, 3348 (1998); R. Proietti Zaccaria and F. Rossi, Phys. Rev. B **67**, 113311 (2003); D. Taj, L. Genovese, and F. Rossi, Europhys. Lett. **74**, 1060 (2006).
[7] See, e.g., *Physics of Quantum Electron Devices*, edited by F. Capasso (Springer, Berlin, 1990).
[8] See, e.g., C. Gmachl *et al.*, Rep. Prog. Phys. **64**, 1533 (2001); R. Koehler *et al.*, Nature **417**, 156 (2002); R.C. Iotti and F. Rossi, Rep. Prog. Phys. **68**, 2533 (2005).
[9] See, e.g., *Semiconductor Macroatoms: Basic Physics and Quantum-device Applications*, edited by F. Rossi (Imperial College Press, London, 2005); Focus on *Solid State Quantum Information*, edited by R. Fazio, New J. Phys. **7** (2005).
[10] See e.g., M. O. Scully and M. S. Zubairy, *Quantum Optics* (Cambridge University Press, Cambridge, 1997)
[11] See, e.g., E.B. Davies, *Quantum Theory of Open Systems* (Academic Press, London, 1976).
[12] G. Lindblad, Commun. math. Phys. **48**, 119 (1976).
[13] See, e.g., S.G. Schirmer and A. I. Solomon, Phys. Rev.A **70**, 022107 (2004).
[14] See, e.g., E. Fermi, *Nuclear Physics* (University of Chicago Press, 1950).
[15] See, e.g., L.P. Kadanoff and G. Baym, *Quantum Statistical Mechanics*, Benjamin, New York (1962).
[16] See, e.g., L.V. Keldysh, Sov. Phys. JETP **20**, 1018 (1965).
[17] See, e.g., H. Haug and A.-P. Jauho, *Quantum Kinetics in Transport and Optics of Semiconductors* (Springer, Berlin, 1996).
[18] See, e.g., R. Lake and S. Datta, Phys. Rev. B **45**, 6670 (1992); C. Rivas *et al.*, Appl. Phys. Lett. **78**, 814 (2002).
[19] See, e.g., F. Rossi, R. Brunetti, and C. Jacoboni, in *Hot Carriers in Semiconductor Nanostructures: Physics and Applications*, edited by J. Shah (Academic Press inc., Boston, 1992), p. 153.
[20] See, e.g., N.N. Bogoliubov, *Lectures on Quantum Statistics* (Gordon and

Breach, New York, 1967).
[21] See, e.g., V.M. Axt and S. Mukamel, Rev. Mod. Phys. **70**, 145 (1998).
[22] See, e.g., R. Brunetti, C. Jacoboni, and F. Rossi, Phys. Rev.B **39**, 10781 (1989); P. Bordone *et al.*, Phys. Rev.B **59**, 3060 (1999).
[23] See, e.g., T. Kuhn, in *Theory of Transport Properties of Semiconductor Nanostructures*, edited by E. Schöll (Chapman and Hall, London, 1998), p. 173.
[24] R.C. Iotti, E. Ciancio, and F. Rossi, Phys. Rev.B **72**, 125347 (2005).
[25] See, e.g., A.A. Budini, Phys. Rev.A **74**, 053815 (2006).

Chapter 12

Transient Four-wave Mixing of Excitons in Quantum Dots from Ensembles and Individuals

Paola Borri

Cardiff University, School of Biosciences, Museum Avenue PO Box 911, Cardiff CF10 3US, United Kingdom

Wolfgang Langbein

Cardiff Univesity, School of Physics and Astronomy, The Parade, Cardiff CF24 3AA, United Kingdom

The dephasing time of an optical excitonic transition in a semiconductor quantum dot is of fundamental interest and of key importance in many recent applications. It sets the time scale during which the coherence of the excitonic transition is preserved and therefore quantum computing operations based on coherent-light matter interaction are possible. The dephasing time scale is strictly related to intrinsic carrier scattering mechanisms. Direct measurement of the exciton dephasing processes in epitaxially grown quantum dots has been realized in recent years using transient four-wave mixing, unveiling a complex dephasing scenario which has stimulated significant progress in related theoretical models. We review here the most important results on transient four-wave mixing of self-assembled InGaAs/GaAs quantum dot ensembles, with particular emphasis on the temperature-dependent dephasing time and related exciton-phonon interaction. The recent development of the heterodyne spectral interferometry technique has also allowed performing transient four-wave mixing on individual, localized excitons. We present results of this technique on excitons localized in monolayer islands of GaAs/AlAs quantum wells and in self-assembled CdTe/ZnTe quantum dots, with emphasis on the subjects of photon echo formation and coherent control and readout of the polarization.

Contents

12.1. Introduction

Nonlinear optical spectroscopy is a powerful technique to investigate the dynamics of charge carriers in semiconductors and semiconductor nanostructures. More specifically, the third-order nonlinearity probed in four-wave mixing (FWM) or spectral hole burning experiments can be used to determine dephasing times even in the presence of large inhomogeneous broadening where linear spectroscopy usually fails. These nonlinear methods were extensively used in the 90's to investigate dephasing of excitons in semiconductor quantum wells (QWs) [1]. However their application to epitaxially-grown semiconductor nanostructures of reduced dimensionality such as quantum wires and quantum dots (QDs) turned out to be quite difficult due to reasons of both signal strength and directional selectivity. Transient FWM on localized excitons in GaAs islands [2, 3] and II-VI epitaxially grown quantum dots [4, 5] was reported in the late 90's, both systems exhibiting stronger oscillator strengths (radiative lifetimes of 100-200 ps) than the strongly-confined InGaAs/GaAs self-assembled QDs.

Recently, with the implementation of a sensitive heterodyne detection, measurements of transient four-wave mixing with a large dynamic range were reported by us, investigating ensembles of self-assembled InGaAs/GaAs QDs with radiative lifetimes in the nanosecond range and substantial inhomogeneous broadening [6]. Among the epitaxially-grown QD systems, InGaAs/GaAs QDs are probably the most widely investigated. Coherent light-matter interaction in these systems is receiving increasing attention due to possible solid-state implementations in the emerging field of quantum information processing, as discussed in this book. Moreover, the application of self-assembled InGaAs/GaAs QDs to optoelectronic devices in the optical telecommunications wavelength window of 1.3-1.5 μm holds the promise of lower costs compared to the present InP-based technology, combined with expected superior performances arising from the reduced dimensionality [7]. Beside its fundamental interest, the knowledge of the dephasing time, inversely proportional to the homogeneous broadening, of an excitonic transition in a QD is of crucial importance for many of these applications. The dephasing time sets the time scale during which the coherence of the excitonic transition is preserved and therefore operations based on coherent-light matter interaction are possible. This time scale is strictly related to intrinsic mechanisms such as the coupling to the radiation field, carrier-phonon scattering and carrier-carrier scattering. Its study yields a direct probe of those carrier dynamics which ultimately are limiting the high-speed performances of QD-based lasers/amplifiers. The direct measurement of the exciton dephasing processes in self-assembled InGaAs/GaAs QDs using transient FWM has lead to several interesting, and sometimes surprising discoveries and has stimulated many other experimental and theoretical studies. We will review here our most important results on transient FWM of self-assembled InGaAs/GaAs QDs, with particular emphasis on the temperature-dependent dephasing time and related exciton-phonon interaction.

Although transient FWM on QD ensembles is a valuable tool to measure dephasing times in inhomogeneously broadened systems, this type of experiments performed on single QDs are of significant benefit to clearly distinguish between processes occurring within one dot from those occurring in different dots and averaged in the ensemble response. With the recent introduction of the heterodyne spectral interferometry (HSI) technique, we were able to perform transient FWM measurements on individual, localized excitons [8–11]. We will review here the results of single–dot transient FWM performed on individual excitonic transitions localized in monolayer islands of GaAs/AlAs QWs and in self-assembled CdTe/ZnTe QDs, with emphasis on the subjects of photon echo formation and coherent control and readout of the polarization. The article is organized as follows. In Section 12.2 we recall the basic concepts of transient FWM in homogeneously and inhomogeneously broadened two-level systems. Section 12.3 is aimed to give a formal description of the heterodyne detection technique, including HSI which is discussed in subsection 12.3.1. Transient FWM in InGaAs/GaAs QD ensembles is discussed in Section 12.4 with particular attention to the aspect of dephasing by acoustic-phonon interactions. These interactions result in a non-Lorentzian homogeneous lineshape discussed in 12.4.1, and in a temperature-dependent broadening of the zero-phonon linewidth discussed in 12.4.2. The recent achievement of measuring transient FWM on single QDs is addressed in Section 12.5. The subjects of photon echo formation and coherent control in the regime of optical Rabi oscillations are discussed in 12.5.1 and 12.5.2, respectively.

12.2. Transient Four-Wave Mixing: Background

The dephasing time, conventionally called T_2, of an optical transition can be measured experimentally in many ways. The homogeneous width of the absorption spectral lineshape is inversely proportional to the dephasing time, thus experiments can be performed either in the time domain to directly address the transient decay of the polarization induced by a pulsed coherent light field or in the spectral domain by measuring the steady-state optical absorption lineshape. Generally, the response of the medium to the incident field depends on the field intensity. For example, only in the linear response limit (i.e. in the first order of the incident field amplitude) the absorption lineshape is Lorentzian with the energy full-width at half-maximum (FWHM) given by $\gamma = 2\hbar/T_2$. At higher orders, effects such as the power broadening [12], the quadratic Stark shift [13], and the optical Stark splitting (the signature of Rabi oscillations in the spectral domain) become important.

One obvious limitation of linear spectroscopy in determining the homogeneous lineshape is inhomogeneous broadening, which has to be deconvoluted from the total line broadening. In the time domain, this translates into an additional decay rate (sometimes indicated as $1/T_2^*$ [14]) of the macroscopic first-order polarization inversely proportional to the inhomogeneous spectral width σ. In strongly inhomogeneously broadened systems, linear spectroscopy usually fails in measuring the dephasing time since $\gamma \ll \sigma$. Note that a recently developed linear technique based on a time-resolved speckle-analysis of the resonant Rayleigh scattering in systems with static disorder allows the measurement of the T_2 time also in the presence of large inhomogeneous broadening [15].

Even when isolating one system from the inhomogeneous ensemble, linear spectroscopy might be severely affected by background light. This is, for example, the case when light is detected in the transmitted or reflected directions (which are the directions of propagation of the first order polarization in planar samples) and only a small part of it contains the effect of the investigated resonance. The analysis of secondary emission, such as resonant Rayleigh scattering or photoluminescence might overcome the problem. Indeed, recent achievements in performing photoluminescence spectroscopy with high spatial and spectral resolutions allowed extracting the homogeneous broadening from isolated excitons weakly confined by the lateral disorder in thin GaAs QWs [16] and in strongly confined single InGaAs QDs [17].

Experiments based on third-order signals, such as four–wave mixing in the transient coherent domain after pulsed excitation [1] or spectral hole-burning in the frequency domain with continuous–wave excitation [18] allow one to overcome the presence of an inhomogeneous distribution and the signal can be detected free of background with appropriate selection in the direction and/or frequency domain. The formal treatment of third–order non-linearities can be found in Ref. [13] while the application of transient FWM to measurement of the dephasing time in semiconductors is reviewed in Refs. [1, 19]. A comprehensive review of laser spectroscopy including the subjects of nonlinear, coherent and time-resolved spectroscopy can be found in Ref. [18].

Let us recall the key properties of transient degenerate two-beam FWM. In this configuration, two exciting pulses are used, spectrally centered at the same optical frequency ω which is usually in or near resonance with the optical transition of interest. The pulses are described by the complex electric fields $\mathcal{E}_{1,2}(t)$ centered at time $t = 0$, and can be delayed by a variable delay time $\tau_{\rm P}$ so that the total exciting field is $\mathcal{E}(t) = \mathcal{E}_1(t) + \mathcal{E}_2(t - \tau_{\rm P})$. To separate effects at the different orders in the incident field amplitude, the population and the polarization (or the diagonal and off-diagonal elements of the density matrix) can be expanded in a Taylor series [1, 13]. Consequently, the optical Bloch equations separate in a series of equations which can be truncated to a desired order. This perturbation approach allows one to describe nonlinear signals associated with the third-order polarization, which for example in media with inversion symmetry is the lowest nonlinear polarization allowed in the electric-dipole approximation.

In extended media, one can discriminate the orders experimentally using defined wavevectors (i.e. directions) $\vec{k}_{1,2}$ of the excitation fields, in which case the nth-order polarization $P^{(\mathbf{n})} \propto \mathcal{E}_1^{n_1} \mathcal{E}_1^{*m_1} \mathcal{E}_2^{n_2} \mathcal{E}_2^{*m_2}$ with $\mathbf{n} = (n_1, m_1, n_2, m_2)$, where the exponents $n_{1,2}$ and $m_{1,2}$ are natural numbers including zero, emits in the direction $l_1\vec{k}_1 + l_2\vec{k}_2$ with $l_{1,2} = n_{1,2} - m_{1,2}$. For example, FWM signal creating an echo for positive $\tau_{\rm P}$ corresponds to $n_1 = 0, m_1 = 1, n_2 = 2, m_2 = 0$ and is emitted in the direction $2\vec{k}_2 - \vec{k}_1$, which can be intuitively understood as the direction into which pulse 2 is diffracted by the density grating created by the interference of pulses 1 and 2. Since phase matching conditions (i.e. energy and momentum conservation) result in a critical thickness for the mixing process to be efficient [13], this directional selection geometry is suited for thin films. For sub-wavelength sized systems like individual excitonic transitions, the broken translational invariance prohibits the use of such a wavevector selection.

On the other hand, we can still use the time invariance provided that there is a temporal stability of the investigated structure over the course of the experiment. In this case, the orders of the nonlinear polarization can be discriminated using a frequency selection scheme. In a degenerate resonant case this is obtained by slightly shifting the optical frequencies of pulses 1 and 2, for example by radio-frequency amounts $\Omega_{1,2}$ using acousto-optics modulators, and by repeating the experiment using pulse trains which then exhibit controlled phase variations given by $\exp(i\Omega_{1,2}t)$. Such a frequency selection scheme, combined with an interferometric detection of the FWM field amplitude, is the essence of the heterodyne detection scheme which we used in our experiments. A detailed description of this technique can be found in the following section 12.3.

The time evolution of the third–order FWM polarization created by $\mathcal{E}(t) = \mathcal{E}_1(t) + \mathcal{E}_2(t - \tau_\mathrm{P})$ can be solved analytically assuming delta-like pulses of amplitudes $\mathcal{E}_{10}$, $\mathcal{E}_{20}$, an approximation physically valid for pulse durations much shorter than the dephasing time of the optical transition under consideration. For a system described as a two-level optical transition without inhomogeneous broadening and with a single exponential decay of the polarization given by T_2 such solution has the form [1]:

$$P^{(3)}_{\mathrm{FWM}} \propto \mu^4 e^{-i\omega t} \mathcal{E}^*_{10}\mathcal{E}^2_{20}\theta(\tau_\mathrm{P})\theta(t-\tau_\mathrm{P})e^{-(g-i\omega)(t-\tau_\mathrm{P})}e^{-(g^*+i\omega)\tau_\mathrm{P}} \,. \qquad (12.1)$$

The following notations are used: μ projection of transition dipole moment along the field polarization direction (assuming co-polarized exciting fields), $g = i\omega_0 + 1/T_2$ (ω_0 resonance frequency of the two-level transition) and θ the Heaviside function. The above solution also assumes that the delta-like pulse approximation is valid compared to the time scale $1/(\omega-\omega_0)$. Apart from the oscillatory terms (which reduce to $e^{-i\omega_0 t}$ at resonance $\omega = \omega_0$) the polarization decays exponentially with a time constant T_2 and is non-zero only when $\tau_\mathrm{P} > 0$ and $t > \tau_\mathrm{P}$. The corresponding time-resolved FWM intensity $\propto |P^{(3)}_{\mathrm{FWM}}|^2$ is shown in the top part of Fig. 12.1. The time-integrated FWM intensity as a function of delay time τ_P is also decaying exponentially as shown in Fig. 12.1, and its decay constant can be used to measure the dephasing time. The phase conjugation of the g-terms in Eq. (12.1) which contain the two-level transition frequency ω_0 is of crucial importance when considering the FWM signal for an inhomogeneously broadened ensemble, i.e. if a distribution of resonance frequencies ω_{0_j} is present. At time $t = 2\tau_\mathrm{P}$ the term $e^{-(g-i\omega)(t-\tau_\mathrm{P})}e^{-(g^*+i\omega)\tau_\mathrm{P}}$ simplifies into $e^{-2\tau_\mathrm{P}/T_2}$ independent of ω_{0_j}. Therefore, the superposition of the third-order

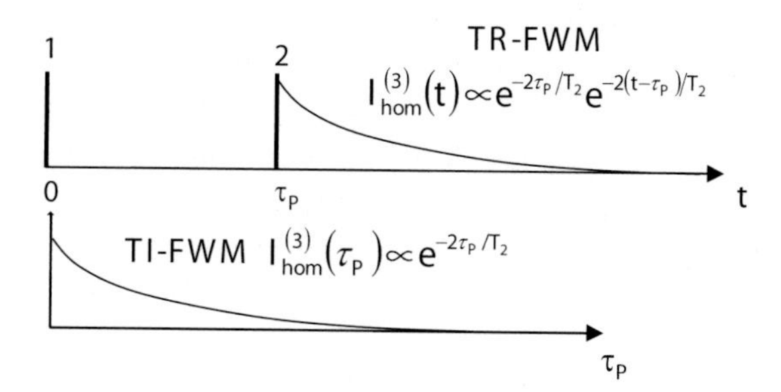

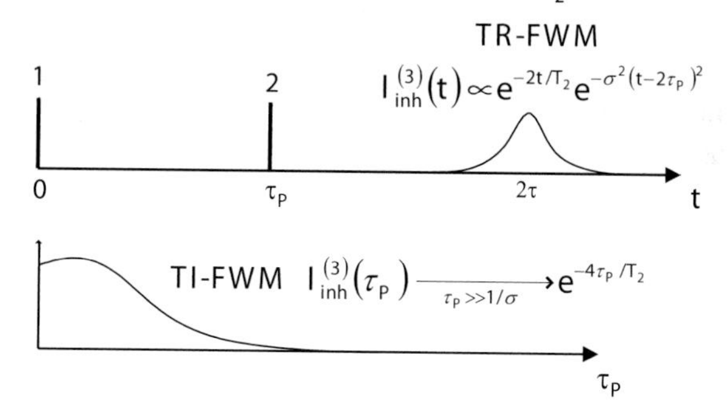

Fig. 12.1. Sketch of the time-resolved four-wave mixing (TR FWM) intensity and of the time-integrated FWM (TI-FWM) intensity versus delay time τ_P between two exciting delta-pulses for both a homogeneously broadened and an inhomogeneously broadened two-level system.

polarizations from N two-level systems with frequencies ω_{0_j} will manifest in a constructive interference at time $t = 2\tau_P$ with a total FWM amplitude N times larger than the individual FWM amplitudes of the single transitions (assuming identical amplitudes of the individual third-order polarizations). For other times, the phases are, in general, randomly distributed due to the distribution of the transition frequencies, and the total FWM amplitude is only about $\sqrt{N}$ times the individual amplitude. In the limit of a large number of systems in the ensemble, the signal at $t = 2\tau_P$ is thus far larger than at other times, and is called a photon echo. A numerical simulation showing the echo formation in a transient FWM experiment for a finite number N of two-level systems with resonance frequencies distributed according to a Gaussian function is shown in Fig. 12.2.

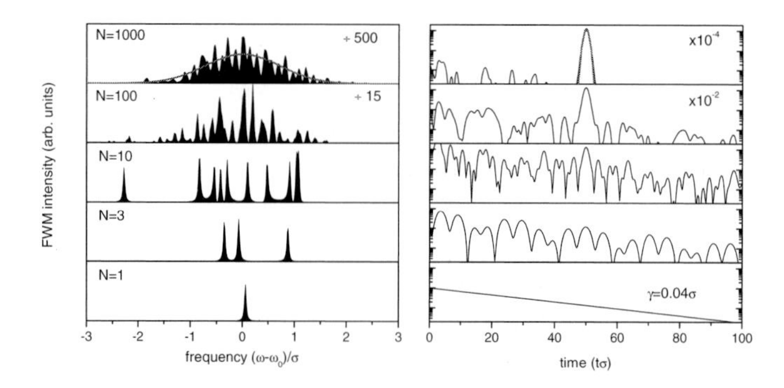

Fig. 12.2. Simulated spectrally resolved FWM intensity (left) and time-resolved FWM intensity (right) for ensembles of N two-level systems with equal transition dipole moment μ and homogeneous linewidth $\gamma = 0.04\sigma$, but transition frequencies ω_k taken from a Gaussian distributed ensemble of standard deviation σ (indicated as red line in the $N = 1000$ FWM spectrum, together with the corresponding FWM echo in the FWM dynamics).

Let us consider an inhomogeneously broadened ensemble with N so large that it can be described by a continuous distribution of transition frequencies according to the Gaussian

$$G(\omega_0') = \frac{1}{\sigma\sqrt{2\pi}} \exp\left[-\frac{(\omega_0' - \omega_0)^2}{2\sigma^2}\right] \tag{12.2}$$

where $NG(\omega_0')d\omega_0'$ is the number of systems with transition frequencies between ω_0' and $\omega_0'+d\omega_0'$. The third-order polarization in Eq. (12.1) summed over the inhomogeneous distribution is then given by the integral:

$$\int_{-\infty}^{+\infty} P^{(3)}_{\mathrm{FWM}}(\omega_0', t) NG(\omega_0') d\omega_0' = NP^{(3)}_{\mathrm{FWM}}(\omega_0, t) \exp\left[-\frac{\sigma^2(t - 2\tau_{\mathrm{P}})^2}{2}\right] \tag{12.3}$$

Eq. (12.3) expresses the result that the FWM of a inhomogeneously broadened large ensemble is a photon echo in real time, i.e. it appears as a pulsed signal after a time interval τ_{P} from the second exciting pulse [1]. Note that with similar considerations as for Eq. (12.3) one can calculate that the time evolution of the *first-order* macroscopic polarization of a inhomogeneously broadened large ensemble is given by the product of an exponential decay with $1/T_2$ time constant and a Gaussian decay inversely proportional to the frequency width of the inhomogeneous distribution. Therefore, when $\sigma \gg 1/T_2$ the dynamic of the macroscopic first-order polarization is dominated by the fast decay from the inhomogeneous broadening and the measure of its time-evolution is not a good probe of the dephasing time T_2. Measuring the third-order FWM signal associated with the solution in Eq. (12.3) overcomes the problem of the inhomogeneous broadening since the photon echo acts as a probe of the homogeneously broadened polarization at the time $t = 2\tau_P$. A sketch summarizing the real-time evolution of the FWM intensity as well as the time-integrated FWM (TI FWM) versus τ_P is shown in Fig. 12.1. Also for the case of a inhomogeneously broadened large ensemble, the TI FWM at long delay times decays exponentially with a time constant proportional to T_2 and can be thus used to measure the dephasing time.

12.3. Heterodyne Detection

An arrangement illustrating the heterodyne technique used in our transient FWM experiments is shown in Fig. 12.3. The pulse train provided by a mode-locked laser source is divided into co–polarized pump (pulse 1), probe (pulse 2), and reference pulse trains. The spectrum of the pulse train consists of a series of frequency modes separated by the repetition rate. Formally, the electric field describing the pulse train can be written as [20]:

$$E(t) = e^{-i\omega_0 t} \sum_n A_n e^{-in\Omega_{\text{rep}} t} + c.c. \tag{12.4}$$

Eq. (12.4) describes a sequence of pulses with a time period of $2\pi/\Omega_{\text{rep}}$. The right side of Eq. (12.4) (without *c.c.*) represents the complex field $\mathcal{E}(t)$ and has the form of a Fourier series with the components A_n. The envelope of the distribution of $|A_n|^2$ gives the spectrum of a single pulse in the train.

In the heterodyne experiment, the pump and probe pulse sequences are frequency upshifted in the range of several tens of MHz by acousto–optic modulators (AOM). We assume that the pulse duration is much smaller than the period of the AOM radio-frequency $(\nu_{\text{AOM}})^{-1} = 2\pi/\Omega_{\text{AOM}}$, i.e. that the spectral width $\Delta\omega$ of the envelop distribution of $|A_n|^2$ is much larger than Ω_{AOM}. This assumption is satisfied for pulses of subpicosecond duration and AOM frequencies in the MHz range. The pump pulses are time-shifted by τ_P relative to the probe pulses, positive for pump pulses leading, by an optical delay line (in the pump beam for Fig. 12.3). After being recombined by a beam splitter (C_2), pump and probe pulses are coupled co–linearly into the investigated sample, which consisted in our

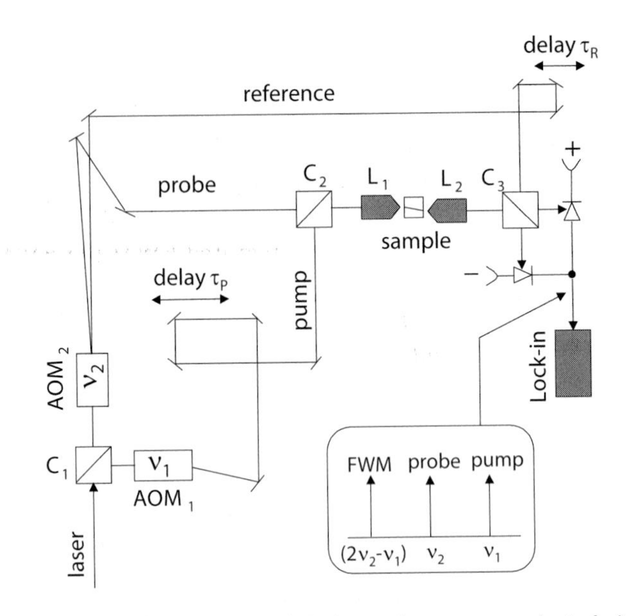

Fig. 12.3. Experimental arrangement of the heterodyne pump-probe technique. C: cube beamsplitter; AOM: acousto–optic modulator; L: lens.

earlier work [6] of InGaAs/GaAs QDs embedded in the active region of a narrow–ridge waveguide with tilted facets to avoid back reflections. In this case, optimum light coupling into the sample was obtained using a lens (L_1) matching the high-numerical aperture of the waveguide. The signal at the output of the waveguide (which can be either the transmitted probe in a differential-transmission experiment or the non-linear combination of pump and probe fields giving rise to the FWM) is collected by a high numerical aperture lens (L_2) and recombined with the reference pulse train in the beamsplitter C_3. The reference pulse sequence is obtained from the undeflected beam at the output of AOM_2 and it is not frequency shifted, but can be time-delayed by τ_R relative to the signal. Let us write formally the expression for the reference and signal pulse trains, respectively:

$$E_r(t) = e^{-i\omega_0(t-\tau_R)} \sum_n A_n^r e^{-in\Omega_{rep}(t-\tau_R)} + c.c. \tag{12.5}$$

$$E_s(t) = e^{-i(\omega_0+\Omega_s)t} \sum_n A_n^s e^{-in\Omega_{rep}t} + c.c. \tag{12.6}$$

In differential transmission, $\Omega_s/2\pi = \nu_2$ (i.e the radiofrequency of AOM_2), while in FWM is $\Omega_s/2\pi = 2\nu_2 - \nu_1$. The presence of the pump influences A_n^s which will be a function of τ_P.

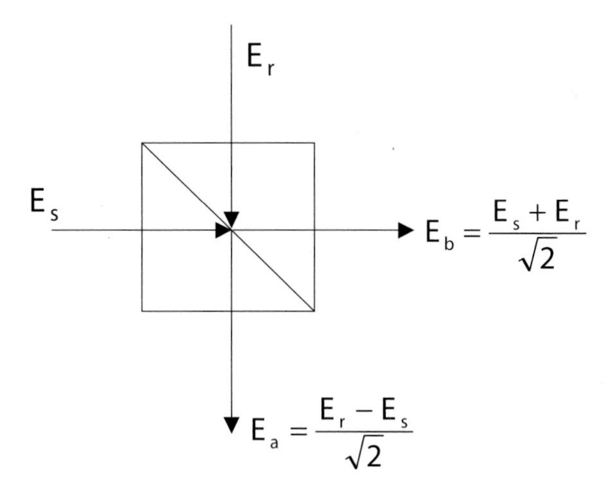

Fig. 12.4. Electric field transmission/reflection scheme for a 50:50 beam splitter.

The interference between signal and reference beams in C_3 can be understood by considering the field transmission/reflection scheme shown for a 50:50 beam splitter in Fig. 12.4. Each photodiode in the detection scheme shown in Fig. 12.3 generates a photo-current proportional to the square of the incoming field time-integrated over the response time T_{det} of the diode. In the differential scheme shown in Fig. 12.3 we detect the photocurrent difference $I_{\text{d}} = I_{\text{b}} - I_{\text{a}}$:

$$\begin{aligned} I_{\text{d}} &\propto \int_{T_{\text{det}}} dt (E_{\text{b}}^2 - E_{\text{a}}^2) \propto \int_{T_{det}} dt E_{\text{r}} E_{\text{s}} \\ &= \int\limits_{T_{\text{det}}} dt \left[e^{-i(2\omega_0 + \Omega_{\text{s}})t + i\omega_0 \tau_{\text{R}}} \sum_{n,m} A_n^{\text{s}} A_m^{\text{r}} e^{-i(n+m)\Omega_{\text{rep}} t + im\Omega_{\text{rep}} \tau_{\text{R}}} \right. \\ &\quad \left. + e^{-i\Omega_{\text{s}} t - i\omega_0 \tau_{\text{R}}} \sum_{n,m} A_n^{\text{s}} A_m^{*\text{r}} e^{-i(n-m)\Omega_{\text{rep}} t - im\Omega_{\text{rep}} \tau_{\text{R}}} + c.c. \right] \end{aligned} \tag{12.7}$$

where Eqs. (12.5,12.6) have been used. The terms in Eq. (12.7) which are rapidly oscillating with $2\omega_0$ average to zero over $T_{\text{det}} \gg 2\pi/\omega_0$. Conversely, terms oscillating at the frequency Ω_{s} or smaller can be time resolved by the detector, i.e. $T_{\text{det}} < 2\pi/\Omega_{\text{s}}$. The lock-in that detects the oscillating photocurrent (see Fig. 12.3) can be synchronized, with an appropriate electrical reference input, to the terms in Eq. (12.7) oscillating with Ω_{s}, i.e. those with $m = n$, or to the terms oscillating with lower frequency

$\Omega_s - l\Omega_{rep}$ with l natural, i.e. those with $m = n + l$. For example, if Ω_s is only slightly larger than Ω_{rep} one can choose $m = n + 1$ for which we find

$$I_d \propto e^{-i(\Omega_s - \Omega_{rep})t} e^{-i(\omega_0 + \Omega_{rep})\tau_R} \sum_n A_n^s A_{n+1}^{*r} e^{-in\Omega_{rep}\tau_R} + c.c. \quad . \tag{12.8}$$

The integration over T_{det} does not appear in Eq. (12.8) since the oscillation with frequency $\Omega_s - \Omega_{rep}$ is slowly varying over T_{det}, i.e. can be time–resolved by the diodes. Note that the τ_R-dependent term which multiplies the term $e^{-i(\Omega_s - \Omega_{rep})t}$ has the form of a Fourier series with the n-th Fourier component being $A_n^s A_{n+1}^{*r}$. The previously imposed condition $\Delta\omega \gg \Omega_{AOM}$ implies that we can approximate A_{n+1}^{*r} with A_n^{*r}. The resulting term $A_n^s A_n^{*r}$ is the n-th Fourier component of the cross–correlation $C(\tau_R)$ between the complex reference and signal fields $\mathcal{E}_{r,s}$ described by the first part of the rhs of Eq. (12.5,12.6), respectively. Detecting the photocurrent with a dual-phase lock-in both the real and the imaginary part of $C(\tau_R)$ can be measured, i.e. one is sensitive to pump–induced amplitude *and* phase changes of the signal.

As shown in Fig. 12.3, the distinction between pump, probe and FWM is made in heterodyne detection as follows. The interference between probe and reference pulses gives rise to a beating signal at the radiofrequency ν_2 of AOM_2 (and its sidebands separated by the repetition rate of the pulse train). The interference of pump and reference beams also gives rise to a beating signal but at the frequency ν_1 of AOM_1 and its sidebands. Similarly, the interference between FWM signal and reference gives rise to a beating signal at the frequency $2\nu_2 - \nu_1$ and its sidebands which again can be distinguished from ν_1 and ν_2.

In the first version of this technique described in Refs. [21, 22], the laser source was at high repetition rate (~ 100 MHz), probe and reference beams were frequency shifted by AOMs with 1 MHz frequency difference, and the frequency filtering of the interference signal between probe and reference beams was achieved by using a high–frequency radio receiver at 1 MHz. For differential measurements, modulations were induced by chopping the pump–beam and their amplitude was measured by a lock–in amplifier after the radio receiver. Switching between amplitude-modulation and frequency-modulation operations of the radio receiver allowed to alternatively detect pump–induced amplitude or phase changes of the transmitted probe field, i.e gain or refractive index changes in a semiconductor optical amplifier with bulk or QW active medium. However, for the measurements of refractive index changes, a calibration with respect to a known optical phase change was necessary. Moreover, in these previous experiments the full amplitude noise from the laser source was detected.

A non–degenerate version of the heterodyne pump-probe was also demonstrated, with pump and probe pulses having independent wavelength and pulse widths. This technique provided both time and spectral–domain information and was utilized on InGaAs/AlGaAs strained-layer single QW optical amplifiers [22, 23].

It was pointed out theoretically [24] and experimentally [25] that the heterodyne detection allows measurement of transient four-wave mixing in active waveguides and overcomes the problem of a critical interaction length that occurs in FWM experiments with spatial selection geometry, since phase–matched conditions are met with co–linear propagation. Measurements of the dephasing time performed with FWM in heterodyne detection were reported on bulk InGaAsP optical amplifiers operating at $1.55\,\mu$m at room temperature [25, 26], and indicated a dephasing time of ~ 100 fs.

We demonstrated a novel version of the detection scheme [26], with a dual-phase lock–in amplifier and a balanced detector for signal recovery. Two photodiodes detect the signal in the differential scheme shown in Fig. 12.3 to reject the common mode laser noise. By creating an electric signal for the reference channel of the lock–in, synchronized with the optical interference signal, the lock–in amplifier directly recovers the interference signal at the proper beat frequency and provides amplitude and phase measurements at the same time, with no need of absolute phase calibration. This novel detection scheme was first demonstrated using a low (300 kHz) repetition rate laser system and measuring the interference signal at the lowest sideband between the AOM frequency and the laser repetition rate, which was detectable by a low–frequency ($\leq$ 100 kHz) lock–in and a low–frequency balanced detector [26]. Using this technique we also pointed out [27] that the coherent nonlinearity affecting the differential transmission signal around zero pump-probe delays (the so–called coherent artifact) can be experimentally separated from the incoherent transient by measuring its mirror component, i.e. the background–free FWM at $2\nu_1 - \nu_2$.

For the experiments which will be discussed in Section 12.4, a high rep-etition rate (76 MHz) laser source was used. In this case, high–frequency photodiodes and a high-frequency lock-in ($\leq$ 200MHz) were necessary. The pump pulse train was frequency shifted by $\nu_1 = 80$ MHz while the probe pulse train was shifted by $\nu_2 = 79$ MHz. Differential transmission and FWM were detected at the lower sideband of ν_2 and $2\nu_2 - \nu_1$ separated by the laser repetition rate, i.e. at 3 MHz and 2 MHz respectively. Moreover, the sample was held in a specially designed cryostat for temperature-dependent measurements, allowing for light–coupling in and out of the waveguide with high numerical aperture.

12.3.1. *Heterodyne spectral interferometry*

We here discuss more in detail the modifications to the heterodyne detection created to allow the measurement of transient FWM of individual quantum dots [9]. A schematic diagram of the experimental setup is given in Fig. 12.5. Similar to the set-up described in the previous subsection, optical pulses of about 150 fs duration at a repetition rate of $\Omega_{\rm rep} = 2\pi \times 76\,$MHz were produced by a mode-locked Ti:sapphire laser and were split into two beams $B_{1,2}$ that were subsequently phase and frequency shifted with AOMs by $\varphi_{1,2} = \Omega_{1,2}t$ with $\Omega_{1,2} = 2\pi \times (79, 80)\,$MHz. They were then recombined in a beam-splitter cube into the same spatial mode $B_{\rm s}$, with a relative delay time $\tau_{\rm P}$, positive for pulse 1 leading. A reference beam $B_{\rm r}$ (the part of B_2 not deflected by the AOM) was propagating close to $B_{\rm s}$ and filled, as with $B_{\rm s}$, the far-field plane (FF) of the microscope objective (MO). By passing through the same optical elements after the beam splitter cube, phase fluctuations between $B_{\rm r}$ and $B_{\rm s}$ were minimized. The path over which the beams travel separately was encapsulated, minimizing phase fluctuations due to air turbulence. The whole setup was temperature stabilized. The beams were coupled into the MO using a beam splitter of only 4% reflectivity in order to reduce the losses in the subsequent transmission in the detection path. The beams were focussed onto the sample plane, called the near-field (NF), by an achromatic MO of 0.85 numerical aperture. This MO was mounted in the cryostat sample chamber on a piezoelectric translation stage, allowing for fine positioning of the focus with 10 nm sensitivity. The response field emitted in the reflected direction was collected by the same MO and spatially filtered at an intermediate image plane to keep only the zeroth order of the Airy diffraction created by the objective aperture for both $B_{\rm r}$ and $B_{\rm s}$ (see NF image in Fig. 12.5). From there, a dual lens system (L2,L3) imaged the FF into a mixing AOM, in which $B_{\rm r}$ and $B_{\rm s}$ spatially overlap. The NF was instead imaged into the directions at the AOM, so that the spatial separation between $B_{\rm r}$ and $B_{\rm s}$ in the NF ($\approx 9\,\mu$m) corresponded to different directions in the AOM. The angle between these directions was chosen to match the AOM Bragg diffraction angle (0.015 rad). By this arrangement, the diffracted beam of $B_{\rm r}$ overlapped with $B_{\rm s}$ and vice versa.

The resulting mixed beams $B_{\mathrm{a,b}}$ passed through a dual lens system (L4,L5) that imaged the NF into the input slit of a high resolution (15 μeV) imaging spectrometer where $B_{\mathrm{a,b}}$ were spatially separated, while the FF was imaged onto the spectrometer grating to avoid vignetting. The spectrally and time-resolved intensities $I_{\mathrm{a,b}}(\omega, t)$ of $B_{\mathrm{a,b}}$ were detected by a liquid nitrogen cooled silicon charge coupled device (CCD) at the output focus plane of the spectrometer. The mixing AOM was driven with an electric field $\propto \cos(\Omega_{\mathrm{D}} t + \varphi)$, so that B_{s} acquired a phase shift $-\Omega_{\mathrm{D}} t - \varphi$ when diffracted into B_{r}, while B_{r} was phase-shifted by $\Omega_{\mathrm{D}} t + \varphi$ when diffracted into B_{s}, where φ is an adjustable phase offset. The diffraction efficiency was adjusted to $\sim 50\%$. This scheme provides a simultaneous measurement of all spectral components of the signal in both amplitude and phase, allowing the determination of the signal in both frequency- and time-domain by Fourier-transform.

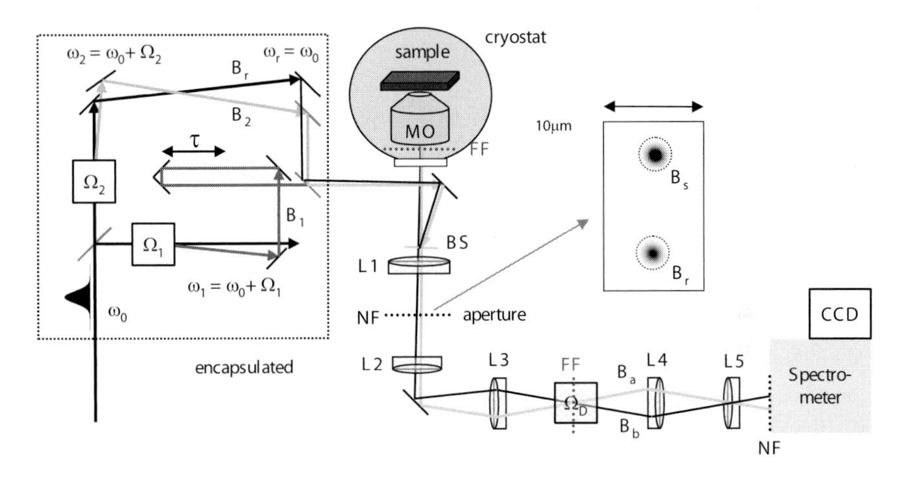

Fig. 12.5. Scheme of the experimental setup. Boxes: Acousto-optical modulators of the indicated frequency. MO: High numerical aperture (0.85) microscope objective, L1-L5 achromatic doublet lenses. Spectrometer: Imaging spectrometer of 15 μeV resolution. NF: near field of sample, FF: far-field of sample.

To describe the interference of the reference and signal fields $E_{\mathrm{r,s}}$ detected in the heterodyne spectral interferometry, we express a pulse train in time domain as a superposition of short pulses with envelope $A(t)$ and central optical frequency ω_0 separated by the time $\tau_{\mathrm{rep}} = 2\pi/\Omega_{\mathrm{rep}}$. This approach is complementary to what shown in the previous subsection, where a Fourier series in frequency domain was used. The unshifted reference field is given in this notation by:

$$E_{\mathrm{r}}(t) = e^{-i\omega_0 t} \sum_n A_{\mathrm{r}}(t - n\tau_{\mathrm{rep}}) + c.c. = \mathcal{E}_{\mathrm{r}} + c.c. \tag{12.9}$$

When frequency shifted by an AOM, we can assume that the pulse duration is much shorter than the inverse of the AOM frequency, so that the AOM generates a phase shift which is constant over the pulse envelope, but is different between the pulses in the train. The signal field E_{s} generated by the exciting pulse trains can be separated into terms proportional to the polarization $P_{\mathrm{s}}^{l_1,l_2}$ which is the sum over all $P^{(\mathbf{n})}$ with equal l_1, l_2. The signal field $E_{\mathrm{s}}^{l_1,l_2}$ (i.e. containing $P_{\mathrm{s}}^{l_1,l_2}$) including the effect of the mixing AOM when diffracted into B_{r} can then be written as:

$$\begin{aligned} E_{\mathrm{s}}^{l_1,l_2}(t) &= e^{-i\omega_0(l_1+l_2)t} \sum_n e^{-i(l_1\Omega_1 + l_2\Omega_2 - \Omega_{\mathrm{D}})n\tau_{\mathrm{rep}}} A_{\mathrm{s}}^{l_1,l_2}(t - n\tau_{\mathrm{rep}}) + c.c. \\ &= \mathcal{E}_{\mathrm{s}}^{l_1,l_2} + c.c \end{aligned} \tag{12.10}$$

The spectrally resolved field $\mathcal{E}(\omega, t)$ detected after a grating spectrometer at the time t for an input field $\mathcal{E}(t)$ is given by [28]

$$\mathcal{E}(\omega, t) = \int_{t-\tau_{\mathrm{m}}}^{t} dt' e^{-i\omega(t-t')} \mathcal{E}(t'). \tag{12.11}$$

where τ_{m} is the maximum time difference in the spectrometer introduced by the grating, which is related to the spectral resolution intensity FWHM by $\Delta\omega \approx 5.6/\tau_{\mathrm{m}}$ and is of the order of 100 ps in our experiment. The detected spectrally and time-resolved intensity $I(\omega, t)$ is then up to a constant factor given by $I(\omega, t) = (\mathcal{E}(\omega, t) + c.c.)^2$, resulting for the intensity of beam B_{a} in

$$\begin{aligned} I_{\mathrm{a}}(\omega, t) &= (\mathcal{E}_{\mathrm{r}}(\omega, t) + \mathcal{E}_{\mathrm{s}}(\omega, t) + c.c.)^2 \\ &= 2\left[|\mathcal{E}_{\mathrm{r}}(\omega, t)|^2 + |\mathcal{E}_{\mathrm{s}}(\omega, t)|^2 + 2\Re(\mathcal{E}_{\mathrm{r}}^*(\omega, t)\mathcal{E}_{\mathrm{s}}(\omega, t))\right] \end{aligned} \tag{12.12}$$

where we have neglected terms oscillating with 2ω which are rapidly averaging to zero. Let us now focus on the interference term $I_{\mathrm{int}} = \mathcal{E}_{\mathrm{r}}^* \mathcal{E}_{\mathrm{s}}$. Since the repetition period τ_{rep} of the pulse train is assumed to be much longer than the duration of the reference pulse A_{r}, the signal field A_{s}, and the spectrometer time difference τ_{m}, only fields corresponding to the same repetition index n in Eq. (12.9) can interfere after the spectrometer, Eq. 12.11. Therefore we find for the contribution of the field $\mathcal{E}_{\mathrm{s}}^{l_1,l_2}$ to this interference term:

$$I_{\rm int}^{l_1,l_2}(\omega,t) = \sum_n \int_{t-\tau_{\rm m}}^{t} dt_{\rm r} e^{i\omega(t-t_{\rm r})} \int_{t-\tau_{\rm m}}^{t} dt_{\rm s} e^{-i\omega(t-t_{\rm s})} \times \tag{12.13}$$
$$A_{\rm r}^*(t_{\rm r}-n\tau_{\rm rep}) A_{\rm s}^{l_1,l_2}(t_{\rm s}-n\tau_{\rm rep}) e^{i\omega_0(t_{\rm r}-(l_1+l_2)t_{\rm s})} e^{-i(l_1\Omega_1+l_2\Omega_2-\Omega_{\rm D})n\tau_{\rm rep}}$$

We detect $I_{\rm int}(\omega,t)$ integrated over the CCD exposure time $T_{\rm e} \approx 50-1000\,{\rm ms}$, resulting in

$$I_{\rm int}^{l_1,l_2}(\omega) = \int_0^{T_{\rm e}} dt I_{\rm int}^{l_1,l_2}(\omega,t) = \tag{12.14}$$
$$I_{\rm single}^{l_1,l_2}(\omega) \times \sum_{n=0}^{N-1} e^{i\omega_0(1-l_1-l_2)n\tau_{\rm rep}} e^{-i(l_1\Omega_1+l_2\Omega_2-\Omega_{\rm D})n\tau_{\rm rep}}$$

with the number of repetitions within the exposure time $N = T_{\rm e}/\tau_{\rm rep}$, and the single repetition interference

$$I_{\rm single}^{l_1,l_2}(\omega) = \int_{-\infty}^{\infty} dt' \int_{t'-\tau_{\rm m}}^{t'} dt_{\rm r} e^{i\omega(t'-t_{\rm r})} \int_{t'-\tau_{\rm m}}^{t'} dt_{\rm s} e^{-i\omega(t'-t_{\rm s})} \times$$
$$A_{\rm r}^*(t_{\rm r}) A_{\rm s}^{l_1,l_2}(t_{\rm s}) e^{i\omega_0(t_{\rm r}-(l_1+l_2)t_{\rm s})} \tag{12.15}$$

Only signal fields with the same carrier frequency as the reference contribute to $I_{\rm int}^{l_1,l_2}(\omega)$, since fields with other carrier frequencies, such as second harmonic generation, create interferences oscillating with multiples of ω_0, averaging rapidly to zero. This leaves only terms with $l_1 + l_2 = 1$, and simplifies the summation in Eq. (12.14) to the frequency-filtering function

$$\sum_{n=0}^{N-1} e^{-in\varphi_0} = \frac{\sin(N\varphi_0/2)}{\sin(\varphi_0/2)} e^{-i(N-1)\varphi_0/2} \tag{12.16}$$

where $\varphi_0 = (l_1\Omega_1 + l_2\Omega_2 - \Omega_{\rm D})\tau_{\rm rep} = \Omega_{\rm i}\tau_{\rm rep}$. This function has maxima at $\varphi_0 = 2\pi m$, m integer, corresponding to $\Omega_{\rm i} = m\Omega_{\rm rep}$, around which it has the behaviour of a low-pass filter $\sin(x)/x$ with $x = T_{\rm e}(\Omega_{\rm i} - m\Omega_{\rm rep})/2$, suppressing signals with frequency differences larger than $4/T_{\rm e}$. By the choice of Ω_1, Ω_2 and $\Omega_{\rm p}$, the $\Omega_{\rm i}$ corresponding to different $l_{1,2}$ are differing in the MHz range. In $I_{\rm int}$, which is a sum over all $I_{\rm int}^{l_1',l_2'}$, we select $I_{\rm int}^{l_1,l_2}$ by using $\Omega_D = l_1\Omega_1 + l_2\Omega_2$, which shifts $\Omega_{\rm i}$ to zero, so that $I_{\rm int}^{l_1,l_2}$ is not suppressed in $I_{\rm int}$, while other $I_{\rm int}^{l_1',l_2'}$ are suppressed by factors of $T_{\rm e}\Omega_{\rm i}' > 10^6$.

To detect the interference term $I_{\rm int}(\omega)$ only, we measure the intensities of both beams $I_{\rm a,b}(\omega)$, and determine $I_{\rm m}(\omega) = I_{\rm a} - I_{\rm b}$. In order to suppress erroneous signal due to systematic variations of detection efficiencies in $I_{\rm a,b}(\omega)$, e.g. due to inhomogeneities in the CCD response, the offset phase φ

of the detection AOM drive is cycled between 0 and π in adjacent exposures, and we use $2I_{\rm d} = I_{\rm m}^{\varphi=0} - I_{\rm m}^{\varphi=\pi}$ to determine $\Re(I_{\rm int}) = I_{\rm d}/2$. In this way, the classical noise and systematic errors from the non-interfering term $|\mathcal{E}_{\rm r}|^2 + |\mathcal{E}_{\rm s}|^2$ are largely suppressed, and the weak interference signal can be detected shot-noise limited. Note that $\mathcal{E}_{\rm s}$ contains also the linear reflection of $\mathcal{E}_{1,2}(t)$, which is typically by far dominating its intensity, and that $|\mathcal{E}_{\rm r}|^2$ has to be significantly larger than $|\mathcal{E}_{\rm s}|^2$ to reach shot-noise limited detection of the interference term. This results in a weak interference due to $\mathcal{E}_{\rm s}^{l_1,l_2}$, typically below 10^{-3} relative modulation in $I_{\rm a,b}(\omega)$, which can only be detected in a well balanced scheme.

$\mathcal{E}_{\rm s}^{l_1,l_2}$ is determined in amplitude and phase by spectral interferometry [29] from the measured interference intensity $I_{\rm d}$, using the experimentally adjusted property of $\mathcal{E}_{\rm r}$ to precede the signal field in time t. Using this property, we can apply $F(\Theta(t)F^{-1}(I_{\rm d}(\omega))) = 2\mathcal{E}_{\rm r}^*(\omega)\mathcal{E}_{\rm s}^{l_1,l_2}(\omega)$ with the Heaviside function $\Theta(t)$, and the Fourier-transform operator F. The reference field amplitude can be determined by blocking $B_{\rm s}$ and measuring $I_{\rm a,b} = 2|\mathcal{E}_{\rm r}|^2$. The time-range of the resulting time-resolved signal field of this technique is limited by the spectral resolution of the spectrometer to $\tau_{\rm m}$, approximately 100 ps in our experiments. Since $\mathcal{E}_{\rm r}$ is the reference pulse reflected by the sample it is helpful to arrange that the reflection is not strongly modifying its spectrum. This can be achieved by a metal coating on the sample with an opening at the signal beam, or by using a surface reflection that is dominated by a non-resonant refractive index. Different $\mathcal{E}_{\rm s}^{l_1,l_2}$ including their relative phase can be measured sequentially during the passive phase stability time of the setup. In order to correct for long-term drifts of the relative phase between signal and reference within a measurement consisting of many CCD exposures, the phase of the measured signal is determined for subintervals of the total acquisition time. If the non-linear signal in the subintervals is too weak to determine the phase, one can monitor the phase drifts of $\mathcal{E}_{\rm s}^{1,0} \propto e^{i\varphi_1}$ and $\mathcal{E}_{\rm s}^{0,1} \propto e^{i\varphi_2}$ instead.

In this way, different material responses can be measured sequentially: For $\Omega_{\rm d} = \Omega_{1,2}$, the reflected excitation pulses 1,2 are measured, while for $\Omega_{\rm d} = 2\Omega_2 - \Omega_1$, the emitted FWM field $\propto \mathcal{E}_1^*\mathcal{E}_2^2$ is measured. Higher-order non-linearities, like six-wave mixing, can be detected analogously.

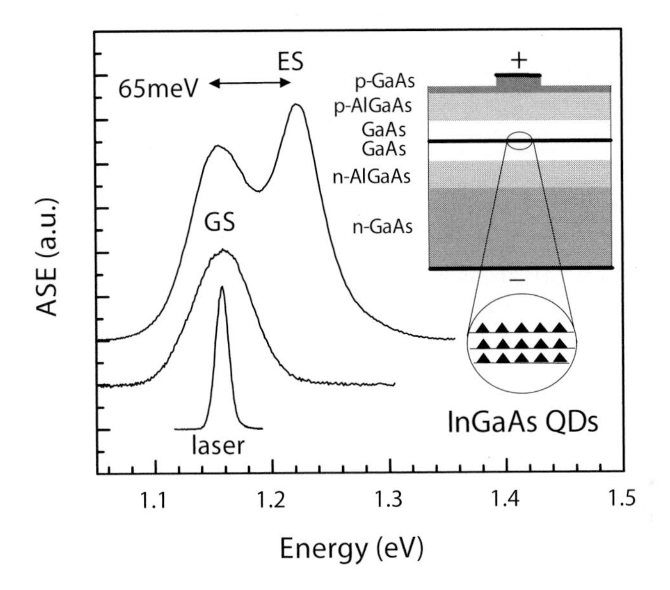

Fig. 12.6. Amplified spontaneous emission (ASE) spectra at 25K at low (0.5mA) and high (20mA) electrical injection current from an ensemble of self-assembled $In_{0.7}Ga_{0.3}As/GaAs$ QDs embedded in the active region of a p-i-n ridge waveguide. GS and ES indicate the ground-state and first optically active excited state transition, respectively. The spectrum of the exiting pulse used in the FWM experiment is also shown. After [6].

12.4. Transient FWM in InGaAs/GaAs Quantum Dot Ensembles

Using transient two-beam FWM in heterodyne detection, we measured the dephasing time and its temperature dependence in an inhomogeneously broadened ensemble of self-assembled $In_{0.7}Ga_{0.3}As/GaAs$ QDs embedded in the active region of a $5\,\mu m$ wide and 0.5 mm long ridge waveguide [6]. The sample consisted of 3 QD layers separated by 35-nm GaAs spacers in the intrinsic region of a p-i-n diode structure allowing for electrical injection and representing a QD-based semiconductor optical amplifier (a sketch of the structure is given in the inset of Fig. 12.6). Low–temperature amplified spontaneous emission (ASE) spectra shown in Fig. 12.6 allowed us to infer a Gaussian inhomogeneous broadening of 60 meV and an energy separation of 65 meV between the excitonic ground–state (GS) transition and the first optically-active excited state(ES) transition, visible for high injection currents for which saturation of the QD ground state occupation is reached. Photoluminescence on a reference sample showed a wetting layer transition ~210 meV above the dot ground-state, a signature of the strong carrier confinement in the QDs.

The FWM experiment was performed using exciting pulses with an optical frequency tuned to the center of the GS transition (as shown in Fig. 12.6) which was ranging from $\hbar\omega$=1.059 eV (1170 nm wavelength) at 300 K to 1.158 eV (1070 nm wavelength) at 7 K. At the center of the GS transition we estimated a negligible contribution (2%) of the ES to the optical density of states. The two exciting pulses were co-polarized and co-linearly propagating into the TE waveguide mode which, compared to TM polarization, maximizes light-matter interaction with the QD excitonic GS transition, according to polarization selection rules in these systems [30].

Transient FWM measurements at 50 K are shown in Fig. 12.7. The experiment was performed without injection current, i.e. the transition from the crystal ground state to the GS exciton was probed. In the figure, $\tau_{\rm R}$ measures the delay of the reference pulse with respect to pulse 2 (probe), i.e. pulse 1 (pump) is at $\tau_{\rm R} = -\tau_{\rm P}$. As expected for a dominantly inhomogeneously broadened ensemble, we observe a FWM signal centered at $\tau_{\rm R} = \tau_{\rm P}$, i.e. the time-resolved FWM is clearly a photon echo (see also Fig. 12.1). The TI FWM versus τ_P is obtained by integrating the TR FWM field amplitude versus τ_R, and it is shown in the right part of Fig. 12.7 for different pulse intensities of the exciting beams. In the figure the intensity of pulse 2, which was twice that of pulse 1, is indicated and I_0 corresponds to an energy per pulse $\leq$ 0.1pJ. Using this excitation intensity as pump intensity in differential transmission we found an absorption bleaching of ~10% relative to the maximum absorption of the investigated waveguide (~ 30 cm^{-1}, 1.5 optical density) corresponding to only 0.1 electron-hole pairs excited per dot on average. With increasing excitation intensity the TI FWM signal scaled as a third-order response and finally saturated at $\geq 8I_0$. Note also the presence of beats in the initial TI FWM dynamics. We interpreted these beats as exciton–biexciton beats [6, 31] and from the beat period we inferred a biexciton binding energy of 3 meV, consistent with other findings in the literature for InGaAs QDs [32]. At high excitation intensity the exciton–biexciton beats change, similar to what is observed in quantum wells when fifth-order contributions are important [33].

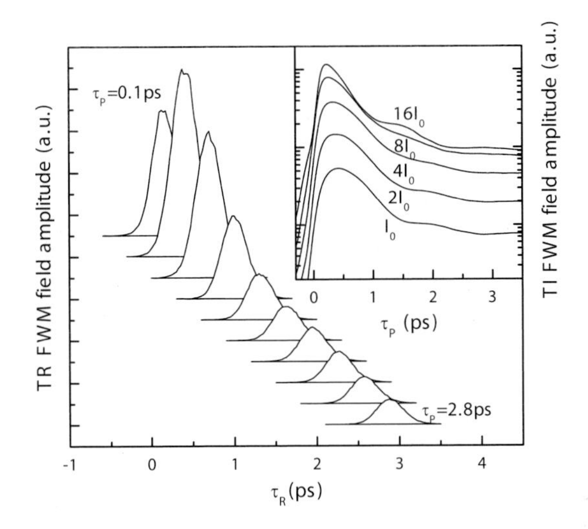

Fig. 12.7. Time-resolved four-wave mixing field amplitude versus reference delay time at 50 K on an ensemble of self-assembled $In_{0.7}Ga_{0.3}As/GaAs$ QDs. The delay time between the exciting pulses varies in 300 fs steps from 0.1ps to 2.8ps. In the inset, the time-integrated four-wave mixing is shown versus delay between the exciting pulses for different excitation intensities. After [6].

12.4.1. *Exciton acoustic-phonon interaction: Non Lorentzian homogeneous lineshape*

Among the physical mechanisms responsible for the dephasing, the interaction of excitons with phonons in QDs is a topic intensively discussed in the literature. Our measurements of the transient FWM decay of the GS excitonic transition and its temperature dependence in an inhomogeneously broadened ensemble of self-assembled $In_{0.7}Ga_{0.3}As/GaAs$ QDs revealed a rather complex, unusual scenario [6]. In Fig. 12.8 the time-integrated FWM is shown as a function of delay time between the exciting pulses, measured for different temperatures in the range from 300 K to 7 K. While from 300 K to 125 K the dynamics is dominated by a dephasing time below 1 ps, for temperatures below 100 K a slow component appears which becomes dominant at smaller temperatures. This component has a mono-exponential decay with a time constant which is increasing with decreasing temperature. An impressive change of the measured dephasing time by more than three orders of magnitude over the investigated temperature range is found. At 300 K the dephasing time is as fast as 200 fs corresponding to 6 meV homogeneous broadening FWHM, while at T=7 K the dephasing time deduced from the mono-exponential FWM decay is 630 ps, equivalent to only 2 μeV homogeneous broadening.

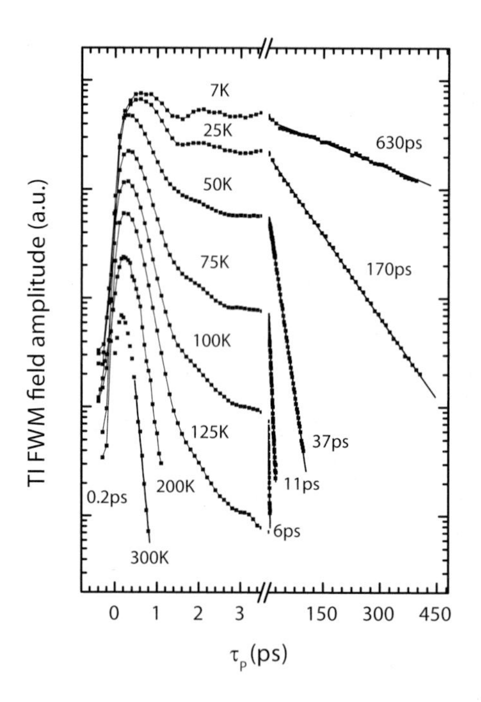

Fig. 12.8. Time-integrated four-wave mixing field amplitude on an ensemble of self-assembled $In_{0.7}Ga_{0.3}As/GaAs$ QDs at different temperatures as indicated. The dephasing times inferred from the exponential decays are indicated together with the exponential fits. After [6].

These results pioneered a new physical understanding of the interaction of excitons with acoustic phonons in self-assembled InGaAs/GaAs QDs. The strongly non-exponential behavior with a fast initial dephasing over few ps followed by a long mono-exponential decay is now understood theoretically within the independent Boson model as the manifestation in time domain of a non–Lorentzian homogeneous lineshape with a narrow zero-phonon line (ZPL) and a broad band from pure dephasing via exciton-acoustic phonon interactions [34–41]. The starting point of this treatment considers that the crystal ground state and the state where the localized exciton is optically created correspond to different adiabatic potentials for the nuclear motion. The modification of the adiabatic potential for the nuclear motion due to the electronic excitation can be expressed in different orders of the nuclear displacement coordinates. A shift of the potential minimum in the lattice displacement coordinates is a linear contribution while a change of the curvature is a quadratic contribution. A relative shift between the potential minima in the absence and in the presence of the localized exciton implies that optical transitions can occur with absorption and emission of phonons, similar to the electronic transitions in molecules showing roto-vibrational bands governed by the Franck-Condon principle. The mathematical treatment of an exciton–phonon interaction which is linear in the phonon displacement operators and does not involve excited excitonic states is given in Refs. [34, 37, 40]. In this case, the form of the exciton-phonon Hamiltonian is reduced to that of the independent Boson model and allows for an analytical solution giving polaronic eigenstates. The exciton–phonon interaction in this form does not lead to a change of the excitonic population and thus accounts only for pure dephasing. Note that the importance of pure dephasing via exciton-phonon interaction in semiconductor quantum dots was already pointed out in earlier works [3, 42] which used perturbation theory as opposed to the analytical solution of the independent Boson model.

When LO phonons are considered, the absorption (emission) spectra calculated for InGaAs QDs exhibit several side lines, separated by the LO phonon frequencies, above (below) a central line, consistent with experiments [43]. When acoustic phonons are taken into account as a continuum of bulk–like modes, which is a good approximation for dots embedded in a matrix with similar elastic properties such as InGaAs QDs in GaAs barriers, the central line decomposes into a sharp zero-phonon line and a broad band caused by the continuum of acoustic–phonon assisted transitions, which at low temperature forms a shoulder toward high (low) energies in the absorption (emission) spectrum. The width of the broad band increases with decreasing size of the dot (i.e. of the excitonic wavefunction) [34], while the weight of the ZPL (ratio of the ZPL area to the total area of the line) decreases with decreasing size of the dot or increasing temperature.

As shown in Fig. 12.8, the amplitude of the fast initial FWM dynamics increases with increasing temperature, which corresponds to a decrease of the weight of the ZPL with increasing temperature well explained theoretically within the independent Boson model. This composite non–Lorentzian line-shape is expected to be more visible in more strongly confined dots and was experimentally observed in the photoluminescence spectra of II–VI epitaxially–grown QDs [34] while is hardly resolved for excitons weakly confined by the lateral disorder in narrow GaAs quantum wells [44–46].

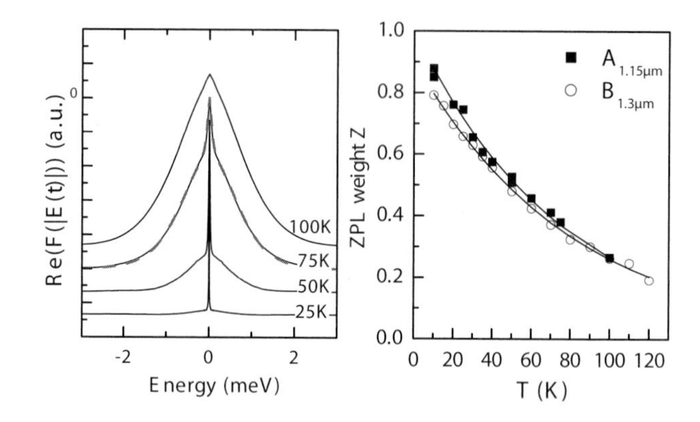

Fig. 12.9. Left: Real part of the Fourier transform of the TI FWM field amplitude measured for different temperatures on an ensemble of self-assembled $In_{0.7}Ga_{0.3}As/GaAs$ QDs (indicated as sample $A_{1.15\mu m}$). A fit (dashed line) to the data at 75 K is also shown. Right: Ratio between the area of the ZPL and the total area of the line (ZPL weight) for sample $A_{1.15\mu m}$ together with results measured on InAs/GaAs QDs emitting near 1.3μm at room temperature (sample $B_{1.3\mu m}$). The solid lines are exponential fits to the data. After [47].

Since the time–integrated photon echo versus the delay τ_P between the exciting pulses is a probe of the time evolution of the first–order polarization, the Fourier transform of the TI-FWM versus τ_P is similar to the homogeneous lineshape [38]. In Fig. 12.9(left) the real part of the Fourier transform of the TI-FWM amplitude is shown, where the long exponential decay results in a sharp Lorentzian ZPL and the fast initial dephasing gives rise to the broad band. It can be shown that the real part of the Fourier transform of the first–order polarization amplitude is identical to the symmetrized absorption lineshape if the ZPL width is small compared to the width of the broad band [48]. By analyzing the lineshape in Fig. 12.9(left) as the sum of a Lorentzian and a $\cosh^{-2}$ function, well fitting the ZPL and the broad band respectively (see red dashed line in Fig. 12.9(left)), we have inferred the ZPL weight Z also taking into account that the TI-FWM amplitude actually yields a weight Z^3, i.e. the third power of the weight obtained with the first-order polarization amplitude [38, 49]. The ZPL weight Z is shown in Fig. 12.9(right) together with results measured on InAs/GaAs QDs emitting near 1.3μm at room temperature and thus more strongly confined [47]. The ZPL weight decreases exponentially with increasing temperature and is lower for the QDs which are more strongly confined, as expected.

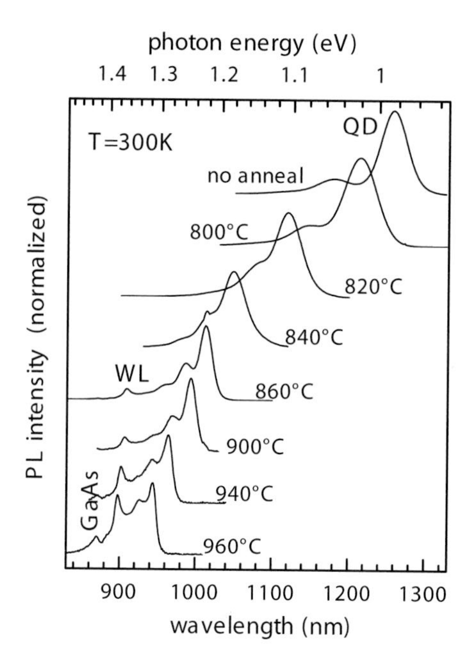

Fig. 12.10. Non-resonantly excited photoluminescence spectra at room temperature of thermally-annealed InAs/GaAs QD ensembles. The nominal annealing temperature is given for each sample. Spectra are vertically displaced for clarity. The barrier (GaAs), wetting-layer (WL) and QD transitions are indicated.

We performed a systematic study of the dependence of the ZPL weight on the extension of the excitonic wavefunction in a series of InAs QDs where the in-plane confinement potential was tuned by thermal annealing over nine samples [49]. The investigated samples were planar structures containing InAs/GaAs self-assembled QDs grown on a GaAs(100) substrate using molecular-beam epitaxy. 10 stacked layers of QDs were grown separated by 100 nm GaAs spacing layers and with $2-3\times10^{10}\,\mathrm{cm}^{-2}$ areal density, i.e. quantum–mechanically uncoupled. QDs were obtained by deposition of a nominal coverage of 2.1 monolayers of InAs at a substrate temperature of 515 °C resulting in the formation of InAs islands that were overgrown with 8 nm GaAs at a reduced substrate temperature of 505 °C. After the growth, nine pieces of the wafer were subjected to rapid thermal annealing of 30 seconds at temperature ranging from 800 °C to 960 °C in order to decrease the depth of the in-plane confining potential by thermally induced In diffusion, most important along the growth direction [50]. Fig. 12.10 shows room–temperature photoluminescence spectra under non-resonant excitation (in the GaAs barrier) for some of the samples. The lowest energy peak corresponds to the GS excitonic transition, showing a sizeable inhomogeneous and also homogeneous broadening at room temperature. We observe that thermal annealing shifts the GS excitonic transition strongly to higher energies, while the wetting layer transition shifts only slightly. Therefore, the energy distance from the GS to the wetting layer transition is tuned in the QD series from 332 meV to 69 meV. This energy distance will be called in the following confinement energy E_c. Similarly, the energy separation between ES and GS is tuned in this series from ~90 meV to ~25 meV.

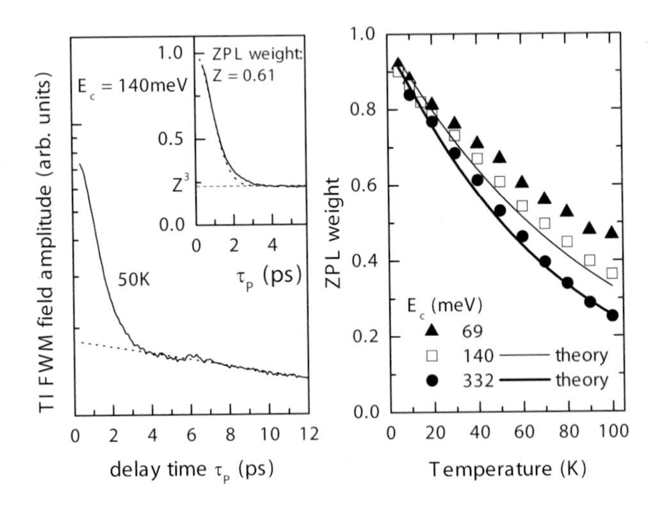

Fig. 12.11. Left: Time-integrated FWM field amplitude at 50 K in thermally-annealed InAs/GaAs QD ensembles with 140 meV confinement energy, together with an exponential fit of the dynamics at long delay times (dotted line). The initial FWM dynamics dived by the exponential fit is shown in the inset together with a Gaussian function (dotted line) fitting the initial decay. The asymptotic value at large delay (dashed line) is used to calculate the ZPL weight, as indicated. Right: ZPL weight versus lattice temperature for three QD samples with different confinement energies, as indicated. Solid lines are theory curved calculated according to the model by Vagov et al. [52]. After [49].

Using the heterodyne technique (combined with a directional selection geometry in these planar samples [51]) we measured transient FWM in resonance with the center of the inhomogeneously broadened GS transition on the sample series [49]. From the non-exponential FWM decay we deduced the ZPL weight, as shown in Fig 12.11. To extract the ZPL weight Z from the FWM data we used the results of Refs. [38, 39], in which the time-resolved FWM field is calculated within the independent Boson model for infinitely short exciting pulses and, assuming an infinitely large inhomogeneous broadening (i.e. an infinitely short photon echo in real time), the TI FWM field envelop $\overline{G}_{\infty}(\tau_{\mathrm{P}})$ in a photon echo experiment (normalized to 1 at $\tau_{\mathrm{P}} = 0$) is derived. Both assumptions are valid approximations if the pulse and the photon echo durations are much shorter than the intrinsic acoustic phonon dynamics. These conditions were met in our experiment, in which the inhomogeneous broadening and the Fourier-limited pulse spectra were much wider than the acoustic-phonon band. We therefore could use the results of Ref. [39] for $\overline{G}_{\infty}(\tau_{\mathrm{P}})$ and for the envelop function of the linear polarization in real time $G_{\mathrm{lin}}(t)$:

$$\overline{G}_{\infty}(\tau_{\mathrm{P}} \to \infty) = |G_{\mathrm{lin}}(t \to \infty)|^3 = Z^3 \,. \tag{12.17}$$

To compare Eq. (12.17) with the experiment, we had to remove from the experimental data the effect of the long time decay of the polarization (i.e. the ZPL width) not included in Eq. (12.17). We have done so by dividing the data through the observed exponential decay of the ZPL (see dotted line in Fig. 12.11(left)), which results in the solid curve shown in the inset. This procedure assumes that the polarization decay can be factorized in the product of the fast decay predicted by the independent Boson model with the long exponential decay [40], i.e. that the same dephasing of the ZPL is present also in the phonon-assisted transitions of the independent Boson model. Moreover, since for $|\tau_{\mathrm{P}}| < 300\,\mathrm{fs}$ the data were dominated by non-resonant FWM in the GaAs barrier, the TI FWM amplitude at $\tau_{\mathrm{P}} = 0$, which is needed to calculate the normalized function $\overline{G}_{\infty}(\tau_{\mathrm{P}})$, was estimated by fitting the initial decay with a Gaussian (dotted curve shown in the inset of Fig. 12.11(left)). The deduced ZPL weight Z is indicated in the inset of Fig. 12.11(left). Using this procedure, the temperature-dependent ZPL weight was determined for three QD samples having different confinement energies as shown in Fig. 12.11(right). We found a systematic decrease of the ZPL weight with increasing temperature and confinement energy, in quantitative agreement with the prediction of the independent Boson model, where the solid lines in Fig. 12.11(right) are calculations of the ZPL weight by Vagov et al. [53].

Not only the ZPL weight but also the time scale of the initial FWM decay was compared with the model by Vagov et al. [53]. For this comparison, the initial FWM dynamics was divided by the exponential decay at long delays, the asymptotic value was subtracted, and the data were

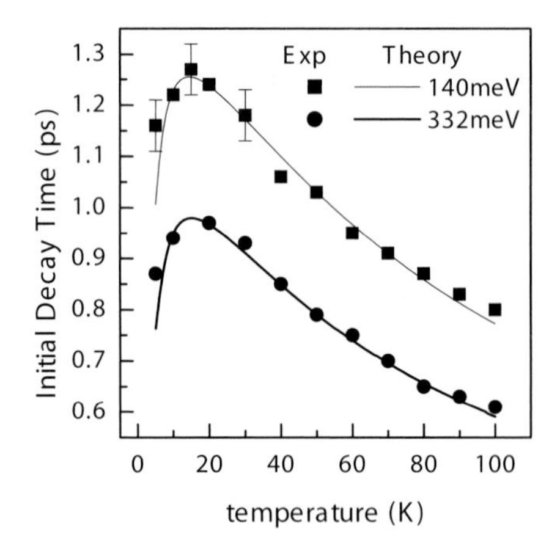

Fig. 12.12. Initial FWM decay time in thermally-annealed InAs/GaAs QD ensembles with 140 meV and 332meV confinement energy as a function of lattice temperature. Solid lines are theory curved calculated according to the model by Vagov et al. [53].

normalized to their maximum value. We then evaluated the delay time at which the normalized signal has dropped by half of its initial amplitude. The extracted times are plotted in Fig. 12.12 as a function of temperature for two samples. Experiments and theory confirm the tendency of a faster decoherence (i.e. a larger width of the acoustic-phonon broad band) for the smaller extension of the excitonic wavefunction, present on the more strongly confined (un-annealed) dots. Moreover, a non-monotonous temperature dependence of the initial decoherence time scale is observed. Such non-monotonous dependence, somewhat unusual, was well explained by the theory and is believed to be a general property of the coupling of localized electronic excitations to a continuum of vibrational modes [53]. It derives from the development of the phonon-assisted transitions from an asymmetric lineshape at low temperatures to a symmetric one at high temperatures for which the thermal energy is larger than the energy of the coupled phonons.

12.4.2. *Temperature-dependent dephasing: Width of the zero-phonon line*

The theoretical description of the exciton–acoustic phonon interaction mentioned in the previous section is able to explain the experimental observation of a decrease of the ZPL weight with increasing temperature. However, within the independent Boson model the ZPL is *un-broadened*, in clear contrast to the experimentally observed broadening of the ZPL which increases with increasing temperature [6, 17]. For a theoretical description of the width of the ZPL, two types of dephasing mechanisms can be distinguished: i) population transfer such as radiative recombination and phonon–assisted transitions toward other confined electronic levels, and ii) pure dephasing processes that do not change the carrier occupation.

From our FWM measurements of the dephasing time on an ensemble of $In_{0.7}Ga_{0.3}As/GaAs$ QDs embedded in the active region of narrow ridge waveguide [6] we deduced a temperature dependence of the width of the ZPL as shown in Fig 12.13. The ZPL FWHM is given by $\gamma = \frac{2\hbar}{T_2}$ where T_2 is the mono-exponential dephasing decay time (see Fig 12.8). Below 100 K the ZPL width measured by us is well fitted by the expression (see solid line in Fig. 12.13) $\gamma = \gamma_0 + aT + b/(\exp(E_A/k_BT) - 1)$ with the following parameters: $\gamma_0 = 0.67\,\mu\text{eV}$, $a = 0.22 \pm 0.01\mu\text{eV/K}$, $E_A = 16 \pm 1\,\text{meV}$, $b = 1.1 \pm 0.2\,\text{meV}$.

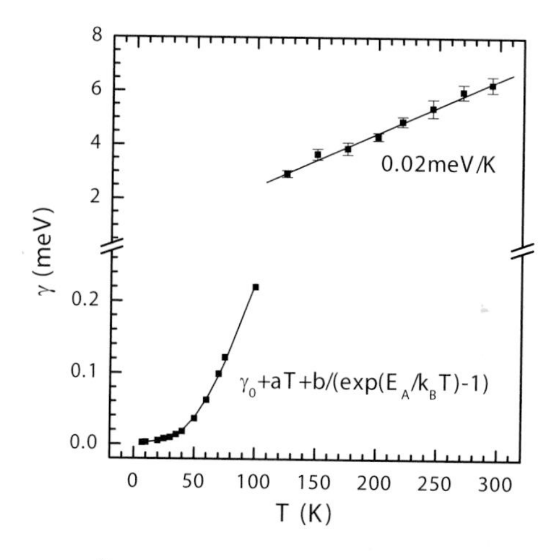

Fig. 12.13. Homogeneous linewidth of the ZPL full–width at half maximum together with fits (solid lines) on an ensemble of self-assembled $In_{0.7}Ga_{0.3}As/GaAs$ QDs. After [6]

The measured broadening was first interpreted in Ref. [6] as a inelastic broadening given by the recombination lifetime (γ_0) and by one–phonon-assisted transitions into other exciton states. A linear temperature dependence is actually *not* expected in QDs [16], contrary to bulk and QWs [54]. In fact, in QDs one-phonon absorption from the excitonic ground state occurs toward energetically well separated excited states. The individual transitions consequently should contribute to the temperature dependence of γ with the phonon occupation Bose functions having activation energies equal to the energy separation between the ground and the involved excited states [16]. If transitions occur into exciton states separated by less than $k_{\mathrm{B}}T$ even at small temperatures, Bose functions reduce to a linear temperature dependence. This linear dependence was therefore attributed to transitions among the fine-structure states of the exciton ground state which are usually separated by only $\sim 100\,\mu$eV or less in InGaAs QDs [31]. However, this interpretation is in conflict with measurements showing an exciton spin–flip time much longer than the radiative lifetime [55]. The activated temperature dependence was attributed to the one-phonon-absorption transition toward the excited exciton state separated by E_{A} involving the first excited hole level, which is consistent with typical energy level spacings calculated in InGaAs QDs [56]. However this interpretation was not supported by the systematic experimental study of the ZPL width which we performed on the series of annealed QDs [49], as we discuss below.

At higher temperatures from 100 K to 300 K the measured homogeneous linewidth of $In_{0.7}Ga_{0.3}As$/GaAs QDs was dominated by a broadening in the few meV range which increases approximately linearly with temperature [6, 17, 57] (see Fig. 12.13) and is of comparable magnitude as the broadening of excitons in (In,Ga)As/(Al,Ga) QWs [54, 58]. Its interpretation is even more complex than for lower temperatures, since the role of the exciton excited states in the modelling of the dephasing (inelastic and pure) cannot be neglected [59]. Moreover, both optical and acoustic phonons have to be taken into account, which still awaits theoretical treatment in the coupled exciton-phonon model including excited states.

The main results of our systematic study of the ZPL width and its temperature dependence in the range 5-100 K in the series of annealed QDs is summarized in Fig. 12.14. The FWHM of the ZPL is shown versus temperature for the unannealed QDs having 332 meV confinement energy, together with a fit to the data in Fig. 12.14(left). We found that a good fit to the temperature dependence of γ for *all* investigated QDs in the series was obtained using two Bose functions with activation energies E_1 and E_2 and coefficients b_1 and b_2:

$$\gamma = \gamma_0 + b_1 \frac{1}{e^{E_1/k_{\mathrm{B}}T} - 1} + b_2 \frac{1}{e^{E_2/k_{\mathrm{B}}T} - 1} \qquad (12.18)$$

where γ_0 represents a temperature independent broadening. Remarkably, the activation energies E_1 and E_2 were found to be constant within error (10%) for all investigated samples, i.e. *independent of the confinement energy*. The parameter b_2 was also roughly constant ($4 \pm 1\,\text{meV}$), while b_1 systematically increased and γ_0 decreased with increasing E_c, as shown in Fig 12.14(right). When fitting the data with Eq. (12.1) including a linear term aT, we find in all samples that $a = 0 \pm 0.02\,\mu\text{eV/K}$, i.e. the linear term is zero within error.

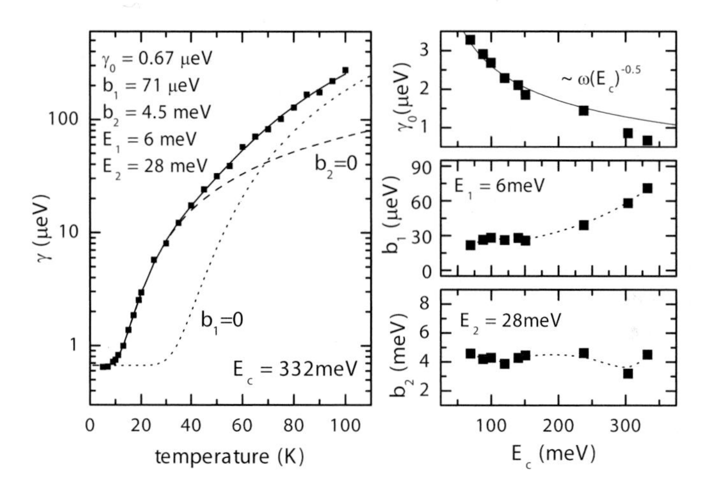

Fig. 12.14. (Left) Homogeneous broadening (ZPL width) versus temperature (square symbols) in the un-annealed InAs/GaAs QDs together with a fit to the data (solid line) according to Eq. (12.18). The contribution of the activated term with E_1 (E_2) in Eq. (12.18) is given, including γ_0, as dashed (dotted) line. (Right) Top: Zero-temperature extrapolated width of the ZPL (square symbols) versus confinement energy in the series of annealed InAs/GaAs QDs compared with a curve (solid line) proportional to the inverse square root of the confinement energy E_c times the optical transition frequency ω. Middle (Bottom): Coefficient b_1 (b_2) of the thermally activated increase of the ZPL width with activation energy E_1 (E_2), versus confinement energy. Dotted lines are guides to the eye. The size of the symbols is equal or bigger than the error bars of the parameters from the fits. After [49]

The zero-temperature dephasing rate γ_0 is given by the *radiative* lifetime T_{rad} (i.e. $\gamma_0 = \hbar/T_{\text{rad}}$), as we experimentally demonstrated [60] by comparing the polarization decay with the strength of the FWM signal and thus of the transition dipole moment. The radiative lifetime in InGaAs QDs depends on the electron-hole wavefunction overlap and on the exciton coherence volume [7]. Both quantities are influenced by the annealing. In particular, the reduction of the in–plane confinement potential by annealing [50] results in an increased extension of the excitonic wavefunction in–plane and thus in an increased exciton coherence area. If the in–plane exciton confinement potential is approximated by a two–dimensional harmonic oscillator potential of depth E_c (neglecting the zero-point quantization energy) reaching the wetting layer at a fixed diameter, the coherence area can be shown [60] to scale like $1/\sqrt{E_c}$. Consequently the radiative decay scales like (see Eqs. (45),(46) in Ref. [61] or Eq.(5.62) in Ref. [7]) $\omega\sqrt{E_c}$ where ω is the frequency of the optical transition, i.e. $\hbar\omega = E_{WL} - E_c$ with E_{WL} the energy of the wetting layer transition. The solid line in the top part of Fig. 12.14 shows this trend which well reproduces the experimental dependence for confinement energies $E_c < 200\,\text{meV}$. The deviation at large confinement energies could be due to an asymmetry of the In distribution in the growth direction, resulting in a decreased electron–hole wavefunction overlap and thus in a decreased radiative rate. This asymmetry is removed in the first stage of annealing. Measurements of the biexciton binding energy in the same sample series are indeed consistent with the presence of an asymmetric In distribution in the growth direction which is symmetrized by annealing. An increase in the electron–hole wavefunction overlap in the initial stage of annealing also reduces the charge separation and thus results in a reduced repulsive exciton–exciton interaction and in an increased biexciton binding energy which we have measured [31].

Concerning the temperature-dependent width of the ZPL, the activation energy $E_2 = 28\,\text{meV}$ is close to the range of optical phonon energies (29-36 meV) observed in InGaAs QDs [62, 63]. Optical phonons feature a density of states peaked at a certain energy. If the involved optical phonon interaction would be a one-phonon absorption process, electronic excited states at the distance of the optical phonon energy would have to be present. While for confinement energies less than 100 meV, this could be due to the continuum of hole states in the wetting layer, at larger confinement energies no such continuum is expected, in contrast to the observation of a rather constant coefficient b_2 versus E_c. An alternative pure dephasing mechanism via elastic interaction with longitudinal optical phonons, which have a finite lifetime due to the decay into acoustic phonons, was discussed in Refs. [36, 59]. For a phonon lifetime smaller than the exciton dephasing time, which is the case in most of our measurements, a ZPL broadening is predicted, having a temperature dependence proportional to $\bar{n}(\bar{n}+1)$ with the phonon occupation $\bar{n}$. Since $\bar{n} \ll 1$ in the measured temperature range, this dependence is similar to the one used in Eq. (12.18). Also the observed value of b_2 is within an order of magnitude equal to the result of Ref. [59] for a similar QD structure. However, in Ref. [64] it was shown that the derivation of a finite dephasing in Ref. [59] is an artefact and that dispersionless LO phonons quadratically coupled to a single QD transition produce no broadening of the ZPL. On the other hand, the same authors of Ref. [64] have recently shown that, when all exciton levels are included in the calculations, dephasing from quadratic-coupling to LO phonons is indeed present [65].

Turning to the activation energy $E_1 = 6\,\mathrm{meV}$, we first note that acoustic phonons do not have a concentrated density of states at this energy. The phonon energy for transitions involving one–acoustic phonon absorption into higher energy states should thus be given by the discrete energy level spacing in the QDs. In the investigated series of QDs, the energy separation from GS to ES transitions was measured to change from $\sim$70 meV to $\sim$25 meV with decreasing E_c from 332 meV to 69 meV. Our finding of a constant activation energy E_1, therefore, clearly contradicts the importance of such phonon absorption processes for the dephasing. Furthermore, acoustic phonons have a long lifetime, so that a discussion similar to the one mentioned above for the optical phonons is not likely. Recent theoretical models have suggested a broadening of the ZPL via pure dephasing mechanisms. In Ref. [40] a temperature-dependent acoustic phonon damping by lattice anharmonicities was introduced semi-phenomenologically in qualitative agreement with our experimental data of Ref. [6]. In Ref. [66] an acoustic phonon damping by surface roughness scattering was introduced, leading to a zero-temperature offset and a linear temperature dependence of the exciton dephasing, in contrast to our experimental observations in the annealed QD series. In Ref. [41] it is shown that phonon–assisted virtual transitions into higher electronic states can be cast into an effective quadratic coupling in the phonon displacement, which is known to lead to a temperature-dependent pure dephasing of the ZPL for a thermal phonon population. This approach predicts a dephasing reproducing the main features of the temperature dependence observed by us, but a quantitative comparison with the dependence on confinement energy was not shown. In synthesis, the discussion about the relevant dephasing mechanisms in quantitative comparisons with experiments is still rather unsettled. We should also point on the experimental finding that the coefficient b_1 is increasing with increasing E_c. This behavior contradicts the intuitive idea that strongly confined QDs should have a homogeneous broadening less sensitive to temperature due to inhibited phonon–assisted transitions between confined electronic states and further supports the importance of pure dephasing mechanisms.

Prior concluding this section, we should remark on the presence of a linear increase with temperature (aT) of the ZPL width which we observed on the $In_{0.7}Ga_{0.3}As/GaAs$ QDs in Ref. [6] but not in the annealed InAs/GaAs QD series. Since the sample investigated in [6] was in a p-i-n diode allowing for electrical injection, we believe that the vicinity of QDs to free carriers in the doped cladding layers and the consequent Coulomb interaction is likely to explain the measured a in that sample.

12.5. Transient FWM in Single Quantum Dots

Recent experiments have circumvented the effect of inhomogeneous broadening by performing linewidth measurements on single localized exciton states in semiconductor quantum dots. Most of these experiments consist of spectrally–resolved single-dot photoluminescence where the measurement time has to be much longer than the dephasing time in order to collect a sufficient number of photons. During this time, fluctuations of the environment lead to slow spectral diffusion of the frequency of the investigated transitions (e.g. by the Stark effect), resulting in an inhomogeneous broadening in the measurement due to the time-ensemble [67]. As a result, the typical line-width of semiconductor quantum dots at low temperatures (5-10 K) measured in single-dot photoluminescence is 10-1000 μeV [68–71], much larger than the homogeneous line-width determined in photon echo experiments [6, 72, 73]. Transient FWM spectroscopy on individual exciton transitions can distinguish between these two broadening mechanisms. Moreover, this technique is a powerful tool to explore quantum state manipulation of single excitonic transitions as a route toward quantum computing. Using the heterodyne spectral interferometry technique discussed in Section 12.3.1 we measured transient FWM on individual excitons localized by lateral disorder in thin GaAs/AlAs quantum wells [8, 9, 11] and in self-assembled epitaxial CdTe/ZnTe quantum dots [10]. In the following subsections we review two main outcomes of this experimental work: i) the evolution of the time-resolved transient FWM from free polarization decay of a single transition to the photon echo formation for many transitions, ii) the coherent control and read-out via Rabi oscillations in the polarization of individual excitonic transitions.

12.5.1. *Photon-echo formation*

As discussed in Section 12.2, the superposition of the third-order FWM polarizations from a inhomogeneously broadened ensemble of N two-level systems results in a constructive interference at the photon echo time after the arrival of the second exciting pulse, which is equal to the delay time between the two exciting pulses. The constructive interference enhances the FWM intensity at the echo time by a factor N compared to other times (see Fig. 12.2) We investigated experimentally the formation of the photon echo with increasing size N of the ensemble by measuring transient FWM on localized excitons in a thin GaAs/AlAs QW grown with a growth interruption [8, 9, 11]. The sample consisted of an AlAs/GaAs/AlAs single QW grown by molecular beam epitaxy with a thickness increasing from approximately 4 nm to 10 nm over a total lateral displacement of 200 mm. The growth was interrupted for 120 s at each interface, allowing for the formation of large monolayer islands on the growth surface. The sample was antireflection coated, and was held in a helium cryostat at a temperature of T=5 K. This type of sample is described in [74] and was also investigated by time-resolved and spectrally resolved speckle analysis [75, 76]. A detailed theoretical model of the interface structure and the excitonic properties is given in [77].

For non-resonant excitation at 1.96 eV, spatially focussed to the diffraction limit of the MO, the confocally detected photoluminescence of an approximately 6 nm thick region of the QW is shown in Fig. 12.15a. In order to select individual states within the optical resolution of the experiment (0.5 μm), we adjusted the fractional monolayer thickness of the QW to be about -0.2 ML, yielding a low density of localized exciton states in the largest monolayer (ML) thickness [77]. This was done by monitoring the PL spectrum while moving the excitation spot along the QW thickness gradient. Individual emission lines are visible in the low-energy part of the spectrum, corresponding to individual localized excitons. Due to the diffusion of the excited carriers prior to recombination, the spatial resolution of the PL spectrum is determined by the resolution of $0.61\lambda/\mathrm{NA}$ in the emission imaging. The same region was investigated by the FWM technique. The excitation spectrum (see Fig. 12.15b) was resonant only to excitons localized in the lower monolayer in order to avoid creating large exciton densities. The measured spectrally resolved FWM for a small positive delay time between the exciting pulses of 1 ps is shown in Fig. 12.15b. It consists of several sharp resonances of 20-30 μeV FWHM. Since the FWM intensity is proportional to the third power of the excitation intensity (in the low excitation intensity limit), the spatial resolution in the FWM can be improved to $0.36\lambda/\mathrm{NA} \approx 320$ nm. FWM resonances of significant strength are only observed at the higher energy side of the PL emission, and the FWM intensities of the resonances are not clearly correlated to the PL intensities.

This is due to the different properties probed by PL and FWM. The FWM intensity is (in the low excitation intensity limit) proportional to the eighth power of the optical transition dipole moment μ of the resonance. The PL intensity instead is determined by the radiative rate, proportional to μ^2, but also by the relaxation dynamics of excitons into the localized state. Additionally, in the PL we use non-resonant excitation, for which charged exciton emission [78], having a binding energy of about 4 meV, could be present, explaining the PL below 1.630 eV in the PL.

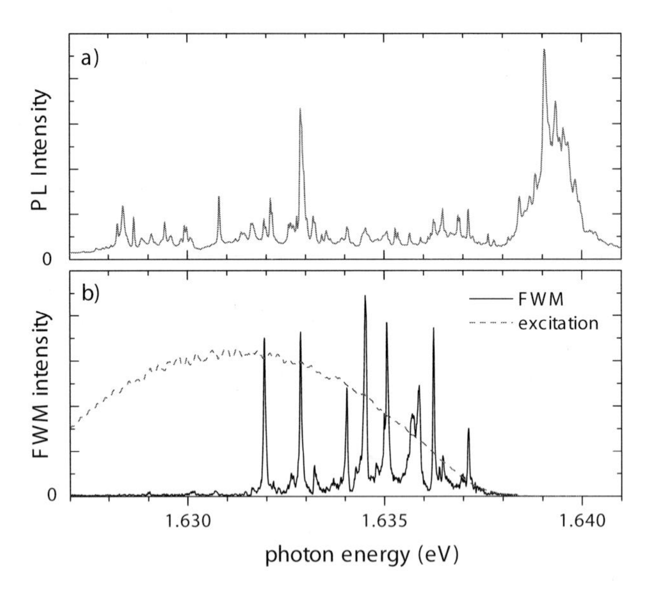

Fig. 12.15. a) Confocally excited / detected PL spectrum of a $(0.5\,\mu\text{m})^2$ area of a AlAs/GaAs/AlAs QW with a thickness of about 20.8 ML (6 nm). b) Spectrally resolved FWM intensity of the same area at 1 ps delay time between the exciting pulses. The spectrum of the excitation pulses is shown as dotted line.

The temporal sequence of the transient FWM experiment, consisting of the arrival of pulse 1, pulse 2, and the subsequent emission of the FWM intensity can be measured in the heterodyne detection by choosing Ω_d equal to Ω_1, Ω_2, or $2\Omega_2 - \Omega_1$, respectively, as discussed in Section 12.3.1. Furthermore, the fact that we can determine, in the spectral domain, the number of states contributing to the time-domain FWM signal allowed us, by appropriate selection of regions of the sample, to control the size of the state ensemble probed in the FWM. By systematically increasing the number of participating transitions from one to many, we followed the formation of the photon echo with increasing ensemble sizes. This evolution is shown in Fig. 12.16 for ensembles of about 1, 4, and 10 transitions. With increasing ensemble size N, the intensity enhancement in the photon echo increases roughly $\propto N$, as expected (see Section 12.2). The peaks off the photon echo time are due to subsidiary constructive interferences of state subgroups, which are prominent due to the small number of participating states.

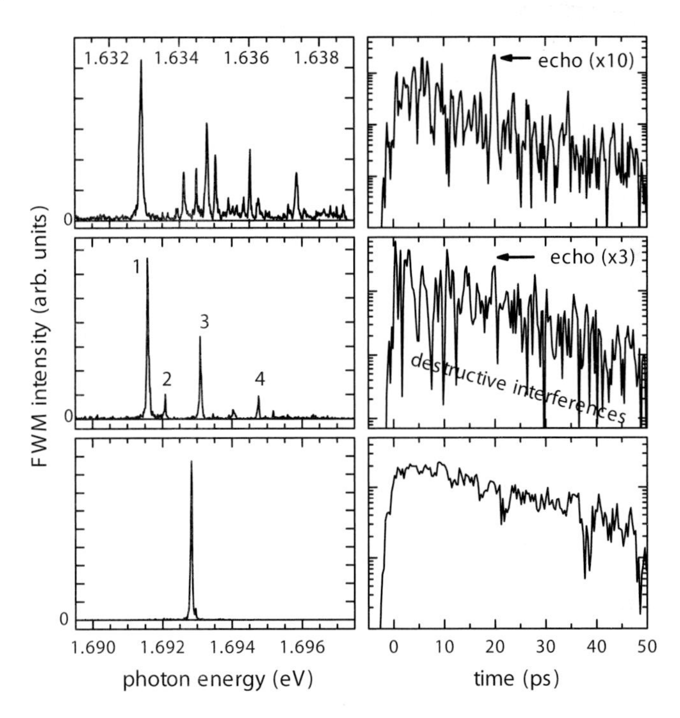

Fig. 12.16. FWM intensity for a delay time between the exciting pulses of 20 ps, spectrally resolved (left) and time-resolved (right), for exciton state ensembles of different size. The estimated fractional enhancement by the constructive interference at the photon echo time $t = 20$ ps is indicated. After [8]

Apart from the evolution from a free polarization decay to a photon echo by increasing the number of individual transitions in the ensemble, we find that even individual transitions can show a photon echo in the FWM when measured by the heterodyne spectral interferometry technique. This is due to the fact that pulsed measurements on single transitions need summation over a large number of repetitions, as the signal created upon an individual pulsed excitation is small: At maximum one can detect a single photon after a single excitation, but typically in average only $10^{-2}..10^{-5}$ photons are detected due to the limited collection efficiencies, detection efficiencies, and also to less than one photon being initially emitted. In the measurements presented here, we average over 1-100 seconds for each FWM spectrum, corresponding to $10^{10}..10^{12}$ repetitions. During this time, slow fluctuations of the environment of the probed transition can lead to a variation of its frequency (e.g. by the Stark effect due to local electric field variations), resulting in an effective inhomogeneous broadening in the time-ensemble probed by the measurement [67]. This manifests itself by a typical line-width of individual excitonic transitions in semiconductor quantum dots at low temperatures measured in single-dot photoluminescence of 100-1000 μeV [71, 79–81], much larger than their homogeneous line-width as determined in large ensemble photon echo experiments in the 1-10 μeV range [6, 72]. Since the heterodyne technique measures the FWM *field* summed over the number of repetitions, also this effective inhomogeneous broadening leads to the formation of a photon echo. In the GaAs/AlAs samples only weak spectral wandering is observed, leading to a negligible photon echo formation from an individual excitonic transition under averaging over multiple repetitions. Much stronger spectral wandering was observed on CdTe quantum dots, and allowed measurements of the photon echo formation due to multiple repetition averaging of the transient FWM field from an individual excitonic transition [10], as will be reviewed in the following.

The investigated sample consisted of self-assembled epitaxial CdTe/ZnTe quantum dots (QDs) grown by molecular beam epitaxy on an undoped [001] ZnTe substrate. Details on the growth and characterization of such structures can be found in Refs. [10, 80]. The quantum dot density in the investigated sample was several $10^{10}/\text{cm}^2$, so that within the spatial resolution typically some 10-100 QDs were probed. Due to the large inhomogeneous broadening (of about 100 meV) the emission energies of individual QDs were in general spectrally well separated, allowing to perform spectral filtering to isolate the FWM from a single QD.

The spectral FWM intensity from a sample region with exciton states of significant line-width (and hence ones expected to show effects from spectral wandering) is given in Fig. 12.17. The inset magnifies a line-shape from a single transition of about 0.35 meV FWHM, which is much larger than the homogeneous linewidth estimated to be less than 0.1 meV as will be discussed below [10]. To investigate the time-evolution of the FWM from this single transition, we spectrally filtered the measured FWM with a Gaussian of 0.62 meV FWHM around the transition (see inset in Fig. 12.17). The resulting time-evolution of the FWM intensity is given in Fig. 12.17 for various delay times between the exciting pulses. A shift of the FWM with increasing delay time is observed. The transition thus does show a photon echo behavior as opposed to a free polarization decay, indicating that it is dominantly inhomogeneously broadened most likely due to spectral wandering originating from the quantum confined Stark effect of fluctuating local electric fields. The homogeneous broadening of the transition is consequently estimated from the decay of the photon echo intensity with delay time $\propto \exp(-4\tau_P/T_2)$, resulting in $T_2 = 20 \pm 5$ ps or a FWHM of $2/T_2 = 66\,\mu$eV. This dephasing time is significantly shorter than the typical PL emission lifetime after non-resonant excitation of 100-200 ps [80, 82], indicating that the dephasing time might not be radiatively limited, as opposed to the dephasing in InGaAs QDs as discussed in Section 12.4.2. However, the radiative lifetime in the investigated CdTe QDs is not well known, and it can in principle take values between the radiative lifetime of excitons in ideal CdTe quantum wells of about 1 ps and the strong confinement limit in the nanosecond range, depending on the size of the localized exciton wavefunction. Additionally, we need to consider that the FWM signal intensity is scaling with the forth power of the radiative decay rate, so that we will preferentially select states with short radiative lifetimes in our experiment. For comparison, excitons in GaAs quantum wells have radiative lifetimes of 10-30 ps and excitons localized in interface fluctuations discussed previously showed dephasing times of 20 - 50 ps, close to the quantum well value indicating a weak confinement.

We should mention that, similar to what measured in InGaAs/GaAs QDs and discussed in Section 12.4.1, excitons localized in CdTe/ZnTe QDs also exhibited a non-Lorentzian homogeneous lienshape consisting of a sharp ZPL superimposed to a broad band from exciton–acoustic phonon interactions [34]. However, its manifestation in the FWM response of a single excitonic transition was challenging to detect. At 7 K a ZPL weight of ~ 0.9 was estimated from the intensity of the spectrally-resolved FWM field, as discussed in [10]. It might be expected that the exciton states best observable in FWM were the ones with the weakest phonon band, since a strong FWM implies a large transition dipole moment, which in turn implies a large coherence volume and thus a large wavefunction size.

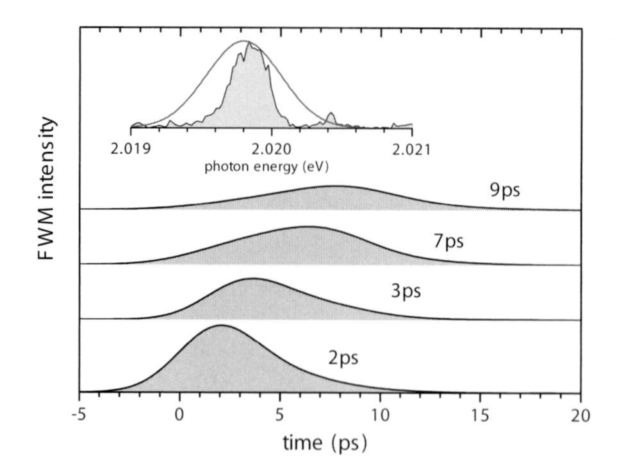

Fig. 12.17. Time-resolved FWM from a single excitonic transition in a CdTe/ZnTe quantum dot at various delays between the exciting pulses. The inset shows the spectrally resolved FWM intensity at 7 ps delay, and the filter function used to isolate the single transition (red solid line). After [10].

12.5.2. *Rabi oscillations*

The coherent control and detection of spin states via nuclear magnetic resonance (NMR) was first demonstrated more than 50 years ago [83]. Shortly thereafter, pulsed techniques were developed to control spin systems, and to measure decoherence and dissipation rates. Applying radio-frequency pulses defined by π or $\pi/2$ pulse areas, a spin state can be switched in phase and orientation. This quantum state manipulation resulted in many proof-of-principle demonstrations of quantum algorithms by liquid state NMR. While conventional NMR approaches build on well-developed experimental capabilities to create and control the quantum properties necessary for quantum computation in ensembles, solid-state approaches towards coherent control can draw on existing nanostructure fabrication technologies for the addressing of individual qubits. By combining the coherent control techniques of NMR with ultrafast pulses in the optical frequency range, engineered exciton and spin-based scalable quantum information processing (QIP) devices are expected to be realizable. The sub-picosecond time scale of the optical pulses allows for a the execution of a significant number of elementary operations within the exciton decoherence time – a feature of fundamental importance for any useful QIP device.

Self-assembled InGaAs/GaAs QDs are good candidates for optical coherent control and possible solid-state implementations of QIP, due to the long dephasing time of the excitonic ground-state transition in the nanosecond range at low temperature [6], the strong confinement energy of 100-300 meV resulting in well separated energy levels and the large transition dipole moments compared to atoms [84, 85]. Using excitonic transitions in self-assembled InGaAs QDs at low temperature, one qubit rotations via Rabi oscillations have been recently demonstrated both in single QDs [86–91] and QD ensembles [84], the latter case being mainly limited by a distribution of transition dipole moments in the ensemble.

Transient FWM combined with heterodyne spectral interferometry offers a powerful way to resonantly drive and coherently manipulate a single exciton state with two excitation pulses, and recover the time-resolved induced polarization in both amplitude and phase. This is particularly relevant since the implementation of quantum computational devices will require the ability to place a qubit in a given configuration, allow it to evolve (with the possible application of additional pulses to further affect the qubit state) and then readout the results of the evolution. Although the applicability of this technique to single InGaAs/GaAs QDs appears to be challenging and has not been demonstrated yet, we succeeded in demonstrating its potential for coherent control and polarization readout of individual states on excitons confined by interface fluctuations in a AlAs/GaAs/AlAs QW [11].

Fig. 12.18a shows the measured FWM intensity and phase spectrally resonant to an individual excitonic transition as function of the pump pulse area Θ_1. The FWM field created by the probe pulse (in this case at a 5 ps delay after the pump pulse) is proportional to the phase–conjugated polarization created by the pump pulse (see Section 12.2). In a two-level system, Rabi oscillations would be expected as the Bloch-vector rotates due to the pump pulse (see sketches in Fig. 12.18a), with the polarization being zero both when the system is purely in the ground or the excited state, i.e. with a period of π in the pulse area. The corresponding polarization is $\propto \sin(\Theta_1)$, and thus the FWM intensity $\propto \sin^2(\Theta_1)$, shown in Fig. 12.18a as solid line. The experimental pulse area unit was adjusted to reproduce the position of the first minimum at $\Theta_1 = \pi$. From the focus size and the pulse duration, we estimated a transition dipole moment of 30-40 Debye. For $\Theta_1 < 1.2\pi$, the data are in agreement with the two-level prediction. Additionally, the predicted π jump in the phase is observed since the polarization changes sign across $\Theta_1 = \pi$. For larger pulse areas, significant deviations from the prediction are present. Such deviations were observed in a pump-probe experiment sensitive to the population inversion [92], and were

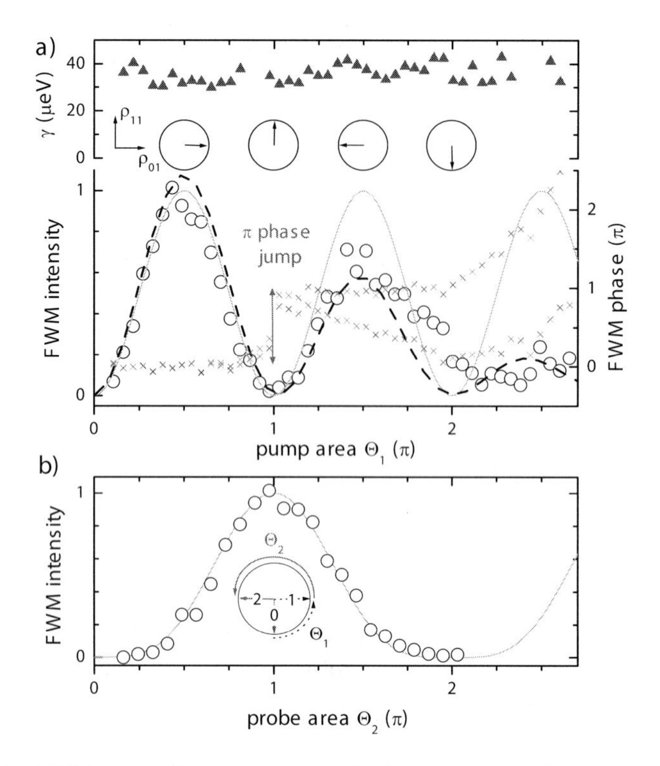

Fig. 12.18. FWM spectrally resonant to a single exciton state localized by interface fluctuations in a thin AlAs/GaAs/AlAs QW. a) As a function of pump pulse area Θ_1 for a probe pulse area $\Theta_2 = \pi$. Circles: measured intensity. Solid line: Expected two-level result $\propto \sin^2(\Theta_1)$. Dashed line: calculated FWM including multiexcitons (see text). The sketches indicate the state vectors after the pump pulse in a Bloch sphere. Crosses: measured FWM phase. The two curves indicate the limits of a systematic error due to a phase drift. Top: line width γ (FWHM) of the exciton state measured in the FWM spectrum. b) Intensity as a function of Θ_2 for $\Theta_1 = \pi/2$. The solid line is the expected two-level result $\propto \sin^4(\Theta_2/2)$. The sketch indicates the corresponding state vector positions 0: initially, 1: after the pump, 2: after pump & probe. The energy density per excitation pulse was $2.2\,\mu\text{J/cm}^2$ for $\Theta_{1,2} = \pi$. After [11]

interpreted as an increased dephasing rate induced by an optically created carrier density, called excitation-induced dephasing (EID), and well known in quantum wells. This mechanism was also discussed for Rabi-oscillations in InGaAs quantum dots [90]. In our experiments, we could directly measure the dephasing of the transition after the application of both excitation pulses using the linewidth γ of its third-order polarization. From the missing photon echo signature in the corresponding time-resolved FWM, we could exclude a significant influence of inhomogeneous broadening due to e.g. spectral wandering. We observed that γ as function of pulse area Θ_1 (see Fig. 12.18a top) was constant within error, inconsistent with the EID hypothesis. We suggested that the deviations from the two-level result might originate from the presence of near-resonant transitions to multi-excitonic states into which the system is driven by the excitation pulses - a coherent process which does not affect the dephasing of the fundamental exciton transition [11]. To support this interpretation, we calculated the FWM intensity (see dashed line in Fig. 12.18a), neglecting dephasing, in presence of 3 transitions to biexcitonic states with binding energies of 0.5, -0.6, -0.8 meV and transition dipole moments 30% of the exciton transition. A decay superimposed to the Rabi-oscillations was observed in the simulation, qualitatively reproducing the measurement.

The FWM polarization phase-conjugated to the pump-induced polarization can be also coherently controlled by the probe pulse area Θ_2. In a two-level system, the FWM intensity created by the probe is $\propto \sin^4(\Theta_2/2)$ [83], with the ideal phase conjugation being achieved by rotating the Bloch-vector by an angle of $\Theta_2 = \pi$ [14]. This prediction is in quantitative agreement with the experiment (see Fig. 12.18b), and represents the verification of a basic two-pulse coherent control on a single excitonic transition. In Ref. [11] we also demonstrated coherent control of the polarization in both amplitude and phase with non-resonant excitation pulses and simultaneous Rabi flopping of several spatially and energetically close exciton states.

12.6. Summary and Outlook

In summary, the ability to perform transient FWM in ensemble and single semiconductor QDs using a sensitive heterodyne detection technique has opened the possibility not only to measure dephasing times and investigate the physical origin of dephasing processes in these systems but also to demonstrate protocols of coherent manipulation and polarization readout, all of which is of key importance for a possible solid-state implementation of quantum computing using excitonic transitions in QDs as semiconductor quantum bits.

From the point of view of the dephasing processes in InGaAs/GaAs QDs, the finding of a radiatively limited dephasing time of the zero-phonon line in the nanosecond range even at finite (5 K) temperature is very promising because, if the homogeneous lineshape would be given by the ZPL only, coherent-light matter interaction in these systems would last as long as the radiative decay, without disruptive effects from pure dephasing. This is also very important to generate indistinguishable photons from single QDs [93]. On the other hand we showed that, in agreement with the independent Boson model, even at low temperatures the zero-phonon line is superimposed to a broad acoustic phonon band, especially for strongly localized dots, corresponding to an initial non-exponential decoherence from pure dephasing by acoustic-phonon interaction. Thermal annealing plays a useful role by reducing the strength of quantum confinement and thus increasing the weight of the ZPL at a given temperature. However, this effect is significant at high temperature (>20 K). At 5 K about 20% of the initial FWM amplitude is already lost after few picosecond, corresponding to a ZPL weight of ~0.9 in all thermally annealed QDs investigated by us. It has been suggested recently that the detrimental effect of such decoherence during an operation performed by a sequence of fixed number of pulses can be significantly reduced by using pulses of finite duration optimized so that the spectral overlap with phonon modes is minimized [94]. Concerning the temperature dependence of the ZPL width, there is clearly the need to go beyond the independent Boson model. However, the comparison of experimental findings with theory is still partly unsettled. Experimental results point towards pure dephasing mechanisms playing a major role, rather than one-phonon absorption processes. A ZPL width from pure dephasing is explained in the theory via a quadratic coupling to both acoustic and optical phonons. However a quantitative modeling is still missing.

The possibility recently opened to perform transient FWM, time and spectrally resolved, in single QDs using heterodyne spectral interferometry is holding great potential. This approach compared to ensemble measurements has the advantage to distinguish between processes occurring within one dot from those occurring in different dots and averaged in the ensemble response. However, the ability to perform this type of measurements in single InAs QDs remains the challenge for future work. The experimental demonstration of this method was shown for excitons localized by interface fluctuations in thin AlAs/GaAs/AlAs quantum wells and in self-assembled CdTe/ZnTe QDs. These measurements allowed to time resolve the formation of photon echo in the third–order polarization response, when more than one localized excitonic transition contributes to the response. They also showed the ability to distinguish the effect of spectral wandering, a well known problem that limits the measurement of the homogeneous linewidth of a single QD with photoluminescence experiments. Spectral wandering, conversely, manifests as a photon echo in the FWM field amplitude summed over many repeats which is detected by the heterodyne spectral interferometry, and therefore does not prevent the measurement of the homogeneous broadening via the decay of the time-integrated photon echo versus delay time. Finally, optical coherent control and polarization readout of individual excitonic states was demonstrated, paving the way for the experimental demonstration of solid-state based quantum computing using optical transitions in quantum dots. Furthermore, the detection of amplitude and phase of the signal allows the implementation of a two-dimensional femtosecond spectroscopy [9], in which mutual coherent coupling between individual quantum dot states can observed and quantified.

Acknowledgments

We acknowledge Stephan Schneider and Brian Patton for their significant contribution to the experimental work reported here. InGaAs/GaAs QDs in a ridge waveguide were provided by Prof. Dieter Bimberg, Technical University of Berlin, Germany. The series of thermally annealed InAs/GaAs QD ensembles was provided by Dirk Reuter and Prof. Andreas Wieck in Bochum University, Germany. The AlAs/GaAs/AlAs thin QW with interface fluctuations was provided by Prof. Jørn Hvam, grown at the III-V Nanolab, a joint laboratory between Research Center COM and the Niels Bohr Institute, Copenhagen University. The CdTe/ZnTe QDs were provided by H. Mariette, CEA-CNRS Grenoble, France. A large part of the experimental work was performed in the Physics Department of Dortmund University, Germany, in collaboration with Prof. Ulrike Woggon.

References

[1] Shah, J. (1996). *Ultrafast Spectroscopy of Semiconductors and Semiconductor Nanostructures* (Springer, Berlin).

[2] Erland, J., Kim, J. C., Bonadeo, N. H., Steel, D. G., Gammon, D. and Katzer, D. S. (1999). Nonexponential photon echo decays from nanostructures: Strongly and weakly localized degenerate exciton states, *Phys. Rev. B* **60**, p. R8497.

[3] Fan, X., Takagahara, T., Cunningham, J. E. and Wang, H. (1998). Pure dephasing induced by exciton-phonon interactions in narrow GaAs quantum wells, *Solid State Commun.* **108**, pp. 857–61.

[4] Gindele, F., Woggon, U., Langbein, W., Hvam, J. M., Leonardi, K., Hommel, D. and Selke, H. (1999b). Excitons, biexcitons, and phonons in ultrathin CdSe/ZnSe quantum structures, *Phys. Rev. B* **60**, pp. 8773–82.

[5] Wagner, H. P., Langbein, W. and Hvam, J. M. (1999). Mixed biexcitons in single quantum wells, *Phys. Rev. B* **59**, p. 4584.

[6] Borri, P., Langbein, W., Schneider, S., Woggon, U., Sellin, R., Ouyang, D. and Bimberg, D. (2001). Ultralong dephasing time in InGaAs quantum dots, *Phys. Rev. Lett.* **87**, p. 157401.
[7] Bimberg, D., Grundmann, M. and Ledentsov, N. N. (1999). *Quantum Dot Heterostructures* (John Wiley and Sons, Chichester).
[8] Langbein, W. and Patton, B. (2005). Microscopic measurement of photon echo formation in groups of individual excitonic transitions, *Phys. Rev. Lett.* **95**, p. 017403.
[9] Langbein, W. and Patton, B. (2006). Heterodyne spectral interferometry for multidimensional nonlinear spectroscopy of individual quantum systems, *Optics Letters* **31**, pp. 1151–1153.
[10] Patton, B., Langbein, W., Woggon, U., Maingault, L. and Mariette, H. (2006). Time- and spectrally-resolved four-wave mixing in single CdTe/ZnTe quantum dots, *Phys. Rev. B* **73**, p. 235354.
[11] Patton, B., Woggon, U. and Langbein, W. (2005). Coherent control and polarization readout of individual excitonic states, *Phys. Rev. Lett.* **95**, p. 266401.
[12] Loudon, R. (1983). *The Quantum Theory of Light* (Oxford Science Publications, Oxford).
[13] Shen, Y. R. (1984). *The Principles of Nonlinear Optics* (John Wiley and Sons, New York).
[14] Allen, L. and Eberly, J. H. (1975). *Optical Resonance and Two-Level Atoms* (Wiley, New York).
[15] Langbein, W., Hvam, J. M. and Zimmermann, R. (1999). Time-resolved speckle analysis: A new approach to coherence and dephasing of optical excitations in solids, *Phys. Rev. Lett.* **82**, p. 1040.
[16] Gammon, D., Snow, E. S., Shanabrook, B. V., Katzer, D. S. and Park, D. (1996). Homogeneous linewidth in the optical spectrum of a single gallium arsenide quantum dot, *Science* **273**, p. 87.
[17] Bayer, M. and Forchel, A. (2002). Temperature dependence of the exciton homogenous linewidth in $in_{0.6}ga_{0.4}as$/gaas self-assembled quantum dots, *Phys. Rev. B* **65**, p. 041308.
[18] Demtröder, W. (1998). *Laser Spectroscopy* (Spinger-Verlag, Berlin).
[19] Klingshirn, C. F. (1995). *Semiconductor Optics* (Spinger-Verlag, Berlin Heidelberg).
[20] Siegman, A. E. (1986). *Lasers* (Oxford).
[21] Hall, K. L., Lenz, G., Ippen, E. P. and Raybon, G. (1992). Heterodyne pump–probe technique for time domain studies of optical nonlinearities in waveguides, *Optics Letters* **17**, pp. 874–876.
[22] Sun, C.-K., Golubovic, B., Fujimoto, J. G., Choi, H. K. and Wang, C. A. (1995b). Heterodyne nondegenerate pump-probe measurement technique for guided-wave devices, *Optics Lett.* **20**, pp. 210–212.
[23] Sun, C.-K., Golubovic, B., Choi, H.-K., Wang, C. A. and Fujimoto, J. G. (1995a). Femtosecond investigations of spectral hole burning in semiconductor lasers, *Appl. Phys. Lett.* **66**, pp. 1650–1652.
[24] Mecozzi, A., Mørk, J. and Hofmann, M. (1996). Transient four-wave mixing

with colinear pump and probe, *Optics Lett.* **21**, p. 1017.
[25] Hofmann, M., Brorson, S. D., Mørk, J. and Mecozzi, A. (1996). Time resolved four-wave mixing technique to measure the ultrafast coherent dynamics in semiconductor optical amplifiers, *Appl. Phys. Lett.* **68**, p. 3236.
[26] Borri, P., Langbein, W., Mørk, J. and Hvam, J. M. (1999b). Heterodyne pump–probe and four–wave mixing in semiconductor optical amplifiers using balanced lock–in detection, *Optics Comm.* **169**, pp. 317–324.
[27] Borri, P., Romstad, F., Langbein, W., Kelly, A. E., Mørk, J. and Hvam, J. M. (2000). Separation of coherent and incoherent nonlinearities in a heterodyne pump–probe experiment, *Optics Express* **7**, pp. 107–112.
[28] Stolz, H. (1994). Time-resolved light scattering from excitons, in G. Höhler, R. D. Peccei, F. Steiner, J. Trümper, P. Wölfle and E. A. Niekisch (eds.), *Springer Tracts in Modern Physics*, Vol. 130 (Springer Verlag Berlin Heidelberg New York).
[29] Lepetit, L., Chériaux, G. and Joffre, M. (1995). Linear techniques of phase measurement by femtosecond spectral interferometry for applications in spectroscopy, *J. Opt. Soc. Am. B* **12**, p. 2467.
[30] Yu, P., Langbein, W., Leosson, K., Hvam, J. M., Ledentsov, N. N., Bimberg, D., Ustinov, V. M., Egorov, A. Y., Zhukov, A. E., Tsatsul'nikov, A. F. and Musikhin, Y. G. (1999). Optical anisotropy in vertically coupled quantum dots, *Phys. Rev. B* **60**, pp. 16680–5.
[31] Langbein, W., Borri, P., Woggon, U., Stavarache, V., Reuter, D. and Wieck, A. D. (2004a). Control of fine-structure splitting and biexciton binding in $\text{in}_x\text{ga}_{1-x}\text{as}$ quantum dots by annealing, *Phys. Rev. B* **69**, pp. 161301(R)–4.
[32] Findeis, F., Zrenner, A., Böhm, G. and Abstreiter, G. (2000). Phonon-assisted biexciton generation in a single quantum dot, *Phys. Rev. B* **61**, p. R10 579.
[33] Albrecht, T. F., Bott, K., Meier, T., Schulze, A., Koch, M., Cundiff, S. T., Feldmann, J., Stolz, W., Thomas, P., Koch, S. W. and Göbel, E. O. (1996). Disorder mediated biexcitonic beats in semiconductor quantum wells, *Phys. Rev. B* **54**, p. 4436.
[34] Besombes, L., Kheng, K., Marsal, L. and Mariette, H. (2001). Acoustic phonon broadening mechanism in single quantum dot emission, *Phys. Rev. B* **63**, p. 155307.
[35] Förstner, J., Weber, C., Danckwerts, J. and Knorr, A. (2003). Phonon-assisted damping of rabi oscillations in semiconductor quantum dots, *Phys. Rev. Letters* **91**, pp. 127401–4.
[36] Goupalov, S. V., Suris, R. A., Lavallard, P. and Citrin, D. S. (2002). Exciton dephasing and absorption line shape in semiconductor quantum dots, *IEEE J. Sel. Topics Q. El.* **8**, pp. 1009–1014.
[37] Krummheuer, B., Axt, V. M. and Kuhn, T. (2002). Theory of pure dephasing and the resulting absorption line shape in semiconductor quantum dots, *Phys. Rev. B* **65**, p. 195313.
[38] Vagov, A., Axt, V. M. and Kuhn, T. (2002). Electron-phonon dynamics in optically excited quantum dots: Exact solution for multiple ultrashort laser pulses, *Phys. Rev. B* **66**, pp. 165312 1–15.

[39] Vagov, A., Axt, V. M. and Kuhn, T. (2003). Impact of pure dephasing on the nonlinear optical response of single quantum dots and dot ensembles, *Phys. Rev. B* **67**, p. 115338.

[40] Zimmermann, R. and Runge, E. (2002b). Dephasing in quantum dots via electron-phonon interaction, in J. H. Davies and A. R. Long (eds.), *Proceedings of the 26th International Conference on the Physics of Semiconductors* (Institute of Physics Publishing, UK), p. M3.1.

[41] Muljarov, E. A. and Zimmermann, R. (2004). Dephasing in quantum dots: Quadratic coupling to acoustic phonons, *Phys. Rev. Lett.* **93**, p. 237401.

[42] Takagahara, T. (1999). Theory of exciton dephasing in semiconductor quantum dots, *Phys. Rev. B* **60**, pp. 2638–2652.

[43] Heitz, R., Mukhametzhanov, I., Stier, O., Madhukar, A. and Bimberg, D. (1999). Enhanced polar exciton-LO-phonon interaction in qunatum dots, *Phys. Rev. Lett.* **83**, pp. 4654–57.

[44] Guest, J., Stievater, T. H., Chen, G., Tabak, E. A., Orr, B. G., Steel, D. G., Gammon, D. and Katzer, D. S. (2001). Near-field coherent spectroscopy and microscopy of a quantum dot system, *Science* **293**, pp. 2224–7.

[45] Mannarini, G. and Zimmermann, R. (2006). Near-field spectra of quantum well excitons with non-markovian phonon scattering, *Phys. Rev. B* **73**, p. 115325.

[46] Peter, E., Hours, J., Senellart, P., Vasanelli, A., Cavanna, A., Bloch, J. and Gérard, J. M. (2004). Phonon sidebands in exciton and biexciton emission from single GaAs quantum dots, *Phys. Rev. B* **69**, pp. 041307(R)–4.

[47] Borri, P., Schneider, S., Langbein, W. and Bimberg, D. (2006). Ultrafast carrier dynamics in InGaAs quantum dot materials and devices, *J. Opt. A: Pure Appl. Opt.* **8**, pp. S33–S46.

[48] Zimmermann, R. and Runge, E. (2002a). *private communication*

[49] Borri, P., Langbein, W., Woggon, U., Stavarache, V., Reuter, D. and Wieck, A. D. (2005). Exciton dephasing via phonon interactions in InAs quantum dots: Dependence on quantum confinement, *Phys. Rev. B* **71**, p. 115328.

[50] Fafard, S. and Allen, C. N. (1999). Intermixing in quantum-dot ensambles with sharp adjustable shells, *Appl. Phys. Letters* **75**, pp. 2374–6.

[51] Borri, P., Langbein, W., U.Woggon, Schwab, M., Bayer, M., Fafard, S., Z.Wasilewski and Hawrylak, P. (2003). Exciton dephasing in quantum dot molecules, *Phys. Rev. Lett.* **91**, p. 267401.

[52] Vagov, A., Axt, V., Kuhn, T., Langbein, W., Borri, P. and Woggon, U. (2004a). Nonmonotonous temperature dependence of the initial decoherence in quantum dots, *Phys. Rev. B* **70**, p. 201305(R).

[53] Vagov, A., Axt, V. M., Kuhn, T., Langbein, W., Borri, P. and Woggon, U. (2004b). Nonmonotonous temperature dependence of the initial decoherence in quantum dots, *Phys. Rev. B* **70**, p. 201305(R).

[54] Borri, P., Langbein, W., Hvam, J. M. and Martelli, F. (1999a). Well-width dependence of exciton-phonon scattering in $In_xGa_{1-x}As/GaAs$ single quantum wells, *Phys. Rev. B* **59**, p. 2215.

[55] Paillard, M., Marie, X., Renucci, R., Amand, T., Jbeli, A. and Gérard, J. M. (2001). Spin relaxation quenching in semiconductor quantum dots,

Phys. Rev. Lett. **86**, p. 1634.

[56] Stier, O., Grundmann, M. and Bimberg, D. (1999). Electronic and optical properties of strained quantum dots modeled by 8-band k·p theory, *Phys. Rev. B* **59**, pp. 5688–701.

[57] Matsuda, K., Ikeda, K., Saiki, T., Tsuchiya, H., Saito, H. and Nishi, K. (2001). Homogeneous linewidth broadening in a $in_{0.5}ga_{0.5}as$/gaas single quantum dot at room temperature investigated using a highly sensitive near–field scanning optical microscope, *Phys. Rev. B* **63**, p. 121304.

[58] Borri, P., Langbein, W., Mørk, J., Hvam, J. M., Heinrichsdorff, F., Mao, M.-H. and Bimberg, D. (1999c). Dephasing in InAs/GaAs quantum dots, *Phys. Rev. B* **60**, pp. 7784–7787.

[59] Uskov, A. V., Jauho, A.-P., Tromborg, B., Mørk, J. and Lang, R. (2000). Dephasing times in quantum dots due to elastic LO phonon–carrier collisions, *Phys. Rev. Lett.* **85**, p. 1516.

[60] Langbein, W., Borri, P., Woggon, U., Stavarache, V., Reuter, D. and Wieck, A. D. (2004c). Radiatively limtied dephasing in InAs quantum dots, *Phys. Rev. B* **70**, p. 033301.

[61] Sugawara, M. (1995). Theory of spontaneous-emission lifetime of wannier excitons in mesoscopic semiconductor quantum disks, *Phys. Rev. B* **51**, p. 10743.

[62] Heitz, R., Born, H., Hoffmann, A., Bimberg, D., Mukhametzhanov, I. and Madhukar, A. (2000). Resonant raman scattering in self-organized InAs/GaAs quantum dots, *Appl. Phys. Lett.* **77**, pp. 3746–8.

[63] Heitz, R., Veit, M., Ledentsov, N. N., Hoffmann, A., Bimberg, D., Ustinov, V. M., Kop'ev, P. S. and Alferov, Z. I. (1997). Energy relaxation by multiphonon processes in InAs/GaAs quantum dots, *Phys. Rev. B* **56**, pp. 10435–10445.

[64] Muljarov, E. A. and Zimmermann, R. (2006). Comment on "dephasing times in quantum dots due to elastic LO phonon-carrier collisions", *Phys. Rev. Lett.* **96**, p. 019703.

[65] Muljarov, E. A. and Zimmermann, R. (2007). LO phonon-induced exciton dephasing in quantum dots: An exactly solvable model, *Phys. Rev. Lett.* **98**, p. 187401.

[66] Ortner, G., Yakovlev, D. R., Bayer, M., Rudin, S., Reinecke, T. L., Fafard, S., Wasilewski, Z. and Hawrylak, P. (2004). Temperature dependence of the zero phonon line width in InAs/GaAs quantum dots, *Phys. Rev. B* **70**, p. 201301(R).

[67] Mukamel, S. (1999). *Principles of Nonlinear Optical Spectroscopy* (Oxford, USA).

[68] Bayer, M., Ortner, G., Stern, O., Kuther, A., Gorbunov, A. A., Forchel, A., Hawrylak, P., Fafard, S., Hinzer, K., Reinecke, T. L., Walck, S. N., Reithmaier, J. P., Klopf, F. and Schäfer, F. (2002). Fine structure of neutral and charged excitons in self-assembled In(Ga)As/(Al)GaAs quantum dots, *Phys. Rev. B* **65**, p. 195315.

[69] Gindele, F., Hild, K., Langbein, W. and Woggon, U. (1999a). Phonon interaction of single excitons and biexcitons, *Phys. Rev. B* **60**, p. R2157.

[70] Leosson, K., Jensen, J. R., Hvam, J. M. and Langbein, W. (2000a). Linewidth statistic of single InGaAs quantum dot photoluminescence lines, *phys. stat. sol. (b)* **221**, pp. 49–53.

[71] Türck, V., Rodt, S., Stier, O., Heitz, R., Engelhardt, R., Pohl, U. W., Bimberg, D. and Steingrüber, R. (2000). Effect of random field fluctuations on excitonic transitions of individual CdSe quantum dots, *Phys. Rev. B* **61**, pp. 9944–7.

[72] Kuribayashi, R., Inoue, K., Sakoda, K., Tsekhomskii, V. A. and Baranov, A. V. (1998). Long phase-relaxation time in CuCl quantum dots: Four-wave-mixing signals analogous to dye molecules in polymers, *Phys. Rev. B* **57**, p. R15084.

[73] Langbein, W., Borri, P., Woggon, U., Stavarache, V., Reuter, D. and Wieck, A. D. (2004b). Radiatively limited dephasing in InAs quantum dots, *Phys. Rev. B* **70**, p. 033301.

[74] Leosson, K., Jensen, J. R., Langbein, W. and Hvam, J. M. (2000b). Exciton localization and interface roughness in growth-interrupted GaAs/AlAs quantum wells, *Phys. Rev. B* **61**, p. 10322.

[75] Kocherscheidt, G., Langbein, W. and Savona, V. (2003). Level-statistics in the resonant rayleigh scattering dynamics of monolayer-split excitons, *phys. stat. sol. (b)* **238**, p. 486.

[76] Langbein, W., Leosson, K., Jensen, J. R., Hvam, J. M. and Zimmermann, R. (2000). Instantaneous rayleigh scattering from excitons localized in monolayer islands, *Phys. Rev. B* **61**, p. R10555.

[77] Savona, V. and Langbein, W. (2006). Realistic heterointerface model for excitonic states in growth-interrupted GaAs quantum wells, *Phys. Rev. B* **74**, p. 75311.

[78] Tischler, J. G., Bracker, A. S., Gammon, D. and Park, D. (2002). Fine structure of trions and excitons in single GaAs quantum dots, *Phys. Rev. B* **66**, p. 081310(R).

[79] Empedocles, S. A., Norris, D. J. and G., B. M. (1996). Photoluminescence spectroscopy of single CdSe nanocrystallite quantum dots, *Phys. Rev. Lett.* **77**, p. 3873.

[80] Marsal, L., Besombes, L., Tinjod, F., Kheng, K., Wasiela, A., Gilles, B., Rouvière, J.-L. and Mariette, H. (2002). Zero-dimensional excitons in CdTe/ZnTe nanostrutures, *J. App. Phys.* **91**, pp. 4936 – 4943.

[81] Patton, B., Langbein, W. and Woggon, U. (2003). Trion, biexciton, and exciton dynamics in single self-assembled CdSe quantum dots, *Phys. Rev. B* **68**, p. 125316.

[82] Couteau, C., Moehl, S., Tinjod, F., G'erard, J. M., Kheng, K., Mariette, H., Gaj, J. A., Romestain, R. and Poizat, J. P. (2004). Correlated photon emission from a single IIVI quantum dot, *Appl. Phys. Lett.* **85**, p. 6251.

[83] Hahn, E. L. (1950). Spin echoes, *Phys. Rev.* **80**, p. 580.

[84] Borri, P., Langbein, W., Schneider, S., Woggon, U., Sellin, R. L., Ouyang, D. and Bimberg, D. (2002). Rabi oscillations in the excitonic ground-state transition of InGaAs quantum dots, *Phys. Rev. B* **66**, p. 081306(R).

[85] Silverman, K. L., Mirin, R. P., Cundiff, S. T. and Norman, A. G. (2003).

Direct measurement of polarization resolved transition dipole moment in InGaAs/GaAs quantum dots, *Appl. Phys. Lett.* **82**, pp. 4552–4554.

[86] Besombes, L., Baumberg, J. J. and Motohisa, J. (2003). Coherent spectroscopy of optically gated charged single InGaAs quantum dots, *Phys. Rev. Lett.* **90**, p. 257402.

[87] Htoon, H., Tagakahara, T., Kulik, D., Baklenov, O., Jr., A. L. H. and Shih, C. K. (2002). Inerplay of rabi oscillations and quantum interference in semiconductor quantum dots, *Phys. Rev. Lett.* **88**, p. 087401.

[88] Kamada, H., Gotoh, H., Temmyo, J., Tagakahara, T. and Ando, H. (2001). Exciton rabi oscillation in a single quantum dot, *Phys. Rev. Lett.* **87**, p. 246401.

[89] Stufler, S., Ester, P., Zrenner, A. and Bichler, M. (2005). Quantum optical properties of a single $In_xGa_{1-x}As$-GaAs quantum dot two-level system, *Phys. Rev. B* **72**, p. 121301(R).

[90] Wang, Q. Q., Muller, A., Bianucci, P., Rossi, E., Xue, Q. K., Takagahara, T., Piermarocchi, C., MacDonald, A. H. and Shih, C. K. (2005). Decoherence processes during optical manipulation of excitonic qubits in semiconductor quantum dots, *Phys. Rev. B* **72**, p. 035306.

[91] Zrenner, A., Beham, E., Stufler, S., Findeis, F., Bichler, M. and Abstreiter, G. (2002). Single quantum dot photodiodes - coherent properties of a two-level system with electric contacts, *Nature* **418**, p. 612.

[92] Stievater, T. H., Li, X., Steel, D. G., Katzer, D. S., Park, D., Piermarocchi, C. and Sham, L. (2001). Rabi oscillations of excitons in single quantum dots, *Phys. Rev. Lett.* **87**, p. 133603.

[93] Varoutsis, S., Laurent, S., Kramper, P., Lemaitre, A., Sagnes, I., Robert-Philip, I. and Abram, I. (2005). Restoration of photon indistinguishability in the emission of a semiconductor quantum dot, *Phys. Rev. B* **72**, p. 041303(R).

[94] Axt, V. M., Machnikowski, P. and Kuhn, T. (2005). Reducing decoherence of the confined exciton state in a quantum dot by pulse-sequence control, *Phys. Rev. B* **71**, p. 155305.

PART 4
Flying Qubits

Chapter 13

Light-matter Interaction in Single Quantum Dot - Micropillar Cavity Systems

S. Reitzenstein, J.-P. Reithmaier*, and A. Forchel

Technische Physik, Universität Würzburg, Am Hubland D-97074 Würzburg, Germany

Technische Physik, INA, Universität Kassel, Heinrich-Plett-Str. 40 D-34132 Kassel, Germany

In this chapter, recent progress in cavity-quantum electrodynamics experiments on single quantum dot microcavity systems is reviewed. Of particular interest is the strong coupling regime, which is characterized by a coherent exchange of energy between a discrete quantum emitter and the cavity mode. Strong coupling is observed when the emitter-photon coupling rate is larger than any dissipative decay rate in the system. Key issues for the optimization of both the emitter and the cavity characteristics are discussed. Differences between weak and strong coupling are illustrated experimentally in the framework of cavity quantum electrodynamics. Furthermore, the demonstration of the quantum nature in a strongly coupled quantum dot micropillar system as well as a coherent photonic coupling of quantum dots mediated by the strong light field in high-Q micropillar cavities are addressed.

Contents

13.1. Introduction

Coherent interaction between bare quantum states is the basis for any type of Qubit. Here the focus will be on the realization of flying Qubits realized by coherent light-matter interaction in solid state. These kind of interaction is described in the framework of cavity quantum electrodynamics (CQED) and can be observed in systems based on high quality microcavities in combination with atom-like emitters [1–3]. CQED addresses in particular a modification of the emitter's spontaneous decay in the presence of confined optical modes in cavities with small mode volumes. In real cavities, photons are not confined for an infinitive amount of time but leave the cavity after a characteristic time, which is inversely proportional to the quality (Q) factor of the latter. The escape of photons through leaky modes introduces dissipation in the system. Now, if the coupling rate between an atom-like emitter and the photonic cavity mode is lower than the photon escape rate irreversible decay dominates, which is described in terms of weak coupling. It is characteristic for the weak coupling regime that spontaneous emission of an emitter can be enhanced or reduced compared with its vacuum value by tuning discrete cavity modes in and out of resonance [4–6]. Recently, due to the enormous progress in nanotechnology processing, it has become feasible to realize quantum dot microcavity systems in which the light-matter coupling rate exceeds any dissipative decay rate [7–9]. In this case, the conditions for strong coupling are fulfilled and vacuum field fluctuations initiate a reversible exchange of energy between the emitter and the cavity. This coherent coupling introduces entanglement as it is associated with the formation of new 'dressed' quantum states and may provide a basis of quantum dot-microcavity systems for future applications in quantum information processing or schemes for coherent control [10–13]. Evidence of strong coupling is usually manifest in the emission spectrum that displays anti-crossing between the quantum dot (QD) exciton and cavity-mode dispersion relations. Strong coupling is characterized by a vacuum Rabi splitting of up to a few 100 μeV in state-of-the-art semiconductor structures. In the following, several aspects of strong coupling will be addressed exemplarily for quantum-dot micropillar cavity systems.

13.2. Theoretical Background

Different theoretical schemes have been applied to describe the interaction of a single quantum dot exciton coupled resonantly to an empty mode of a microcavity. These range from a simple two coupled harmonic oscillator model to a fully quantum mechanical theory [14–16]. In the ideal case for which the emitter's relaxation and cavity losses can be neglected, spontaneous emission becomes reversible, i.e. the emitted photon is reabsorbed and reemitted again and again. This gives rise to vacuum Rabi oscillations which frequency is proportional to the strength of the atomfield coupling, and hence to the vacuum field amplitude at the emitter's location.

The parameter g is the so-called coupling constant, which is defined as a scalar product of the transition matrix element of the dot dipole moment $\mathbf{d}$ with the local value of the electric field $\mathbf{E}$ at the dot position in the cavity ($g = |\langle \mathbf{d} \cdot \mathbf{E} \rangle|$). For the case of the emitter (exciton dipole) placed at the maximum of the photon electric field distribution in the microcavity, the coupling constant can be expressed in terms of the oscillator strength (OS) $f = 2m\omega_0 d^2/(e^2\hbar)$ as [15]

$$g = \left(\frac{1}{4\pi\epsilon_r\epsilon_0} \frac{\pi e^2 f}{m\hat{V}} \right)^{1/2} \tag{13.1}$$

where ϵ_0 and ϵ_r are the vacuum and relative medium permittivity, m is the free electron mass, f is the oscillator strength of the QD transition, and $\hat{V}$ corresponds to the effective cavity mode volume.

The nonperturbative dynamics of the QD transition coupled to a cavity mode can be described by quantum-mechanical theory. The relevant Hamiltonian is

$$H = \hbar\omega_0\hat{\sigma}_3 + \hbar\omega_\mu \left(\hat{a}_\mu^\dagger \hat{a}_\mu + \frac{1}{2} \right) + i\hbar\omega_\mu \left(\hat{\sigma}_- \hat{a}_\mu^\dagger + \hat{\sigma}_- \hat{a}_\mu^\dagger \right) \tag{13.2}$$

where $\hat{\sigma}_-$, $\hat{\sigma}_+$,$\hat{\sigma}_3$ are pseudospin operators for the two-level system with ground (excited) state $|g>$ ($|e>$). The spectrum of this Hamiltonian consists of a ground state $|g, 0>$ and a ladder of doublets $|e, n>$, $|g, n{+}1>$, $n = 0, 1, ...$ which in the resonance case $\omega_0 = \omega_\mu$ give rise to dressed states split by $2\hbar g\sqrt{n+1}$.

In real cavities photons leak out of the cavity and spontaneous emission is coupled into a continuum of non-resonant modes. If decoherence processes affecting either the photon or the emitter, are slow enough on the scale of the Rabi period, the coupled system is still in a strong coupling regime and experiences damped Rabi oscillation. Faster decoherence processes lead to overdamping; in this so-called weak coupling regime, the emitter relaxes monotonically down to its ground state. When the main decoherence process is due to cavity damping, characterized by the resonance quality factor Q of the mode it is still possible to modify the spontaneous emission of the this emitter to a large extent and in particular its spontaneous emission rate $1/\tau_{cav}$, by varying the strength of the emitter-field coupling (i.e. $\hat{V}$) or the cavity losses (i.e. Q).

Basically, the energies of these two modes (exciton and photon) at resonance can be expressed by the following formula [14, 15]:

$$E_{1,2} = E_0 - i(\gamma_c + \gamma_X)/4 \pm \sqrt{g^2 - (\gamma_c - \gamma_X)^2/16)} \tag{13.3}$$

where E_0 is the energy of the uncoupled modes, and γ_C and γ_X are the full widths at half maximum of the cavity mode and exciton line, respectively.

From Eq. 13.3 the vacuum Rabi splitting for the system under consideration can be obtained directly as

$$\Delta E = 2\hbar\Omega = 2\sqrt{g^2 - (\gamma_c - \gamma_X)^2/16} \tag{13.4}$$

According to Eq. 13.3, any mode splitting due to strong coupling appears when

$$g^2 - ((\gamma_c - \gamma_X)^2/16 > 0 \tag{13.5}$$

Since for present microcavities the QD exciton linewidth is usually much smaller (by an order of magnitude at least) than the photon mode linewidth, the condition in Eq. 13.5 can be simplified to [7]

$$g > \gamma_c/4 \tag{13.6}$$

This is the threshold condition for strong coupling of a single QD exciton and a microcavity photon. Using Eq. 13.1 which tells that $g \propto \sqrt{f/\hat{V}}$ and considering that the cavity quality factor is defined by $Q = E_c/\gamma_c$, one obtains from Eq. 13.6 the figure of merit $Q\sqrt{f/\hat{V}}$ for the observation of strong coupling. For pillar microcavities, the mode volume scales approximately with the square of the cavity diameter d_C, which leads to $Q\sqrt{f}/d_c$ as a quantity for technological optimization. Thus, it is necessary to maximize the oscillator strength of the QD exciton transition and the Q-factor to micropillar diameter ratio simultaneously. The optimization of both, the quantum dot properties as well as the cavity finesse, will be described in detail in the subsequent sections.

13.3. Sample Fabrication

13.3.1. *Growth of quantum dot structures*

Pronounced light-matter coupling effects are observed for QDs with a large oscillator strength in low mode volume, high-Q microcavity systems. Therefore, optimization of the quantum dot growth aims at the realization of large dot structures with large dipole-moment and large oscillator strength, respectively. Self-assembled semiconductor quantum dots provide a three-dimensional confinement potential for the carriers and consequently have a discrete energy spectrum with δ-like density of states and an ultra-narrow gain spectrum [17]. They have already been utilized in electronic and optoelectronic devices, such as semiconductor lasers [18], and more recently also for single photon sources [19, 20].

In the well known Stranski-Krastanov growth mode [21], the formation of quantum dots is initiated by strain relaxation during epitaxial growth of GaInAs on a GaAs substrate as the deposited layer exceeds a critical thickness. As a result, the growth characteristic switches from a two-dimensional to a three-dimensional growth mode. In this way, high quality low dimensional nanostructures are formed as the strain is elastically relaxed without introduction of crystal defects. The dot size, area density, and optical properties can be widely controlled by growth parameters, such as growth temperature, growth rate and III/V-ratio. The $In_xGa_{1-x}As$ quantum dot structures are typically grown in solid source molecular beam epitaxy (MBE) systems on (0 0 1) undoped GaAs substrates. Hereby, varying systematically the In content allows one to address the influence of strain on the shape and the optical properties of the QDs [22]. For such studies, a 300 nm GaAs buffer is grown on the substrate before the QD layer is formed using submonolayer deposition of InAs and GaInAs.

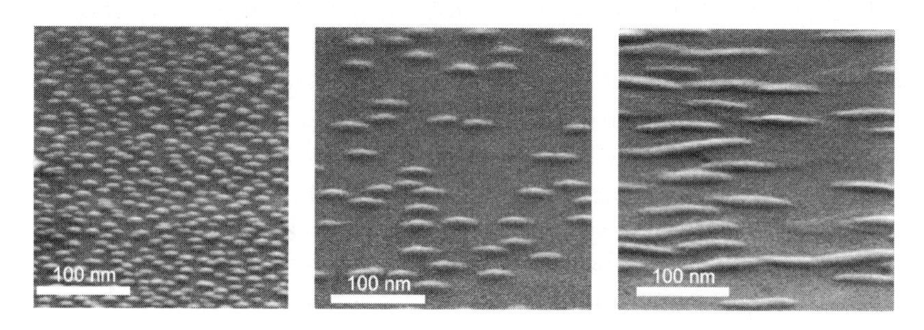

Fig. 13.1. Surface scanning electron microscope (SEM) images of uncapped $In_{0.6}Ga_{0.4}As$ (left panel), $In_{0.45}Ga_{0.55}As$ (center panel) and $In_{0.3}Ga_{0.7}As$ (right panel) quantum dot samples. The surface is tilted by 70° to enhance the height contrast.

Strongly strained $In_{0.6}Ga_{0.4}As$ dots (QD1) form by depositing a nominally 1.4 nm thick InGaAs layer of 7 cycles of 0.1 nm InAs and 0.1 nm $In_{0.20}Ga_{0.80}As$. The strain is reduced for $In_{0.45}Ga_{0.55}As$ dots (QD2) grown by depositing a nominally 2.1 nm thick layer via submonolayer deposition of 11 cycles of 0.07 nm InAs and 0.12 nm $In_{0.125}AsGa_{0.875}$. In addition, $In_{0.30}Ga_{0.70}As$ dots (QD3) can be realized by depositing a nominally 4.5 nm thick layer via deposition of 30 cycles of 0.03 nm InAs and 0.12 nm $In_{0.12}Ga_{0.88}As$. A growth temperature of 590^oC is used for the GaAs buffer, 470^oC for sample QD1 and migration enhanced growth at 510^oC for the low-strain dot layers QD2 and QD3. A cap of 20 nm GaAs is deposited in case of the overgrown samples used for optical studies at the same growth temperature as used for the QD growth.

The influence of the strain on dot nucleation and morphology, is investigated on uncapped dot samples with different In content. Surface scanning electron microscope (SEM) images of the samples labelled QD1 (a), QD2 (b) and QD3 (c) are shown in Fig. 13.1 under an angle of 70° to enhance the height contrast. High strain results in diameters between $10-15$ nm and a rather high dot density of about $1-2\times10^{11}cm^{-2}$ for the $In_{0.6}Ga_{0.4}As$ QDs. For CQED experiments on the single dot level the dot density should be much lower. A lower dot density of about $1 - 2 \times 10^{10}$cm^{-2} and larger dot diameters of 20–25 nm are realized for sample QD2 for which the In content was reduced to 45%. The larger diameter of these QDs is partly related to the lower strain due to the lower In content and to a higher growth temperature of 510°C in comparison to 470°C used for QD1. The higher growth temperature results in a larger migration length on the surface which reduces the dot density and enlarges the dot diameter [23].

Still larger dot dimensions were realized for sample QD3 by decreasing the In content to 30% as can be seen in 13.1. Due to a further decrease of strain, the island growth is mainly initiated by crystal steps on the surface. As a consequence elongated dot structures with typical lengths of $50 - 100$ nm and widths of about 30 nm form preferentially orientated along the $[0\bar{1}1]$ direction. Due to the larger dimensions, the dot volume is increased by one order of magnitude in comparison with the state of the art circular shaped high In-containing GaInAs or InAs dots with diameters of $15 - 20$ nm. In addition, a further decrease of the dot density down to about $6 - 9 \times 10^{9}$cm^{-2} is achieved for $In_{0.3}Ga_{0.7}As$ QDs.

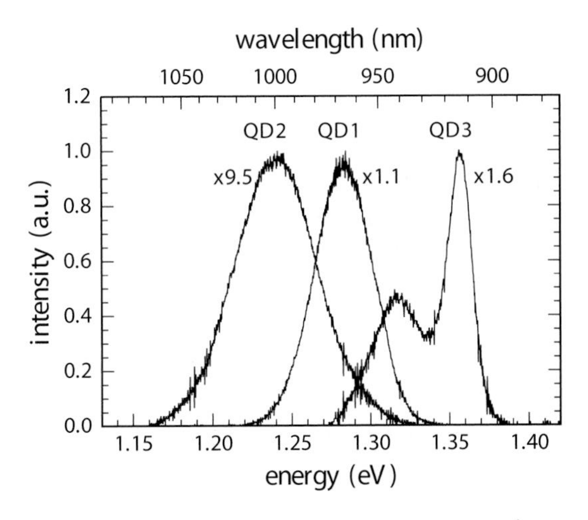

Fig. 13.2. Low temperature PL spectra of the GaInAs dot samples with an In content of 60% (QD1), 45% (QD2), 30% (QD3).

PL spectra of GaInAs dot samples with In contents of 60% (QD1), 45% (QD2), and 30% (QD3) are summarized in Fig. 13.2. The PL emission of QD1 is centered at 1.283 eV with an inhomogeneous linewidth of 45 meV. The larger dot sizes of sample QD2, shifts the emission energy to lower energies (1.242 eV) even though less In was supplied during growth. A very inhomogeneous size distribution and a broadening of the linewidth to 62 meV is attributed to the reformation of some circular shaped dots to elongated dash-like structures during the overgrowth process. Sample QD3 with 30% In content and a deposited nominal layer thickness of 4.5 nm is just above the critical condition for dot formation and shows a double peak emission. In fact, an In content of 27% seems to be already too low to reach the critical thickness for dot formation. Related to this observation the high energy peak of QD3 is interpreted as a quantum well transition. The second peak at lower energy (1.315 eV) is associated with transitions of large elongated QDs. Here, the much lower In content of QD3 shifts the QD emission to significantly higher energies in comparison to QD1 and QD2 samples.

13.3.2. *Realization of active high-Q micropillar cavities*

The interaction of electromagnetic radiation with single quantum states, e.g., an electron state in an atom or in a solid state quantum dot structure, is rather weak due to the low oscillator strength. However, this interaction can be strongly increased by several orders of magnitude by embedding the atom or the QD in a high quality (Q) resonator. Semiconductor systems in particular have the advantage of allowing one to embed QDs into the fabrication process of high-Q cavities. Different approaches were investigated and published in the literature to improve the Q-factor in semiconductor microcavities. Promising candidates for high-Q low mode volume microcavities are photonic crystal membrane structures [24], semiconductor microdisk structures [25] and micropillar cavities [26, 27], respectively.

Micropillars have some advantages compared with other microcavity approaches in terms of mode control and overlapping between light field and quantum dots. In micropillars large overlap of the central fundamental mode with the quantum dot states can be achieved and the confined mode spectrum can be directly controlled by the post diameter [28, 29]. In addition, the possibility of optical excitation and detection perpendicular to the sample surface facilitates optical experiments.

Of special interest are micropillars with large Q-factors and embedded low density self-assembled quantum dots with high oscillator strength. Cavity structures are usually grown by solid source molecular beam epitaxy GaAs substrates and consist of bottom distributed Bragg reflector (DBR) followed by a one λ GaAs cavity and a top DBR. The DBRs are realized

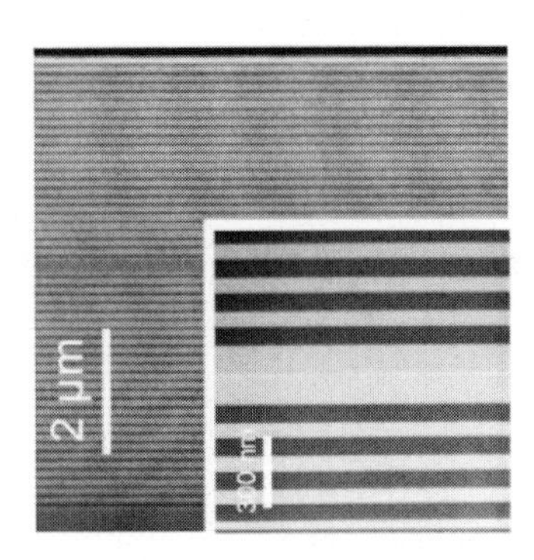

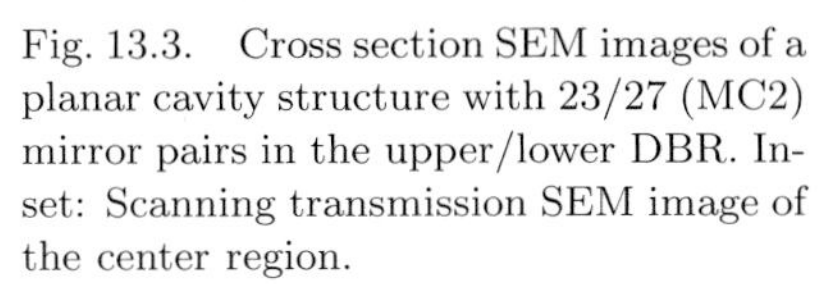
Fig. 13.3. Cross section SEM images of a planar cavity structure with 23/27 (MC2) mirror pairs in the upper/lower DBR. Inset: Scanning transmission SEM image of the center region.

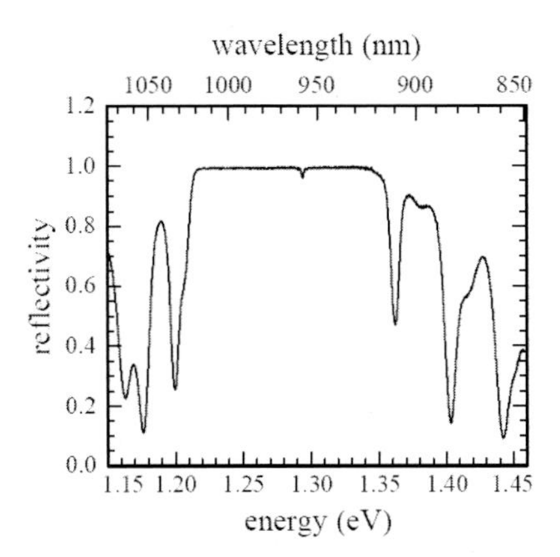

Fig. 13.4. Reflectivity spectrum of MC1 measured at 300 K with a stopband of 113 nm and a resonance peak at about 1.29 eV.

by a stack of alternating quarter-wavelength thick GaAs and AlAs mirror pairs. The reflectivity of the DBRs increases with the number of mirror pairs and high-Q cavities are designed with nearly identical reflectivities of the bottom and the top DBR. A cross section SEM image of a structure with 23/27 mirror pairs in the upper/lower DBR is shown in Fig. 13.3. The DBRs are separated by a one λ thick GaAs cavity. A QD layer is placed in the middle of the cavity, i.e., at the antinode of the electrical field.

A rather low dot density of $n_{QD} \lesssim 1 \times 10^{10}\text{cm}^{-2}$ is chosen in order to allow single dot investigations and to reduce damping of the optical modes by bandgap absorption in the active layer. The planar cavity structures show a pronounced stopband which can be characterized by reflectivity measurements, usually performed at room temperature. In reflectivity measurements of planar cavity structures, the spectral resolution is limited by the finite acceptance angle because of the continuous in-plane mode dispersion. A typical reflectivity spectrum of a planar microcavity structure with 20/24 mirror pairs in the lower (upper) DBR is shown in Fig. 13.4. Here, the cavity resonance can be seen at 1.29 eV as a small dip within the stopband with a width of about 113 nm.

Based on planar microcavity structures, high quality, low mode volume micropillars with circular cross-sections and a diameter range between 1 μm

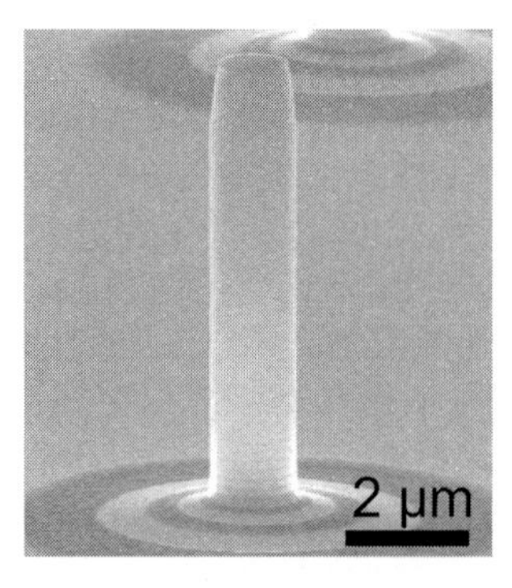

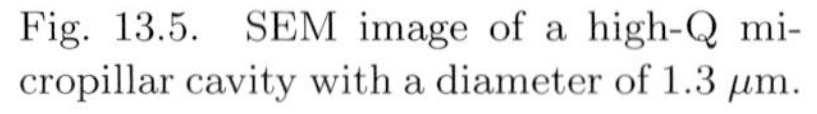
Fig. 13.5. SEM image of a high-Q micropillar cavity with a diameter of 1.3 μm.

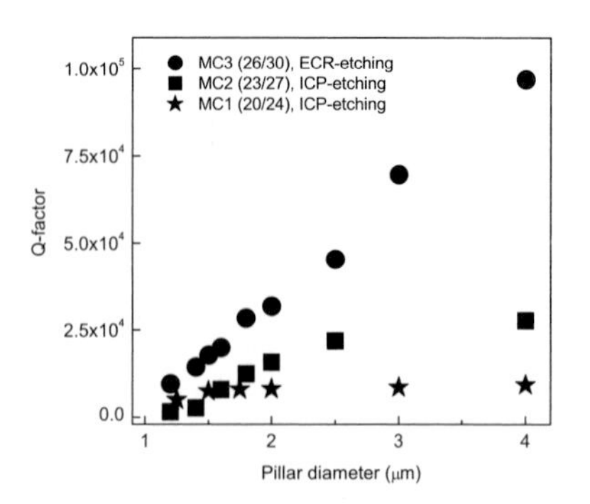

Fig. 13.6. Pillar Q-factor as a function of pillar diameter for three microcavity layouts with different numbers of mirror pairs in the DBRs.

and about 4.0 μm are fabricated by means of modern semiconductor nanotechnology. As a first step, micropillars are defined into resist (PMMA) by high resolution electron beam lithography. After evaporation of a metal mask (Ni) and a subsequent lift-off process, the semiconductor material is removed at unexposured regions down to few remaining layers of the lower DBR by reactive ion etching. This process is performed either with inductively coupled plasma (ICP) or electron cyclotron resonance (ECR) etching. Optimization of the plasma etching allows the realization of very smooth sidewalls and avoids a degeneration of the top mirror. This reduces optical losses due to photon scattering at surface irregularities, and provides strong and uniform optical confinement in the micropillars. Here, ECR etching has the advantage that undercutting of the lower DBR observed for ICP etching is almost negligible [27]. An SEM image of a micropillar with a diameter of 1.3 μm and a height of about 6 μm is shown Fig. 13.5. In this example, the bottom mirror was removed except for the last 2 - 3 GaAs/AlAs mirror pairs.

For practical reasons, it is interesting to study the influence of the micropillar geometry on the resonance Q-factor defined as the ratio E_c/γ_c of the emission energy E_c and the mode linewidth γ_c (FWHM). This allows one to address the influence of the number of mirror pairs in the DBRs on the resonator quality and to choose micropillars with the maximum Q/d_c ratio for investigations of strong light-matter interaction. the measured Q-factors are depicted in Fig. 13.6 versus the pillar diameter in the range between 1 and 4 μm for three microcavity designs with 20/24 (MC1), 23/27

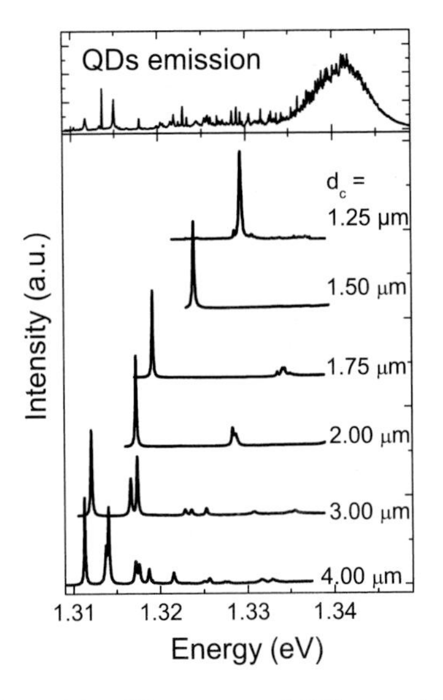

Fig. 13.7. PL spectra of QD ensemble emission and micropillar cavities of different diameter.

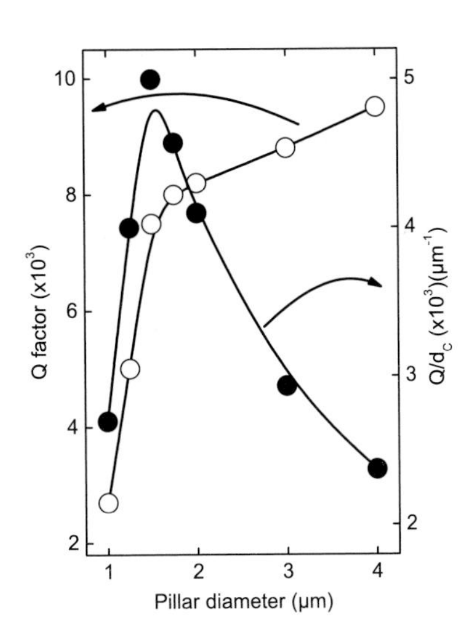

Fig. 13.8. Dependence of the pillar Q-factor and the Q/d_c on the pillar diameter.

(MC2) and 26/30 (MC3) mirror pairs in the upper/lower DBR.

Some interesting features can be identified in the data displayed in Fig. 13.6. First of all, it is seen that the Q-factor drops with decreasing cavity diameter d_c. For sample MC1 with 20/24 mirror pairs, which was processed by ICP etching, the Q-factor saturates above $d_c \approx 2.0$ μm at a value of about 10000 and decreases down to 5000 at $d_c \approx 1.0$ μm. Higher Q-factors of up to 30000 were determined for MC2 with 20/24 mirror pairs as a result of the higher reflectivity of the DBRs [27]. However, because of strong underetching a strong decrease of the Q-factor is observed due to diffraction losses in the underetched lower mirror. An increase of the mirror pairs to 26/30 in case of MC3 further reduces the losses through the DBRs in vertical direction leading to Q-factors of up to 90000 for a pillar diameter of 4 μm. For such high number of mirror pairs, the Q-factor depends strongly on the influence of residual bandgap absorption in the limit of large diameters. In the low diameter limit, the Q is mainly limited by leaky modes and diffraction losses [30]. The latter could be strongly decreased due to optimized pillar processing using ECR etching with almost no underetching. This certainly contributes to the significantly higher Q-factors of MC3 exceeding 20000 below $d_c = 2$ μm in comparison with MC2.

For the experimental realization of strong coupling, large QDs providing a high oscillator strength are embedded in high-Q micropillar cavities with a large number of mirror pairs in the lower and upper DBR. Here it is crucial, to have the fundamental mode of the micropillar cavity located in the low energy tail of the QD emission band with a low spectral density of QD exciton lines in order to allow single dot investigations. A comparison of the microcavity mode spectra with the ensemble QD emission of a structure with an removed upper DBR is shown in Fig. 13.7. For the shown micropillar diameters, the modes match the range dominated by single dot lines seen on the low energy tail of the QD emission band as required for single dot investigations. For a given cavity design, pillars with the maximum Q/d_c ratio have the highest coupling constant and are therefore most suitable for investigations of strong coupling. Fig. 13.8 shows a diameter dependence of typical Q-factors and Q/d_c values for pillars based on MC1. The highest Q/d_c ratio of about 5000 μm^{-1} is obtained for pillars with a diameter of about 1.5 μm which are the best candidates for the observation of strong coupling. In addition to the spectral matching, spatial matching with the field in the cavity is required to obtain the maximum coupling strength. This requires the QD to be located laterally in the center of the cavity which introduces some randomness, since it cannot be controlled in the present technology due to the random distribution of the QDs in the plane of the active layer. Thus, further technological optimization will probably aim at the controlled positioning of QDs in the field maximum of micropillar cavities in order to ensure a maximum coupling strength [31].

13.4. Experimental Techniques

Light-matter interaction in individual micropillar cavity structures is usually studied by means of micro-photoluminescence techniques. For optical excitation, the laser-beam is focused into a few micron diameter spot, centered at the pillar by a microscope objective. Usually, continuous wave (CW), non-resonant excitation at either 514 nm or 532 nm is used for optical pumping. However, for photon correlation experiments described in section 13.6 it is necessary to work with pulsed resonant excitation using a mode-locked pulse Ti:sapphire laser in conjunction with an Hanbury-Brown and Twiss setup [32]. The micropillar geometry is very advantageous with respect to in- and outcoupling of light and the microscope objective applied for excitation can also be used to collect the emission of the sample. The PL is usually detected by a high resolution spectrometer in combination with a nitrogen cooled silicon CCD camera. For the tuning through resonance the different temperature variation of the band gap and of the refractive index n is exploited. The temperature dependence of the emission energy of the quantum dot excitons (typically -0.04 meV/K at 25 K) is governed by the temperature dependence of the band gap, whereas the temperature dependence of the cavity modes (about -0.005 meV/K at 25 K) is determined by the much weaker temperature dependence of the refractive index. By selecting micropillars with single QD exciton lines near, i.e., within a range of 1 meV above the energy of the cavity resonance, it is possible to tune the QD excitons in and out of resonance with the cavity mode.

13.5. Weak Coupling and Strong Coupling in Micropillar Cavities

Light-matter interaction of single quantum dots in micropillar cavities was first investigated about a decade ago [4, 5]. In the early works enhancement and inhibition of spontaneous emission was observed in the weak coupling regime. However, the onset of strong coupling was not reached in these studies. Due the strong improvement in semiconductor technology it has become feasible to fulfill the condition for strong coupling, i.e., the rate of energy exchange exceeds any dissipative decay rate. This becomes possible by optimizing both the quantum dot as well as the micropillar properties as discussed in the previous sections [7].

13.5.1. *Weak coupling*

If the rate of exchange of energy between an quantum dot exciton and the cavity mode is smaller than the underlying decay rates, then the exciton-photon system is in the regime of weak coupling which is usually manifested by an enhancement of the exciton recombination rate accompanied by an increase of the QD emission intensity at resonance with the cavity mode. The enhancement of recombination rate is described in terms of the Purcell-effect [33] and can be observed in the PL spectra shown in Fig. 13.9. Here, the temperature tuning of a single QD line through resonance with the fundamental mode of a 1.8 μm micropillar cavity is presented. The energy of the resonance has been placed at zero detuning. In this example standard (small volume) $In_{0.6}Ga_{0.4}As$ self-assembled quantum dots were used in combination with a cavity exhibiting a rather low $Q/d_c = 3600\ \mu m^{-1}$ ratio.

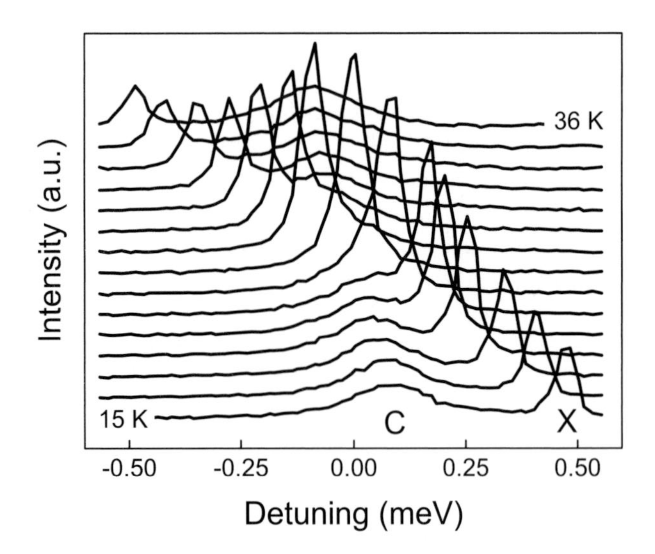

Fig. 13.9. Light-matter interaction in the weak coupling regime. Temperature tuning between 15 and 35 K is used to shift a single QD exciton X into resonance with the fundamental mode of a 1.8 μm diameter micropillar cavity with a Q of 6500. On resonance an enhancement of the emission intensity is observed due to the Purcell-effect.

The variation of the mode energies of the spectra shown in Fig.13.9 is depicted in Fig. 13.10 (upper panel) as a function of detuning. As expected for the weak coupling regime simple, crossing of the exciton line (X) and the cavity mode (C) occurs at resonance, which can be explained in terms of Eq. 13.3 when $g^2 < (\gamma_c - \gamma_x)^2/16$ and both eigenenergies have the same real part. However, the emitter relaxation into its ground state which is characterized by an exponential decay, is enhanced by the Purcell-factor F_p. A maximum Purcell-factor is expected for an ideal, resonant emitter located at the field maximum with its dipole moment aligned along the local electrical field. In this case F_p is given by

$$F_p = \frac{\tau_{free}}{\tau_{cav}} = \frac{3Q(\lambda_c/n)^3}{4\pi^2 \hat{V}}, \tag{13.7}$$

where τ_{free} and τ_{cav} denotes the spontaneous lifetime of the emitter in free space and in the cavity, respectively [33]. In a real system spatial mismatching and emission into leaky modes reduces the experimental Purcell-factor [4]. In the present case, the QD emission intensity is enhanced by

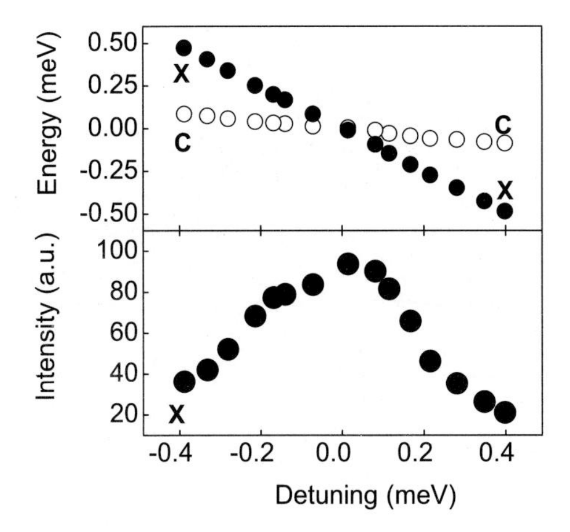

Fig. 13.10. Mode energies (upper panel) and intensity of the QD exciton (lower panel) as a function of detuning between the exciton (X) and photon mode (C). The weak coupling regime is characterized by a simple crossing behavior of the mode energies and an enhancement of the exciton intensity at resonance.

a factor of about 5 at resonance compared with its off-resonance value for large detunings. This behavior typical for the weak coupling regime is illustrated in the lower panel of Fig. 13.10, where the exciton intensity is plotted versus detuning.

13.5.2. *Strong coupling*

Considerably different spectra have been recorded for an optimized pillar structure with $d_c = 1.5$ μm and $Q/d_c = 4900$ μm^{-1}, and low density layer of enlarged volume $\text{In}_{0.3}\text{Ga}_{0.7}\text{As}$ quantum dots embedded in the center of the λ cavity. Temperature tuning of this optimized quantum dot - micropillar structure is shown in Fig. 13.11 in a range of PL spectra 5 - 30 K. In this example a clear qualitative difference in the emission properties is observed in comparison with the data depicted in Fig. 13.9. First of all, an increase of intensity at resonance is almost absent. Secondly, no simple line crossing typical for the weak coupling but a well pronounced anticrossing (polaritonic) behavior is observed near resonance. Furthermore, the exciton and photon modes exchange their properties, i.e., linewidths and intensities at resonance. This is a clear signature of coherent coupling of two modes as expected in the strong coupling regime [7].

In Figure 13.12 the measured peak energies and FWHM values for a 1.5 μm micropillar are plotted as function of the temperature [7]. For

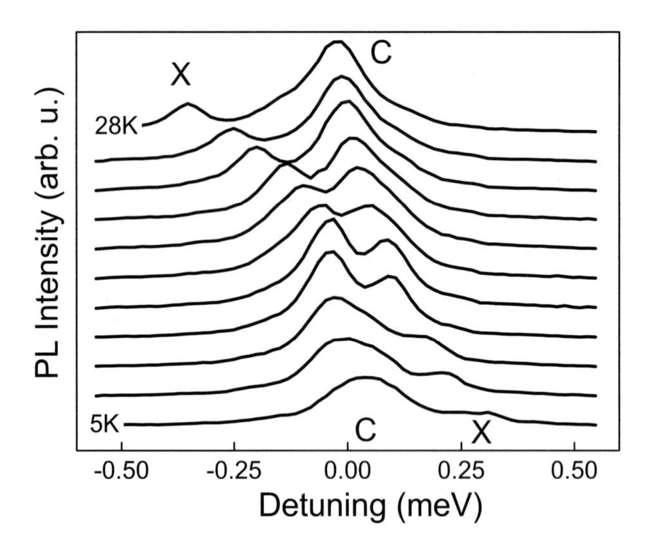

Fig. 13.11. Light-matter interaction in the strong coupling regime. Temperature tuning between 5 and 28 K is used to shift a single QD exciton X through resonance with the fundamental mode of a 1.5 μm diameter micropillar cavity with a Q of 7350. On resonance at about 20 K pronounced anticrossing is observed as a fingerprint of the strong coupling regime.

the strongly coupled system, the energies of the higher lying mode show a rather strong variation before reaching resonance at about 20 K and a much weaker dependence below 20 K. On the other hand, the lower laying mode displays only a weak temperature variation at low temperature and changes to a stronger temperature dependence above 20 K, i.e., after passing the resonance. Here, the strong variation with temperature is associated with excitonic behavior while the weaker temperature dependence identifies the cavity mode. Therefore the anticrossing leads to a change from an exciton like dispersion to a cavity mode like one for the upper branch, whereas the lower branch changes from a cavity mode dispersion to that of an exciton with increasing temperature. The energy separation of the coupled modes on resonance of about 140 μeV corresponds to the vacuum Rabi splitting of this coupled single QD exciton solid state cavity system. Strong coupling leads to a normal mode mixing at resonance which is reflected in equal mode linewidths on resonance. For increasing temperature up to 30 K two distinct modes are identified in the emission spectrum. However, at 30 K the components of the emission have exchanged their properties in a way such that the low energy line has a FWHM similar to that of the excitonic mode located at 5 K at higher energy. Simultaneously, at 30 K the high energy line displays a FWHM which correspond to the values for the cavity mode at low temperature.

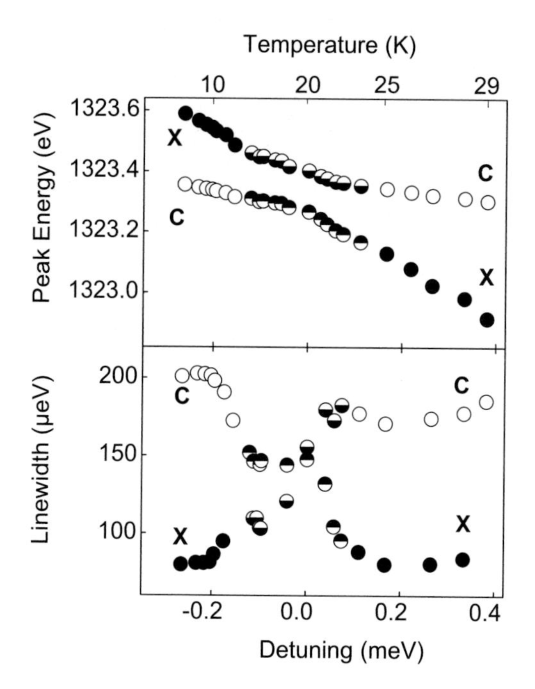

Fig. 13.12. Emission energy and linewidth (FWHM) as a function of detuning between the exciton (X) and photon mode (C). The strong coupling regime with a vacuum Rabi splitting of 140 μeV is reflected in the anticrossing behavior of the mode energies and the exchange of linewidth at resonance.

The discussed features are clear fingerprints of strong coupling between the single QD exciton and the single photon mode. It is interesting to estimate the coupling constant g from the experimentally obtained value for vacuum Rabi splitting ($\Delta E_r = 140$ μeV). Taking the measured cavity linewidth into account the coupling constant equals to 80 μeV according to Eq. 13.4. This value exceeds clearly the threshold $g_c/4 = 45$ μeV (Eq. 13.6) for the observation of strong coupling and confirms again that the strong coupling regime has been reached. It is interesting to estimate the oscillator strength according to Eq. 13.1 for the present coupling constant and a mode volume of about 0.3 μm^3 calculated numerically for the 1.5 μm diameter micropillar cavity. One obtains an effective oscillator strength of about 50 which is about five times higher than the values reported for self assembled InAs QDs. This indicates that the larger extension of the $In_{0.3}Ga_{0.7}As$ QDs results in higher dipole moment. Similar oscillator strengths are reported for natural quantum dots realized by monolayer fluctuations in GaAs/AlGaAs quantum wells. It is clear that dots with large oscillator strengths are crucial for the observation of strong coupling as the threshold value $g = \gamma_c/4$ would not be reached for self assembled InAs dots with a typical oscillator strength of about 10 for the present Q/d_C values.

It is important to note that strong coupling refers here to the empty cavity (vacuum) Rabi splitting and the strong coupling occurs between a single QD exciton and single photon. A careful adjustment of the experimental parameters is required for the demonstration of vacuum Rabi splitting. For example, low excitation powers were applied in order to ensure that multi-exciton effects can be ruled out. Still, the QD line could be a superposition of two or more degenerated single exciton transitions related to different dots. However, due to the low spectral density of QD emission lines which is typically about one line per meV, the probability to find 2 lines emitting within the spectral resolution of 50 μeV is less than 1 %. This estimation shows that multi-emitter contributions are very unlikely for the present example of strong coupling. Nevertheless, it is interesting to prove the single emitter regime experimentally by studying the photon statistics of the emitted light which will be addressed in the next section of this review. Also, the average photon number in the cavity mode must be much lower than unity for the investigation of vacuum field Rabi splitting. Since most of the applied laser power (2 μW, 532 nm) is absorbed in the GaAs sections of the upper Bragg reflector it can be estimated that only a small fraction of about 1 nW reaches the cavity. In the worst case estimation, all the photons reaching the cavity are absorbed there, and the resulting carrier pairs all relax in the dot under consideration. This results in an average photon number of less than 1% in the present cavity with a photon lifetime of about 5 ps. Therefore, effects due to the population of the cavity with more than one photon at a time are considered to be negligible as required for the observation of vacuum Rabi splitting.

13.6. Photon Antibunching in the Single QD Strong Coupling Regime

In this section evidence will be presented that the emission from a strongly coupled single QD-microcavity system is dominated by a single quantum emitter. In particular, it will be shown that the strong light matter interaction originates from a single quantum emitter coupled to the cavity mode. It is clear that this regime corresponds to the most fundamental situation of light-matter interaction in semiconductors and needs to be distinguished from the semiclassical case of vacuum Rabi-splitting in quantum well-mirocavity structures for which a large number of quantum well excitons contribute to the coupling [29, 34]. Proofing the quantum nature of the strongly coupled QD-microcavity system exploits the fact that photon emission from a single quantum emitter is antibunched [19]. Such a non-classical emission behavior is expected for a single quantum dot strongly coupled to a microcavity mode. The key to the observation of antibunching in such a system is to resonantly pump a selected QD via an excited state. In this case background population of the cavity mode by non-resonant emitters which are usually excited by an above-band pump [7–9], is strongly reduced. This is crucial for photon correlation experiments aiming at the proof of the single dot regime since background population by non-resonant emitter leads to uncorrelated photon emission of the cavity mode which masks the antibunching property of a single resonant emitter.

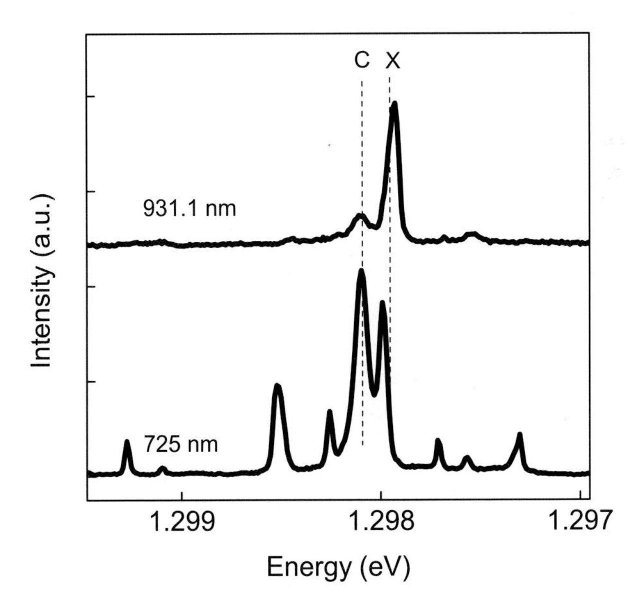

Fig. 13.13. PL spectra of a 1.8 μm micropillar under nonresonant and resonant p-shell excitation at 725 nm and 931.1 nm, respectively.

Studies of the photon statistic of a strongly coupled single quantum dot-micropillar cavity system were performed on a sample based on a planar cavity with 26/30 mirror pairs in the upper/lower DBR and InGaAs QDs with an In content of about 40% in the active layer. For this structures the highest Q/d_c-ratios are observed for 1.8 μm diameter pillars, which typically exhibit Q-factors of 10000 - 20000, for a mode volume of $V_m = 0.43\ \mu\text{m}^{-3}$. Usually, quantum dots in microcavities are excited non-resonantly, i.e., above the bandgap of the surrounding GaAs cavity layer. In this case, no cavity emission is expected if no QD level is resonant with the mode. However, studies of QD high-Q microcavity systems show strong cavity emission even when no QD is resonant with the cavity mode [7]. Here the coupling mechanism is most probably related to energy transfer by acoustic phonons but no detailed study of the underlying processes has been reported so far. Population of the cavity mode by non-resonant emitters can be strongly suppressed by tuning the laser to resonantly pump the excited state, i.e. p-shell exciton, of a selected QD [35] which quickly thermalizes to the QD ground state (s-shell). The effect of resonant pumping on the cavity emission is illustrated in Fig. 13.13. CW above-band pumping shown in the lower trace in Fig. 13.13 results in a strong cavity mode emission even though no resonant QD is present. In contrast, resonant pumping of an excited state for a specific QD (937.1 nm in this case) allows a selective excitation of a chosen QD and reduces background cavity emission by almost an order of magnitude.

Strong coupling of a resonantly excited QD with the mode of a 1.8 μm with a Q of 15200 is demonstrated in Fig. 13.14. Similar to Fig. 13.11 clear anticrossing is observed by temperature tuning of a selected QD mode through resonance with the cavity mode. The mode dispersion and linewidths of this resonantly pumped QD-cavity system show again an exchange of properties of the two lines where almost identical linewidths are observed at the resonance temperature. Here, the vacuum Rabi splitting amounts to 56 μeV which corresponds to a coupling constant of $g = 35\ \mu$eV, significantly higher than the threshold value $\gamma_c/4 = 21\ \mu$eV for strong coupling. This coupling constant corresponds to an estimated oscillator strength of about 20 which is notably lower than the one observed for $In_{0.3}Ga_{0.7}As$ QDs. The difference in oscillator strength is related to a lower dipole moment of the 40% In QDs in consistence with their smaller lateral extension. The smaller size of QDs is also reflected in a higher splitting between the p- and s-shell of 25–30 meV in comparison to about 5 meV reported for $In_{0.3}Ga_{0.7}As$ QDs [36]. Even though the coupling constant is

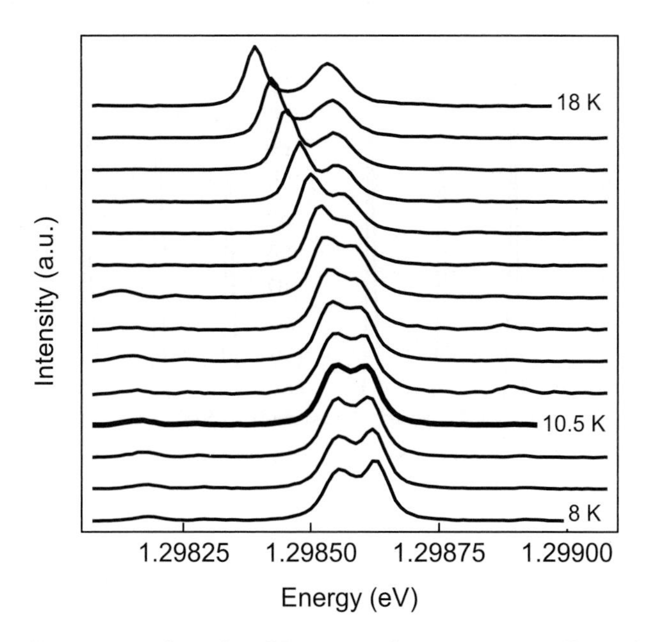

Fig. 13.14. Temperature dependent PL spectra of a 1.8 μm micropillar under resonant p-shell excitation at 936.25 - 936.45 nm. Resonance characterized by a vacuum Rabi-splitting of 56 μeV is observed at 10.5 K.

lower for the smaller 40% In QDs they are more suitable for photon correlation experiments since the larger sp-splitting facilitates resonant p-shell pumping.

The quantum nature of the strongly coupled QD-microcavity system and the single emitter regime is verified by measuring the photon autocorrelation function $g^2(\tau) = \langle: I(t)I(t+\tau) :\rangle / \langle I(t) \rangle^2$ of the PL emission. On resonance, the lifetime of the emitter is strongly reduced to about 15 ps which is twice the cavity lifetime. Photon antibunching at such short time scales can not be resolved by conventional photon counting modules under CW excitation. In fact, pulsed excitation is required to circumvent the problem of limited temporal resolution. Under pulsed resonant excitation the photons collected from the coupled QD-cavity system at resonance show pronounced antibunching which is presented in Fig. 13.15. The observed value of the autocorrelation function $g^2_{r,r}(0) = 0.18 < 1/2$ proves that the emission from the coupled QD cavity is indeed dominated by the single QD emitter. Further insight of the single quantum dot-microcavity system is obtained by cross-correlation experiments under off-resonance condition. First of all, for a detuning of 0.4 nm to lower energies the emission of the

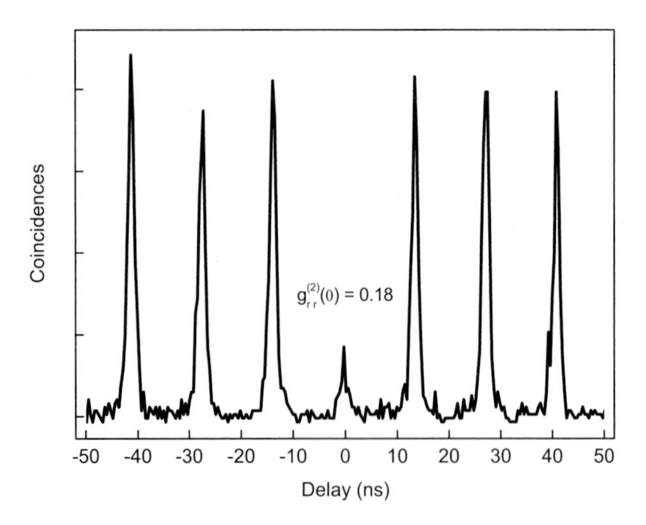

Fig. 13.15. Photon correlation function of the strongly coupled single quantum dot-micropillar cavity system on resonance.

resonantly excited QD is antibunched as expected with $g^{(2)}_{x,x}(0) = 0.19$ as can be seen in the left panel of Fig. 13.16. Interestingly, even under off-resonance conditions the cavity emission with $g^{(2)}_{c,c}(0) = 0.39 < 1/2$ shows antibunching, which means that the cavity emission is dominated by a single quantum emitter (Fig. 13.16, center panel). However, the slightly higher value $g^{(2)}_{c,c}(0) = 0.39$ indicates that even under resonant excitation some background emitters contribute weakly to the cavity emission. Most interestingly, the cross-correlation function between the QD exciton and cavity emission $g^{(2)}_{x,c}$ reveals strong antibunching with $g^{(2)}_{x,c} = 0.22$ (Fig. 13.16, right panel) and conclusively proves that the single QD emitter is responsible for both peaks in the PL spectrum. This strong correlation for the largely detuned system implies that the QD emission is coupled efficiently into the cavity mode which cannot be explained by radiative coupling. Further studies are necessary to determine the exact mechanism responsible for this coupling but most probably it is mediated by the absorption or emission of thermally populated acoustic phonons.

13.7. Coherent Photonic Coupling of Quantum Dots

Coherently coupled solid state systems formed by interacting semiconductor quantum dots are of considerable interest for fundamental studies and

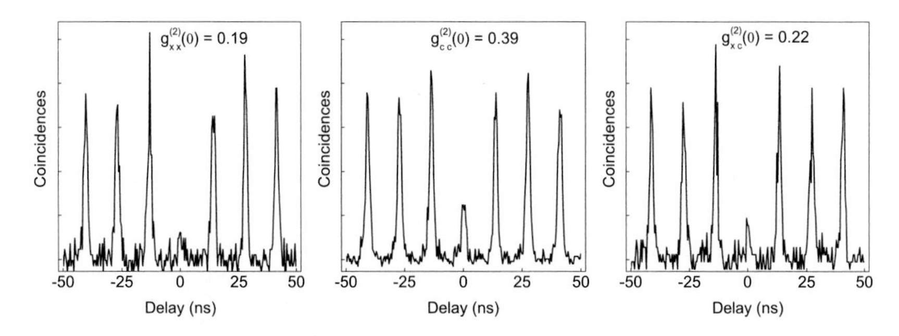

Fig. 13.16. Photon correlation functions of the single quantum dot-micropillar system for a detuning of the exciton line by 0.4 nm from the cavity mode. Left panel: Autocorrelation function $g^{(2)}_{x,x}(\tau)$ of the exciton. Center panel: Autocorrelation function of the cavity mode $g^{(2)}_{c,c}(\tau)$. Cross-correlation function $g^{(2)}_{x,c}(\tau)$ of the exciton and cavity mode.

for applications including quantum information processing [11]. For applications in quantum information processing systems, energetically close entangled transitions such as in excitonic molecules, where excitons are located in two electronically coupled quantum dots, or biexciton transitions in a single quantum dot have been proposed. These investigations of coherent coupling between quantum dots have focused on electronic coupling by Coulomb interaction or by quantum mechanical tunnelling of carriers [37–39]. The latter coupling mechanism has stringent technological requirements primarily due to the short interaction length of the tunnelling. In order to show significant tunnelling, the two quantum dots should be located within a few nanometers of each other and their energy levels should match. A long range photonic coupling of quantum dots can be mediated by the optical mode of a micropillar cavity providing three dimensional optical confinement. Here, the coupling of the dots results in a large splitting of 250 μeV at resonance due to the coherent interaction with two emitters.

In this section, it will be shown that a photonic coupling can be used to provide a coherent interaction between two quantum dots and an optical mode of a micropillar cavity. This system represents a new type of quantum dot molecule with greatly reduced requirements for the spatial separation of the dots (on the order of the wavelength of light, i.e., a few 100 nm) in comparison with electronically coupled molecules (a few nanometers). A system in which two quantum dot excitons are coupled coherently by the optical mode can be realized only if the quantum dot exciton transition energies are sufficiently close, i.e., if they have a spectral separation smaller than or comparable to the exciton - photon coupling energy. The transition energies of the quantum dot excitons formed by self assembly in epitaxy of non-lattice matched semiconductors vary due to size and composition fluctuations. This allows us to observe two QDs with spectrally distinct but energetically close exciton emission lines. The large photon field in a micropillar cavity can then be used to couple the emitters coherently in the regime of strong coupling. In particular, a light - coupled coherent QD system is realized when two quantum dot excitons are tuned on resonance with the optical mode and interact jointly with photons in the optical mode of the cavity.

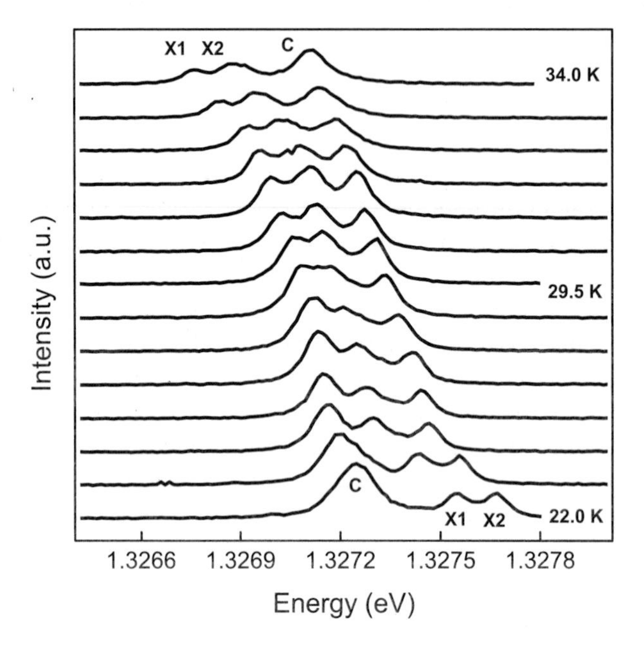

Fig. 13.17. Temperature dependent PL spectra of a 1.8 μm micropillar showing coherent photonic coupling of two QD excitons (labeled X1 and X2) mediated by the strong light field of the cavity mode C. Resonance with the cavity mode is observed for 29.5 K.

Coherent photonic coupling is demonstrated in a micropillar with a diameter of 1.6 μm and a Q of 9600 for which two almost degenerate $In_{0.3}Ga_{0.7}As$ QDs show strong coupling with the cavity mode. The low temperature (22 K) emission of the respective structure is shown in Fig. 13.17 and consists of the cavity line (C) located at about 1.3273 eV and two higher lying excitonic transitions of two different QDs, which are labelled X1 (exciton at about 1.3276 eV) and X2 (exciton at about 1.3277 eV) [40].

As will be shown below, it is crucial for the observation of coherent photonic coupling that the energetic spacing between the two X lines (about 120 μeV) is smaller than the linewidth (FWHM) of the cavity mode of about 140 μeV in the present case. Again, temperature tuning is applied in order to shift the QD excitons through resonance with the common cavity mode. At low temperatures the cavity mode is located at lower energy than both X1 and X2. At above 30 K the cavity mode has become the highest lying component of the spectrum. At low or at high temperatures, i.e. for an uncoupled system, the cavity mode is characterized by a higher intensity and a larger spectral width in comparison with the dot lines. No crossing of the lines is observed in the whole temperature range and all lines have similar intensities and widths near resonance, as expected for strong coupling. Unlike to the situations discussed in the previous sections, strong coupling involves simultaneously two QD excitons and the cavity mode and mode mixing of three modes occurs.

More information on the underlying coupling mechanisms and the respective coupling constants is obtained by studying the dispersion relations. The respective energy dependencies of the QD excitons and the cavity mode are obtained from evaluations of the experimental data using Lorentzian lineshape fits and are plotted in Fig. 13.18 for coherent (upper panel) coupling as a function of detuning. The quantum dot excitons X1 and X2 (open triangles and squares) and the cavity emission (full circles) have characteristic temperature dependences out of resonance, which can be used to identify them as cavity mode (weak temperature (T) dependence) and quantum dot excitons (stronger T-dependence). In case of the coherent coupling, the temperature dependencies of two of the three lines clearly change character when going through resonance at about 29.5 K. Above the resonance temperature the high energy line shows a weak temperature dependence, like that of the cavity at low temperatures, and the two lines now at low energy have acquired the characteristic strong temperature dependence observed for the QD excitons below resonance temperature. The smallest separation between the two outermost lines of the triplet is reached at resonance with a splitting of 250 μeV. In addition to the experimental data, results of calculations which have been obtained by using a model of three coupled oscillators are presented in Fig. 13.18 as solid lines. The calculations are based on an extension of earlier two oscillator calculations and allow the determination of the coupling constants under the assumption that the dots are positioned in the maximum of the electromagnetic field of the fundamental mode in the micropillar cavities.

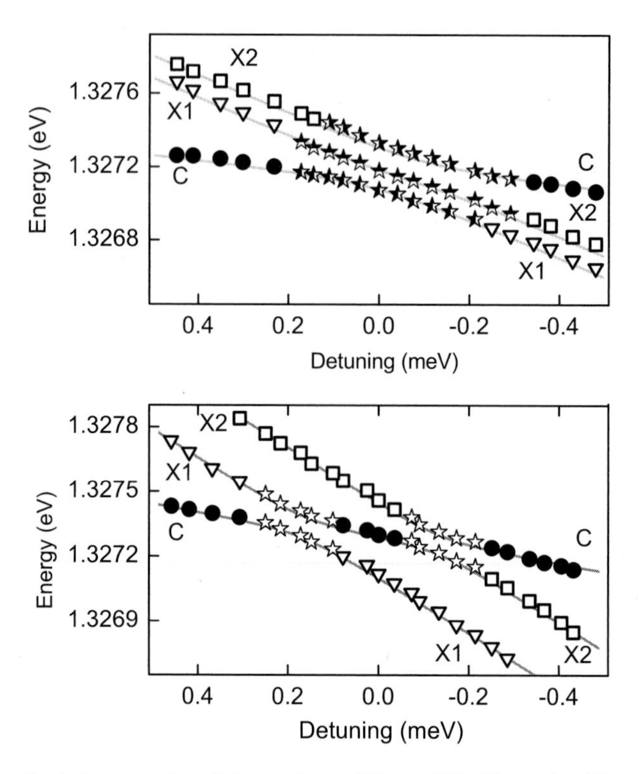

Fig. 13.18. Emission energies of the excitonic (X1: ▽, X2: □), cavity (C:•) and mixed modes (⋆) of the experimental spectra shown in Fig. 13.17 and Fig. 13.19, respectively. The upper panel corresponds to coherent photonic coupling of X1 and X2 while sequential strong coupling is illustrated in the lower panel. Results of modelling the experimental data by a three oscillator model are plotted in solid lines.

This is ensured for the vertical direction by the position of the QD layer in the center of the cavity while the positions of the self-assembled QDs are random in the plane of the dots as discussed in section 13.3.2. The Rabi splitting is characterized by the scalar product of the dipole moment and the electrical field. Thus, the above assumption results in a tendency to underestimate the dipole moments of the excitons in the evaluation. Using the the experimental Q-values and the experimental values for the spectral separation of the two dot excitons as input, the dispersion relations can be modelled in good agreement with experiment. The modelling yields values for the exciton - photon coupling constants of the two excitons of $g_1 = 66\,\mu$eV (X1), $g_2 = 76\,\mu$eV (X2). The individual values are consistent with the oscillator strength 40 - 50 of the large $In_{0.3}Ga_{0.7}As$ QDs used in the present investigation.

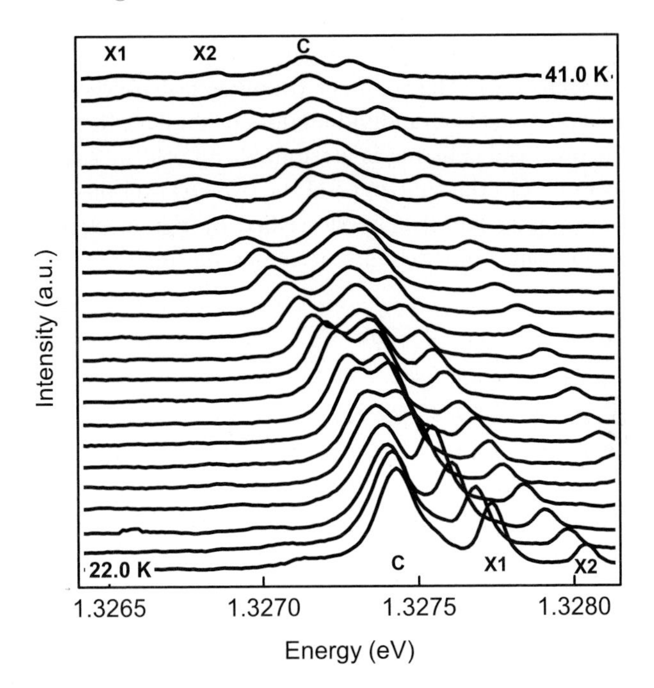

Fig. 13.19. Temperature dependence of photoluminescence spectra for a micropillar cavity ($d_c = 1.8$ μm) showing a sequential strong coupling of two excitons (X1 and X2) with the cavity mode (C). Resonance with the cavity mode is observed for 33 K (X1) and 37 K (X2), respectively.

Coherent photonic coupling needs to be distinguished from a sequential strong coupling which can appear when the energetic separation δ between two excitonic transitions is significantly larger than the cavity mode linewidth. Resonance tuning of such a system is presented in Fig. 13.19 which shows temperature dependent PL spectra of another micropillar with a cavity mode C and two nearby QD excitons X1 and X2. In this case, δ is notably larger then the cavity mode linewidth of 160 μeV and amounts to 315 μeV. When the temperature is increased from 22 K to 41 K, first the exciton X1 goes into the strong coupling with the cavity mode. The corresponding vacuum Rabi splitting is about 100 μeV, which is significantly smaller than in the previous case of two QDs. For higher temperature, the cavity mode has a separate anticrossing of a similar magnitude with the second dot exciton X2. These two strong interactions are well separated and correspond to the behaviour of an individual QD exciton in strong coupling with an optical mode. The solid lines show results again from calculations which provide coupling coefficients of the two quantum dot excitons of $g_1 = 62$ μeV, and $g_2 = 56$ μeV.

The total splitting in the coherent coupling case exceeds values for the vacuum Rabi splittings from individual dots in similar pillars by a factor of about 2. The observed splitting is attributed to both the cavity-exciton interactions as well as to the splitting δ of the two QD excitons at low T. Here, the coherent coupling part can be described by the effective Hamiltonian (neglecting dephasing) of three interacting oscillators:

$$H = \sum_{\alpha=C,X1,X2} E_\alpha \hat{a}_C^\dagger \hat{a}_\alpha - (1/2) \sum_{\alpha=X1,X2} g_\alpha \left[\hat{a}_C^\dagger \hat{a}_\alpha + \hat{a}_\alpha^\dagger \hat{a}_C \right] + C_0 \quad (13.8)$$

where E_C, E_{X1} and E_{X2} are the energies of the cavity mode, and excitons X1 and X2. $\hat{a}_\alpha$ and $\hat{a}_\alpha^\dagger$ denote the corresponding destruction and creation operators, g_1 and g_2 represent the two exciton-photon coupling constants, and $C_0 = (E_C + E_{X1} + E_{X2})/2$. This results in the following determinant for their eigenvalues:

$$\begin{vmatrix} E_c - E & g_1 & g_2 \\ g_1 & E_{X1} - E & 0 \\ g_2 & 0 & E_{X2} - E \end{vmatrix} = 0 \quad (13.9)$$

In the degenerate case with $E_{X1} = E_{X2}$ and $g_1 = g_2$, the vacuum Rabi splitting arises due to the coherent interaction with two equal emitters and the effective coupling $\sqrt{g_1^2 + g_2^2} = \sqrt{2}g$ increases by a factor of $\sqrt{2}$ compared to the single dot case which is well known for strongly coupled systems [41]. In the general case, the splitting is also influenced by the energetic splitting between the two excitons and one obtains

$$\Delta E \approx 2\sqrt{2g^2 + (\delta^2/4) + \left[(\Delta + \delta/2)^2\right]/3} \quad (13.10)$$

for the total splitting of the three modes, where $\Delta = E_{X1} - E_C$, and $g_1 = g_2 = g$ is used for simplicity. On resonance, i.e. $\Delta = 0$, both the coherent part g as well as the splitting δ give contributions to the total splitting. In case of the data shown in Fig. 13.17 the coherent part dominates while $\delta \approx 120$ μeV contributes only by approximately 20% to the increase of the splitting in comparison with to the single dot case. For large enough energy differences between the two quantum dot excitons, they will come into resonance with the cavity mode in sequence rather than simultaneously as shown in Fig. 13.19.

13.8. Conclusion

In conclusion, single quantum dot - micropillars have been proven to be an attractive system for CQED experiments on fundamental light-matter interaction. Due to the enormous progress in semiconductor technology in recent years, both the properties of the single quantum emitters, i.e. quantum dots, as well as the optical properties of low mode volume micropillar cavities have been improved considerably to allow now studies on the single particle level. In particular, the long sought solid state implementations of strongly coupled cavity mode - two level emitter systems has been demonstrated by using single quantum dots in high quality factor cavities with small mode volumes. In this article we have also presented proof of the quantum nature of the strongly coupled quantum dot-micropillar cavity system which is nicely reflected in pronounced photon antibunching of the emitted photons. Also the feasibility of coherent photonic coupling of quantum dots has been demonstrated which could be a versatile basis of quantum information building blocks in the solid state.

References

[1] Y. Yamamoto and R. Slusher, Optical processes in microcavities, *Physics Today*. pp. 66–73, (1993).
[2] J. M. Gérard and B. Gayral, InAs quantum dots: artificial atoms for solid-state cavity-quantum electrodynamics, *Physica E*. **9**, 131–139, (2001).
[3] K. J. Vahala, Optical microcavities, *Nature*. **424**, 839–846, (2003).
[4] J. M. Gérard, B. Sermage, B. Gayral, B. Legrand, E. Costard, and V. Thierry-Mieg, Enhanced spontaneous emission by quantum boxes in a monolithic optical microcavity, *Phys. Rev. Lett.* **81**, 1110–1113, (1998).
[5] M. Bayer, T. L. Reinecke, F. Weidner, A. Larionov, A. McDonald, and A. Forchel, Inhibition and enhancement of the spontaneous emission of quantum dots in structured microresonators, *Phys. Rev. Lett.* **86**, 3168–3171, (2001).
[6] G. S. Solomon, M. Pelton, and Y. Yamamoto, Single-mode spontaneous emission from a single quantum dot in a three-dimensional microcavity, *Phys. Rev. Lett.* **86**, 3903–3906, (2001).
[7] J. P. Reithmaier, G. Sek, A. Löffler, C. Hofmann, S. Kuhn, S. Reitzenstein, L. V. Keldysh, V. D. Kulakovskii, T. L. Reinecke, and A. Forchel, Strong coupling in a single quantum dotsemiconductor microcavity system, *Nature*. **432**, 197–200, (2004).

[8] T. Yoshie, A. Scherer, J. Hendrickson, G. Khitrova, H. M. Gibbs, G. Rupper, C. Ell, O. B. Shchekin, and G. Deppe, Vacuum rabi splitting with a single quantum dot in a photonic crystal nanocavity, *Nature.* **432**, 200–203, (2004).
[9] E. Peter, P. Senellart, D. Martrou, A. Lemaitre, J. Hours, J. M. Gérard, and J. Bloch, Exciton-photon strong-coupling regime for a single quantum dot embedded in a microcavity, *Phys. Rev. Lett.* **95**, 067401–1–4, (2005).
[10] C. Monroe, Quantum information processing with atoms and photons., *Nature.* **416**, 238–246, (2002).
[11] A. Imamoglu, D. D. Awschalom, G. Burkard, D. P. DiVincenzo, D. Loss, M. Sherwin, and A. Small, Quantum information processing using quantum dot spins and cavity qed, *Phys. Rev. Lett.* **83**, 4204–4207, (1999).
[12] X. Q. Li and Y. J. Yan, Quantum computation with coupled quantum dots in optical microcavities, *Phys. Rev. B.* **65**, 205301–1–5, (2002).
[13] A. Kiraz, M. Atatüre, and A. Imamoglu, Quantum- dot single-photon sources: Prospects for applications in linear optics quantum-information processing, *Phys. Rev. A.* **69**, 032305–1–10, (2004).
[14] S. Rudin and T. L. Reinecke, Oscillator model for vacuum rabi splitting in microcavities, *Phys. Rev. B.* **59**, 10227–10233, (1999).
[15] L. Andreani, G. Panzarini, and J.-M. Gérard, Strong-coupling regime for quantum boxes in pillar microcavities: Theory, *Phys. Rev. B.* **60**, 13267–13279, (1999).
[16] G. Khitrova, H. Gibbs, F. Jahnke, M. Kira, and S. Koch, Nonlinear optics of normal-mode-coupling semiconductor microcavities, *Mod. Phys. 71.* **71**, 1591–1639, (1999).
[17] Y. Arakawa and H. Sakaki, Multidimensional quantum well laser and temperature dependence of its threshold current, *Appl. Phys. Lett.* **40**, 939–941, (1982).
[18] D. L. Huffaker, G. Park, Z. Zou, O. B. Shchekin, and D. G. Deppe, 1.3 μm room-temperature gaas-based quantum-dot laser, *Appl. Phys. Lett.* **73**, 2564–2566, (1998).
[19] P. Michler, A. Kiraz, C. Becher, W. V. Schoenfeld, P. M. Petroff, L. Zhang, E. Hu, and A. Imamoglu, A quantum dot single-photon turnstile device, *Science.* **290**, 2282–2285, (2000).
[20] C. Pagesi, D. Fattal, J. Vucovic, G. S. Solomon, and Y. Yamamoto, Indistinguishable photons from a single-photon device, *Nature.* **419**, 594–597, (2002).
[21] L. N. Stranski and L. Krastanov, *Akad. Wiss. Lit. Mainz Math. Naturwiss. K1 IIb.* **146**, 797, (1939).
[22] A. Löffler, J.-P. Reithmaier, A. Forchel, A. Sauerwald, D. Peskes, T. Kümmell, and G. Bacher, Influence of the strain on the formation of GaInAs/GaAs quantum structures, *J. Crys. Growth.* **286**, 6–10, (2005).
[23] W. Ma, R. Nötzel, H. P. Schönherr, and K. H. Ploog, Shape transition of coherent three-dimensional (In, Ga)As islands on GaAs (100), *Appl. Phys. Lett.* **79**, 4219–4221, (2001).
[24] Y. Akahane, T. Asano, B. S. Song, and S. Noda, High-q photonic nanocavity in a two-dimensional photonic crystal, *Nature.* **425**, 944–947, (2003).

[25] B. Gayral, J. M. Gérard, A. Lemaitre, C. Dupuis, L. Manin, and J. L. Pelouard, High-q wet-etched GaAs microdisks containing InAs quantum boxes, *Appl. Phys. Lett.* **75**, 1908–1910, (1999).

[26] J. M. Gérard and B. Gayral, Strong purcell effect for InAs quantum boxes in three-dimensional solid-state microcavities, *J. Lightwave Technol.* **17**, 2089, (1999).

[27] A. Löffler, J. P. Reithmaier, G. Sek, C. Hofmann, S. Reitzenstein, M. Kamp, and A. Forchel, Semiconductor quantum dot microcavity pillars with high-quality factors and enlarged dot dimensions, *Appl. Phys. Lett.* **86**, 111105–1–3, (2005).

[28] J. P. Reithmaier, H. Z. M. Röhner, F. Schäfer, and A. Forchel, Size dependence of confined optical modes in photonic quantum dots, *Phys. Rev. Lett.* **78**, 378–381, (1997).

[29] T. Gutbrod, M. Bayer, A. Forchel, J. Reithmaier, T. Reinecke, S. Rudin, and P. Knipp, Weak and strong coupling of photons and excitons in photonic dots, *Phys. Rev. B.* **57**, 9950–9956, (1998).

[30] T. Rivera, J.-P. Debray, J. M. Gérard, B. Legrand, L. Manin-Ferlazzo, and J. L. Oudar, Optical losses in plasma-etched AlGaAs microresonators using reflection spectroscopy, *Appl. Phys. Lett.* **74**, 911–913, (1999).

[31] A. Badolato, K. Hennessy, M. Atatüre, J. Dreiser, E. Hu, P. M. Petroff, and A. Imamoglu, Deterministic coupling of single quantum dots to single nanocavity modes, *Science.* **308**, 1158–1161, (2005).

[32] R. H. Brown and R. Q. Twiss, The question of correlation between photons in coherent light rays, *Nature.* **178**, 1447–1415, (1956).

[33] E. M. Purcell, Spontaneous emission probabilities at radio frequencies, *Phys. Rev.* **69**, 681, (1946).

[34] R. Houdre, R. Stanley, U. Oesterle, M. Ilgems, and C. Weisbuch, Room-temperature cavity polaritons in a semiconductor microcavity, *Phys. Rev. B.* **49**, 16761–16764, (1994).

[35] C. Santori, G. S. M. Pelton, Y. Dale, and Y. Yamamoto, Triggered single photons from a quantum dot, *Phys. Rev. Lett.* **86**, 1502–1505, (2001).

[36] T. Mensing, S. Reitzenstein, A. Löffler, J. Reithmaier, and A. Forchel, Magnetooptical investigations of single self assembled $In_{0.3}Ga_{0.7}As$ quantum dots, *Physica E.* **32**, 131–134, (2006).

[37] M. Bayer, P. Hawrylak, K. Hinzer, S. Fafard, M. Korkusinski, Z. R. Wasilewski, O. Stern, and A. Forchel, Coupling and entangling of quantum states in quantum dot molecules, *Science.* **291**, 451–453, (2001).

[38] X. Q. Li, Y. W. Wu, D. Steel, D. Gammon, T. H. Stievater, D. S. Katzer, D. Park, C. Piermarocchi, and L. J. Sham, An all-optical quantum gate in a semiconductor quantum dot, *Science.* **301**, 809–811, (2003).

[39] H. J. Krenner, M. Sabathil, E. C. Clark, A. Kress, D. Schuh, M. Bichler, G. Abstreiter, and J. J. Finley, Direct observation of controlled coupling in an individual quantum dot molecule, *Phys. Rev. Lett.* **94**, 057402–1–4, (2005).

[40] S. Reitzenstein, A. Löffler, C. Hofmann, A. Kubanek, M. Kamp, J. P. Reithmaier, A. Forchel, V. D. Kulakovskii, L. V. Keldysh, I. V. Ponomarev,

and T. L. Reinecke, Coherent photonic coupling of semiconductor quantum dots, *Opt. Lett.* **31**, 1738–1740, (2006).

[41] G. Khitrova, H. M. Gibbs, M. Kira, S. W. Koch, and A. Scherer, Vacuum rabi splitting in semiconductors, *Nature Physics.* **2**, 81–90, (2006).

Chapter 14

Entangled Photons via Biexciton-Resonant Hyperparemetric Scattering

Keiichi Edamatsu

Research Institute of Electrical Communication, Tohoku University
2-1-1 Katahira, Aoba-ku, Sendai 980-8577, Japan

Generation of polarization-entangled photons from a biexciton via resonant hyperparametric scattering is reviewed and discussed. The quantum state of the generated photon pair is characterized using quantum state tomography.

Contents

14.1. Introduction

Entanglement is one of the essential resources of quantum information and communication technology. The non-local nature of entanglement conflicts with classical local realism, as pointed out by Einstein, Podolsky and Rosen [1]. The existence criterion of the non-local quantum correlation was formalized by Bell's inequality [2] and by its generalization known as CHSH-Bell's inequality [3]. From the 1960s through the 80s several experiments [4–9] had been carried out to test Bell's inequality; they had demonstrated the violation of Bell's inequality and thus manifested the existence of non-local quantum correlations brought by entanglement. In these early experiments, photon pairs emitted from atomic cascades are used to

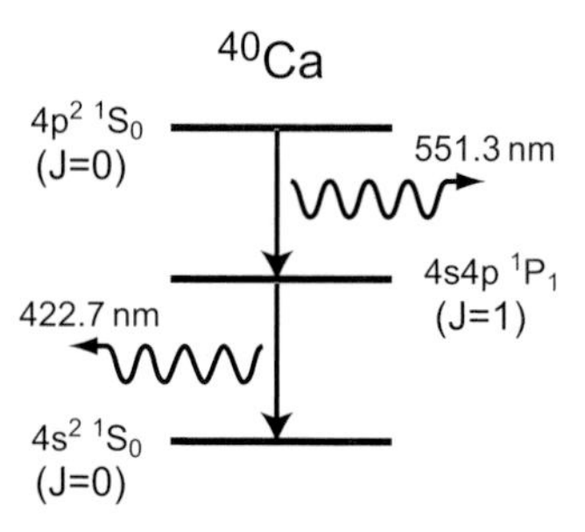

Fig. 14.1. Schematic diagram of entangled photon pair emission via atomic cascade in calcium.

hold entanglement in their polarization states. Following to such pioneering works, a number of methods to generate entangled photons have been proposed. To date, parametric down-conversion (PDC) has been being the most popular and powerful method to obtain entangled photons [10, 11]. Most recently, entangled photon generation using semiconductor materials [12–16] attracts attention because of its potential for the realization of entangled photon emitting diodes in the near future. In these experiments, cascaded photon emission from a biexciton, a solid state analogue of the atomic cascade, was used to generate polarization entanglement [17]. In this chapter, we review the experimental results on the generation of polarization-entangled photons from a biexciton in a CuCl crystal via resonant hyperparametric scattering [12, 16], and discuss the quantum state of the generated photon pairs.

14.2. Entangled Photon Generation via Atomic and Biexciton Cascades

Prior to describing the entangled photon generation via biexciton in semiconductor materials, it is worth noting atomic cascade emission; both have the same principle in the entangled photon generation. Entangled photons emitted from atomic cascades were investigated using calcium [4, 5, 8, 9] and mercury [6, 7] atoms. For example, a diagram of the atomic cascade in calcium is illustrated in Fig. 14.1. The atom is excited from its ground state $|g\rangle=(4s^2\ {}^1S_0)$ to the excited state $|e\rangle=(4p^2\ {}^1S_0)$ by two-photon excitation [8]. The excited state consists of two excited 4p electrons forming zero total angular momentum (J=0). Thus, the angular momentum part of the

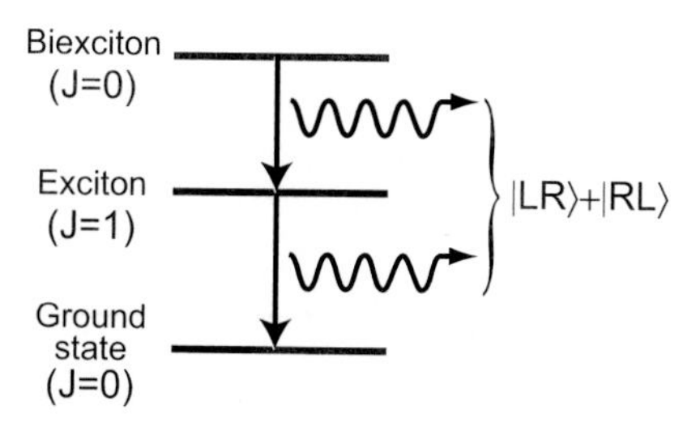

Fig. 14.2. Schematic diagram of entangled photon generation via cascade emission from a biexciton.

excited state is

$$|\mathrm{e}\rangle = \frac{1}{\sqrt{3}} \left(|+1\rangle_{\mathrm{A}} |-1\rangle_{\mathrm{B}} - |0\rangle_{\mathrm{A}} |0\rangle_{\mathrm{B}} + |-1\rangle_{\mathrm{A}} |+1\rangle_{\mathrm{B}} \right), \tag{14.1}$$

where $|m\rangle_i$ denotes the state of each excited electron (i=A or B) represented by the orbital magnetic quantum number m . The excited state decays to the intermediate state $|\mathrm{i}\rangle$=(4s4p $^1\mathrm{P}_1$) by emitting a visible photon (wavelength $\lambda = 551.3$ nm), and the intermediate state decays back to the ground state $|\mathrm{g}\rangle$ by emitting another photon ($\lambda = 422.7$ nm). In this scheme, the total angular momentum J of the atom changes as $J = 0 \to 1 \to 0$. Assuming that the two photons are emitted along the quantization axis, the second term in Eq. (14.1) does not contribute to the optical transition. As a result, if we observe the two photons emitted to opposite directions, the photon pair is entangled so that their polarization state is the maximally entangled state

$$|\psi\rangle = \frac{1}{\sqrt{2}} \left(|\mathrm{L}\rangle_{\mathrm{A}} |\mathrm{L}\rangle_{\mathrm{B}} + |\mathrm{R}\rangle_{\mathrm{A}} |\mathrm{R}\rangle_{\mathrm{B}} \right) \equiv \frac{1}{\sqrt{2}} \left(|\mathrm{LL}\rangle + |\mathrm{RR}\rangle \right), \tag{14.2}$$

where L and R denote the polarization state (L: left circular, R: right circular) of each photon. The polarization entangled state (14.2) can be rewritten in linear polarization bases as

$$|\psi\rangle = \frac{1}{\sqrt{2}} \left(|\mathrm{HH}\rangle - |\mathrm{VV}\rangle \right), \tag{14.3}$$

where H and V denote horizontal and vertical linear polarizations, respectively. Thus, from this atomic cascade, generation of entangled photons in the triplet Bell state is expected. For instance, Aspect *et al.* performed the test of CHSH-Bell's inequality [3] utilizing the photons thus generated [8, 9].

The excitonic system in a semiconductor quantum dot (QD) can also be used to generate entangled photon pairs. Benson *et al.* pointed out that cascaded two-photon emission from a biexciton, i.e., a bound state of two excitons, can be used to generate polarization entanglement [17]. Similar to the atomic cascade described above, the biexciton decays into the ground state via an exciton state emitting two sequent photons. The lowest bound state of the biexciton has zero total angular momentum ($J = 0$). The optically-allowed (bright) exciton is $J = 1$ and the ground state is $J = 0$. Thus, the change in angular momentum of the biexciton cascade is the same as that of the atomic cascade, resulting that the emitted photon pair has the polarizatioin entanglement:

$$|\psi\rangle = \frac{1}{\sqrt{2}}\left(|\mathrm{LR}\rangle + |\mathrm{RL}\rangle\right) = \frac{1}{\sqrt{2}}\left(|\mathrm{HH}\rangle + |\mathrm{VV}\rangle\right). \tag{14.4}$$

Note that we assume here that we observe the two photons emitted to the same direction, whereas we assumed to observe the two photons emitted to opposite directions in Eqs. (14.2) and (14.3)

In spite of the theoretical prediction described above, entangled photon generation from semiconductor QD has not been successful until most recently, mainly for the following reason. Most of self-organized QDs have small asymmetry in shape that induces the splitting of the exciton energy levels depending on their polarizations. In practice, it is quite common that the spactra of single QD emission (biexciton as well as exciton emission) splits into two components depending on the polarizations [18, 19]. The splitting causes the distinguishability between the photon pairs emitted from the biexciton cascade depending on the polarization. Let the two principal axes of the polarization be set as H and V. Note that the two-photon polarization state is no longer the superposition of the two as shown in Eq. (14.4) but the statistical mixture of $|\mathrm{HH}\rangle$ and $|\mathrm{VV}\rangle$. The density matrix representation of the state is

$$\rho = \frac{1}{2}\left(|\mathrm{HH}\rangle\langle\mathrm{HH}| + |\mathrm{VV}\rangle\langle\mathrm{VV}|\right), \tag{14.5}$$

which is the mixed state that has only a classical correlation. The state shows classical correlation when the polarization is observed along their principal axes; it however does not exhibits any correlation when observed, for instance, in circular polarization or $\pm 45^\circ$ polarization bases. Thus, the state (14.5) is not entangled. Such mixed state of the two-photon polarization was observed experimentally [20]. Most recently, two experimental groups reported the generation of polarization entangled photons from a single QD [13–15]. The two groups took different approaches to realize the entanglement. One used QDs having reduced asymmetry and

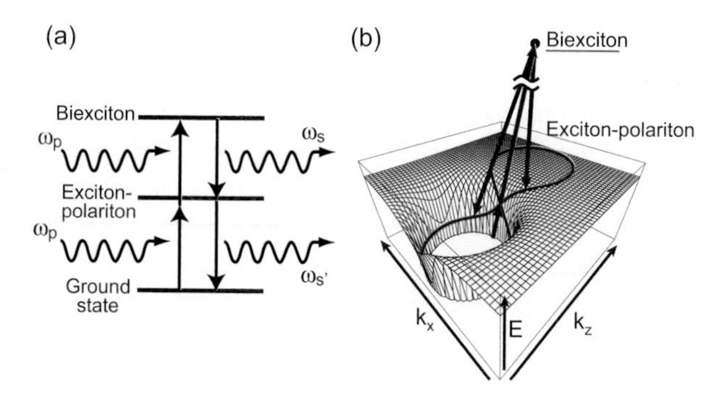

Fig. 14.3. Schematic diagram of biexciton-resonat hyper parametric scattering (RHPS). (a) A simple picture that explains two pump photons (frequency: ω_p) are converted to the two scattered photons (ω_s, $\omega_{s'}$). (b) The diagram taking account of the exciton-polariton dispersion drawn in two dimensions of momentum space. The red curve on the polariton-dispersion surface indicates the states on which the phase-matching condition can be satisfied.

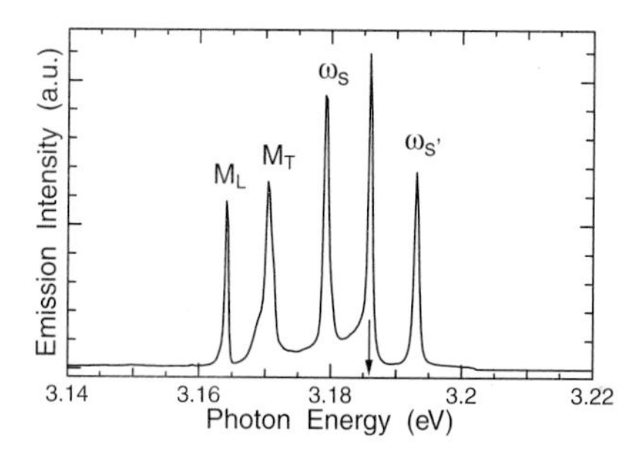

Fig. 14.4. Observed RHPS spectrum of CuCl crystal at 4 K. The downward arrow indicates the photon energy of the pump light ($\hbar\omega_p$=3.1861 eV). The peaks indicated by ω_s and $\omega_{s'}$ are the lights emitted via the RHPS. The peaks M_L and M_T are resonant biexciton luminescence leaving longitudinal and transverse excitons, respectively.

added in-plane magnetic field to match the spectra for H and V polarizations [13, 14]. The other used spectral filtering of the emission spectra selecting identical spectra from the overlapping spectral range of the two polarized emissions [15].

Entangled photon pairs can also be generated using biexcitons in bulk semiconductor crystal, via the process called biexciton-resonat hyper parametric scattering (RHPS) [12, 16]. In the process, as illustrated in Fig. 14.3, two pump photons (frequency: ω_p) create the biexciton by resonant two-photon excitation. The biexciton decays into the ground states emitting two scattered photons (ω_s, $\omega_{s'}$). Figure 14.4 presents an example of the

RHPS spectra observed in a CuCl crystal. The two peaks indicated by ω_s and $\omega_{s'}$ are the signals emitted via the RHPS. The photon pairs thus generated also hold the polarization entanglement (14.4), reflecting the zero angular momentum of the biexciton. Although the RHPS, or two-photon resonant Raman scattering in other words, has been extensively studied since the 1970s [21–28], the applicability of the process to the generation of entangled photons was theoretically pointed out only recently [29]. RHPS is a $\chi^{(3)}$ nonlinear process concerning four photons, whereas PDC is a $\chi^{(2)}$ nonlinear process concerning three photons. In the RHPS, the photons concerned must obey a certain phase matching condition, as in PDC. Since the photons in the crystal forms exciton-polaritons as a consequence of strong exciton-photon interaction, the phase matching condition must take the dispersion of the exciton-polariton into account [21–28]. The result is that the two scattered photons are emitted along the two almost identical conical surfaces, as in the type-I PDC's case [11]. In addition, recent theoretical works have predicted that RHPS will be dramatically enhanced in a microcavity [30, 31] or films having specific thickness [32]. In this context, RHPS in CuCl film has been demonstrated experimentally [33].

14.3. Observation of Entanglement

14.3.1. *Polarization correlation measurement*

In order to observe and evaluate entanglement held by photon pairs, we must perform (i) simultaneous measurement of physical quantities of constituent photons and (ii) analysis of correlation between them. In the observation of polarization entanglement, projection measurement of the two-photon polarization state incorporating coincidence measurement is necessary. Let the polarization states of photons selected by the two polarization analyzers be α and β, respectively. For instance, α and β may be H, V, L, or R. The two-photon polarization sate onto which the state is projected is $|\alpha\beta\rangle$. The probability $p_{\alpha\beta}$that finds the system in the polarization state $|\alpha\beta\rangle$ is

$$p_{\alpha\beta} = \mathrm{Tr}\{\rho|\alpha\beta\rangle\langle\alpha\beta|\}, \tag{14.6}$$

where ρ is the density matrix of the two-photon polarization state. If the polarization state of the photon pair is in a pure state $|\psi\rangle$, $\rho = |\psi\rangle\langle\psi|$ and

$$p_{\alpha\beta} = |\langle\alpha\beta|\psi\rangle|^2. \tag{14.7}$$

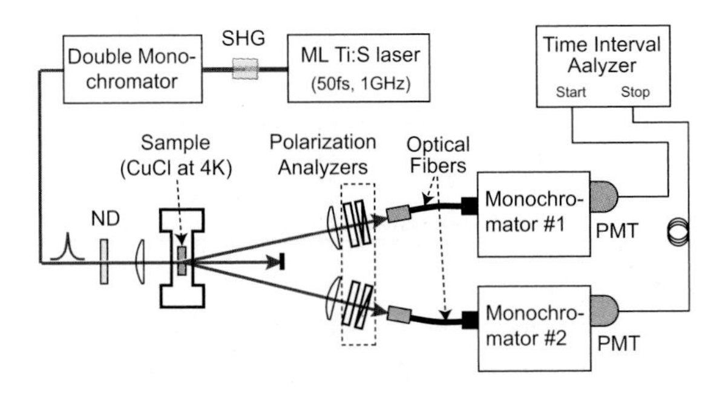

Fig. 14.5. Schematic illustration of the experimental setup for observing entangled photon pairs generated from RHPS in CuCl crystal. SHG: second harmonic generation. ND: neutral density filter. PMT: photon counting photomultiplier.

The rate of coincidence measurement of the photon pairs in which the polarization state is projected onto $|\alpha\beta\rangle$ is proportional to $p_{\alpha\beta}$. For example, if the state is in the entangled state (14.4), it is expected that the projection measurement gives equal probabilities 1/2 for LR, RL, HH and VV polarization combinations. Note that if the state is in the mixed state (14.5), which has only a classical correlation, it shows perfect correlation in the linear polarization bases (H, V) but no polarization correlation in the circular bases (L, R).

Figure 14.5 presents the experimental setup used to observe entangled photon pairs generated from RHPS in a CuCl crystal [16]. In this experiment, a femtosecond mode-locked Ti:sapphire laser having a 1 GHz repetition rate was used as a pump source. The high repetition pumping is essential for the reduction of accidental background signals. The second harmonics of the pump light was then spectrally filtered by a zero-dispersion double monochromator to be in the two-photon resonance with the lowest biexciton ($\hbar\omega_p$=3.1861 eV). The output light from the monochromator was focused on the sample through an ND filter. In order to suppress the accidental coincidence of uncorrelated photons, the pump power was set to around 10 μW. The emitted photons from the sample were fed into the optical multi-mode fibers connected to the two monochromators. Using the polarization analyzers consisting of a quarter wave plate and a polarizer in front of each fiber, we measured the polarization projection measurement as described above. The photons having appropriate frequencies, ω_s and $\omega_{s'}$ (See Figs. 14.3 and 14.4), were selected by the monochromators #1 and #2, and were detected by the two photomultipliers. The time-interval analyzer recorded the difference in arrival time between the photons.

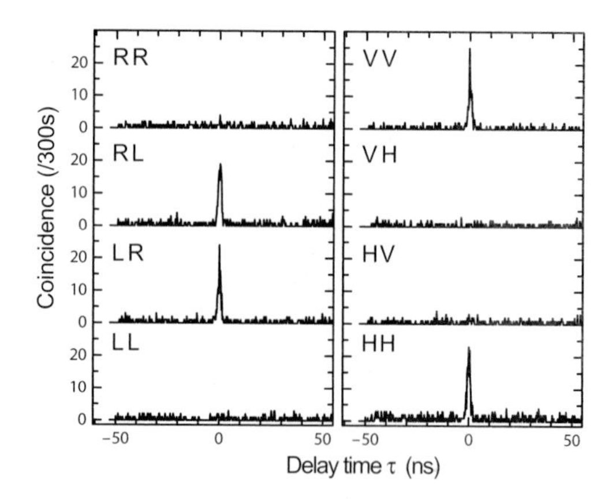

Fig. 14.6. Time correlation histograms of the photon pairs generated from RHPS in CuCl crystal. Polarization correlation measurements were made in circular (L, R) and linear (H, V) polarization bases. The central peak at $\tau = 0$ is the correlated two-photon coincidence signal originating from one biexciton. The background signal at $\tau \neq 0$ results from the uncorrelated photons.

Figure 14.6 presents an example of the polarization correlation measurements thus obtained [16]. In the figure, histograms of the coincidence count rate are shown for a set of polarization combinations, as a function of delay time τ between two photons detected. In the histogram, the central peak at τ=0 is the correlated two-photon coincidence signal originating from a biexciton. Thus, the polarization correlation of this central peaks presents the quantum correlation. On the other hand, the noisy background signals that appear at $\tau \neq 0$ originate from uncorrelated photons generated by independent pump pulses (interval: 1 ns) and have no polarization correlations. In these histograms, it is clear that the coincidence signals at τ=0 appear for counter circular (LR and RL) or parallel (HH and VV) polarization combinations, in consistent with the theoretical prediction in Eq. (14.4).

14.3.2. *Violation of Bell's inequality*

An explicit and quantitative prove of the non-local quantum correlation brought by entanglement is to examine the inequality proposed by Clauser, Horne, Shimony and Holts (CHSH) [3]. According to the CHSH theory, Bell's inequality can be written as

$$S = |E(\theta_{\mathrm{A}}, \theta_{\mathrm{B}}) - E(\theta_{\mathrm{A}}, \theta'_{\mathrm{B}}) + E(\theta'_{\mathrm{A}}, \theta_{\mathrm{B}}) + E(\theta'_{\mathrm{A}}, \theta'_{\mathrm{B}})| \leq 2, \qquad (14.8)$$

and $E(\theta_A, \theta_B)$ is given by

$$E(\theta_A, \theta_B) = \frac{C(\theta_A, \theta_B) + C(\theta_A^\perp, \theta_B^\perp) - C(\theta_A^\perp, \theta_B) - C(\theta_A, \theta_B^\perp)}{C(\theta_A, \theta_B) + C(\theta_A^\perp, \theta_B^\perp) + C(\theta_A^\perp, \theta_B) + C(\theta_A, \theta_B^\perp)}, \quad (14.9)$$

where $C(\theta_A, \theta_B)$ is the coincidence count for each polarization angle and $\theta_i^\perp \equiv \theta_i + 90°$. The maximally entangled state (14.4) violates Bell's inequality; we obtain $S = 2\sqrt{2} > 2$ at $\theta'_A - \theta_A = 45°$, $\theta_B - \theta_A = 22.5°$, and $\theta'_B - \theta_A = 67.5°$.

In order to confirm experimentally that the generated photon pairs actually violate CHSH-Bell's inequality (14.8), we must measure the polarization correlation for 16 combinations of analyzer setting ($\theta_A = 0°$, $45°$, $90°$, $135°$; $\theta_B = 22.5°$, $67.5°$, $112.5°$, $157.5°$). From the measurements similar to those in Fig. 14.6, coincidence counts $C(\theta_A, \theta_B)$ were recorded for the 16 polarization combinations. From the result, the S value of $2.34 \pm 0.10 > 2$ was obtained. It is clear that the S value apparently violates Bell's inequality by more than 3 times the standard deviation. Although the obtained S value violates Bell's inequality, the obtained S value was smaller than the ideal value $2\sqrt{2}$. This indicates the degradation of entanglement to some extent, as discussed below.

14.3.3. *State tomography*

Polarization entanglement between two photons is evaluated from their density matrix. The density matrix for the entangled state (14.4) is expressed as

$$\rho = \frac{1}{2} (|\mathrm{HH}\rangle + |\mathrm{VV}\rangle) (\langle \mathrm{HH}| + \langle \mathrm{VV}|) = \frac{1}{2} \begin{pmatrix} 1 & 0 & 0 & 1 \\ 0 & 0 & 0 & 0 \\ 0 & 0 & 0 & 0 \\ 1 & 0 & 0 & 1 \end{pmatrix}. \quad (14.10)$$

In this matrix representation, $|\mathrm{HH}\rangle$, $|\mathrm{HV}\rangle$, $|\mathrm{VH}\rangle$, and $|\mathrm{VV}\rangle$ are used as the basis set. The maximally entangled state holds full coherence between the constituent bases and thus the diagonal ($|\mathrm{HH}\rangle\langle \mathrm{HH}|$ and $|\mathrm{VV}\rangle\langle \mathrm{VV}|$) and off-diagonal ($|\mathrm{HH}\rangle\langle \mathrm{VV}|$ and $|\mathrm{VV}\rangle\langle \mathrm{HH}|$) elements have the same absolute value 1/2 in the density matrix (14.10).

In order to obtain experimentally the density matrix of two-photon polarization states, we must observe the polarization correlation described in Sec. 14.3.1 for various polarization combinations. Then, using the procedure called state tomography [34], we can reconstruct the density matrix that is estimated from the experimental data. The number of bases, or the dimension of Hilbert space, of a 2 qubit system is $2 \times 2 = 4$, and thus the minimal number of polarization combinations in the correlation measurement required to reconstruct the density matrix is $4^2 = 16$. Mathematically, we can reconstruct the density matrix from the 16 combinations of the experimental data by linear algebra. In practice, however, errors and noises in the experimental data usually make the reconstructed density matrix unreliable; sometimes it produces unphysical density matrices that have, for instance, negative eigenvalues. In such cases, we should reconstruct the density matrix from more than 16 experimental data by maximum likelihood estimation [34] so that the resultant density matrix is not to be unphysical.

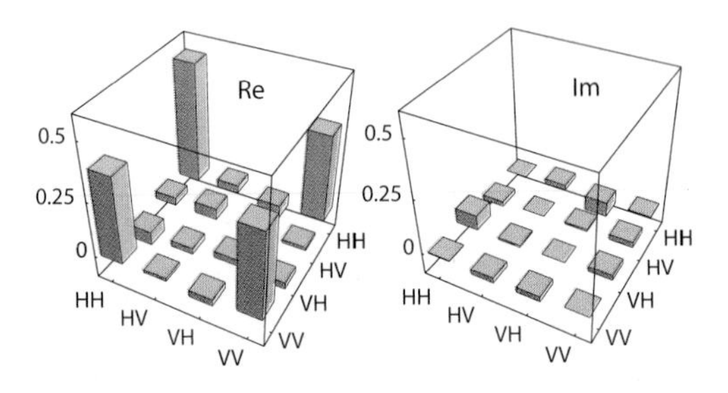

Fig. 14.7. Reconstructed density matrix of the two-photon polarization state for the photon pairs generated from RHPS in CuCl.

Figure 14.7 shows the histogram of reconstructed density matrix of the two-photon polarization state of the photon pairs generated by RHPS in a CuCl crystal [16]. To obtain the density matrix, data of 22 combinations in polarization correlation measurements, including those shown in Fig. 14.6, were used. The diagonal elements reflect the polarization correlation in the linear polarization bases (HH, HV, VH, and VV) in Fig. 14.6 and the off-diagonal elements show the coherence between them. We see that the values of diagonal elements ($|\mathrm{HH}\rangle\langle\mathrm{HH}|$ and $|\mathrm{VV}\rangle\langle\mathrm{VV}|$) and the off-diagonal elements ($|\mathrm{HH}\rangle\langle\mathrm{VV}|$ and $|\mathrm{VV}\rangle\langle\mathrm{HH}|$) are close to 0.5 and other elements are almost zero. Thus we see that the experimentally reconstructed density matrix is quite close to the expected one shown in Eq. (14.10).

Strictly speaking, however, there are some deviations from the ideal one indicating that the entanglement is degraded to some extent by, for instance, imperfection of the source and disturbance from environment. The first signature of the degradation is an imbalance between the coefficients of $|\mathrm{HH}\rangle\langle\mathrm{HH}|$ and $|\mathrm{VV}\rangle\langle\mathrm{VV}|$. This degradataion comes from the experimental condition in which we expect the geometrical imbalance between the intensity of H and V polarizations of the generated photons [16, 25, 26]. The second signature is the appearance of diagonal elements $|\mathrm{HV}\rangle\langle\mathrm{HV}|$ and $|\mathrm{VH}\rangle\langle\mathrm{VH}|$ that are not expected originally to appear. This may happen when some noises are mixed up to the original state. Actually, such a noise comes from the accidental coincidence signals originating from uncorrelated photons. The third signature of the degradation, which is not significant in our reconstructed density matrix, is a disordered phase relationship between $|\mathrm{HH}\rangle$ and $|\mathrm{VV}\rangle$. In this case, the amount of off-diagonal elements ($|\mathrm{HH}\rangle\langle\mathrm{VV}|$ and $|\mathrm{VV}\rangle\langle\mathrm{HH}|$) becomes smaller. The extreme case of the phase disorder results in the classical mixed state such as Eq. (14.5), in which all the off-diagonal elements are 0. Thus, by analyzing the density matrix, we can make an estimation of the degree of entanglement and the origins of its degradation.

The fidelity (F) defined by

$$F \equiv \langle\psi|\rho|\psi\rangle \tag{14.11}$$

is a measure that evaluates the closeness of the state expressed by the density matrix ρ to the ideal pure state $|\psi\rangle$. More generally, the fidelity of the state ρ to the state expressed by a density matrix ρ_0 is defined by

$$F \equiv \left(\mathrm{Tr}\sqrt{\sqrt{\rho_0}\rho\sqrt{\rho_0}}\right)^2 . \tag{14.12}$$

F takes a value in the range from 0 to 1. It is obvious that F=1 for the ideal state: $\rho = \rho_0$. The fidelity of the reconstructed density matrix shown in Fig. 14.7 to the ideal maximally entangled state (14.10) is calculated to be F=0.85. The fidelity of the classically correlated states (14.5) to the maximally entangled state is F=0.5; this is the upper bound of the fidelity of any classical state to the maximally entangled state. Thus the result F=0.85 manifests that the observed two-photon polarization state has true entanglement beyond the classical limit. If one consider a state that models the degradation of entanglement assuming the imbalance of the polarization at the source and the mixture of noise originating from uncorrelated photons, the fidelity of the reconstructed density matrix to the model state is 0.94. This means that the degradation of entanglement in the observed sate is almost expressed by the two origins modeled in the analysis. It is noteworthy that a part of such degradation can be compensated by local filtering of the polarization, i.e., by entanglement distillation [35]. Entanglement of formation (E_F), or ebit, is a measure of entanglement; 1

ebit corresponds to a pair of qubit in a maximally entangled state and E_F quantifies how much ebit can be constructed from the state concerned, or how much ebit is necessary to construct the state, by use of local operation and classical communication (LOCC) [36]. Concurrence (C) and tangle ($T \equiv C^2$) are also used as quantitative measures of entanglement. In a 2-qubit system, C and T can be calculated from a density matrix and then we can obtain E_F [37]. The value of E_F obtained from the density matrix shown in Fig. 14.7 is E_F=0.65. It means that if we have 100 sets of the two-photon state, we can get 65 pairs of maximally entangled photons via LOCC. Taking account of the fact that E_F=0 for the classically correlated state (14.5), our state shown in Fig. 14.7 has considerably high degree of entanglement, even though it is not perfect.

14.4. Summary and Outlook

We have discussed the generation of entangled photons via biexciton-resonant hyperparametric scattering in a CuCl crystal. From the first realization of entangled photons using atomic sources, a lot of works on the generation of entangled photons have been carried out. Biexciton cascades in semiconductor materials will open the way of realizing entangled photon emitting diodes. Such a practical source of entangled photons is one of the key technologies in the development of quantum information and communication technology.

Acknowledgments

Our experimental work [12, 16] described in this chapter was done with Drs. G. Oohata and R. Shimizu. The author is grateful to Professors T. Itoh, H. Ishihara, M. Nakayama, and H. Ohno for valuable discussion.

References

[1] A. Einstein, B. Podolsky, and N. Rosen, Can quantum-mechanical description of physical reality be considered complete?, *Phys. Rev.* **47**(10), 777–780, (1935).
[2] J. Bell, On the Einstein-Podolsky-Rosen paradox, *Physics.* **1**, 195–200, (1964).
[3] J. F. Clauser, M. A. Horne, A. Shimony, and R. A. Holt, Proposed experiment to test local hidden-variable theories, *Phys. Rev. Lett.* **23**(15), 880–884 (1969).

[4] C. A. Kocher and E. D. Commins, Polarization correlation of photons emitted in an atomic cascade, *Phys. Rev. Lett.* **18**(15), 575–577 (1967).
[5] S. J. Freedman and J. F. Clauser, Experimental test of local hidden-variable theories, *Phys. Rev. Lett.* **28**(14), 938–941 (1972).
[6] J. F. Clauser, Experimental investigation of a polarization correlation anomaly, *Phys. Rev. Lett.* **36**(21), 1223–1226 (1976).
[7] E. S. Fry and R. C. Thompson, Experimental test of local hidden-variable theories, *Phys. Rev. Lett.* **37**(8), 465–468 (1976).
[8] A. Aspect, P. Grangier, and G. Roger, Experimental tests of realistic local theories via Bell's theorem, *Phys. Rev. Lett.* **47**(7), 460–463 (1981).
[9] A. Aspect, P. Grangier, and G. Roger, Experimental realization of Einstein-Podolsky-Rosen-Bohm gedankenexperiment: A new violation of Bell's inequalities, *Phys. Rev. Lett.* **49**(2), 91–94 (1982).
[10] P. G. Kwiat, K. Mattle, H. Weinfurter, A. Zeilinger, A. V. Sergienko, and Y. Shih, New high-intensity source of polarization-entangled photon pairs, *Phys. Rev. Lett.* **75**(24), 4337–4341 (1995).
[11] P. G. Kwiat, E. Waks, A. G. White, I. Appelbaum, and P. H. Eberhard, Ultrabright source of polarization-entangled photons, *Phys. Rev. A.* **60**(2), R773–R776 (1999).
[12] K. Edamatsu, G. Oohata, R. Shimizu, and T. Itoh, Generation of ultraviolet entangled photons in a semiconductor, *Nature.* **431**(7005), 167–170, (2004).
[13] R. M. Stevenson, R. J. Young, P. Atkinson, K. Cooper, D. A. Ritchie, and A. J. Shields, A semiconductor source of triggered entangled photon pairs, *Nature.* **439**(7073), 179–182, (2006).
[14] R. J. Young, R. M. Stevenson, P. Atkinson, K. Cooper, D. A. Ritchie, and A. J. Shields, Improved fidelity of triggered entangled photons from single quantum dots, *New Journal of Physics.* **8**(2), 29, (2006).
[15] N. Akopian, N. H. Lindner, E. Poem, Y. Berlatzky, J. Avron, D. Gershoni, B. D. Gerardot, and P. M. Petroff, Entangled photon pairs from semiconductor quantum dots, *Phys. Rev. Lett.* 96(13):130501, (2006).
[16] G. Oohata, R. Shimizu, and K. Edamatsu, Photon polarization entanglement induced by biexciton: Experimental evidence for violation of Bell's inequality, *Phys. Rev. Lett.* 98(14):140503, (2007).
[17] O. Benson, C. Santori, M. Pelton, and Y. Yamamoto, Regulated and entangled photons from a single quantum dot, *Phys. Rev. Lett.* **84**(11), 2513–2516, (2000).
[18] D. Gammon, E. S. Snow, B. V. Shanabrook, D. S. Katzer, and D. Park, Fine structure splitting in the optical spectra of single gaas quantum dots, *Phys. Rev. Lett.* **76**(16), 3005–3008 (1996).
[19] R. J. Young, R. M. Stevenson, A. J. Shields, P. Atkinson, K. Cooper, D. A. Ritchie, K. M. Groom, A. I. Tartakovskii, and M. S. Skolnick, Inversion of exciton level splitting in quantum dots, *Phys. Rev. B.* 72(11):113305, (2005).
[20] C. Santori, D. Fattal, M. Pelton, G. S. Solomon, and Y. Yamamoto, Polarization-correlated photon pairs from a single quantum dot, *Phys. Rev. B.* 66(4):045308, (2002).
[21] M. Inoue and E. Hanamura, Emission spectrum from the bose-condensed

excitonic molecules, *J. Phys. Soc. Jpn.* **41**(4), 1273–1284 (1976).
[22] F. Henneberger and J. Voigt, Polariton effects in optical-spectra of excitonic molecules, *phys. stat. sol. (b).* **76**(1), 313–323, (1976).
[23] F. Bechstedt and F. Henneberger, 2-photon resonance Raman-scattering by excitons via an intermediate excitonic molecule .1. bare-exciton approach, *phys. stat. sol. (b).* **81**(1), 211–220, (1977).
[24] F. Henneberger, K. Henneberger, and J. Voigt, 2-photon resonance Raman-scattering by excitons via an intermediate excitonic molecule .2. polarization properties and polarization effects, *phys. stat. sol. (b).* **83**(2), 439–450, (1977).
[25] T. Itoh and T. Suzuki, Excitonic polariton-polariton resonance scattering via excitonic molecules in CuCl, *J. Phys. Soc. Jpn.* **45**(6), 1939–1948 (1978).
[26] F. Henneberger, J. Voigt, and E. Gutsche, Optical orientation effects in the secondary radiation of excitonic molecules, *J. Lumin.* **20**(3), 221–225, (1979).
[27] B. Hönerlage, R. Lévy, J. B. Grun, C. Klingshirn, and K. Bohnert, The dispersion of excitons, polaritons and biexcitons in direct-gap semiconductors, *Physics Reports.* **124**(3), 161–253, (1985).
[28] M. Ueta, H. Kanzaki, K. Kobayashi, Y. Toyozawa, and E. Hanamura, *Excitonic Processes in Solids.* (Springer-Verlag, Berlin, New York, 1986). and references therein.
[29] S. Savasta, G. Martino, and R. Girlanda, Entangled photon pairs from the optical decay of biexcitons, *Solid State Commun.* **111**, 495–500, (1999).
[30] H. Ajiki and H. Ishihara, Enhanced generation of entangled-photon pairs from a cavity system, *physica status solidi (c).* **3**(7), 2440–2443, (2006).
[31] H. Ajiki and H. Ishihara, Entangled-photon generation in biexcitonic cavity QED, *J. Phys. Soc. Jpn.* **76**(5), 053401 (2007).
[32] M. Bamba and H. Ishihara, Entangled-photon generation via biexcitons in nano-structures, *physica status solidi (c).* **3**(10), 3460–3463, (2006).
[33] M. Nakayama, T. Nishioka, S.Wakaiki, G. Oohata, K. Mizoguchi, D. Kim, and K. Edamatsu, Observation of biexciton-resonant hyper-parametric scattering in SiO_2/CuCl layered structures, *Jpn. J. Appl. Phys.* **46**, L234–L236, (2007).
[34] D. F. V. James, P. G. Kwiat, W. J. Munro, and A. G. White, Measurement of qubits, *Phys. Rev. A.* **64**(5), 052312 (2001).
[35] P. G. Kwiat, S. Barraza-Lopez, A. Stefanov, and N. Gisin, Experimental entanglement distillation and 'hidden' non-locality, *Nature.* **409**(6823), 1014–1017, (2001).
[36] C. H. Bennett, D. P. DiVincenzo, J. A. Smolin, and W. K. Wootters, Mixed-state entanglement and quantum error correction, *Phys. Rev. A.* **54**(5), 3824–3851 (1996).
[37] W. K. Wootters, Entanglement of formation of an arbitrary state of two qubits, *Phys. Rev. Lett.* **80**(10), 2245–2248 (1998).

Chapter 15

Entangled Photon Pair Emission and Interference from Single Quantum Dots

R. M. Stevenson, R. J. Young, and A. J. Shields

Toshiba Research Europe Ltd., Cambridge Research Laboratory, 208 Cambridge Science Park, Milton Road, Cambridge CB4 0GZ, United Kingdom

We review in this chapter our recent realization and development of entangled light emission from individual InAs quantum dots. We describe how energetic which-path information is circumvented by attention to the confinement energy of the dot, resulting in generation of entangled photon pairs with up to 76% fidelity. Improvements in device design are detailed that reveal the importance of controlling the charge of the dot in order to maximize both emission intensity and fidelity. Finally we present interferometric measurements of the entangled biphoton state that shows a two-fold enhancement of the de Broglie wavelength compared to single photons.

Contents

15.1. Introduction

Entanglement between two qubits is a fundamental requirement for quantum computing, and for optical qubits this resource takes the form of a pair of entangled photons. The importance of entangled light states extends into other quantum optical technologies, such as secure quantum communication over long distances [1], and quantum interferometry and metrology [2–5].

Generating entangled photon pairs with semiconductor devices is very appealing, the technology is convenient, well-established and lends itself well to low cost mass production. A more subtle advantage for semiconductor light sources is the possibility to link solid-state quantum operations with optical qubits, which are otherwise only weakly interacting.

We review here the first experimental realization of a semiconductor source of triggered entangled photons. The triggered nature of the device limits the emission to no more than one photon pair per emission period. This is essential to avoid errors in applications, and is a precursor to an on-demand source required for efficient quantum computing. We go on to identify the factors limiting the performance of the entangled photon source. Of particular interest, we find that operation of the device is close to the limit of slow spin scattering, and this situation can be furthered by the control of the exciton-charging probability. In addition, we find that the entangled biphoton is surprising robust against dephasing, in stark contrast to the devastating effect of decoherence upon single photons emitted by similar structures.

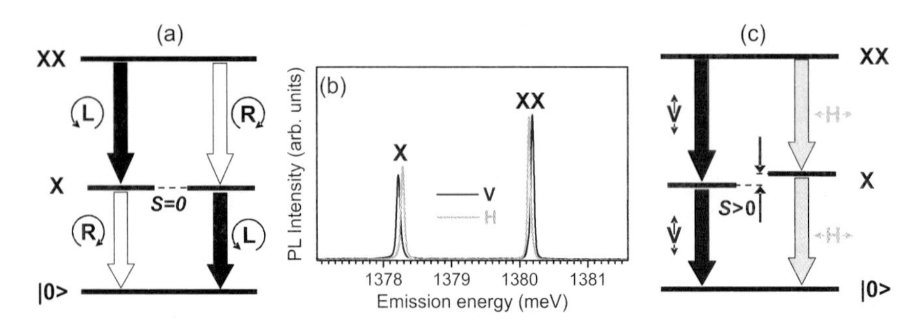

Fig. 15.1. (a) An energy level diagram illustrating the optical transitions involved in the decay from the biexciton state (XX) through the intermediate exciton states (X) in an ideal quantum dot. (b) Polarized photoluminescence spectra from a single quantum dot, the peaks correspond to emission from the XX and X states as indicated. (c) The energy level diagram for the optical decay of the biexction state modified by a splitting between the bright exciton states. Colors indicate photon polarizations; black (white) corresponds to left (right) hand circular and red (grey) corresponds to vertical (horizontal).

The device design employs a single quantum dot as the active region, which was previously proposed as a source of entangled photon pairs by Benson et al. [6]. In the ideal case, the emission sequence is represented by figure 15.1(a), and proceeds as follows. The quantum dot is excited into the biexciton state (XX) consisting of two electrons and two heavy-holes. Only recombination of an electron and hole with opposing spins can couple to the optical modes. Recombination of the spin-up electron, followed by the spin-down electron, emits a σ^- polarized photon followed by a σ^+ polarized photon, or, vice versa. Provided that there is no additional way to distinguish the polarization of a photon, the order of recombination is unknown until measured. Thus the emitted pair exists in a superposed, maximally entangled state $(|\mathrm{R}_{XX}\mathrm{L}_X\rangle + |\mathrm{L}_{XX}\mathrm{R}_X\rangle)/\sqrt{2}$, where L and R denote the left and right hand polarizations of the circular basis, and the XX and X subscripts denote the biexction and exciton photons respectively.

Unfortunately, the situation in reality is generally not ideal [7–9]. The expected entangled state $(|\mathrm{R}_{XX}\mathrm{L}_X\rangle + |\mathrm{L}_{XX}\mathrm{R}_X\rangle)/\sqrt{2}$ can be re-written as $(|\mathrm{H}_{XX}\mathrm{H}_X\rangle + |\mathrm{V}_{XX}\mathrm{V}_X\rangle)/\sqrt{2}$ in the rectilinear basis, and $(|\mathrm{D}_{XX}\mathrm{D}_X\rangle + |\mathrm{A}_{XX}\mathrm{A}_X\rangle)/\sqrt{2}$ in the diagonal polarization basis, where H, V, D, and A represent horizontal, vertical, diagonal, and anti-diagonal polarizations. Thus the polarization of the second photon is expected to correlate with the first, and have the same linear, or opposite circular, polarization. Initial experiments showed that only rectilinear polarization correlation is observed, and thus the emission from the quantum dots is not entangled [7–9]. The lack of entanglement is attributed to distinguishability of the orthogonally polarized two photon states by energy, and photoluminescence reveals polarization dependent splitting of both the biexciton and exciton photons as shown in figure 15.1(b). [8, 10]. Due to the common initial (biexciton) and final (ground) states, the origin of the splitting is governed by the level structure of the intermediate exciton state. As shown in figure 15.1(c), the degeneracy of the exciton state is lifted in ordinary dots, resulting in fine structure splitting of the exciton S. Structural asymmetries in-plane, such as the strain, composition and shape, unbalance the exchange interaction, which in turn results in the hybridisation of the exciton spin states to produce linearly polarized eigenstates [11, 12].

Control of the polarization splitting S is crucial in order to emit entangled photon pairs. The strategy we use for the devices measured here is based on using a growth process capable of directly producing quantum dots with zero polarization splitting. For this, we rely upon the discovery that the splitting is strongly dependent on the emission energy of the dot [15]. Figure 15.2(a) plots the polarization splitting S as a function of the emission energy for a number of quantum dots. As the emission energy increases, the splitting decreases. At $\sim 1.4\,\mathrm{eV}$ the splitting is zero, so we design our devices to operate at this wavelength. The origin of the depen-

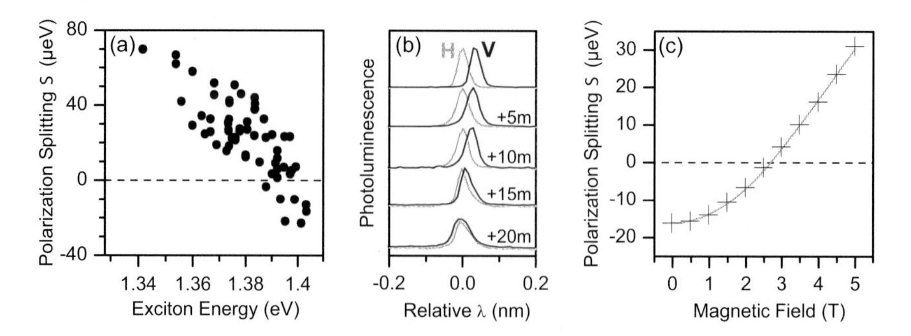

Fig. 15.2. (a) Exciton emission energy plotted as a function of the splitting between the bright exciton states for a number of quantum dots. (b) Several polarized photoluminescence spectra from the exciton state in the same dot, between each set of spectra from top to bottom the dot was annealed for 5 mins at 675°C. (c) The exciton splitting for a single dot plotted as a function of an applied in-plane magnetic field. The solid line show a fit according to expected behaviour.

dence of S with emission energy is attributed to the changing confinement of the quantum dot. Previous measurements have indicated that the area of the exciton increases with decreasing emission energy, which would result in weaker overlap of the electron and hole, and hence smaller exchange energies and splitting [15].

Alternative schemes exist to reduce the splitting. Another permanent approach is to anneal the quantum dot, which also has the effect of reducing the confinement [13–17]. A similar relationship between the splitting and emission energy after anneal is found compared to that of growth alone [15]. Furthermore, the spitting of a selected dot can be tuned to zero after fabrication [17]. This is achieved by first protecting the device from desorption of As by encapsulating with an insulator such as silicon nitride, and repeatedly annealing the device for 5 mins at 675°C. The results of such a process are illustrated in figure 15.2(b), which shows the relative energies of the H and V polarized exciton emission, as a function of the annealing time. After a total anneal of 20 min, the final splitting is -1.8$\pm$2 μeV, i.e. zero within error. Schemes allowing the reversible control of the splitting include the application of external electric [18, 19], strain [22] and magnetic fields [23]. The latter has been shown to enable entangled photon emission from suitable dots [20], and its control over S is demonstrated in figure 15.2(c), which plots the splitting as a function of field. The splitting increases approximately quadratically with field, as expected [23]. However, in this work we use the in-plane magnetic field only to increase the splitting S, and deactivate entanglement. A final strategy to generate entangled photons with quantum dots does not control splitting, and instead employs energetic post-selection of the emitted photons [24].

15.2. Sample and Experimental Design

We now describe the devices and experiments that demonstrate the emission of entangled photon pairs from single quantum dots. The sample used was grown by MBE on a GaAs substrate, and included a single self-assembled quantum dot layer, with the thickness of InAs optimised to achieve the desired quantum dot density of $1\,\mu m^{-2}$ [20]. Background light from the wetting layer is suppressed by a $20°$ C increase in the growth temperature, which results in a blue-shift of the wetting layer due to intermixing of the InAs wetting layer with the surrounding GaAs [21].

AlAs/GaAs distributed Bragg reflectors were grown above (2 repeats) and below (14 repeats) the dot layer to form a planar microcavity, resonant with the optimum quantum dot energy of 1.4 eV. The cavity enhances the light collection efficiency from the top of the sample by an order of magnitude.

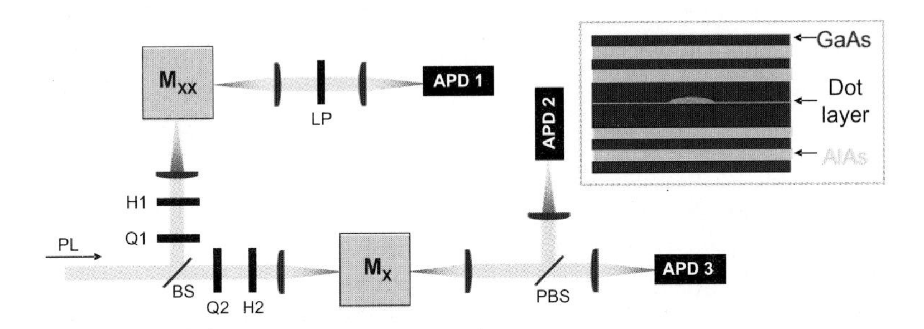

Fig. 15.3. A schematic of experimental setup used to measure polarized cross correlation between the biexciton (XX) and exciton (X) photons. Quarter-wave plates (Q1 and Q2) and Half-wave plates (H1 and H2) were inserted in front of the entrance slits to the two spectrometers (M_X and M_{XX}). A beam-splitter (BS) was used to divide the photoluminescence into the two spectrometers. A polarizing beam-splitter was placed prior to Avalanche Photodiodes (APD's) following M_X and a linear polarizer (LP) between M_{XX} and APD 1. The inset illustrates the basic sample design, the InAs quantum dots are grown in the center of an optical cavity formed by alternating layers of AlAs (light) and GaAs (dark).

Photoluminescence (PL) was excited non-resonantly at $\sim$ 10 K, using a 635 nm laser diode emitting 100 ps pulses at 80 MHz. The properties of photon pairs emitted by a selected quantum dot were analyzed by polarization and time dependent correlations between the XX and X photons, as shown schematically in figure 15.3. Emission was isolated by two spectrometers, tuned to the XX and X emission energies. The insertion of appropriately oriented quarter-wave or half-wave plates preceding each of the spectrometers allows any polarization measurement basis to be selected. The spectrally filtered XX, and X emission passed through a linear polariser, and polarising beam splitter respectively, and was detected by three silicon avalanche photo-diodes (APDs). The time between detection events on different APD's was measured, to determine the second order correlation functions. Finally, the number of counts was integrated over each quantum dot decay cycle.

The probability of detecting coincident photons is dependent on the excitation rate, which drifts and fluctuates during the integration time of our experiments. This can be compensated for by comparing the simultaneous correlation measurement of the XX detection channel with each of the orthogonally polarised X detection channels [20, 21]. For an unpolarised source, as verified for our dots, we define the degree of correlation as $C=(g^{(2)}_{XX,X}-g^{(2)}_{XX,\overline{X}})/(g^{(2)}_{XX,X}+g^{(2)}_{XX,\overline{X}})$, where the second order correlation functions, $g^{(2)}_{XX,X}$ and $g^{(2)}_{XX,\overline{X}}$ are the simultaneously measured, normalised coincidences of the biexciton photon with the co-polarised exciton and orthogonally-polarised exciton photons respectively. We use the degree of correlation C as the primary measure of polarization correlation, due to its robust nature against systematic errors. C takes the values 1, -1, and 0 for perfect polarization correlation, anti-correlation and no polarization correlation, respectively. Note that the probability that a photon pair is correlated has a linear relationship to C.

15.3. Correlations for Entangled and Classical Photons

Figure 15.4(a) shows the degree of correlation between H polarized XX and rectilinearly polarized X photons, as a function of the time delay between the photons, in cycles, for a quantum dot with S$\sim$0. For non-zero time delays, C$\sim$0, indicating that there is no polarization memory from one cycle to the next. The noise on the data is caused by the random errors associated with the finite number of counts. At zero time delay, the large peak of 70%

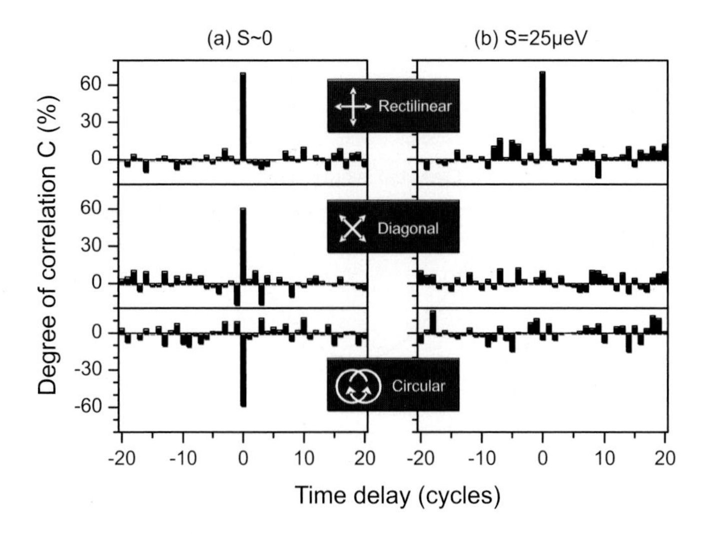

Fig. 15.4. The degree of correlation C between the biexciton and exciton photons (defined in the text) in three different polarization bases as labelled as a fucntion of the delay between the two photons for (a) a dot with S$\sim$0 and (b) S=25μeV.

demonstrates a strong polarization correlation between H polarized excitons and biexcitons in this dot. A similar magnitude of polarization correlation is present for the diagonal, and circular measurement bases, as shown. These results are consistent with the expected polarization correlation in the linear bases, and polarization anti-correlation in the circular basis. In fact, the total magnitude of the correlation in the three measured independent bases can only be explained by entanglement.

In contrast, the situation for a quantum dot with finite splitting is very different. Figure 15.4(b) shows the results of a similar experiment on a dot with $S = 25\,\mu$eV, induced by an in-plane magnetic field. A similar degree of rectilinear polarization correlation is observed to the entangled dot shown in (a). However, no discernable polarization correlation is observed in the diagonal and circular polarized detection bases. This is precisely as expected for a quantum dot with finite splitting, where the photons are distinguishable by energy in addition to polarization.

A feature of entangled photon pairs is that measurement of one photon changes the polarization of the other photon. This is demonstrated by figure 15.5, which plots the degree of correlation for the entangled quantum dot as the function of the rotation of the detection polarization, achieved by adjusting the angle of a single half wave plate, placed directly after the

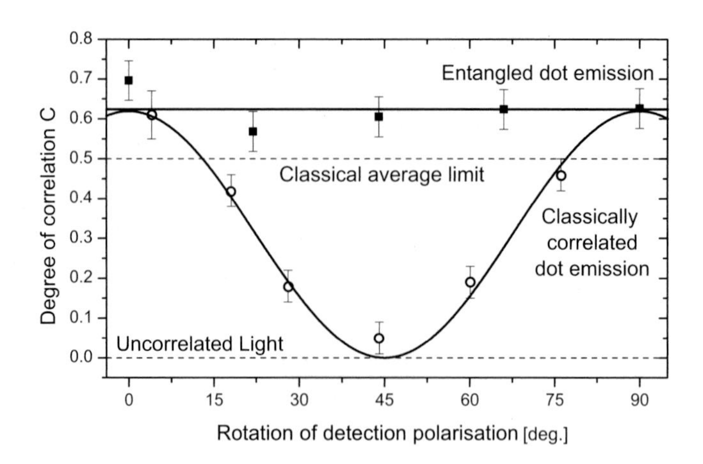

Fig. 15.5. The degree of linear polarization correlation between the biexciton and exciton photons plotted as function of the rotation of the linear measurement basis for a dot emitting classically correlated photons (open) and for a dot emitting polarization-entangled photons (closed). Dashed lines indicate the expected average for an uncorrelated light source and the maximum average for a classical light source.

microscope objective lens. Closed points show that the degree of polarization correlation C is approximately independent of the half wave plate angle. This is an expected result for photon pairs being emitted in the entangled $(|\mathrm{R}_{XX}\mathrm{L}_X\rangle + |\mathrm{L}_{XX}\mathrm{R}_X\rangle)/\sqrt{2}$ state, since the linear polarization measurement of the first photon defines the linear polarization of the second photon. For classically polarization correlated photon pairs, the degree of correlation varies sinusoidally with wave-plate angle, as shown for a similar dot with finite splitting S. The entangled dot has average linear correlation of 62.4±2.4%, which is five standard deviations above the 50% limit for classical pairs of photons and proves that the quantum dot emits polarization entangled photon pairs.

To fully characterise the two photon state emitted by the dot, the two photon density matrix can be constructed from correlation measurements, using quantum state tomography [25]. Our formulism requires the measurement of C for all combinations of the V, H, L, and D biexciton polarizations, with the rectilinear, diagonal and circular polarized exciton detection bases [21]. The resulting density matrix representing the emission from the quantum dot is shown in figure 15.6.

The strong outer diagonal elements in the real matrix (blue bars, figure 15.6) demonstrate the high probability that the photon pairs have the same linear polarization. The inner diagonal elements represent the probability

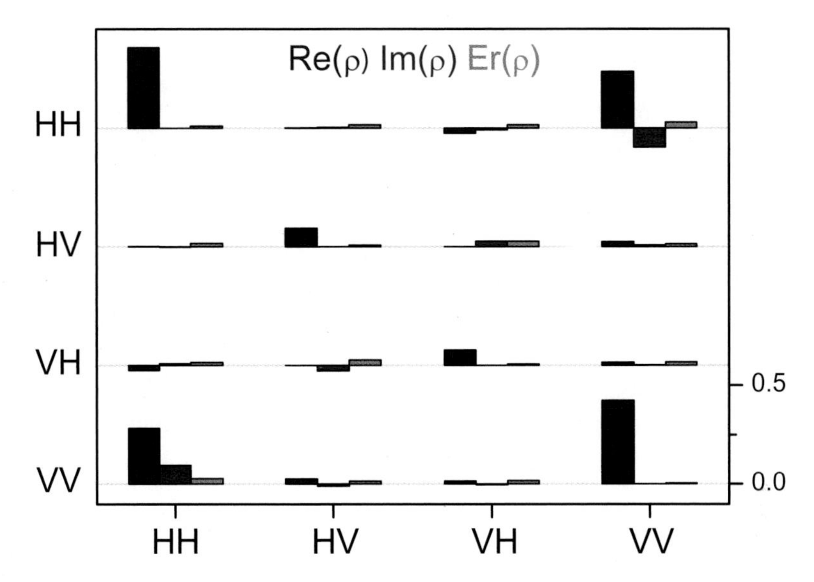

Fig. 15.6. The two-photon density matrix (ρ) in the rectilinear basis for the biexciton decay from a dot with no resolvable bright exciton splitting. Red (blue) bars indicate the real (imaginary) components of the matrix and green bars indicate the magnitude of random error associated with each element.

of detecting oppositely linearly polarized photons, which is attributed to roughly equal parts of background light and exciton spin scattering in the dot [20].

The outer off-diagonal elements in the real matrix (blue bars) are strong, and clear indicators of entanglement. Small imaginary off-diagonal elements (red bars) are additionally seen, which indicates that there may be a small phase difference between the $|H_{XX}H_X\rangle$ and $|V_{XX}V_X\rangle$ components of the entangled state for this particular dot. All other elements are close to zero, given the errors associated with the procedure (green bars).

The measured two photon density matrix projects onto the expected $(|\mathrm{H}_{XX}\mathrm{H}_X\rangle+|\mathrm{V}_{XX}\mathrm{V}_X\rangle)/\sqrt{2}$ entangled state with fidelity 0.702 ± 0.022. This proves that the photon pairs we detect are entangled, since for pure or mixed classical un-polarized states, the fidelity cannot exceed 0.5. Other tests including the tangle and eigenvalue are also positive for entanglement, by many standard deviations [21]. Improvements in the device structure have recently increased the fidelity to 0.76 ± 0.04. This improvement is encouraging, and next we consider how to continue the trend of increasing performance from the entangled photon pair source.

15.4. Improving Device Characteristics

To further improve the properties of the entangled photon source, we now consider how to increase the rate of photon pair generation and collection, and strategies to maximize the fidelity of the entangled state.

First we consider improving the efficiency at which photon pairs are collected. Two important factors limiting the detected entangled light intensity are; emission blocking due to charged exciton states, and the directionality of the emission. The latter can be improved by optical confinement using microcavities, and the device already includes an unbalanced planar microcavity which improves collection efficiency 10 fold compared to a simple bulk device. More complex designs such as micropillars and photonic crystal cavities are likely to lead to further improvements. For example, with full optical confinement in three dimensions, it is possible to achieve Purcell enhancement of the radiative decay rate for single photon emission in a desired direction [27, 28]. Quantum dots typically show significant charged exciton photon emission, due to the presence of background p-doping in the substrate. Charged excitations block emission from the neutral biexciton for a complete cycle, it is thus important to eradicate them. In section 15.5, we control the charge state of a single dot, and show the implications for entangled photon generation.

Secondly we consider improving the fidelity of the emission. Fidelity is sensitive to several major factors, including distinguishability of the emitted photons, presence of uncorrelated light, and decoherence of the entangled state. As shown above, the problem of distinguishability has been solved by controlling the growth of single quantum dots such that the intermediate exciton level is degenerate [15]. Uncorrelated light is attributed to emission originating from device layers other than the dot, especially the wetting layer, and to photon pairs emitted after the intermediate exciton spin has scattered. Uncorrelated light affects both classically polarization correlated and entangled light equally, and we have demonstrated that blue-shifting the wetting layer away from the dot resonance reduced the degree of uncorrelated light by a factor of 3 [21]. Further improvements are to be expected from the use of resonant excitation. In contrast, decoherence only affects the entangled photon pair state. For entangled photon pairs created by two-photon interference, decoherence is extremely limiting, and necessitates the use of resonant excitation [7, 29, 30]. Decoherence in our system is investigated using interferometry in section 15.6, from which we conclude that the device is robust against such problems.

15.5. Charge Control

To enable the exciton charge to be controlled, dots were grown in the intrinsic region of a p-i-n diode, incorporating a cavity design similar to that described in section 15.2. The top two layers of the bottom mirror were doped n-type with silicon and the top mirror was p-doped with carbon. In addition, an AlAs/GaAs superlattice was formed 5nm below the dots, consisting of 7 repeats of 2nm layers [31]. The superlattice forms a barrier, preventing electron tunneling out of the quantum dots. Charging and discharging of the dots is therefore governed by the number of holes in the dot, which is controlled by the bias, the device design enabling this is shown schematically in the inset to figure 15.7.

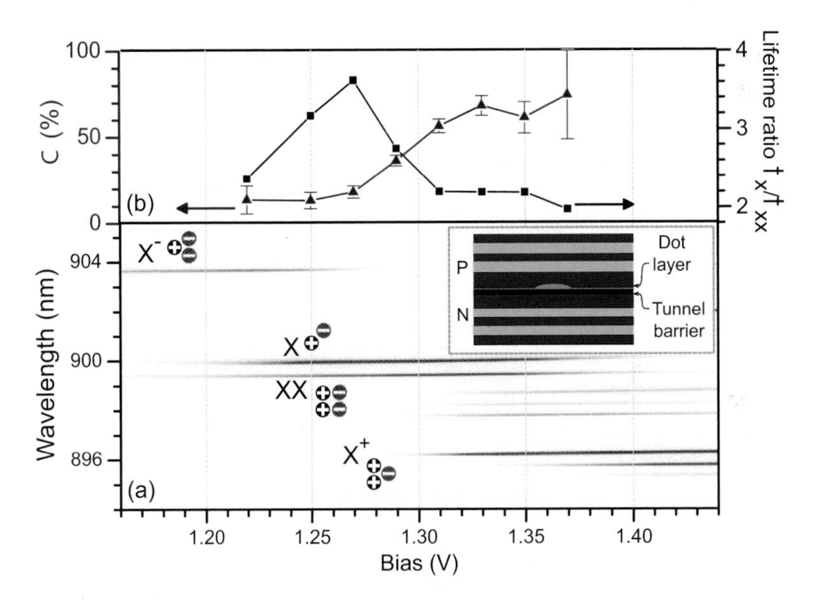

Fig. 15.7. (a) Photoluminescence from a single quantum dot as a function of the bias applied to the dot layer. Emission from the negatively charged (X^-), positively charged (X^+) and neutral exciton states is labelled. (b) The degree of rectilinear polarization correlation between the biexciton and exciton photon and the ratio of the states' lifetimes as a function of the applied bias. The inset illustrates the sample design, the top (bottom) mirror is p-type (n-type) doped. A barrier beneath the dot layer inhibits electrons from tunneling out of the dots.

Figure 15.7 (a) shows PL from a single quantum dot measured as a function of the bias applied between the p- and n-type regions of the device. Below a bias of ~1V no PL was measured from the quantum dots, indicating that the rate at which the holes tunnel from the dots was much greater than the radiative lifetime. Between 1 and 1.2 V PL is predominantly emitted from the negatively charged exciton (X^-). As the bias increases further, emission from the neutral exciton (X) and biexciton (XX) states dominates, followed by emission from the positively charged exciton (X^+). This demonstrates that we can use this device to carefully control the charge of the exciton complexes which are dominant in the photoluminescence spectra of a single quantum dot [32, 33]. The maximum photon pair emission rate corresponds to a bias of $\sim$ 1.30 V.

Another question is what effect the charging probability has upon the fidelity of the entanglement. The quantum dot in the charge tuning diode has finite splitting, and is thus not entangled. However, as discussed in section 15.4, background and spin-scattering degrades entanglement, and equally affects classical correlation. Therefore, the degree of polarization correlation in the rectilinear basis C was measured as a function of applied bias. The results are shown in figure 15.7(b). The degree of correlation increases at higher biases from 56±4% at 1.31 V to 68±6% at 1.33 V. Furthermore the correlation drops sharply at lower voltages to 13±5% at 1.25 V. This sharp degradation is likely due to exciton spin scattering, induced by excess electrons in the vicinity of the dot, as now discussed with reference to time resolved measurements.

The exciton and biexciton radiative lifetimes were measured as a function of bias, from fitting the exponential decay of PL as a function of time. The ratio of the biexciton lifetime to that of the exciton is plotted as a function of bias in figure 15.7(b), together with the degree of correlation C. At high biases the polarization correlation between the pair of photons is high and the ratio of lifetimes is close to 2. This result is expected in the limit of slow exciton dephasing, and is the typical ratio of lifetimes observed in InAs quantum dots. As the bias is reduced, and the degree of correlation drops, the peak lifetime ratio is found to be 3.62±0.02. Theoretical studies of the lifetime ratio predict a value of $\sim$4 in the fast spin-flip limit. A long-lived [34] dark exciton state is formed when the electron in an optically active exciton state undergoes a spin-flip. Such event would therefore increase the measured lifetime of the exciton and explain the large drop-off in correlation measured. At the lowest biases the lifetime ratio is found to decrease again, as the hole tunneling time out of the dot becomes comparable to the exciton radiative lifetime.

An interesting feature of this device is that the bias required to maximize the pair rate differs slightly from the bias under which the largest proportion of pairs are polarization-correlated. Thus selecting a bias for maximum efficiency, or maximum fidelity, would generate less entangled photon pairs than choosing an intermediate value.

15.6. Biphoton Interference and Decoherence

Biphoton interferometry is a powerful technique, that typically employs coincident detection of pairs of photons to probe key characteristics of non-classical light. Such properties include the coherence length and the de Broglie wavelength. The nature of two-photon interference can be remarkably different from that of the constituent single photons. For example, higher frequency interference fringes have been observed using entangled photons generated by parametric down conversion [5, 35], or by post-selected measurements using classical light [36], which is the basis of quantum imaging applications such as quantum lithography [4], and low-cell-damage biomedical microscopy [2].

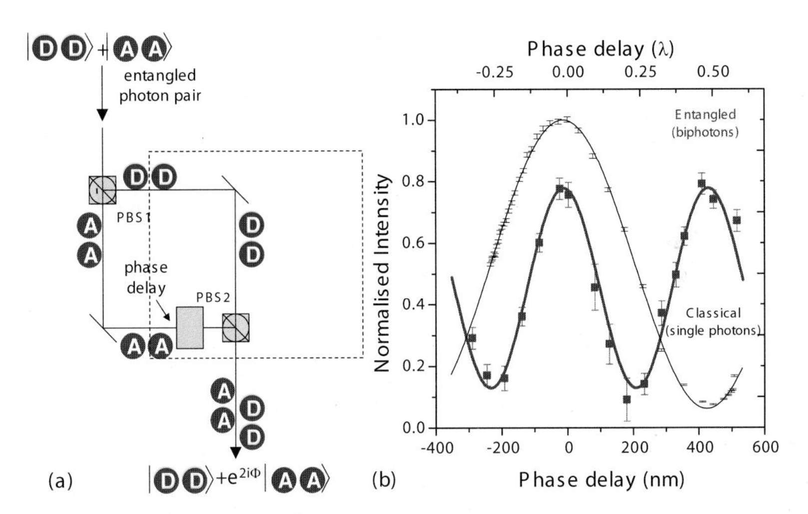

Fig. 15.8. (a) A schematic diagram of the biphoton interferometer. (b) The normalised intensity of classical single photons (black) and normalised biphoton intensity of entangled photon pairs (blue) as a function of the phase delay.

Here we measure biphoton interference for the quantum dot device using an interferometer similar to that shown in figure 15.8(a). The phase control required to measure an interferogram is supplied by a polarization dependent phase delay. A combination of appropriately configured polarizing beam-splitters PBS1 and PBS2 force D and A polarized photons to take separate paths through the interferometer. Both A polarized photons are delayed by phase ϕ, so that output of the interferometer is $(|\mathrm{D}_{XX}\mathrm{D}_X\rangle + e^{2i\phi}|\mathrm{A}_{XX}\mathrm{A}_X\rangle)/\sqrt{2}$. Interference of the two-photon amplitudes of $|D_{XX}D_X\rangle$ and $|A_{XX}A_X\rangle$ is achieved by measurement in the rectilinear basis. The biphoton intensity variation with ϕ results in an interferogram, with an expected period π, corresponding to a de Broglie wavelength of $\lambda/2$, where is the average wavelength of the single biexciton and exciton photons. For the single photon input state $|V\rangle = (|D\rangle + |A\rangle)/\sqrt{2}$, the output from the interferometer is $(|D\rangle + e^{i\phi}|A\rangle)/\sqrt{2}$. Measuring the intensity in the vertical polarization would yield an interferogram with period 2π, corresponding to a de Broglie wavelength λ.

The required polarization dependent phase delay was realized by a liquid crystal with voltage dependent birefringence. The collinear nature of the interferometer provides exceptional stability compared to a typical independent path Mach-Zehnder design.

Single photon interference fringes from light emitted by a quantum dot, were measured by inserting a linear polarizer before the interferometer, to select only vertically polarized photons. Using a single APD, the intensity of the resulting single exciton photon state $|V\rangle$ was measured as a function of the phase delay, and normalized to the maximum. The results are shown in figure 15.8(b) as black points. Clear interference fringes are seen, and the intensity varies as a function of the phase delay in agreement with the fit to expected sinusoidal behavior, shown by the solid line. The period of the oscillations is determined to be 877±35nm (0.99±0.03)λ, approximately equal to the wavelength of the quantum dot emission of ~885nm, as expected.

The linear polarizer set before the interferometer was removed, so that entangled photon pairs emitted by the quantum dot could be analyzed. The normalized biphoton intensity is equal to $(g^{(2)}_{VV} + g^{(2)}_{HH})/(g^{(2)}_{VV} + g^{(2)}_{VH} + g^{(2)}_{HV} + g^{(2)}_{HH})$, where the denominator and numerator and are proportional to the biphoton generation and detection rates respectively, and g represents the second order correlation for coincident detection of two photons with polarization denoted by subscripts. For an unpolarized source, such as this dot, this is approximated to $g^{(2)}_{VV}/(g^{(2)}_{VV} + g^{(2)}_{HV})$. The second order correlation functions for the $|V_{XX}V_X\rangle$ and $|H_{XX}H_X\rangle$ two-photon detection bases were measured simultaneously as described above.

The measured normalised biphoton intensity, indicated by blue points in figure 15.8(b), shows strong interference fringes. The difference in the period of oscillations compared to the classical single photon case is very striking. The fringes fit well to the predicted sinusoidal behavior shown as a solid line, from which we determine the period of the oscillations to be 442$\pm$36nm $(0.50\pm0.03)\lambda$. The period is equivalent to the de Broglie wavelength of the biphoton, which is in excellent agreement with the two-fold reduction from 885nm to 443nm expected for an entangled photon pair source.

The shorter de Broglie wavelength for the entangled state compared to the single photon implies that up to a two-fold enhancement of the imaging resolution is possible using biphoton detection. The visibility of the measured biphoton interferogram compared to that for single photon detection demonstrates enhanced phase resolution of the interferometer with a biphoton source [2]. Note that the exciton and biexciton have different wavelengths, but this does not affect the visibility of biphoton interference, as it is the total energy of the two-photon state that defines the de Broglie wavelength.

From the fit to sinusoidal behavior we also directly determine the phase difference between the $|D_{XX}D_X\rangle$ and $|A_{XX}A_X\rangle$ components of the entangled state emitted by the source to be $(0.02\pm0.03)\lambda$. To our knowledge, this represents the most direct measurement of phase offset reported for entangled photons generated by a quantum dot, and shows that the diagonally polarized two-photon amplitudes are in-phase within error [26].

Control of the phase offset can manipulated the entangled state. For example, a lambda/4 delay transforms the state $(|\mathrm{H}_{XX}\mathrm{H}_X\rangle + |\mathrm{V}_{XX}\mathrm{V}_X\rangle)/\sqrt{2}$ into $(|\mathrm{H}_{XX}\mathrm{V}_X\rangle + |\mathrm{V}_{XX}\mathrm{H}_X\rangle)/\sqrt{2}$, and the photons are then polarization anti-correlated in the rectilinear basis.

Uncorrelated light reduces the amplitude of the interference maxima for entangled or mixed classical light, and the interference minima for entangled light. We measure interference for mixed, polarization correlated light, by increasing the splitting S with an in-plane magnetic field of 4T (not shown). From this we determine that uncorrelated light contributes towards $(37\pm7)\%$ of the biphoton intensity. The corresponding maximum possible fidelity and interference amplitudes for the entangled state are 0.73 ± 0.05 and 0.63 ± 0.07 respectively, in close agreement with measured values of 0.76 ± 0.04 and 0.65 ± 0.04 respectively.

The agreement between the expected maximum, and measured values for both fidelity and interference amplitude suggests that biphoton decoherence does not degrade the entangled state. In contrast, the single photon coherence time limits the visibility of two-photon interference using two sequential photons from a quantum dot, and necessitates the use of resonant excitation.

15.7. Conclusion

The use of quantum dots to generate entangled photon pairs has promised a very attractive route to mainstream technological implementations of quantum computing, quantum optics and quantum information applications. Here we have shown that despite initial difficulties with fine structure splitting in quantum dots, there is every possibility that they will realize their full potential. The $> 70\%$ fidelity of the entangled photons measured here shows that quantum dot technology offers a realistic solution to the problem of generating entangled light. In addition we have demonstrated that factors limiting performance are well understood. Tuning the charging probability of the dot has demonstrated control not only over the emission intensity, but also the degree of correlation, which suggest the future possibility to eradicate exciton spin scattering. Similarly, biphoton interference suggests that decoherence does not limit existing devices. With further improvements to reject background light and improve collection efficiency, the prospect of developing a practical, electrically driven, LED-like source of entangled photons, is very promising indeed.

Acknowledgments

We would like to acknowledge partial funding by the EU projects QAP and SANDiE, and by the EPSRC through the IRC for Quantum Information Processing.

References

[1] H.-J. Briegel, W. Dür, J. I. Cirac, P. Zoller: Phys. Rev. Lett. **81** 5932 (1998)
[2] V. Giovannetti, S. Lloyd, L. Maccone: Science **306**, 1330 (2004)
[3] E. J. S. Fonseca, P. H. Souto Ribeiro, S. Pádua, C. H. Monken: Phys. Rev. A **60**, 1530 (1999)
[4] A. N. Boto, P. Kok, D. S. Abrams, S. L. Braunstein, C. P. Williams, J. P. Dowling: Phys. Rev. Lett. **85**, 2733 (2000)
[5] E. Edamatsu, R. Shimizu, T. Itoh: Phys. Rev. Lett. **89**, 213601 (2002)

[6] O. Benson, C. Santori, M. Pelton, T. Yamamoto: Phys. Rev. Lett. **84**, 2513 (2000)
[7] C. Santori, D. Fattal, M. Pelton, G. S. Solomon, Y. Yamamoto: Phys. Rev. B **66**, 045308 (2002)
[8] R. M. Stevenson, R. M. Thompson, A. J. Shields, I. Farrer, B. E. Kardynal, D. A. Ritchie, M. Pepper: Phys. Rev. B(R) **66**, 081302 (2002)
[9] S. M. Ulrich, S. Strauf, P. Michler, G. Bacher, A. Forchel: Appl. Phy. Lett. **83**, 1848 (2003)
[10] D. Gammon, E. S. Snow, B. V. Shanabrook, D. S. Katzer, D. Park: Phys. Rev. Lett. **76**, 3005 (1996)
[11] V. D. Kulakovskii, G. Bacher, R. Weigand, T. Kümmell, A. Forchel, E. Borovitskaya, K. Leonardi, D. Hommel: Phys. Rev. Lett. **82**, 1780 (1999)
[12] O. Stier, M. Grundmann, D. Bimberg: Phys. Rev. B **59**, 005688 (1999)
[13] W. Langbein, P. Borri, U. Woggon, V. Stavarache, D. Reuter, and A. D. Wieck: Phys. Rev. B **69**, 161301 (2004)
[14] A. I. Tartakovskii, M. N. Makhonin, I. R. Sellers, J. Cahill, A. D. Andreev, D. M. Whittaker, J-P. R. Wells, A. M. Fox, D. J. Mowbray, M. S. Skolnick, K. M. Groom, M. J. Steer, H. Y. Liu, and M. Hopkinson, Phys. Rev. B **70**, 193303 (2004)
[15] R. J. Young, R. M. Stevenson, A. J. Shields, P. Atkinson, K. Cooper, D. A. Ritchie, K. M. Groom, A. I. Tartakovskii, M. S. Skolnick: Phys. Rev. B **72** 113305 (2005)
[16] R. Seguin, A. Schliwa, T. D. Germann, S. Rodt, K. Pötschke, A. Strittmatter, U. W. Pohl, D. Bimberg, M. Winkelnkemper, T. Hammerschmidt, and P. Kratzer: Appl. Phys. Lett. **89**, 263109 (2006)
[17] D. J. P. Ellis, R. M. Stevenson, R. J. Young, A. J. Shields, P. Atkinson, D. A. Ritchie: Appl. Phys. Lett. **90**, 011907 (2007)
[18] K. Kowalik, O. Krebs, A. Lematre, S. Laurent, P. Senellart, P. Voisin, and J. A. Gaj: Appl. Phys. Lett. **86**, 041907 (2005)
[19] B. D. Gerardot, S. Seidl, P. A. Dalgarno, R. J. Warburton, D. Granados, J. M. Garcia, K. Kowalik, O. Krebs, K. Karrai, A. Badolato, P. M. Petroff: Appl. Phys. Lett. **90**, 041101 (2007)
[20] R. M. Stevenson, R. J. Young, P. Atkinson, K. Cooper, D. A. Ritchie, A. J. Shields: Nature **439**, 179 (2006)
[21] R. J. Young, R. M. Stevenson, P. Atkinson, K. Cooper, D. A. Ritchie, A. J. Shields: New J. Phys. **8**, 29 (2006)
[22] S. Seidl, M. Kroner, A. Högele, K. Karrai, R. J. Warburton, A. Badolato, and P. M. Petroff: Appl. Phys. Lett. **88**, 203113 (2006)
[23] R. M. Stevenson, R. J. Young, P. See, D. Gevaux, K. Cooper, P. Atkinson, I. Farrer, D. A. Ritchie, A. J. Shields: Phys. Rev. B **73**, 033306 (2006)
[24] N. Akopian, N. H. Lindner, E. Poem, Y. Berlatzky, J. Avron, D. Gershoni, B. D. Gerardot, P. M. Petroff: Phys. Rev. Lett. **96**, 130501 (2006)
[25] D. F. V. James, P. G. Kwiat, W. J. Munro, A. G. White: Phys. Rev. A **64**, 052312 (2001)
[26] R. M. Stevenson, A. J. Hudson, R. J. Young, P. Atkinson, K. Cooper, D. A. Ritchie, A. J. Shields: Opt. Express **15**, 6507 (2007)

[27] A. Bennett, D. Unitt, P. Atkinson, D. Ritchie, A. J. Shields: Opt. Express **13**, 50 (2005)
[28] D. G. Gevaux, A. J. Bennett, R. M. Stevenson, A. J. Shields, P. Atkinson, J. Griffiths, D. Anderson, G. A. C. Jones, D. A. Ritchie: Appl. Phys. Lett. **88**, 131101 (2006)
[29] D. Fattal, K. Inoue, J. Vuckovic, C. Santori, G. S. Solomon, Y. Yamamoto: Phys. Rev. Lett. **92**, 037903 (2004)
[30] A. J. Bennett, D. C. Unitt, P. Atkinson, D. A. Ritchie, A. J. Shields: Opt. Express, **13**, 7772 (2005)
[31] R. J. Young, S. J. Dewhurst, R. M. Stevenson, A. J. Shields, P. Atkinson, K. Cooper, D. A. Ritchie, Appl. Phys. Lett. **91**, 011114 (2007)
[32] R. J. Warburton, C. Schäflein, D. Haft, F. Bickel, A. Lorke, K. Karrai, J. M. Garcia, W. Schoenfeld, P. M. Petroff: Nature **405**, 926 (2000)
[33] J. J. Finley, P. W. Fry, A. D. Ashmore, A. Lematre, A. I. Tartakovskii, R. Oulton, D. J. Mowbray, M. S. Skolnick, M. Hopkinson, P. D. Buckle, P. A. Maksym: Phys. Rev. B **63**, 161305 (2001)
[34] R. M. Stevenson, R. J. Young, P. See, I. Farrer, D. A. Ritchie, A. J. Shields: Physica E **21**, 381 (2004)
[35] P. Walther, J. -W. Pan, M. Aspelmeyer, R. Ursin, S. Gasparoni, A. Zeilinger: Nature **429**, 158 (2004)
[36] G. Khoury, H. S. Eisenberg, E. J. S. Fonesca, D. Bouwmeester: Phys. Rev. Lett. **96**, 203601 (2006)

PART 5
Qubit Applications

Chapter 16

Telecom-wavelength Single-photon Sources Based on Single Quantum Dots

Martin B. Ward and Andrew J. Shields

Toshiba Research Europe Limited, 208 Cambridge Science Park, Milton Road, Cambridge CB4 0GZ, United Kingdom

Blandine Alloing, Carl Zinoni, Christelle Monat, and Andrea Fiore

Ecole Polytechnique Fédérale de Lausanne (EPFL), Institute of Photonics and Quantum Electronics, Station 3, CH-1015 Lausanne, Swizerland

In this chapter we will discuss the use of InAs quantum dots for telecom wavelength single photon applications. We will describe two molecular beam epitaxial growth techniques to achieve low density distributions of quantum dots that are large enough to emit telecom wavelength photons at low temperatures. Single dot spectroscopy in conjunction with time-resolved measurements provide information on the origin of the main emission lines. Correlation spectroscopy using both InGaAs avalanche photodiodes and superconducting single photon detectors will be discussed and used to characterize the photon emission statistics. Ungated measurements at telecom wavelengths using the superconducting single photon detectors offer ready access to additional details in the second-order correlation function. Progress towards practical electrically driven sources will be presented including oxide aperture and planar cavity LEDs. We will see how the inclusion of electrical contacts enables additional flexibility in the temporal output of a device by tuning the emission wavelength with time-varying Stark shifts.

Contents

16.1. Introduction

Photons are appealing carriers of quantum information due to the long distances they can travel with small disturbances to their quantum states. Optical fibers enable nodes of a network to be linked without a direct line of sight and the possibility of practical distributed quantum information processing and communications. However, for photons to be transferred over relatively long distances their wavelengths should fall within one of the low-loss optical fiber wavelength bands, i.e., around 1.31 μm or 1.55 μm.

Semiconductor quantum dots, with their line emission spectra, are known to generate single photons [1–4] and entangled photon pairs, [5–7] as discussed elsewhere in this volume. Quantum dots have also been extensively developed for use in telecom wavelengths lasers and LEDs. However, for single photon applications the requirement for low dot densities (the opposite requirement from most laser applications) presents a particular growth challenge. We discuss two very different approaches to the growth of dots suitable for single photon work at telecom wavelengths and consider device architectures to isolate and collect photons from the resulting dot distributions. We also consider architectures to enable electrical injection, while satisfying the photon collection requirements. Characterization of the photons emitted must be performed using detectors that are sensitive at telecom wavelengths. We will describe the current state-of-the-art in time-resolved and correlation spectroscopy measured with both InGaAs APDs and superconducting single photon detectors.

The application that is most technologically developed is that of quantum key distribution, in which a digital key is exchanged between two parties on a stream of single photons. [8, 9] Information is encoded on some property such as the polarization, phase or time and the system arranged such that any attempt by an eavesdropper to extract information from the photons in the communication channel can be detected as an increased error rate. When single photons (i.e., Fock states of the electromagnetic field with photon number exactly equal to one [10]) are used it can be proven that a key can be shared about which no information could have been extracted from the photons while in transit. Current implementations are mainly fiber-based and use attenuated lasers to generate photons. Due to the Poissonian nature of laser emission there is always some probability that a pulse will contain more than one photon. In basic systems this significantly limits the lengths and bit rates that can be achieved. More advanced systems using decoy pulses have been demonstrated to reduce the impact on performance [11–13] but the highest possible performance would still derive from using a pure stream of single photons. Tests using heralded photons from down conversion sources [14] have been undertaken, but a convenient and practical semiconductor source of telecom wavelength photons would benefit these potential security applications.

16.2. Quantum Dot Growth

The usability of quantum dots for device applications requires a high crystalline quality, a low areal density (in order to isolate a single dot in a device such as a microcavity or an LED structure) and good size homogeneity for reproducibility of the emission wavelength. The InAs/GaAs system is by far the most studied of all quantum-dot systems since these materials can be easily embedded in a microcavity. A common approach to decrease the dot density is to grow a thin layer of InAs close to the critical thickness of the 2D to 3D growth mode transition. [15] This results in a lower QD density, but also in a decreasing QD size, which leads to a blueshift of the emission wavelength. Most studies using this approach have concentrated on QDs emitting in the $\lambda < 1000\,\mathrm{nm}$ range, due to the difficulty of growing sufficiently sparse layers of dots that are large enough to emit at longer wavelengths and also due to the lower sensitivity of detection systems in the near-infrared. Alternative approaches were needed and in this section we will discuss two molecular beam epitaxy (MBE) techniques that have been used to produce QDs layers that are shown to emit single photons at $1.3\,\mu\mathrm{m}$. Metallorganic vapor-phase epitaxy (MOVPE) has also been used (Ref. 16–20) to achieve low densities of InAs QDs on (In)GaAs and InP.

16.2.1. *Ultra-low growth rate technique*

16.2.1.1. *Growth optimization*

A technique to reduce the overall dot density while maintaining long-wavelength emission has been developed by using a combination of ultra-low InAs growth rates (<0.002 ML/s) and capping with an InGaAs layer. [21]

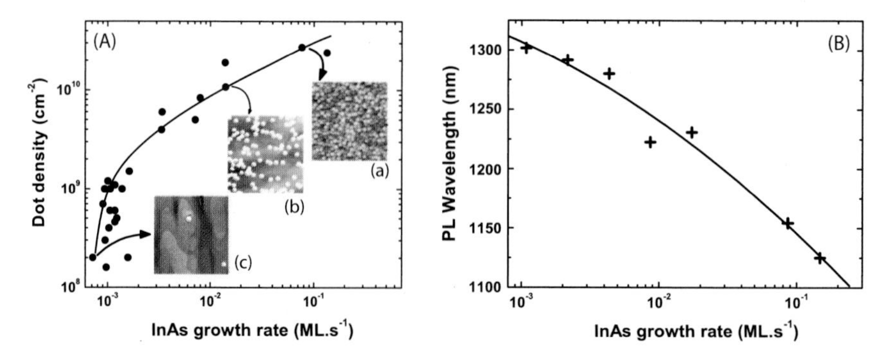

Fig. 16.1. (A) Dot density, as measured from AFM images plotted as a function of InAs deposition rate. The solid line is a guide to the eye. AFM images: 2.1 ML InAs deposed with following growth rates: (a) 0.16 ML/s, (b) 0.08 ML/s and (c) 0.0015 ML/s. (B) Evolution of the PL emission wavelength at room temperature as function of the InAs growth rate. The solid line is a guide to the eye. After Ref. 21.

Figure 16.1 A and 16.1 B present the effect of the InAs growth rate on the QD density and emission wavelength. 2.1 monolayers (MLs) of InAs were deposited after a GaAs buffer growth at different rates in the range 0.16–0.0012 ML/s at fixed growth temperature and As pressure. Two series of samples were grown for photoluminescence (PL) measurements and atomic force microscope (AFM) imaging. By decreasing the InAs growth rate the QD density is shown to decrease strongly while the PL emission is red-shifted (Fig. 16.1 A and B). A dot density as low as 2×10^8 dots/cm^2 is obtained when the InAs growth rate is reduced to 0.001 ML/s. For the same density, the PL emission wavelength reaches 1300 nm at room temperature (see Fig. 16.1 B), i.e., $\sim$1200 nm at low temperature.

To reach the target wavelength of 1300 nm at low temperature it was necessary to extend the room temperature emission wavelength to 1400 nm and this was achieved by capping the quantum dots with a 5 nm-thick $In_{0.15}Ga_{0.85}As$ capping layer after the QD deposition, mainly due to reduced In segregation out of the QD during the capping. Figure 16.2 shows the PL emission spectra of low-density InAs QDs capped with GaAs and InGaAs layers. A strong redshift of the PL wavelength and an enhancement of the PL intensity is observed due to the InGaAs capping layer deposition. Furthermore, the full width at half maximum (FWHM) value of around 25 meV implies very good QD size homogeneity. It should be mentioned that reaching the target emission wavelength at low density in this way requires two In cells in the growth chamber, since QDs and InGaAs capping

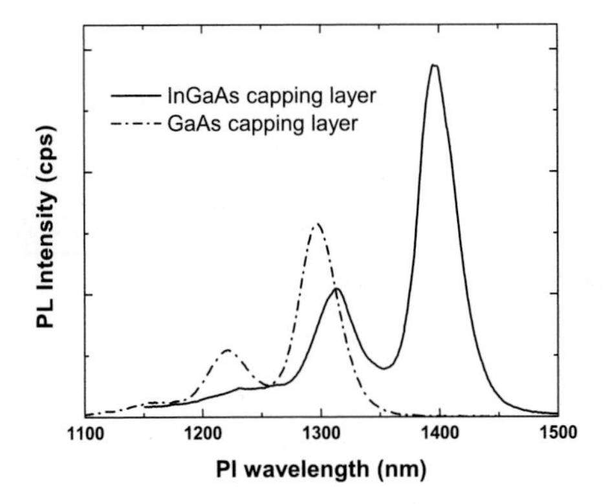

Fig. 16.2. Room temperature PL spectra of InAs QDs grown at 0.0015 ML/s and capped by GaAs or a 5 nm-thick $In_{0.15}Ga_{0.85}As$ layer. After Ref. 21.

layer deposition occur at very different growth rates, necessitating different In cell temperatures.

16.2.1.2. *Structural characterization*

Transmission electron microscope (TEM) images recorded on a sample containing QDs grown at a growth rate of 0.0015 ML/s with an InGaAs capping layer provide information on the structure of the resulting quantum dots. [22] Figure 16.3 a shows cross-sectional (002) dark-field TEM images of the QDs, indicating a lens shape with a 9 nm height. This value is larger than the typical height around 5 nm observed for QDs grown at higher InAs growth rates. [23–25] Using the technique proposed in Ref. 26 the In content in the QDs was estimated from TEM contrast in dark-field (002) images. Figure 16.3 b shows the In composition profile of the QDs capped by InGaAs along the $\langle 001 \rangle$ direction, through the QD center. QDs present a strong In gradient with a maximum In composition of 65% in the center of the dots and 35% at the bottom. The combination of this very high In composition with large QD sizes enables the emission wavelength to reach 1300 nm at low temperature.

Plan-view TEM images (Fig. 16.3 c) reveal the QDs to have a square base and a mean width of 37.5 nm (36 nm) with a standard deviation of 0.3 nm (1.4 nm) along the $\langle 100 \rangle$ ($\langle 010 \rangle$) axes, respectively. This shows a remarkable size homogeneity.

A fraction of the QDs are dislocated, forming large clusters with nonuniform shapes as seen in Fig. 16.3 c. The QD size appears to reach a critical

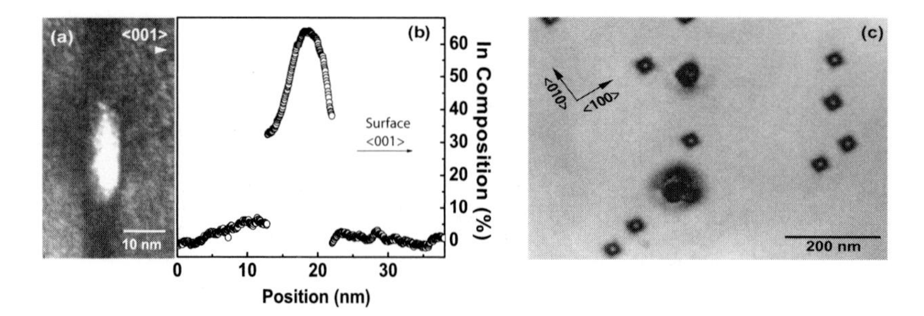

Fig. 16.3. (a) (002) Dark-field images of QDs grown at low InAs growth rate (0.0015 ML/s) and capped by InGaAs. (b) InAs composition variation in the ⟨001⟩ direction of the QDs. (c) plan view image of QDs grown at low growth rate (0.0015 ML/s) and capped by GaAs. After Ref. 22.

value above which plastic relaxation occurs. Then, under these growth conditions, attachment of adatoms to existing islands is favored as opposed to nucleation of new islands, even if it leads to the formation of relaxed clusters. The self size-limiting effect that has been observed in Refs. 27, 28 seems to be overcome here by faster aggregation of adatoms.

16.2.2. *Bimodal dot growth*

Another approach to isolate emission from single telecom wavelength dots is to appreciate that emission from smaller dots can be removed by a spectral filter. This relaxes somewhat the requirement on the overall dot density, making the density of telecom-wavelength dots the important parameter. Emission from the wetting layer and other dot emission lines is commonly removed by spectral filtering in work at shorter wavelengths. The tail of the wetting layer emission sometimes extends to the wavelength of short wavelength dot emission, where it can form a significant contribution to the residual background emission from some devices. At telecom wavelengths the wetting layer emission is spectrally separated further from the operating wavelength. However, attempting to select emission from a dot in the long-wavelength tail of the inhomogeneous linewidth of a standard quantum dot distribution would place stringent requirements on the spectral filter and would only be possible for quite limited total dot densities. Fortunately, one of the subtleties of the self-assembled dot formation process can be used to increase the wavelength separation of the largest quantum dots enabling them to be used for single photon generation. [29] There exists a second critical InAs thickness above which it is favorable for dots of a

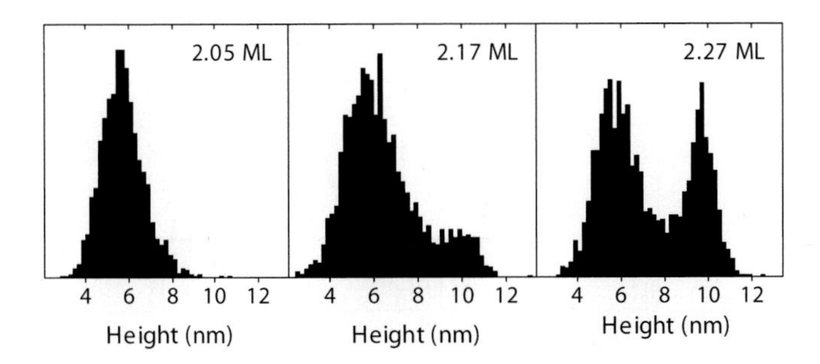

Fig. 16.4. Distribution of quantum dot heights (from AFM), for different InAs coverages on GaAs expressed in monolayers (ML). The dots were grown by molecular beam epitaxy on a GaAs substrate at a rate of ~0.018–0.019 ML/s and at a substrate temperature of ~505–510 °C. After Ref. 29.

new taller shape to form. [30] The analysis of AFM surface profile images from wafers with a range of InAs deposition thicknesses (Fig. 16.4) reveals the onset of a bimodal distribution. For deposition of ~2.05 ML of InAs the quantum dots appear relatively uniform in size with an average height of ~5.7 nm. Deposition of ~2.27 ML leads to the formation of a clear bimodal distribution with many taller dots of a distinct average height ~9.6 nm. Shape transitions in this material system and the importance of the thermodynamic stability of various facets have been discussed in the literature [31, 32]. Kinetic considerations are critical in determining the final dot shape and the details remain sensitive to growth conditions, with significant differences reported between the observations of different groups.

The larger dots which form above the second critical coverage threshold emit light at longer wavelengths than the small dots, as is apparent from the PL spectra of Fig. 16.5 recorded for InAs thicknesses below and above the second critical thickness. The sample with greater InAs coverage shows an additional PL peak close to 1258 nm due to emission from the larger dots. Through careful control of the InAs deposition around the second critical coverage, an arbitrarily low density of these large, long-wavelength quantum dots can be produced. InGaAs capping can be used to extend the emission wavelength (as in Sec. 16.2.1 above). More recently, a modified growth technique has enabled the growth of dot distributions where the number of smaller dots has been reduced substantially, while maintaining the long-wavelength emission. [33]

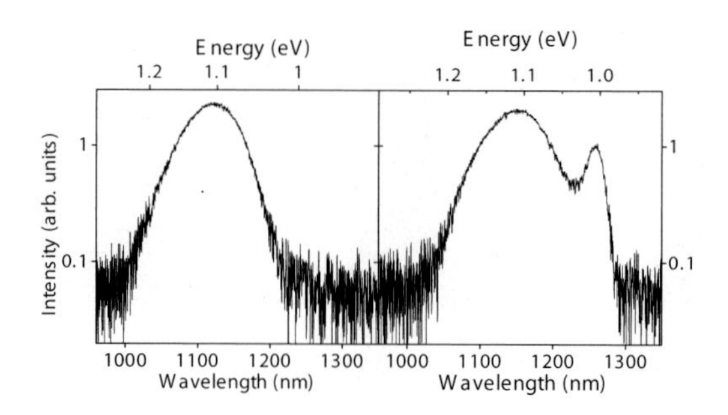

Fig. 16.5. PL spectra (left) below and (right) above the second critical InAs thickness. After Ref. 29.

16.3. Single Quantum Dot Spectroscopy

The performance of standard detectors for telecom wavelengths falls significantly below that achieved by good silicon detectors, both in terms of detection efficiency and noise. However, using liquid-nitrogen-cooled InGaAs array detectors, high-quality spectra can be recorded from single telecom wavelength quantum dots. [21, 34–36] The integration times required can be substantially longer but the design of a sample can influence the extraction efficiency and embedding the quantum dot in a resonant cavity can be highly beneficial. Cavities with relatively low quality factors (Q) can help match the emission pattern to the numerical aperture of practical collection optics. [37] High-Q cavities can make studies of the full line structure of dots more difficult, since the collection efficiency can be reduced outside the linewidth of the cavity.

The samples measurement system must also be able to spatially isolate the emission from ideally a single long-wavelength quantum dot. The spatial resolution required depends on the QD density but in order to achieve a high extraction efficiency into an objective lens, many designs of cavity structure cannot be arbitrarily reduced in scale, since scattering and diffraction losses would become unacceptable. The achievable spatial resolution for a given sample design can therefore place an upper limit on the acceptable telecom-wavelength QD density. In samples without a cavity and in those incorporating weak planar cavities (with no more than a few periods in any upper DBR), wet / dry etched mesas and apertures in metal layers have been used successfully. Reactive ion etched pillar microcavities, with or without oxide apertures, have been used for higher-Q cavities.

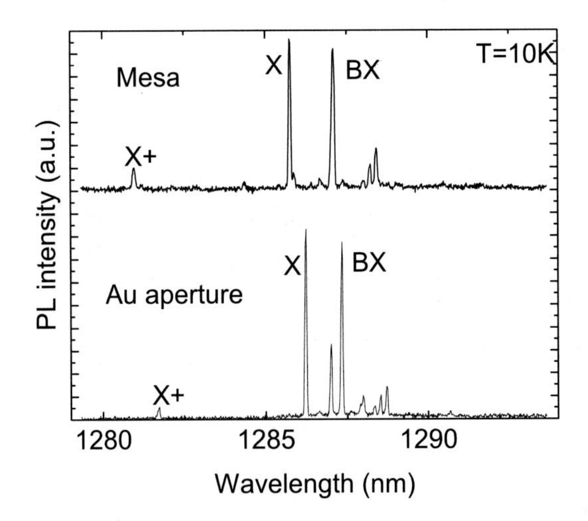

Fig. 16.6. Spectra from two single QDs spatially isolated in different devices demonstrating a repeatable spectral signature. After Ref. 38.

16.3.1. *Exciton/Biexciton identification at 1300 nm*

The QDs studied in this section were grown by MBE using the ultra-low growth rate technique described above in Sec. 16.2.1. The dots were embedded in a planar microcavity, designed to achieve an extraction efficiency of 9% into an external numerical aperture (NA) of 0.5. The intensity of PL from QDs embedded in the micro-cavity showed more than an order of magnitude increase over QDs grown under the same conditions but embedded in bulk GaAs. [21]

Both mesas and apertures can be considered to provide spatial isolation with this wafer structure. In Fig. 16.6 we compare the ground state spectra of single QDs emitting at the same energy but isolated by different techniques. The apertures of $\sim 1\,\mu\text{m}^2$ in gold masks were obtained by lift-off using a chemically amplified resist (UVIII) patterned by 100 kV e-beam lithography, and $\sim 1\,\mu\text{m}^2$ etched mesas realized by reactive-ion-etching using a SiO_2 mask patterned by e-beam lithography. In both devices the same spectral signature is evident. It was observed that mesas provide improved spectral isolation, since, with the metallic aperture, light is collected from dots in the vicinity of the aperture contributing to a higher background. As compared to mesas, metallic apertures can have a lower device yield due to problems related to the lift-off process. Measuring a repeatable spectral signature on devices containing single QDs is a first important step towards the identification of the spectral lines and their attribution to specific charge

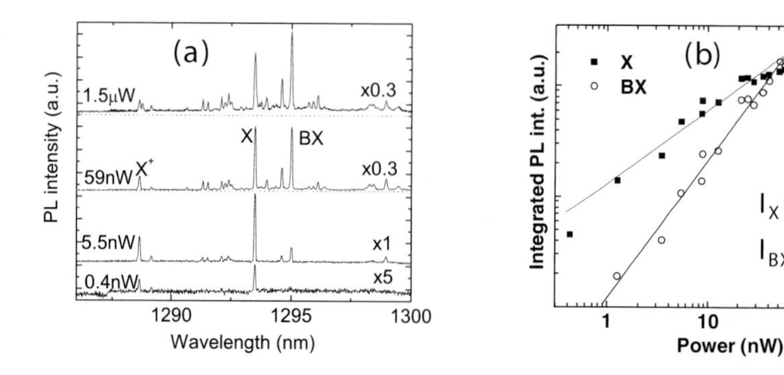

Fig. 16.7. (a) PL of a single QD for different pump powers. (b) Integrated PL intensity for the X and BX lines plotted as a function of pump power.

configurations of the QD.

Non-resonant excitation provides a practical tool for investigating the confined energy structure of the QDs; by increasing the excitation intensity of the laser the sequential filling of the QD states can be controlled. Figure 16.7 a shows spectra from a single QD for increasing power densities by exciting carriers in the GaAs bulk using a pulsed diode laser emitting at 750 nm with a repetition rate of 80 MHz. At the lowest excitation density a single line dominates the spectrum, corresponding to the QD populated by a single exciton (X) in the ground state. The highest energy peak is attributed to the recombination of a positive trion; its large negative binding energy and its appearance at low power densities justifies the assignment. [39, 40] As the power is increased the intensity of the X increases due to the increased probability of finding an exciton in the QD, and more spectral lines begin to appear on the low energy side of the X transition. These lines correspond to the recombination of an electron–hole pair in the ground state in the presence of other charges in the QD and the shift in energy of the new transition lines is due to Coulomb interactions, including exchange and correlation, between the carriers. At 59 nW the spectrum becomes more complex and both the X and BX lines reach a maximum and saturate. Increasing the excitation power by about 2 orders of magnitude no significant change is observed in the spectra except for an increase in the background emission. In Fig. 16.7 b are plotted the integrated PL intensities of the X and BX lines as a function of the excitation intensity. At low excitation power, the PL intensity dependence on the laser power P can be fitted by the relation $I_{X,BX} \propto P^n$, with $n = 0.70 \pm 0.05$ and 1.35 ± 0.05 for the X and BX lines, respectively. (The uncertainty on the fit having been estimated from fitting the data from several QDs.) The fact that the ratio of the exponents is equal to 2 ($n_{BX}/n_X = 2$) suggests that the BX line corresponds to the biexciton emission: the recombination of an exciton in the ground state of the QD in the presence of another electron–hole pair in the ground state. The PL intensity ratio BX/X greater than 1 could be explained by the dark X state acting as a channel for depleting the exciton state, either by non-radiative decay or by allowing additional time for charging of the dot (following BX emission) and subsequent emission at a different energy.

16.3.1.1. *Time resolved PL*

Further support for the identification of the X and BX spectral lines comes from their time evolution since the BX will always recombine before the X and will display a shorter radiative lifetime. The options for time resolved measurements on single dots at telecom wavelengths are more limited than at shorter wavelengths. Efficient Streak cameras are not readily available and upconversion techniques are not sufficiently sensitive. Time-correlated fluorescence spectroscopy is the technique of choice for PL decay measurements at ultralow light levels. In this approach, histograms of photon arrival times on a single-photon detector are recorded on a correlation card as a function of delay from the laser pulse. However, at this wavelength it is not currently possible to get good single photon detection performance from APDs under continuous operation. However, a gated InGaAs APD can be used to record the decay curve of a single telecom-wavelength quantum dot, enabling the rates of depopulation of the excitonic states to be studied. The arrival times of photons on the detector have to be synchronized with the opening of the gate and care taken to maintain afterpulsing at an acceptable level. A fiber-coupled tunable narrow band-pass filter can be used to isolate a single X or BX transition for measurement, as shown in Fig. 16.8. The continuous lines represent least square fits to the decaying part of the PL for the X and BX emission after dark noise subtraction. The least square fits are calculated from the convolution between a two (one) term exponential decay function for the X (BX) and the setup response function. (Setup response was measured with a sample of GaNInAs quantum wells, emitting at 1300 nm at room temperature, with a lifetime previously measured to be 50 ps, [41] which is well below the ~600 ps temporal resolution of the detector. It was measured at a count rate similar to the experimental one since the time resolution achieved with an APD can depend on the rate.)

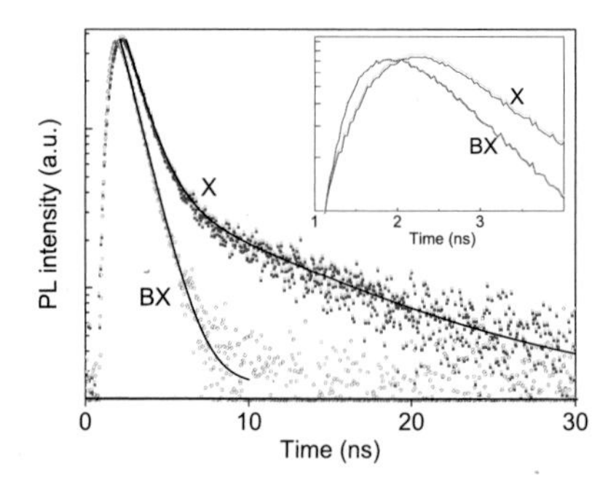

Fig. 16.8. Time dependence of the PL intensity of the single X and BX transitions, the solid line is the least squares fit to the decaying PL. Inset: blowup for short time delays. After Ref. 42.

The error on the fits here are estimated at ± 0.2 ns.

The delay in the start of the decay of the X emission compared to the BX, is evidence of cascade emission and validates the peak assignment (inset Fig. 16.8). The exciton lifetime is characterized by a double exponential decay of 1.1 ns and 8.6 ns: the fast component derives from the recombination of a bright exciton while the slower decaying part suggests the presence of a dark exciton state that is repopulating the allowed exciton transition. [43, 44] The BX decay time was measured to be 1.0 ns; the ratio of 1.1 between the exciton and biexciton lifetime is consistent with previous studies on single QDs. [45] When the QD size is comparable to the Bohr radius, the Coulomb effects are not a small perturbation to the dominating quantization of the kinetic energy. For holes, the changes in the wavefunction spatial distribution are more significant, as compared to electrons, due to their higher effective mass. As a consequence, a larger QD size results in an increased spatial separation of the hole wavefunction, a reduction of the overlap integral and an increased BX radiative lifetime. From cross-sectional TEM images (Fig. 16.3) these QDs have base dimensions of 37.5 nm which is comparable to the exciton Bohr radius in bulk InAs (34 nm). [46]

16.3.2. *Measuring $g^{(2)}(\tau)$ at telecom wavelengths*

Correlation measurements are important in assessing the quality of the emission from quantum dots. The second-order intensity correlation function $g^{(2)}(\tau)$ is measured putting two single-photon detectors in the two arms of a Hanbury Brown and Twiss interferometer, and counting the coincidences between detections as a function of an electrical delay between the two detectors. For classical light, $g^{(2)}(\tau) \geq 1$, while for an ideal single-photon state $g^{(2)}(0) = 0$. [10] This type of measurement is now widely used at wavelengths accessible with silicon APDs for making detailed measurements on quantum dots (as reported extensively elsewhere in this volume) as well as finding wider applications in areas such as the study of biological systems. The coincidence rate of a correlation measurement relates to square of the singles count rate and, for low $g^{(2)}(0)$, the zero-delay background relates to twice the dark count fraction. Detectors at telecom wavelengths exhibit both lower efficiencies and larger dark count rate probabilities and correlation measurements remain challenging on anything but the best samples. The majority of correlation measurements to date have used InGaAs APDs operating in Geiger mode. In order to achieve the best signal-to-noise ratios, most measurements have used narrow gates widths of the order of 3 ns with no attempt to time resolve within the gates. This results in correlation measurements where the time delays are represented by integer numbers of excitation periods, giving rise to discrete histograms, rather than full representations of $g^{(2)}(\tau)$. While the measurement of $g^{(2)}(0)$ is important for the characterization of the light statistics for single photon emitters, many important physical properties of the system are hidden in the correlation function $g^{(2)}(\tau)$ for any delay. [3, 47, 48] Superconducting single photon detectors (SSPDs) are capable of continuous operation and have recently been used to perform time-resolved correlation measurements at these wavelengths. [49]

Measurements in which dark counts are not very significant are now being achieved with both types of detectors but in some early measurements dark counts and counts arising from background emission formed the majority of detection events. [18] Care is needed in comparing experimental results as data is sometimes presented after large corrections for dark counts, and sometimes also for background emission estimated from the spectrum. While correcting for dark counts can be valid, care is needed to ensure that the rates are stable and that they can be estimated sufficiently accurately. Stray counts are normally assumed to be Poissonian in these corrections. The correlation data in this chapter are plotted as measured with the influence of dark counts and their correction discussed in the associated comments.

16.3.2.1. *Correlation measurements with InGaAs APDs*

Most telecom wavelength single photon applications are in single-mode fiber and many commercially available InGaAs APDs are supplied coupled to single-mode fiber. It is often sensible to use a fiber-optic version of the Hanbury Brown and Twiss measurement system [50], since this offers good stability and low losses, once initially coupled to fiber. A wavelength division multiplexer (WDM) can be used to conveniently couple an excitation laser into a fiber without introducing large losses to the single photons being collected via the same fiber and a fiber-coupled tunable spectral filter can select emission from a particular dot line.

Fig. 16.9 a shows the results of a correlation measurement on a quantum dot taken from a bimodal distribution of dots, embedded in a pillar microcavity to enhance the photon collection efficiency. [29] The pillar had a nominal diameter of 2 μm with 25-period (11-period) lower (upper) DBR mirrors giving typical Q-values ~ 600. The excitation was at 2 MHz from a ps-pulsed laser diode at 785 nm and the selected emission line has an exciton-like power dependence. The height of the zero-delay bin is clearly below that of the average, indicating a suppression of multiphoton emission. By measuring out to long time delays it is possible to estimate the average height of the finite delay peaks very precisely (e.g., ± 250 periods used here) giving an average height of 365 with a standard deviation of 20 (close to $\sqrt{365}$ as expected). There were just 67 counts in the zero-delay bin giving a measured suppression of multiphoton pulses to 18.3% of the value expected for a Poissonian source of the same intensity. Dark counts in the detectors accounted for just $\sim 1.4\%$ and 2.8% of the signals on the APDs and light leaking through the spectral filter in this early measurement contributed $\sim 5\%$ on a single APD. Accounting for these two factors the residual two-photon rate of the source was expected to be below 10%. A post-selection technique [51] was employed in these correlation measurements in which any avalanches in an APD occurring within 150 μs of another count in the same APD were ignored. This reduces the impact of afterpulsing and we note that the deadtime built into many packaged silicon APDs means that a photon closely following a previous count in the same APD would not be recorded in the majority of experiments at shorter wavelengths either. An antibunching effect extending to one period (500 ns) away from zero-delay with a height of $\sim 85\%$ of the other finite-delay peaks can be seen. Similar effects have been observed ~ 900 nm with low excitation powers of light above the GaAs bandgap. [52] Fig. 16.20 b shows similar results at a

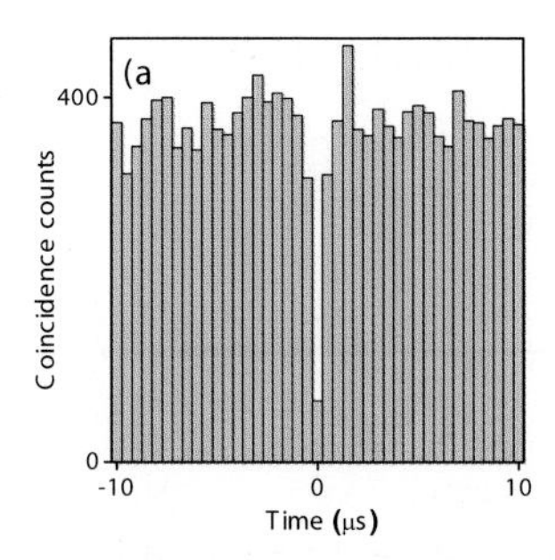

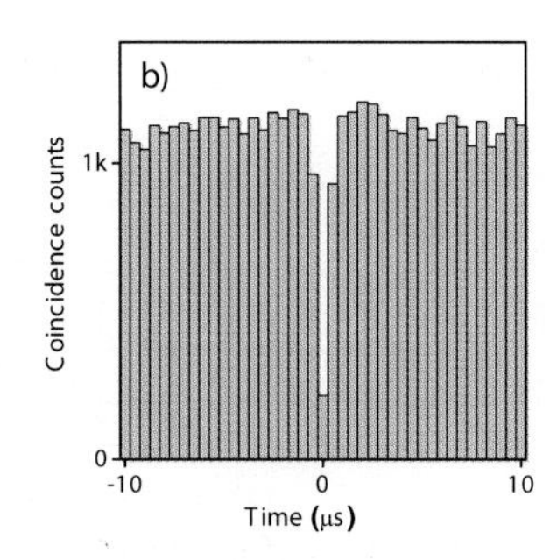

Fig. 16.9. Results of correlation measurements using InGaAs APDs on an X emission line at (a) 5.2 K and (b) the same line at 32 K. For details see Ref. 29.

temperature of 32 K, where the dot line was close to being on resonance with the cavity mode. Clear optically-excited single photon emission has also been demonstrated using dots grown using the ultra-low growth rate technique. [42]

16.3.2.2. *Superconducting single photon detectors*

The recent development of single-photon detectors based on NbN superconducting nanostructures (SSPDs), [53] promises orders-of-magnitude improvement over InGaAs APDs in sensitivity, dark count, jitter and repetition frequency. In particular, fiber-coupled SSPDs with quantum efficiency $> 5\%$, dark count rate around 10 Hz, detection time jitter < 100 ps and maximum counting rate > 50 MHz have been recently demonstrated. [49] Figure 16.10 a shows results from the testing of a fiber-coupled Hanbury Brown and Twiss setup based on a pair of SSPDs illuminated by 750 nm laser pulses. [49] The bias current through the SSPDs was adjusted to obtain a dark count of just 10–30 Hz. As expected, the histogram is characterized by a series of peaks whose integrated areas are all the same within the statistical fluctuation of the measurement; this is the signature of the intensity correlation function for a coherent pulsed source. The FWHM = 440 ps of the peaks corresponds to twice the jitter of the system.

Using the same setup we measured $g^{(2)}(0)$ for a single X line from a QD. A preliminary measurement is shown in Fig. 16.10 b made at a power of 50 nW at 750 nm and repetition frequency of 80 MHz. We remark that this repetition frequency is unachievable with most existing APD setups and that high speed operation of APDs (> 100 MHz) is still under optimization. [54, 55] The coincidence histogram is characterized by periodic peaks

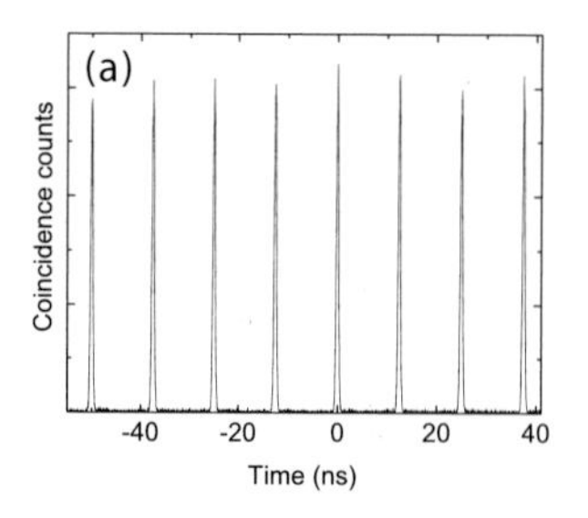

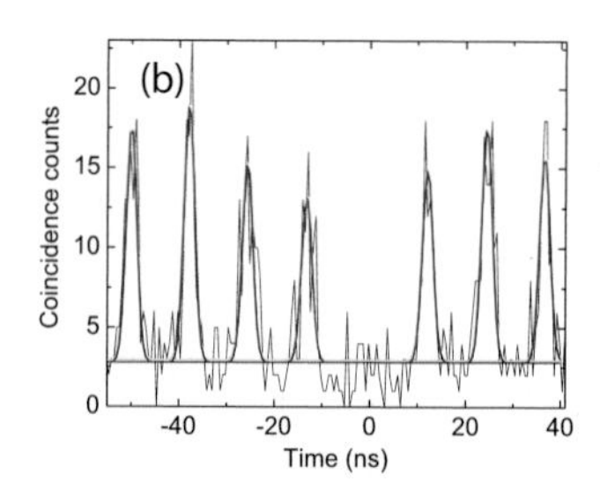

Fig. 16.10. (a) Measured intensity correlation function for a pulsed coherent source. (b) Antibunching on a single exciton line from a QD similar to Fig. 16.6. After Ref. 49.

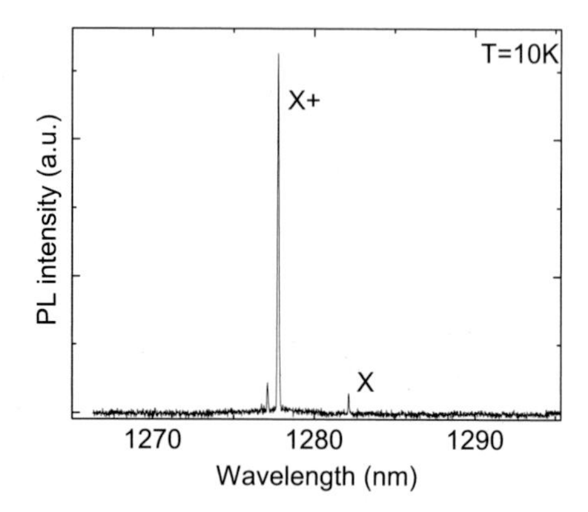

Fig. 16.11. PL under quasi-resonant from a single QD showing the positively charged exciton line.

separated by the laser repetition period except for zero delay; this is the signature of a single photon emitter under pulsed excitation. The peaks are well fitted by Gaussian time distributions with an offset of 3 coincidences and a FWHM = 2.2 ns that corresponds to twice the X lifetime. [42] The coincidences between the peaks are due to uncorrelated light entering the system and detector dark counts.

A more interesting measurement which reveals the dynamics of the charge population of the QD is the measurement of $g^{(2)}(\tau)$. The measurement was made on the positively charged exciton emission (Fig. 16.11) under CW excitation by pumping resonantly in the excited state of the trion. The resulting histogram, measured for a pump power of 0.2 mW, is shown in Fig. 16.12 a for long delays (0.5 μs). For time delays between 3 ns and 200 ns an increase of the correlation function is observed: this

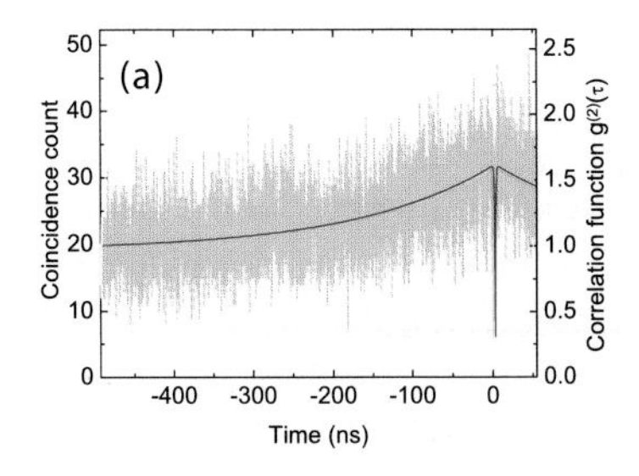

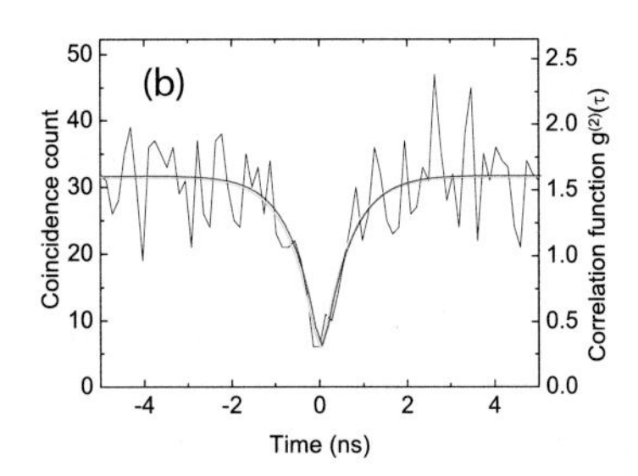

Fig. 16.12. Intensity correlation function measured on the charged exciton line (Fig. 16.11) at (a) long time delays and (b) short time delays. After Ref. 49.

bunching behavior, already studied for short-wavelength QDs, [52] shows that after emission of a photon from the positive trion, the QD remains charged allowing re-excitation of the charged exciton state. For short time delays $< 3\,\text{ns}$, (Fig. 16.12 b), an antibunching dip is observed, confirming the sub-Poissonian statistics of the light emitted by the trion line.

The bunching and antibunching behavior can be modeled in a three level system with the following expression [56]: $g^{(2)}(\tau) = 1 - (1 + a) \times \exp(-\tau/\tau_1) + a \times \exp(-\tau/\tau_2)$. To account for the limited setup resolution, detection of uncorrelated photons and dark counts, the experimental data was fitted by convolving a gaussian time distribution (FWHM = 220 ps) with the correlation function corrected for noise [3]: $g_n^{(2)}(\tau) = 1 + \rho^2(g^{(2)}(\tau) - 1)$. The fit provides the values, a = 0.8, $\tau_1 = 0.62\,\text{ns}$, $\tau_2 = 170.8\,\text{ns}$ and $g^{(2)}(0) = 0.18 \pm 0.02$.

16.4. Single Quantum Dot Light Emitting Diodes

If possible, a practical single photon source should operate under electrical excitation to minimize production costs. The main challenge is to inject current into a small number of QDs (ideally one), while maximizing the fraction of excitation pulses that result in a photon being successfully coupled to the output single-mode fiber. These two requirements are, to a degree, related since collecting light from structures with dimensions smaller than the wavelength of the emission — as often needed to isolate emission from a single QD — can lead to significant diffraction losses. Typical surface densities of ~ 10–$1000\,\mu\text{m}^{-2}$ (ultra-low growth rate technique) imply a p–n junction-based design at the nanometer scale. Using a submicron aperture to post select emission from a single dot would seriously degrade the collection efficiency and etching a mesa of this diameter could reduce the radiative efficiency, because of defects produced at the mesa sidewalls, as well as diffraction losses. The use of an oxide aperture to limit the current to such a small region is considered in the following section. The ability to intentionally seed the growth of a single dot at an intended location [57–59] within a cavity or growing arbitrarily low densities of long-wavelength dots with the bimodal technique could relax this size requirement.

The first electrical devices [4] did not contain a cavity and therefore had efficiencies limited to $\sim 0.5\%$. Individual telecom-wavelength emission lines were recorded from similar structures around $1.3\,\mu$m, [35] but the incorporation of a cavity is necessary for a practical source. Roughly an order of magnitude improvement has been demonstrated using an apertured planar cavity [60] and we will discuss a telecom wavelength device in Sec. 16.4.2. Several other methods to limit the number of excited QDs have been considered. [61–65]

16.4.1. *Current aperture LEDs*

16.4.1.1. *Device design*

A promising approach to producing high efficiency single photon LEDs is to incorporate a current blocking layer of an electrical insulator that is patterned to define a selective current path towards a small number of QDs. The extent of the rest of the cavity can then be larger since only the dots in the region through which the current flows will be excited. This approach (referred to as "current aperture") was first developed in the context of VCSELs [66] and proposed later for single photon sources. [67] Devices operating at wavelengths around 900 nm (with and without cavities) have been reported in the literature. [68–71]

In this section we will describe telecom wavelength bottom-emitting current aperture LEDs. A schematic cross-section of the device is presented in Fig. 16.13 and the fabrication process is described thoroughly in Refs. 67, 72. The heterostructure includes a sparse array of self-organized InAs/GaAs QDs in the middle of a GaAs *p-i-n* diode. The low density QD layer is surrounded by a bottom *n*-doped distributed $Al_{0.9}Ga_{0.1}As$/GaAs Bragg reflector (3.5 pairs) and a top hole injector composed of a 32 nm undoped AlGaAs layer. Above this hole injector, an $Al_{0.85}Ga_{0.15}As$ layer and a heavily *p*-doped GaAs cap complete the structure. After the creation of a shallow mesa by standard optical lithography and dry etching, the sidewalls of the $Al_{0.85}Ga_{0.15}As$ layer located on top of the QDs are exposed. The crucial step of the process is the oxidation of this $Al_{0.85}Ga_{0.15}As$ layer. The oxidation starts at the exposed surface and then penetrates laterally under the GaAs cap layer, leaving an unoxidized aperture region whose diameter scales with the initial mesa size and can be controlled by the oxidation time with 100 nm accuracy. The final evaporation of the broad metal *p*-contact does not require any critical alignment with the mesa since the Al_2O_3 layer forces the current to flow through the unoxidized aperture of the mesa. Figure 16.13 displays a top-view scanning electron micrograph of a 600 nm wide current aperture LED created through this process. The weak microcavity defined by the top Au mirror and the bottom Bragg mirror is designed to enhance the light extraction from the device bottom (the extraction efficiency is estimated theoretically to be $\sim$4.5% around 1300 nm into a numerical aperture NA = 0.5).

16.4.1.2. *Current scaling and EL measurements*

There are several challenges in the realization of current aperture LEDs with high external quantum efficiency. The analysis of current–voltage (I–V) characteristics points out the critical role of current spreading and carrier diffusion. These distinct mechanisms act as parasitic and undesirable current channels that tend to enlarge the actual width of the current injection area beyond the current aperture diameter. On the one hand, current spreading, which is a current flow parallel to and around the active region, can be suppressed by reducing the doping level of the layers between the oxide and the active layer [73]. On the other hand, as carrier diffusion in the active area has been shown to strongly depend on the dimensionality of the active region, or in other words on the number of remaining degrees of freedom offered to the carriers, it should be entirely suppressed in an ideal QD. Such a trend has been observed in similar LEDs with high density QDs. [73] However, when combined with a low density QD active region, the current aperture LED appears to undergo noticeable carrier diffusion. This effect must occur in the wetting layer prior to the carrier capture in the QDs and is enhanced by the slow capture time involved in low density QDs. Figure 16.14 a shows the exponential diode characteristic I–V curves obtained at room temperature for a series of current aperture LEDs with various diameters (between 0.6 and 10 μm) and low density QDs. A diffusion length of 1.2 μm is extracted from the scaling of the current density J versus V (Fig. 16.14 b). However, the possibility to fit the J–V curves over almost 4 orders of magnitude of current with a single parameter precludes the existence of current spreading. The external quantum efficiency for the

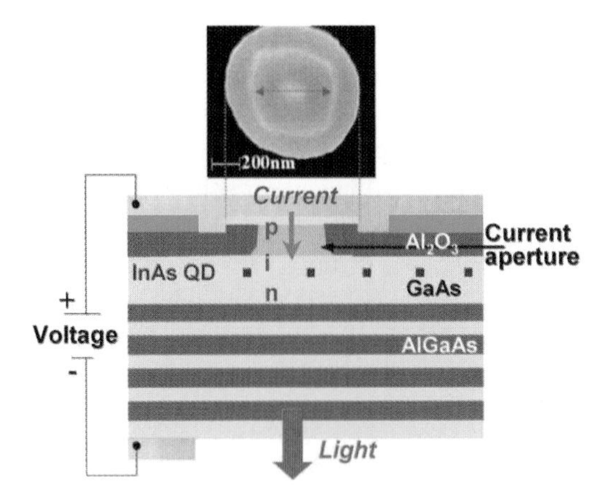

Fig. 16.13. Schematic cross-section of the current aperture LED device. The scanning electron micrograph shows a top view of a LED with a 600 nm-wide current aperture after the oxidation processing step. After Ref. 72.

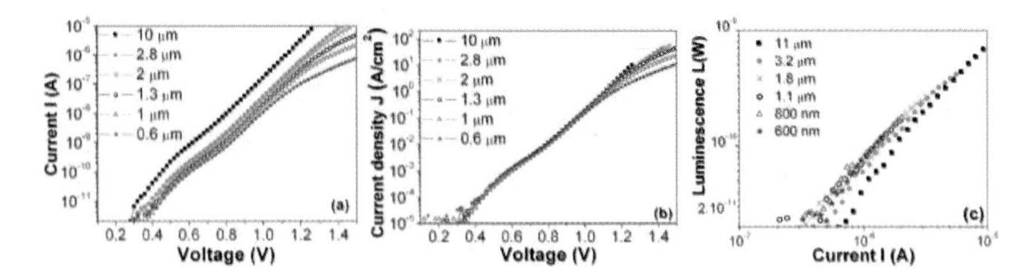

Fig. 16.14. (a) Current–voltage (I–V) characteristics of current aperture LEDs with different diameters as indicated in the legend. (b) Corresponding Current density–Voltage (J–V) obtained from the current aperture diameter increased by a diffusion length fit parameter of 1.2 μm. (c) Luminescence vs current (L–I) characteristics at room temperature for LED devices with current–aperture diameters ranging from 600 nm to 11 μm. After Ref. 72.

same series of devices has been obtained at room temperature through the measurement of light–current (L–I) curves (Fig. 16.14 c). [72] Although the measured efficiency is limited to $\sim 10^{-4}$ (mainly due to the mismatch between the cavity resonance and the QD emission at room temperature), it appears to be roughly independent on the current aperture diameter. The fact that it is not degraded for the smallest devices demonstrates that the oxide does not introduce any defects in the active region.

Proof that only a few QDs are electrically pumped in the device is given by electroluminescence (EL) measurements at low temperature ($\sim$40 K). Figure 16.15 displays EL spectra that have been measured on a 600 nm-

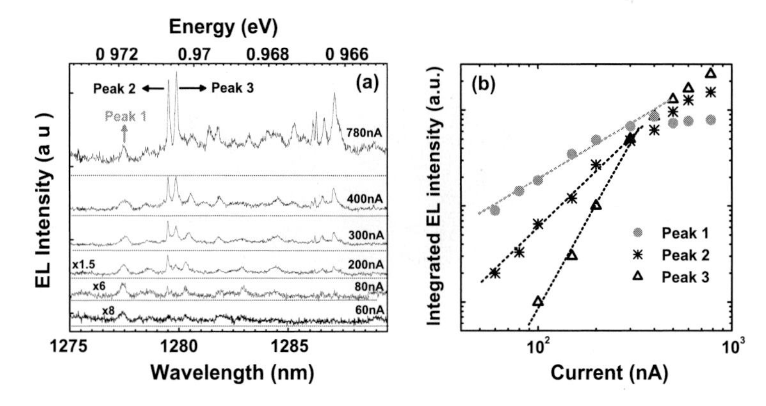

Fig. 16.15. (a) Low temperature electroluminescence (EL) spectra for a 600 nm wide current aperture LED and increasing current injection. Different offsets, which are indicated by a dotted line, have been added to the curves for clarity. (b) Integrated intensity of three different lines as a function of the injected current. After Ref. 72.

diameter current aperture LED for different currents. At low current, a single line (at 1277.3 nm) dominates the spectrum in a defined spectral region. An increase of the current results in a growing number of narrow lines, accompanied by the complex dynamics of their relative intensities. This is the signature of the gradual population of the QD states by the increasing number of injected carriers. In Fig. 16.15 b, the integrated intensities of three peaks are plotted versus the injected current. The intensity I_1 of peak 1 at 1277.3 nm exhibits a roughly linear dependence ($I_1 \propto I^{1.2}$) on the current until it saturates around 300 nA. Such behaviour enables us to assign this peak to the exciton state. The low amount of current required to saturate the QD exciton state attests the high injection efficiency. The intensity I_2 of peak 2 at 1279.9 nm shows a nearly quadratic dependence on I ($I_2 \propto I^{1.95}$), suggesting emission from a biexciton state. The other peaks (like the peak 3) that appear at higher current may be attributed to multiexcitonic states (possibly charged) on the basis of their nonlinear dependence on the current.

These results show the capability of the current aperture approach to selectively inject current into a restricted area that is correlated to the unoxidized aperture diameter and without compromising the radiative efficiency of the device. Although the carrier diffusion length along the active region remains significant at room temperature relative to the small device diameter, the narrow lines measured from the QD EL at low temperature demonstrate the effectiveness of the approach to inject current into one or

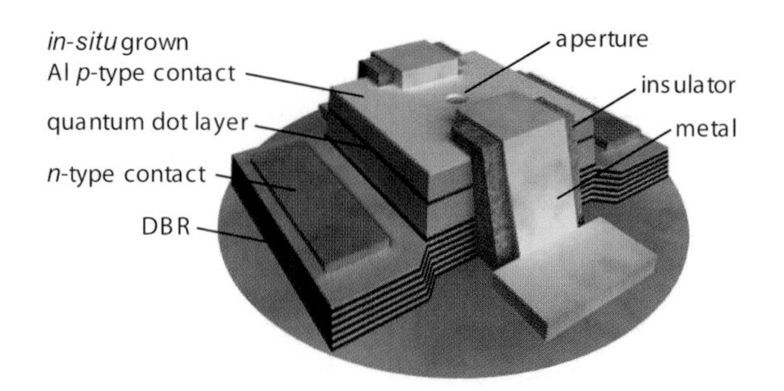

Fig. 16.16. Schematic diagram of a telecom wavelength planar cavity LED. For clarity the diagram is not to scale but typical lateral dimensions of the mesa would be $\sim 60\,\mu\text{m} \times 40\,\mu\text{m}$. After Ref. 33.

few QDs at telecom wavelengths. Eventually, the extraction efficiency of the LED could be further improved by combining the device with a wavelength scale high-Q microcavity, using the oxide current aperture to also provide lateral optical confinement. [74]

16.4.2. *Planar cavity LEDs*

16.4.2.1. *Device design*

Planar cavity LEDs can be fabricated reliably using standard photolithography and have provided some of the highest performance electrically-driven single photon emission to date. A telecom wavelength device design is shown schematically in Fig. 16.16 and described in detail in Ref. 33. The device is grown by MBE and consists of a 3-wavelength GaAs cavity containing an InAs dot layer at a cavity antinode. A cavity is formed between a distributed Bragg reflector (DBR) and the surface GaAs–air interface. A bimodal dot growth technique provided a suitably low density of telecom-wavelength dots to enable the emission from one dot to be isolated by an aperture of diameter one to a few μm. A thin Be-doped layer topped by an Al cap deposited *in-situ* forms a non-annealed p-type ohmic contact. Electrical connections run to the Al layer on the top of an etched mesa and to Si-doped layers in the DBR via diffused AuGeNi n-type ohmic contacts.

16.4.2.2. *Electroluminescence spectroscopy*

Fig. 16.17 a shows the EL spectrum from a dot in a planar cavity LED. The intensity of the emission line 1 increases roughly linearly with current as shown. This line is likely to be due to emission from a neutral or charged

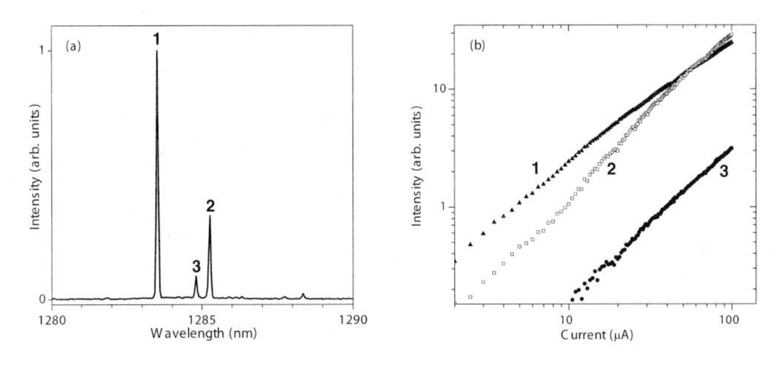

Fig. 16.17. (a) EL spectrum under a constant current of $\sim 10\,\mu$A at ~ 10 K. (b) Current dependence (d. c.) of the intensities of the lines in a. After Ref. 33.

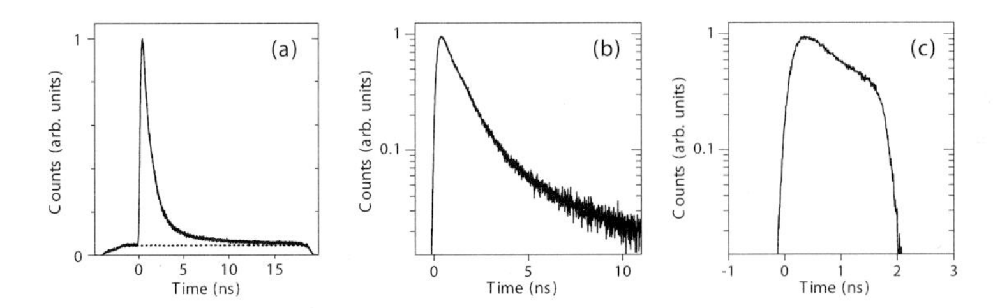

Fig. 16.18. Time-resolved EL data for line 1 in Fig. 16.17 (a) before correction with detector response shown as dotted line and (b) on a logarithmic scale after correction for dark counts. Approx 25 ns APD gate pulses were used. (c) Truncation of the output on the addition of negative-going voltage pulses. After Ref. 33.

single exciton. Lines 2 and 3 increase more rapidly and are likely to arise from more complex (charged) excitonic states.

The time resolved EL from line 1 is shown in Fig. 16.18 a. The response of the InGaAs APD to the ~ 25 ns gate is shown by the dotted line and is seen to be very flat during the gate period. In Fig. 16.18 b the data is plotted on a logarithmic scale after correction for the dark counts. A fit to the steep part of the decay (0.5–2.5 ns) yields a decay time ~ 1.3 ns. However, the real radiative lifetime of the emission line will be closer to 1 ns after the system response (the full-width at half-maximum (FWHM) of the peak measured from a 1300 nm ps-pulsed distributed feedback laser diode was ~ 0.26 ns) and the finite excitation pulse length are considered. A slower component is also visible in the data with an apparent lifetime ~ 10 ns. This is similar to the long-lived component discussed in Sec. 16.3.1.1 and is tentatively attributed to the presence of a dark exciton that is repopulating the allowed exciton [42].

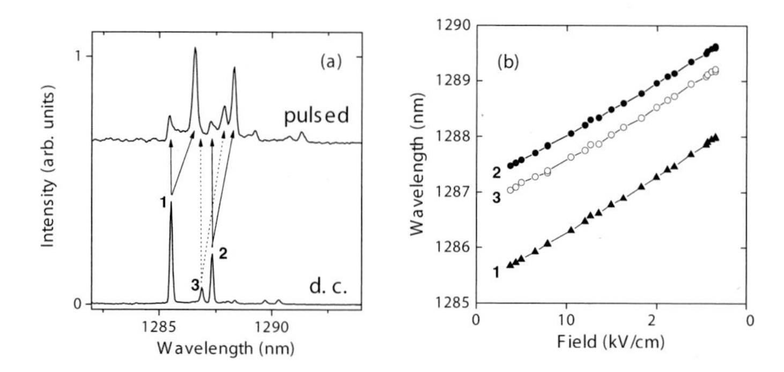

Fig. 16.19. (a) Time averaged EL spectra for d. c. and pulsed excitation. (b) Wavelength shifts of the dominant emission peaks from a. The internal electric fields are roughly estimated here by subtracting the applied voltage from the GaAs bandgap energy and dividing by the length of the intrinsic region. After Ref. 33.

Electrical contacts allow considerable additional flexibility in the operation of such devices. This is illustrated in Fig. 16.18 c where negative-going pulses (0.5 V of duration 10 ns) are applied to the device shortly after each positive-going excitation pulse enabling the tail of the output to be truncated after $\sim$2 ns here. This could be beneficial for applications where it is important to control the photon jitter to within typical detector gate widths. Application of secondary pulses in this way need not reduce the number of photons emitted within the detector gate period.

The primary mechanism responsible for the output truncation is believed to be Stark shifting of the emission wavelength outside the passband of the spectral filter used to select photons from a single emission line. Fig. 16.19 a shows the influence of such time-varying Stark shifts on the time-averaged EL spectrum. When a constant current is driven, the emission is dominated by three lines. Under pulsed excitation each of these lines splits into two components as indicated by the arrows on the figure. The weak component that remains at a roughly constant wavelength is due to emission during the excitation pulse itself. The dominant component shifts to longer wavelengths when the device is maintained at lower bias voltages between excitation pulses. This component arises from emission occurring during the periods between the excitation pulses when the dot experiences higher internal electric fields. [75] The wavelength shift of the dominant component of each line as a function of the estimated internal electric field between excitation pulses is plotted in Fig. 16.19 b.

Strong carrier confinement in telecom wavelength quantum dots allows the emission to be Stark shifted out of the passband of the spectral filter (FWHM $\sim 0.5\,\text{nm}$) without strong carrier loss. At shorter wavelengths the removal of carriers from the dots by the internal electric field can be more significant. [76] In future, modified multi-pulse biasing conditions could create time varying Stark shifts to eliminate multi-photon output due to re-excitation during finite-length electrical pulses. If the excitation pulse is sufficiently above the voltage maintained during the emission period, the associated Stark shift could be large enough for emission during the excitation pulse to be removed with a simple spectral filter. This would relax the requirement on the length of the excitation pulse that it is much less than the radiative lifetime. This can otherwise be difficult to satisfy fully, particularly where a cavity is used to enhance the spontaneous emission rate. If the photon energy was chosen to be resonant with a high-Q cavity during the desired emission period and off-resonance during the excitation pulse, this technique may be possible with little reduction in efficiency, due to the associated variations in the spontaneous emission rate on- and off-resonance.

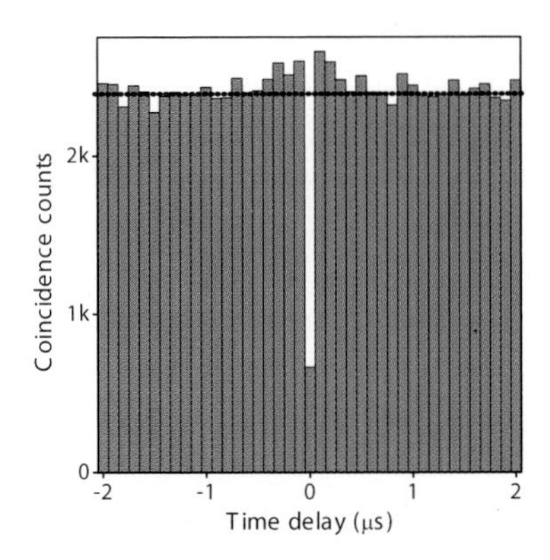

Fig. 16.20. Histogram of delays between avalanches measured in each of the two APDs in a correlation measurement. Voltage pulses were applied to excite the device. A small bunching effect [52] is visible in the first few finite-delay bins. After Ref. 33.

16.4.2.3. *Demonstration of single photon emission*

Figure 16.20 shows a histogram representing the distribution of delays between detection events in each of the two APDs of a fiber optic Hanbury Brown and Twiss measurement (on line 1 from Fig. 16.17). The source and detectors were both operated at 10 MHz and the zero-delay bin is suppressed strongly below the finite-delay bins giving $g^{(2)}(0) = 0.28 \pm 0.01$. Better performance can be inferred for the source itself since dark counts in the detectors contribute to the residual coincidence counts at zero-delay, suggesting $g^{(2)}(0) \sim 0.19$ for the source itself. [33].

16.5. Conclusion

With their high radiative efficiency, short radiative lifetime and good thermal stability, quantum dot (QD) nanostructures offer several advantages among other light sources for single photon emission. Moreover, they can be easily inserted into microcavities and electrical devices to enhance their emission characteristics and practicality. Dispersion is a notorious problem for fiber systems using broadband light sources but the naturally narrow linewidth of quantum dot emission limits the effect of chromatic and polarization mode dispersion over typical photon transmission distances. The larger dots used for telecom wavelength emission offer substantially stronger confinement of carriers than the small dots more widely used at shorter wavelengths, easing the path to higher operating temperatures.

Using quantum dots for single photon applications at telecom wavelengths presents new challenges for growers and spectroscopists alike. We have seen demonstrations of optically-excited and electrically-driven single photon emission performed using QDs grown by two very different techniques to overcome the problem of growing large dots at low densities: ultra-low growth rates and bimodal dot growth. Detailed single dot spectroscopy is possible using cooled InGaAs array detectors and can be combined with information from time-resolved measurements to help understand the origins of the simplest emission lines. However, the subtleties of the dot formation processes and richness of the QD emission line structures are only beginning to be understood in detail.

Due to their high operating temperature, InGaAs APDs are currently the single photon detector used in the majority of telecom wavelength applications and they have been used for most time resolved and correlation measurements to date. Superconducting single photon detectors offer continuous operation with improved signal to noise ratios and look very promising for future investigations.

Oxide apertures provide a convenient way to limit the excitation current in an electrical device to a small area without requiring etched structures of such limited size. Planar cavity LEDs provide a simple way to increase the collection efficiency of an LED by at least an order of magnitude. Electrical contacts provide new flexibility to tune the emission wavelength on time scales shorter than the natural radiative lifetime of the QDs via time-varying Stark shifts on the application of voltage pulses.

Telecom wavelength single photon emitters have advanced very substantially in the last three years and further improvements to the performance of both emitters and detectors will benefit network applications of quantum information processing and computing.

Acknowledgments

MBW and AJS gratefully acknowledge P. See, P. Atkinson (UC), T. Farrow, Z. L. Yuan, O. Z. Karimov, A. J. Bennett, D. G. Gevaux, D. C. Unitt, K. Cooper (UC) and D. A. Ritchie (UC) for their contributions to the work presented here (where UC is the Cavendish Laboratory of the University of Cambridge) and the UK Department of Trade and Industry (LINK 'Optical Systems for the Digital Age' Q-LED), the UK Engineering and Physical Sciences Research Council and the European Commission (SECOQC, QAP and SANDiE) for their support of this work.

BA, CZ, CM and AF gratefully acknowledge A. Gerardino (CNR-IFN), G. Goltsman (MSPU), L. Lunghi (CNR-IFN), G. Patriarche (CNRS-LPN) and V. Zwiller (EPFL) for their contribution to the work presented here (where CNR-IFN is the Institute of Photonics and Nanotechnology of the Italian National Research Council, MSPU is the Moscow State Pedagogical University, and CNRS-LPN is the Laboratoire de Photonique et de Nanostructures of the French CNRS). Work at EPFL was funded by the Swiss National Science Foundation (Professeur boursier and NCCR Quantum Photonics programs), and the European Commission (projects QAP, SINPHONIA and ePIXnet).

References

[1] P. Michler, A. Kiraz, C. Becher, W. V. Schoenfeld, P. M. Petroff, L. Zhang, E. Hu, and A. Imamoğlu, A quantum dot single-photon turnstile device, *Science.* **290**(5500), 2282–2284, (2000).

[2] V. Zwiller, H. Blom, P. Jonsson, N. Panev, S. Jeppesen, T. Tsegaye, E. Goobar, M.-E. Pistol, L. Samuelson, and G. Björk, Single quantum dots emit single photons at a time: Antibunching experiments, *Appl. Phys. Lett.* **78** (17), 2476–2478, (2001).

[3] C. Becher, A. Kiraz, P. Michler, A. Imamoğlu, W. V. Schoenfeld, P. M.

Petroff, Lidong Zhang, and E. Hu, Nonclassical radiation from a single self-assembled InAs quantum dot, *Phys. Rev. B.* 63(12):121312, (2001).

[4] Z. L. Yuan, B. E. Kardynal, R. M. Stevenson, A. J. Shields, C. J. Lobo, K. Cooper, N. S. Beattie, D. A. Ritchie, and M. Pepper, Electrically driven single-photon source, *Science.* **295**(5552), 102–105, (2002).

[5] R. M. Stevenson, R. J. Young, P. Atkinson, K. Cooper, D. A. Ritchie, and A. J. Shields, A semiconductor source of triggered entangled photon pairs, *Nature.* **439**(7073), 179–182, (2006).

[6] R. J. Young, R. M. Stevenson, P. Atkinson, K. Cooper, D. A. Ritchie, and A. J. Shields, Improved fidelity of triggered entangled photons from single quantum dots, *New J. Phys.* **8**(2), 29, (2006).

[7] N. Akopian, N. H. Lindner, E. Poem, Y. Berlatzky, J. Avron, D. Gershoni, B. D. Gerardot, and P. M. Petroff, Entangled photon pairs from semiconductor quantum dots, *Phys. Rev. Lett.* 96(13):130501, (2006).

[8] C. H. Bennett and G. Brassard. Quantum cryptography: public key distribution and coin tossing. In *Proc. of the IEEE Int. Conf. on Comp., Syst. and Sig. Process., Bangalore, India, 1984*, pp. 175–179. IEEE, Piscataway, NJ, (1984).

[9] N. Gisin, G. Ribordy, W. Tittel, and H. Zbinden, Quantum cryptography, *Rev. Mod. Phys.* **74**(1), 145–195, (2002).

[10] R. Loudon, *The Quantum Theory of Light.* (Oxford University Press, Oxford, 2000), third edition. ISBN 019-850177-3.

[11] Q. Wang, X.-B. Wang, and G.-C. Guo, Practical decoy-state method in quantum key distribution with a heralded single-photon source, *Phys. Rev. A.* 75(1):012312, (2007).

[12] T. Schmitt-Manderbach, H. Weier, M. Fürst, R. Ursin, F. Tiefenbacher, T. Scheidl, J. Perdigues, Z. Sodnik, C. Kurtsiefer, J. G. Rarity, A. Zeilinger, and H. Weinfurter, Experimental demonstration of free-space decoy-state quantum key distribution over 144 km, *Phys. Rev. Lett.* 98(1):010504, (2007).

[13] Z. L. Yuan, A. W. Sharpe, and A. J. Shields, Unconditionally secure one-way quantum key distribution using decoy pulses, *Appl. Phys. Lett.* 90(1): 011118, (2007).

[14] A. Soujaeff, T. Nishioka, T. Hasegawa, S. Takeuchi, T. Tsurumaru, K. Sasaki, and M. Matsui, Quantum key distribution at 1550 nm using a pulse heralded single photon source, *Opt. Express.* **15**(2), 726–734, (2007).

[15] D. Leonard, K. Pond, and P. M. Petroff, Critical layer thickness for self-assembled InAs islands on GaAs, *Phys. Rev. B.* **50**(16), 11687–11692, (1994).

[16] N. I. Cade, H. Gotoh, H. Kamada, H. Nakano, S. Anantathanasarn, and R. Nötzel, Optical characteristics of single InAs/InGaAsP/InP(100) quantum dots emitting at 1.55 μm, *Appl. Phys. Lett.* 89(18):181113, (2006).

[17] N. I. Cade, H. Gotoh, H. Kamada, T. Tawara, T. Sogawa, H. Nakano, and H. Okamoto, Charged exciton emission at 1.3 μm from single InAs quantum dots grown by metalorganic chemical vapor deposition, *Appl. Phys. Lett.* 87 (17):172101, (2005).

[18] K. Takemoto, Y. Sakuma, S. Hirose, T. Usuki, N. Yokoyama, T. Miyazawa, M. Takatsu, and Y. Arakawa, Non-classical photon emission from a single

InAs/InP quantum dot in the 1.3-μm optical-fiber band, *Jpn. J. Appl. Phys., Part 2.* **43**(7B), L993–995, (2004).
[19] T. Miyazawa, K. Takemoto, Y. Sakuma, S. Hirose, T. Usuki, N. Yokoyama, M. Takatsu, and Y. Arakawa, Single-photon generation in the 1.55-μm optical-fiber band from an InAs/InP quantum dot, *Jpn. J. Appl. Phys.* **44** (20), L620–622, (2005).
[20] K. Takemoto, M. Takatsu, S. Hirose, N. Yokoyama, Y. Sakuma, T. Usuki, T. Miyazawa, and Y. Arakawa, An optical horn structure for single-photon source using quantum dots at telecommunication wavelength, *J. Appl. Phys.* 101(8):081720, (2007).
[21] B. Alloing, C. Zinoni, V. Zwiller, L. H. Li, C. Monat, M. Gobet, G. Buchs, A. Fiore, E. Pelucchi, and E. Kapon, Growth and characterization of single quantum dots emitting at 1300 nm, *Appl. Phys. Lett.* 86(10):101908, (2005).
[22] B. Alloing, C. Zinoni, L. H. Li, A. Fiore, and G. Patriarche, Structural and optical properties of low-density and in-rich InAs/GaAs quantum dots, *J. Appl. Phys.* 101(2):024918, (2007).
[23] P. B. Joyce, T. J. Krzyzewski, G. R. Bell, and T. S. Jones, Surface morphology evolution during the overgrowth of large InAs–GaAs quantum dots, *Appl. Phys. Lett.* **79**(22), 3615–3617, (2001).
[24] P. Offermans, P. M. Koenraad, R. Nötzel, J. H. Wolter, and K. Pierz, Formation of InAs wetting layers studied by cross-sectional scanning tunneling microscopy, *Appl. Phys. Lett.* 87(11):111903, (2005).
[25] D. M. Bruls, J. W. A. M. Vugs, P. M. Koenraad, H. W. M. Salemink, J. H. Wolter, M. Hopkinson, M. S. Skolnick, Fei Long, and S. P. A. Gill, Determination of the shape and indium distribution of low-growth-rate InAs quantum dots by cross-sectional scanning tunneling microscopy, *Appl. Phys. Lett.* **81**(9), 1708–1710, (2002).
[26] G. Patriarche, L. Largeau, J.-C. Harmand, and D. Gollub, Morphology and composition of highly strained InGaAs and InGaAsN layers grown on GaAs substrate, *Appl. Phys. Lett.* **84**(2), 203–205, (2004).
[27] I. Mukhametzhanov, Z. Wei, R. Heitz, and A. Madhukar, Punctuated island growth: An approach to examination and control of quantum dot density, size, and shape evolution, *Appl. Phys. Lett.* **75**(1), 85–87, (1999).
[28] T. Kaizu and K. Yamaguchi, Self size-limiting process of InAs quantum dots grown by molecular beam epitaxy, *Jpn. J. Appl. Phys., Part 1.* **40**(3B), 1885–1887, (2001).
[29] M. B. Ward, O. Z. Karimov, D. C. Unitt, Z. L. Yuan, P. See, D. G. Gevaux, A. J. Shields, P. Atkinson, and D. A. Ritchie, On-demand single-photon source for 1.3 μm telecom fiber, *Appl. Phys. Lett.* 86(20):201111, (2005).
[30] H. Lee, R. R. Lowe-Webb, W. Yang, and P. C. Sercel, Formation of InAs/GaAs quantum dots by molecular beam epitaxy: Reversibility of the islanding transition, *Appl. Phys. Lett.* **71**(16), 2325–2327, (1997).
[31] M. C. Xu, Y. Temko, T. Suzuki, and K. Jacobi, Shape transition of InAs quantum dots on GaAs(001), *J. Appl. Phys.* 98(8):083525, (2005).
[32] P. Kratzer, Q. K. K. Liu, P. Acosta-Diaz, C. Manzano, G. Costantini, R. Songmuang, A. Rastelli, O. G. Schmidt, and K. Kern, Shape transi-

tion during epitaxial growth of InAs quantum dots on GaAs(001): Theory and experiment, *Phys. Rev. B.* 73(20):205347, (2006).

[33] M. B. Ward, T. Farrow, P. See, Z. L. Yuan, O. Z. Karimov, A. J. Bennett, A. J. Shields, P. Atkinson, K. Cooper, and D. A. Ritchie, Electrically driven telecommunication wavelength single-photon source, *Appl. Phys. Lett.* 90(6): 063512, (2007).

[34] S. Kaiser, T. Mensing, L. Worschech, F. Klopf, J. P. Reithmaier, and A. Forchel, Optical spectroscopy of single InAs/InGaAs quantum dots in a quantum well, *Appl. Phys. Lett.* **81**(26), 4898–4900, (2002).

[35] M. B. Ward, D. C. Unitt, Z. L. Yuan, P. See, R. M. Stevenson, K. Cooper, P. Atkinson, I. Farrer, D. A. Ritchie, and A. J. Shields, Single quantum dot electroluminescence near 1.3 μm, *Physica E.* **21**(2–4), 390–394, (2004).

[36] N. I. Cade, H. Gotoh, H. Kamada, H. Nakano, and H. Okamoto, Fine structure and magneto-optics of exciton, trion, and charged biexciton states in single InAs quantum dots emitting at 1.3 μm, *Phys. Rev. B.* 73(11):115322, (2006).

[37] H. Benisty, H. De Neve, and C. Weisbuch, Impact of planar microcavity effects on light extraction—Part I: basic concepts and analytical trends, *IEEE J. Quant. Electron.* **34**(9), 1612–1631, (1998).

[38] C. Zinoni, B. Alloing, C. Monat, L. H. Li, L. Lunghi, A. Gerardino, and A. Fiore, Time resolved measurements on low-density single quantum dots at 1300 nm, *Physica Status Solidi C.* **3**(11), 3717–3721, (2006).

[39] G. Bester and A. Zunger, Compositional and size-dependent spectroscopic shifts in charged self-assembled $In_xGa_{1-x}As$/GaAs quantum dots, *Phys. Rev. B.* 68(7):073309, (2003).

[40] M. Ediger, P. A. Dalgarno, J. M. Smith, B. D. Gerardot, R. J. Warburton, K. Karrai, and P. M. Petroff, Controlled generation of neutral, negatively-charged and positively-charged excitons in the same single quantum dot, *Appl. Phys. Lett.* 86(21):211909, (2005).

[41] A. Markus, A. Fiore, J. D. Ganière, U. Oesterle, J. X. Chen, B. Deveaud, M. Ilegems, and H. Riechert, Comparison of radiative properties of InAs quantum dots and GaInNAs quantum wells emitting around 1.3 μm, *Appl. Phys. Lett.* **80**(6), 911–913, (2002).

[42] C. Zinoni, B. Alloing, C. Monat, V. Zwiller, L. H. Li, A. Fiore, L. Lunghi, A. Gerardino, H. de Riedmatten, H. Zbinden, and N. Gisin, Time-resolved and antibunching experiments on single quantum dots at 1300 nm, *Appl. Phys. Lett.* 88(13):131102, (2006).

[43] O. Labeau, P. Tamarat, and B. Lounis, Temperature dependence of the luminescence lifetime of single CdSe/ZnS quantum dots, *Phys. Rev. Lett.* 90 (25):257404, (2003).

[44] B. Patton, W. Langbein, and U. Woggon, Trion, biexciton, and exciton dynamics in single self-assembled CdSe quantum dots, *Phys. Rev. B.* 68 (12):125316, (2003).

[45] G. Bacher, R. Weigand, J. Seufert, V. D. Kulakovskii, N. A. Gippius, A. Forchel, K. Leonardi, and D. Hommel, Biexciton versus exciton lifetime in a single semiconductor quantum dot, *Phys. Rev. Lett.* **83**(21), 4417–4420,

(1999).

[46] H. T. Grahn, *Introduction to semiconductor physics.* (World Scientific Publishing, Singapore, 1999). ISBN 981-02-3302-7.

[47] D. V. Regelman, U. Mizrahi, D. Gershoni, E. Ehrenfreund, W. V. Schoenfeld, and P. M. Petroff, Semiconductor quantum dot: A quantum light source of multicolor photons with tunable statistics, *Phys. Rev. Lett.* 87(25):257401, (2001).

[48] E. Moreau, I. Robert, L. Manin, V. Thierry-Mieg, J. M. Gérard, and I. Abram, Quantum cascade of photons in semiconductor quantum dots, *Phys. Rev. Lett.* 87(18):183601, (2001).

[49] C. Zinoni, B. Alloing, L. H. Li, F. Marsili, A. Fiore, L. Lunghi, A. Gerardino, Yu. B. Vakhtomin, K. V. Smirnov, and G. N. Gol'tsman, Single-photon experiments at telecommunication wavelengths using nanowire superconducting detectors, *Appl. Phys. Lett.* 91(3):031106, (2007).

[50] R. Hanbury Brown and R. Q. Twiss, Correlation between photons in two coherent beams of light, *Nature.* **177**(4497), 27–29, (1956).

[51] A. Yoshizawa, R. Kaji, and H. Tsuchida, After-pulse-discarding in single-photon detection to reduce bit errors in quantum key distribution, *Opt. Express.* **11**(11), 1303–1309, (2003).

[52] C. Santori, D. Fattal, J. Vučković, G. S. Solomon, E. Waks, and Y. Yamamoto, Submicrosecond correlations in photoluminescence from InAs quantum dots, *Phys. Rev. B.* 69(15):205324, (2004).

[53] G. N. Gol'tsman, O. Okunev, G. Chulkova, A. Lipatov, A. Semenov, K. Smirnov, B. Voronov, A. Dzardanov, C. Williams, and R. Sobolewski, Picosecond superconducting single-photon optical detector, *Appl. Phys. Lett.* **79**(6), 705–707, (2001).

[54] N. Namekata, S. Sasamori, and S. Inoue, 800 MHz single-photon detection at 1550-nm using an InGaAs/InP avalanche photodiode operated with a sine wave gating, *Opt. Express.* **14**(21), 10043–10049, (2006).

[55] Z. L. Yuan, B. E. Kardynal, A. W. Sharpe, and A. J. Shields, High speed single photon detection in the near infrared, *Appl. Phys. Lett.* 91(4):041114, (2007).

[56] S. C. Kitson, P. Jonsson, J. G. Rarity, and P. R. Tapster, Intensity fluctuation spectroscopy of small numbers of dye molecules in a microcavity, *Phys. Rev. A.* **58**(1), 620–627, (1998).

[57] H. Z. Song, T. Usuki, S. Hirose, K. Takemoto, Y. Nakata, N. Yokoyama, and Y. Sakuma, Site-controlled photoluminescence at telecommunication wavelength from InAs/InP quantum dots, *Appl. Phys. Lett.* 86(11):113118, (2005).

[58] P. Atkinson, M. B. Ward, S. P. Bremner, D. Anderson, T. Farrow, G. A. C. Jones, A. J. Shields, and D. A. Ritchie, Site-control of InAs quantum dots using *ex-situ* electron-beam lithographic patterning of GaAs substrates, *Jpn. J. Appl. Phys., Part 1.* **45**(4A), 2519–2521, (2006).

[59] P. S. Wong, G. Balakrishnan, N. Nuntawong, J. Tatebayashi, and D. L. Huffaker, Controlled InAs quantum dot nucleation on faceted nanopatterned pyramids, *Appl. Phys. Lett.* 90(18):183103, (2007).

[60] A. J. Bennett, D. C. Unitt, P. See, A. J. Shields, P. Atkinson, K. Cooper, and D. A. Ritchie, Microcavity single-photon-emitting diode, *Appl. Phys. Lett.* 86(18):181102, (2005).
[61] M. H. Baier, C. Constantin, E. Pelucchi, and E. Kapon, Electroluminescence from a single pyramidal quantum dot in a light-emitting diode, *Appl. Phys. Lett.* **84**(11), 1967–1969, (2004).
[62] Xiulai Xu, D. A. Williams, and J. R. A. Cleaver, Electrically pumped single-photon sources in lateral *p–i–n* junctions, *Appl. Phys. Lett.* **85**(15), 3238–3240, (2004).
[63] T. Miyazawa, J. Tatebayashi, S. Hirose, T. Nakaoka, S. Ishida, S. Iwamoto, K. Takemoto, T. Usuki, N. Yokoyama, M. Takatsu, and Y. Arakawa, Development of electrically driven single-quantum-dot device at optical fiber bands, *Jpn. J. Appl. Phys., Part 1.* **45**(4B), 3621–3624, (2006).
[64] R. Schmidt, U. Scholz, M. Vitzethum, R. Fix, C. Metzner, P. Kailuweit, D. Reuter, A. Wieck, M. C. Hübner, S. Stufler, A. Zrenner, S. Malzer, and G. H. Döhler, Fabrication of genuine single-quantum-dot light-emitting diodes, *Appl. Phys. Lett.* 88(12):121115, (2006).
[65] O. Benson, C. Santori, M. Pelton, and Y. Yamamoto, Regulated and entangled photons from a single quantum dot, *Phys. Rev. Lett.* **84**(11), 2513–2516, (2000).
[66] D. L. Huffaker, L. A. Graham, and D. G. Deppe, Ultranarrow electroluminescence spectrum from the ground state of an ensemble of self-organized quantum dots, *Appl. Phys. Lett.* **72**(2), 214–216, (1998).
[67] A. Fiore, J. X. Chen, and M. Ilegems, Scaling quantum-dot light-emitting diodes to submicrometer sizes, *Appl. Phys. Lett.* **81**(10), 1756–1758, (2002).
[68] D. J. P. Ellis, A. J. Bennett, A. J. Shields, P. Atkinson, and D. A. Ritchie, Electrically addressing a single self-assembled quantum dot, *Appl. Phys. Lett.* 88(13):133509, (2006).
[69] D. J. P. Ellis, A. J. Bennett, A. J. Shields, P. Atkinson, and D. A. Ritchie, Oxide-apertured microcavity single-photon emitting diode, *Appl. Phys. Lett.* 90(23):233514, (2007).
[70] A. Lochmann, E. Stock, O. Schulz, F. Hopfer, D. Bimberg, V. Haisler, A. Toropov, A. Bakarov, and A. Kalagin, Electrically driven single quantum dot polarised single photon emitter, *Electron. Lett.* **42**(13), 774–775, (2006).
[71] M. Scholz, S. Büettner, O. Benson, A. I. Toropov, A. K. Bakarov, A. K. Kalagin, A. Lochmann, E. Stock, O. Schulz, F. Hopfer, V. A. Haisler, and D. Bimberg, Non-classical light emission from a single electrically driven quantum dot, *Opt. Express.* **15**(15), 9107–9112, (2007).
[72] C. Monat, B. Alloing, C. Zinoni, L. H. Li, and A. Fiore, Nanostructured current-confined single quantum dot light-emitting diode at 1300 nm, *Nano Lett.* **6**(7), 1464–1467, (2006).
[73] A. Fiore, M. Rossetti, B. Alloing, C. Paranthoen, J. X. Chen, L. Geelhaar, and H. Riechert, Carrier diffusion in low-dimensional semiconductors: A comparison of quantum wells, disordered quantum wells, and quantum dots, *Phys. Rev. B.* 70(20):205311, (2004).

[74] C. Zinoni, B. Alloing, C. Paranthoën, and A. Fiore, Three-dimensional wavelength-scale confinement in quantum-dot microcavity light-emitting diodes, *Appl. Phys. Lett.* **85**(12), 2178–2180, (2004).

[75] P. W. Fry, I. E. Itskevich, D. J. Mowbray, M. S. Skolnick, J. J. Finley, J. A. Barker, E. P. O'Reilly, L. R. Wilson, I. A. Larkin, P. A. Maksym, M. Hopkinson, M. Al-Khafaji, J. P. R. David, A. G. Cullis, G. Hill, and J. C. Clark, Inverted electron-hole alignment in InAs-GaAs self-assembled quantum dots, *Phys. Rev. Lett.* **84**(4), 733–736, (2000).

[76] A. J. Bennett, D. C. Unitt, P. See, A. J. Shields, P. Atkinson, K. Cooper, and D. A. Ritchie, Electrical control of the uncertainty in the time of single photon emission events, *Phys. Rev. B.* 72(03):033316, (2005).

Chapter 17

Quantum Information Processing with Quantum Dots in Photonic Crystals

Jelena Vučković, Dirk Englund, Andrei Faraon, Ilya Fushman, and Edo Waks

Ginzton Laboratory, Stanford University, Stanford CA 94305, USA

In this chapter, we review our recent progress on implementing quantum information processing with quantum dots in photonic crystals. We describe a recently developed technique for selective tuning of quantum dots on chip, which addresses the problem of large quantum dot inhomogeneous broadening - usually considered the main obstacle in employing such platform in practical quantum information processing systems. We also present experimental results on cavity QED with a single quantum dot in photonic crystal, both in the weak and strong coupling regimes, which is critical for construction of such systems. The last part of the chapter is dedicated to our preliminary experimental work on integrated quantum information processing circuits, where we demonstrate the generation and transfer of single photons on a photonic crystal chip. Finally, we briefly discuss future prospects and a proposal for a quantum repeater that can be implemented using the described quantum dot-photonic crystal platform.

Contents

17.1. Introduction and Overview

Cavity QED-based approaches have been considered some of the most promising for practical implementation of quantum information processing [1, 2]. Such quantum networks would combine the ease of storing and manipulating quantum information in atoms, ions [3, 4], or quantum dots [5], with the advantages of transferring information between nodes via photons, using coherent interfaces [6–8]. So far, the remarkable demonstrations of basic building blocks of such networks have relied on atomic systems [9–12]. Without a doubt, a successful solid-state implementation of such an approach would open new opportunities for scaling of practical quantum information processing systems [5]. Among the proposed solid-state implementations, systems comprising of quantum dots in photonic crystals are emerging as likely the most favorable, because of the small volumes of photonic crystal cavities which enable strong quantum dot-cavity field interaction [13–15], as well as because of the ability to integrate photonic crystal components monolithically on chip [16].

In this chapter, we review our recent progress on implementing quantum information processing based on quantum dots in photonic crystals. We start by very briefly reviewing photonic crystal cavities and their fabrication in Section 2. Due to the space limitations, we do not describe quantum dot growth properties and photonic crystals designs, which have been described elsewhere [17–21]. In Section 3, we describe our recently developed technique for selective tuning of quantum dots on chip, which addresses the problem of large quantum dot inhomogeneous broadening [22]. In Section 4, we present experimental results on cavity QED with a single quantum dot in photonic crystal, both in the weak and strong coupling regimes, showing the ability to suppress spontaneous emission into all modes other than the resonant cavity mode [13], as well as to initiate vacuum Rabi oscillation between a single quantum dot and the cavity field. In particular, we show that the selective quantum dot tuning technique (presented in Section 3) can be used to tune a quantum dot into the strong coupling regime with a cavity [22]. In Section 5, we show our preliminary work on integrated quantum information processing circuits, where we demonstrate the generation and transfer of single photons on a photonic crystal chip via a waveguide between two cavites [16]. Finally, in Section 6 we briefly discuss a proposal for a quantum repeater that can be implemented using the described quantum dot-photonic crystal platform [23, 24].

17.2. Photonic Crystal Circuits: Design and Fabrication

All experiments presented in this chapter are based on planar photonic crystal (PC) devices in GaAs with incorporated InAs quantum dots (QDs). QD wafers were grown by molecular beam epitaxy by our collaborators. They generally contained an active region consisting of a 150 – 160 nm thick GaAs layer with a centered InAs/GaAs QD layer. The active layer was grown on an AlGaAs sacrificial layer (with high Al content), which is in turn grown on a GaAs substrate. The suspended PC structures were fabricated on such a wafer by a combination of electron beam lithography and dry etching (which creates a PC pattern in the top GaAs layer containing QDs), and a final wet etching step which dissolves a sacrificial layer underneath PC components. An example of fabricated GaAs PC cavities using the described procedure is shown in Fig. 17.1. In the same figure we also show the photoluminescence spectrum from the same type of cavity, obtained by exciting InAs QDs embedded inside of it, and indicating the cavity quality (Q) factor of 20000.

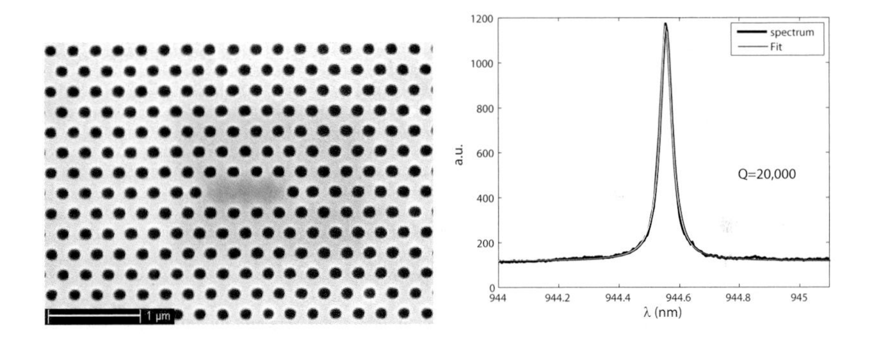

Fig. 17.1. Left: Photonic crystal cavity fabricated in GaAs slab containing InAs QDs. Right: Photoluminescence spectrum from the same type of cavity, indicating Q=20000.

17.3. Local Quantum Dot Tuning on a Photonic Crystal Chip

A major challenge facing proposals for solid-state quantum information processing with quantum dots coupled to optical cavities is spatial and spectral matching of distinct inhomogeneously broadened quantum dots. Spatial alignment can be achieved either by positioning the PC cavity on an already identified QD [15, 25], or by relying on chance. For spectral alignment, there are a few techniques that can be used to modify the emission wavelength of InAs quantum dots: Stark shift [26], Zeeman shift [27], temperature tuning [28] and strain tuning [29]. In this Section, we describe a recently developed technique which enables independent control of QDs, employing structures with high-Q cavities whose temperature is controlled by laser beams [22]. Our in-situ technique allows extremely precise spectral tuning of InAs quantum dots by up to 1.8nm and of cavities of up to 0.4nm (4 cavity linewidths). The technique is crucial for spectrally aligning distinct quantum dots on a photonic crystal chip and forms an essential step toward creating on-chip quantum information processing devices.

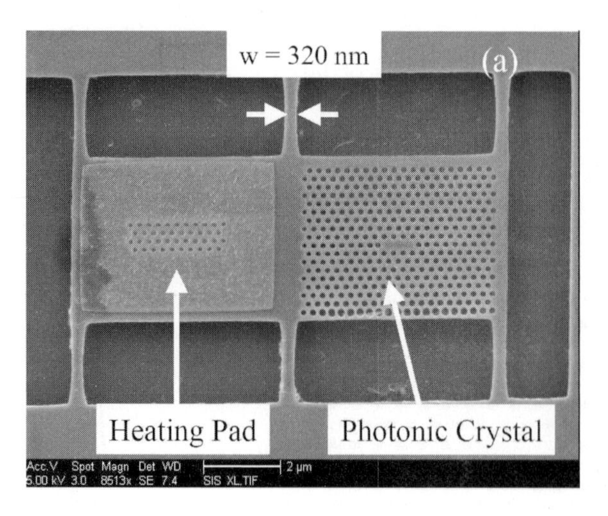

Fig. 17.2. Scanning electron microscope image of the fabricated structure showing the PC cavity, the heating pad and the connection bridges. The temperature of the structure was controlled with a laser beam (960nm) focused on the heating pad.

To achieve independent on-chip tuning, distinct regions containing the quantum dots of interest must be kept at different temperatures. Since GaAs is a good thermal conductor, on-chip local thermal insulation must be provided to achieve significant local heating. For this reason, we fabricated suspended PC structures with minimal thermal contact to the rest of the chip, as shown in Fig. 17.2.

The fabricated structures (12μm long, 4μm wide, 150nm thick) consist of a PC cavity and a heating pad (Fig. 17.2). To provide the thermal insulation needed for efficient device heating, the structure was connected to the rest of the chip by only six narrow bridges. The thermal conductivity of narrow (≈100nm), cold (4K - 10K) GaAs bridges is reduced by up to four orders of magnitude with respect to the bulk GaAs [30], thus improving the thermal insulation. We tested two devices with connection bridges of the same length (2μm) but different widths: w = 320nm and w = 800nm. The temperature of the device was controlled by using a focused laser beam to heat up the pad next to the photonic crystal cavity. To minimize background photoluminescence in single quantum dot measurements, the heating laser is tuned below the QD absorption frequency. A metal layer (20nm Cr/15nm Au) was deposited on the heating pad to increase heat absorption. The thermal conductivity of GaAs beams with cross sections on the order of 100nm/100nm, and the absorption coefficient of the metal layer are not well known. As measured by Fon et al. [30], the thermal conductivity of GaAs beams with dimensions 100nm/200nm/6μm is about 3×10^{-2}WK^{-1}cm^{-1} at 10K, three orders of magnitude lower than the bulk value. Because of the size similarity, we assume that the connection bridges from our device have a similar thermal conductivity. Assuming that $10^{-2}\,mW$ of heat is absorbed in the heating pad and considering that the device has the same thermal conductivity as the bridges, we expect the temperature of the membrane to increase by a few tens of degrees Kelvin.

The local quantum dot tuning measurements were performed in a continuous flow liquid helium cryostat maintained at 10 K. A Ti:Sapph laser tuned at 855 nm was used to excite the quantum dots while a 960 nm laser diode acted as the heating laser. Using a pinhole, we collected photoluminescence from a quantum dot located inside the photonic crystal slab. By increasing the power of the heating laser, QD emission was observed to redshift (Fig. 17.3(a)). The QD linewidth broadens with increasing heating pump power, as expected from experiments where the full sample is heated. We were able to tune the quantum dot by 1.4nm while the linewidth broadened from 0.04nm to 0.08nm. The quantum dot could be further shifted by 1.8nm but the photoluminescence (PL) intensity dropped rapidly. To show the compatibility of this local tuning technique with single photon measurements and quantum information processing, we proved antibunched single photon emission from an exciton transition using a Hanbury Brown-Twiss interferometer while the emission line was shifted by 0.8nm (Fig. 17.3(b)). Antibunching can be measured as long as the thermal energy is not larger than the confining energy of the QD, which in our case corresponds to a detuning of ≈1.4nm.

To investigate the thermal properties of the fabricated devices, we compared the shift of quantum dots located on structures with different bridge

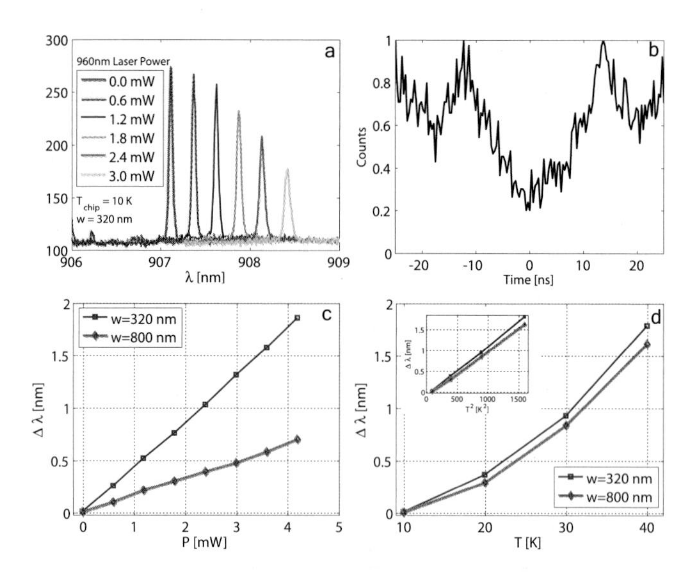

Fig. 17.3. (a) Quantum dot tuning vs. heating pump power. The structure is connected to the substrate by bridges measuring 320nm in width. The quantum dot emission shifts by 1.4nm while increasing the heating laser power to 3mW. Only a small fraction of the heating laser power is absorbed in the metal pad. (b) Autocorrelation measurement showing single photon anti bunching while the QD was detuned by 0.8nm using the local tuning technique. (c) Dependence of the QD detuning on the heating laser power. The two data sets correspond to structures with different thermal contact to the substrate (320nm and 800nm bridges). (d) QD temperature tuning by changing the temperature of the entire chip in the cryostat. The inset shows that the detuning is linear in T^2.

widths, $w = 320$nm and $w = 800$nm. The thermal conductance of the bridges is proportional to their width w. Under the same pump conditions the temperature of the structure is inversely proportional to w so we would expect the QD to shift 2.5 (i.e., 800/320) times further for the structure with thinner bridges. The QD shift observed on the two structures is plotted in Fig. 17.3(c). For the same pump power, the QD shifts 2.65 times further for $w = 320$nm than for $w = 800$nm, in good agreement with the expected result.

The temperature dependence of the QD shift was determined by changing the temperature of the entire chip in the cryostat. The results are plotted in Fig. 17.3(d) and indicate that a shift of 1.8nm corresponds to a temperature of 40K. This implies that during the local temperature tuning experiment, the structure was also heated to 40K. The QD shift shows a quadratic dependence with temperature (Fig. 17.3(d) inset), which is expected since the band gaps of GaAs/InAs have a quadratic temperature dependence in this temperature interval [31]. Our experimental data shows a linear dependence of the QD shift with the heating laser power (Fig. 17.3(c)), which implies a linear relation between the power of the heating laser and T^2. The local temperature gradient also induces strain which can be responsible for shifting the QD emission. To release the strain in the suspended membrane, we used a focused laser beam to cut some of the connection bridges next to the QD. After the strain release, we still observed the same shift of the QD with the heating power which indicates that the shift is mainly due to temperature.

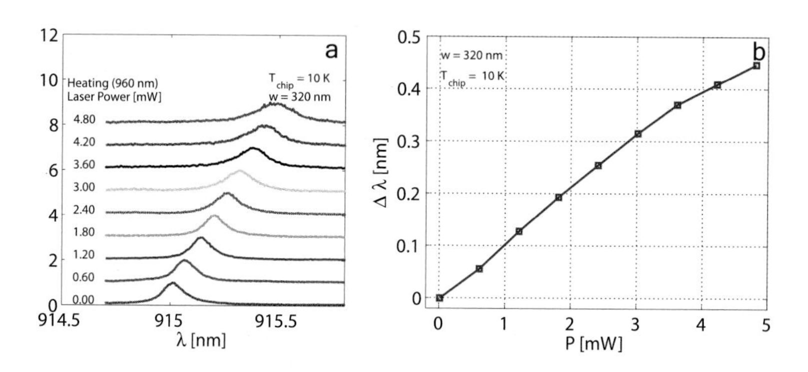

Fig. 17.4. (a) Detuning of the PC cavity resonance with increasing temperature due to local heating. (b) Dependence of the PC cavity resonance wavelength on the local heating power.

Not only the QDs but also the PC cavities shift their resonant frequency with temperature. The local heating technique was used to shift a PC cavity located on the $w = 320$nm structure. Using the same heating power as for the QD tuning, we observed the cavity resonance redshift by up to 0.48nm (Fig. 17.4(a) and (b)), about 3 times less than the QD shift. The quality factor of the cavity dropped from 7600 to 4900. Beside temperature tuning, photonic crystal cavities can be tuned using chemical digital etching [32] or by deposition of molecular layers on top of the PC membrane [33] .

We emphasize that our local temperature tuning technique is completely reversible and does not affect the structure of the PC cavities or the QDs. Another tuning technique that relies on locally heating microcavities to permanently change the structure of the resonator and the QDs has been reported by Rastelli et al. [34].

17.4. Cavity QED with Quantum Dots in Photonic Crystals

Let us assume that a single quantum dot is isolated in a microcavity, and is coupled to the cavity field with the position-dependent coupling strength $g(\vec{r}) = g_0 \psi(\vec{r}) \cos(\xi)$, where $g_0 = \frac{\mu}{\hbar}\sqrt{\frac{\hbar\omega}{2\epsilon_M V}}$, $\psi(\vec{r}) = \frac{E(\vec{r})}{|E(\vec{r}_M)|}$, and $\cos(\xi) = \frac{\vec{\mu}\cdot\hat{e}}{\mu}$ [17]. In this expression, $\vec{r}_M$ denotes the point where the electric field energy density $\epsilon(\vec{r})|E(\vec{r})|^2$ is maximum and ϵ_M is the dielectric constant at this point ($\epsilon_M = \epsilon(\vec{r}_M)$). (We assume that the cavity is designed in such a way that $\epsilon_M = n^2$, where n is the refractive index of the semiconductor used to make it, as desired for cavity QED with QDs embedded in semiconductor). $E(\vec{r})$ is the electric field magnitude, and V is the cavity mode volume, defined as $V = \frac{\iiint \epsilon(\vec{r})|E(\vec{r})|^2 d^3\vec{r}}{\epsilon_M |E(\vec{r}_M)|^2}$. The electric field orientation at the location $\vec{r}$ is denoted as $\hat{e}$, $\vec{\mu}$ is the dipole moment of the dot, and $\mu = |\vec{\mu}|$. Therefore, the coupling $|g(\vec{r})|$ reaches its maximum value of g_0 when the QD is located at the point $\vec{r}_M$ where the electric field energy density is maximum, and when its dipole moment is aligned with the electric field.

The losses of the system can be described in terms of the *cavity field decay rate* κ, equal to $(\omega/2Q)$, and the *dipole decay rate* γ. κ is the decay rate of the resonant cavity mode, while γ includes QD dipole decay to modes other than the cavity mode and to nonradiative decay routes. Depending on the ratio of the coupling parameter $|g(\vec{r})|$ to the decay rates κ and γ, we can distinguish two regimes of coupling between the exciton and the cavity field: the *strong-coupling regime*, for $|g(\vec{r})| > \kappa, \gamma$, and the *weak-coupling regime*, for $|g(\vec{r})| < \kappa, \gamma$ [35].

In the strong-coupling case, the time scale of coherent coupling between the QD and the cavity field is shorter than that of the irreversible decay into various radiative and nonradiative routes. *Vacuum Rabi oscillation* occurs in this case, and the time evolution of the system can be described by oscillation at frequency $2|g(\vec{r})|$ between the state of an excited dot and no photons in the cavity, and the ground-state quantum dot with one photon in the cavity. We will assume that QD dipole decay rates into modes other than the resonant cavity mode and into nonradiative routes are not modified in the presence of a cavity, and estimate γ from the homogeneous linewidth of the QD without a cavity.

On the other hand, in the weak-coupling case, irreversible decay rates dominate over the coherent coupling rate; in other words, the QD-cavity field system does not have enough time to couple coherently before dissipa-

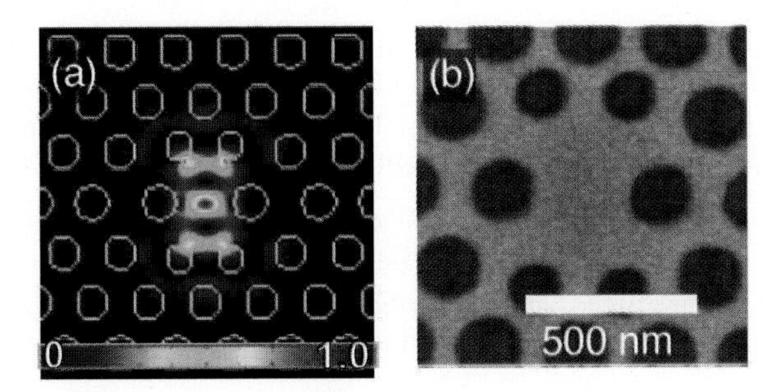

Fig. 17.5. PC cavity used in the weak coupling experiment with a single QD, and the corresponding E-field pattern.

tion occurs. In the systems that we consider in the weak coupling regime, $\kappa > |g(\vec{r})|$, but $\kappa >> \gamma$. In this "bad cavity limit", the spontaneous emission decay rate of the QD into the cavity mode is [36]

$$\Gamma = |g(\vec{r})|^2 \frac{4Q}{\omega} \tag{17.1}$$

which is enhanced relative to the spontaneous emission rate of the same QD in bulk semiconductor by the so-called Purcell factor [37]:

$$\frac{\Gamma}{\Gamma_0} = \frac{3}{4\pi^2}\frac{Q}{V}\left(\frac{\lambda}{n}\right)^3\left(\frac{E(\vec{r})}{E(\vec{r}_M)}\right)^2\left(\frac{\vec{\mu}\cdot\hat{e}}{\mu}\right)^2\frac{\Delta\lambda_c^2}{\Delta\lambda_c^2 + 4(\lambda-\lambda_c)^2} \tag{17.2}$$

In the previous expression, λ is the QD emission wavelength, and λ_c is the cavity mode wavelength.

Therefore, for a QD positioned at the maximum of the electric field energy density, with dipole moment aligned with the electric field, and on resonance with the cavity mode, the Purcell factor is maximized and equal to

$$F = \frac{3}{4\pi^2}\frac{Q}{V}\left(\frac{\lambda}{n}\right)^3. \tag{17.3}$$

For a QD off resonance with the photonic crystal cavity (and inside of photonic band gap), $f = \Gamma/\Gamma_0 < 1$, as a result of the reduced density of optical states inside photonic band gap. For example, in the triangular PC lattice shown in Fig. 17.5, $f \approx 0.2$ [13].

The fraction of emission from a QD that is coupled into one particular cavity mode is known as the spontaneous emission coupling factor β, and is approximately related to the Purcell factor via $\beta = \frac{F}{f+F}$. Therefore, if the QD emission rate is strongly enhanced by its interaction with a cavity mode, and its emission into other modes suppressed by photonic band gap, then $\beta \approx 1$. This regime is extremely important for the construction of highly efficient sources of single photons.

The strong and weak coupling regimes in the solid state are not only of academic interest; they can also be used in the construction of efficient sources of indistinguishable single photons and numerous quantum information processing devices, as described later in this chapter.

17.4.1. *Weak coupling regime*

Over the past decade, optical resonators have been exploited to enhance emission rate for improving numerous quantum optical devices [38, 39]. Single photon sources in particular promise to see large improvements [40]. While more attention has been given to *increasing* emission rate, the reverse is also possible in an environment with decreased density of optical states.

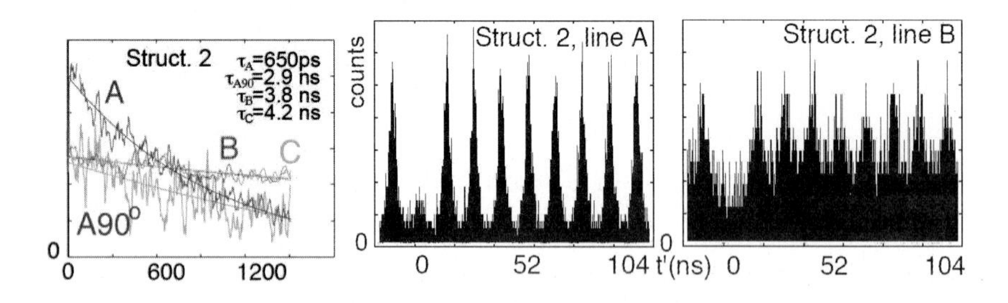

Fig. 17.6. Left: Lifetime measurements on different QDs positioned at various points inside a PC cavity. Right: Measurements of the photon autocorrelation function on the same QDs, indicating antibunching and that single emitters are probed.

Here we describe our experiment confirming that by designing a photonic crystal cavity, we can significantly increase or decrease the spontaneous emission rate of embedded self-assembled InAs QDs [13]. For QDs coupled to a PC cavity shown in Fig. 17.5, we observe up to eight times faster spontaneous emission rate from time resolved photoluminescence measurements and demonstrate antibunching, confirming that single QDs are probed (see Figs. 17.6 and 17.7). Due to spatial misalignment between the QD and cavity, most of the spectrally resonant dots in this experiment have moderate emission rate enhancement. (In the following section, a 12-fold enhancement is observed for a QD in another type of PC cavity [16]). With better spatial and spectral matching, Eq. 17.2 predicts that a rate enhancement of up to $\sim$ 100 is possible in such a cavity, as a result of a moderate Q-factor ($Q = 1000$) and a very small mode volume ($V \approx 0.5(\lambda/n)^3$). In contrast, for individual off-resonant QDs in the PC, we show up to five-fold spontaneous emission rate quenching as a result of the diminished density of optical states in the photonic bandgap. As described above, the simultaneous enhancement of coupled and suppression of uncoupled spontaneous emission rates results in very high cavity coupling efficiency ($\beta \approx 1$), even for moderate Purcell enhancement.

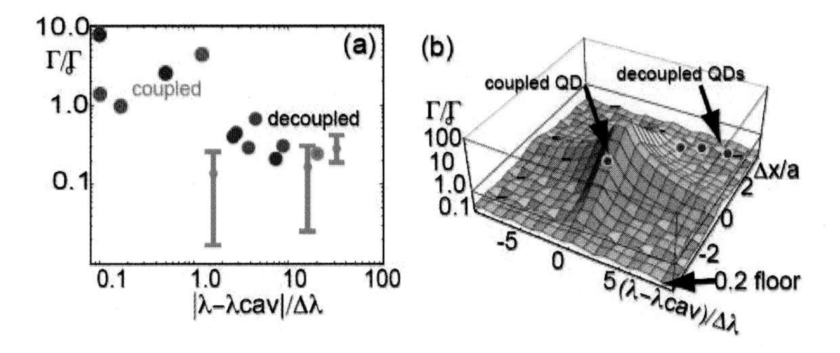

Fig. 17.7. Spontaneous emission rate modification. (a) Experimental (circles) and calculated (bars) data of Γ/Γ_0 of single QD lines vs. spectral detuning (normalized by the cavity linewidth $\Delta\lambda$). Coupling was verified by spectral alignment and polarization matching. (b) Illustration of the predicted spontaneous emission rate modification in the PC as function of normalized spatial and spectral misalignment from the cavity (a is the lattice periodicity). This plot assumes $Q = 1000$ and polarization matching between the emitter dipole and cavity field. The actual spontaneous emission rate modification varies significantly with exact QD location.

17.4.2. *Strong coupling regime*

In the "strong coupling regime," when the coupling strength between the quantum dot and the cavity field g dominates over the loss rates of the system, the quantum dot and the cavity can no longer be treated as distinct entities and entangled modes of the dot and the cavity field are formed. As described previously in this chapter, the quantum dot exchanges photons with the cavity at a very fast rate (Rabi oscillation at the frequency $2g$), and a splitting equal to $2g$ occurs in the cavity resonance, which we refer to as the vacuum Rabi splitting. Our recent result indicating strong coupling

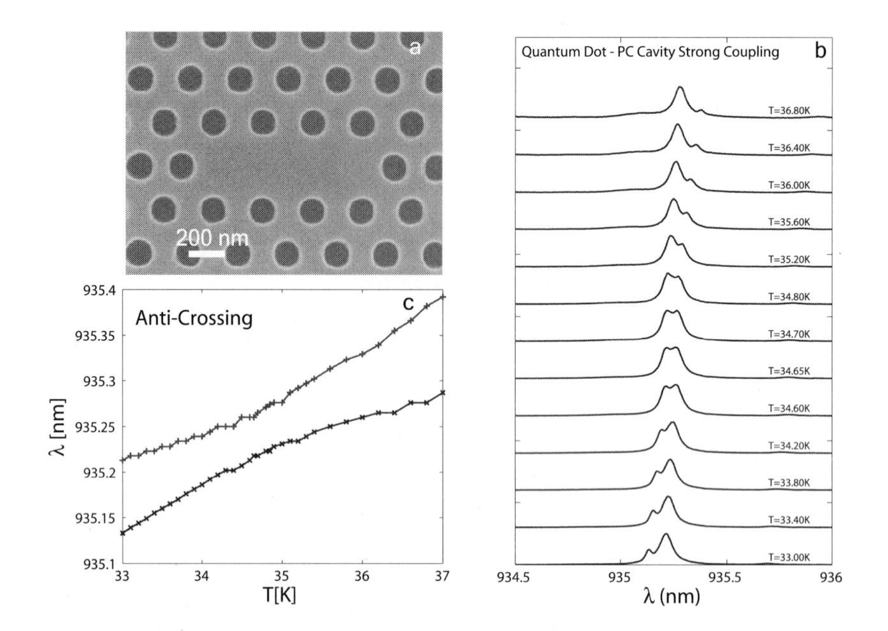

Fig. 17.8. (a) Scanning electron microscope (SEM) image of a photonic crystal cavity. (b) Photoluminescence of a quantum dot strongly coupled to a photonic crystal cavity, using temperature to tune the quantum dot energy through resonance with a cavity mode. Cold-cavity Q-factor is around 10500, and the mode volume is around $(\lambda/n)^3$. (c) Peak positions of photoluminescence in Panel (b) as a function of temperature, exhibiting anti-crossing suggestive of strong coupling behavior.

behavior between a high quality factor cavity mode and a quantum dot emission line is shown in Fig. 17.8. Figure 17.8(a) shows a scanning electron microscope (SEM) image of the cavity, which is a linear three hole-defect photonic crystal cavity. Photoluminescence measurements indicate that the quality factor of this cavity is 10500. The emission line of a quantum dot is tuned into and out of resonance with the cavity using temperature, as shown in Fig. 17.8(b), and the positions of the cavity and QD polariton peaks are plotted as a function of temperature in Fig. 17.8(c). As is consistent with strong coupling behavior, the two peaks exhibit anti-crossing, with the energy separation between the two peaks corresponding to the vacuum Rabi splitting. We have observed such Rabi splitting with a single quantum dot in a number of photonic crystal cavities on different chips, with Q factors ranging from 10000 to 15000.

In the experiment shown in Fig. 17.8, we tuned the temperature of the whole chip by heating the cryostat. However, for practical implementations of quantum information processing, where several quantum gates are integrated on the same chip, it is necessary to individually tune QDs into such regime (as different QDs on chip exhibit different misalignments from the cavity resonances). We used our technique described in Section 3 to locally tune a QD into resonance with a PC cavity mode with $Q = 10000$, as shown in Fig. 17.9(a). We observe splitting and anti-crossing of the polariton states, the signature of strong coupling regime, as the QD is tuned into resonance with the cavity (Fig. 17.9(a-b)). A polariton splitting of 0.05nm is observed. Therefore, our local tuning technique could also be used to tune a single QD into the strong coupling with the cavity.

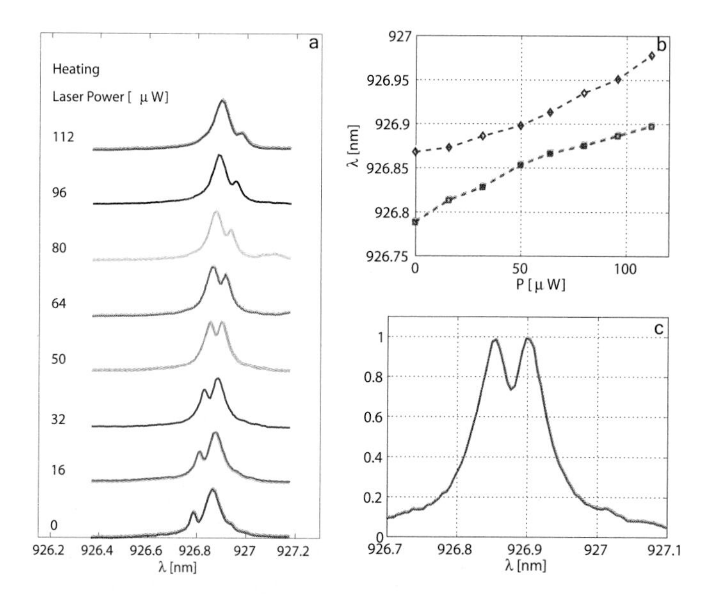

Fig. 17.9. (a) Spectra showing the tuning of a single QD into resonance with the cavity mode using the local heating technique. (b) Anticrossing between the polariton lines of the strongly coupled PC cavity - QD system. (c) Spectrum showing the 0.05 nm splitting of the polaritons.

17.5. Generation and Transfer of Single Photons on a Photonic Crystal Chip

One essential element of quantum information processing systems is a source of indistinguishable single photons, which is required in quantum teleportation [41], linear-optics quantum computation [42], and several schemes for quantum cryptography [43]. Sources have been demonstrated from a variety of systems [44], including semiconductor quantum dots [38, 45–50]. This particular approach for generation of single photons on-demand is based on a combination of pulsed excitation of a single self-assembled semiconductor quantum dot and spectral filtering [17]. When such a QD is excited with a short (ps) laser pulse with the frequency tuned above the band gap of the surrounding semiconductor, electron-hole pairs are created in the surrounding semiconductor matrix; carriers diffuse towards the dot, where they are rapidly trapped and quickly relax to the lowest confined states. The carriers created inside a QD recombine in a radiative cascade, leading to the generation of several photons for each laser pulse. At the moment of the generation of each of the photons in this cascade, the charge configuration of the QD is different. Since the total energy of the QD system is modified by the Coulomb interaction among carriers, all of the photons emitted in the cascade have slightly different frequencies. The last emitted photon for each pulse (corresponding to a single exciton state) thus has a unique frequency, and can be spectrally isolated.

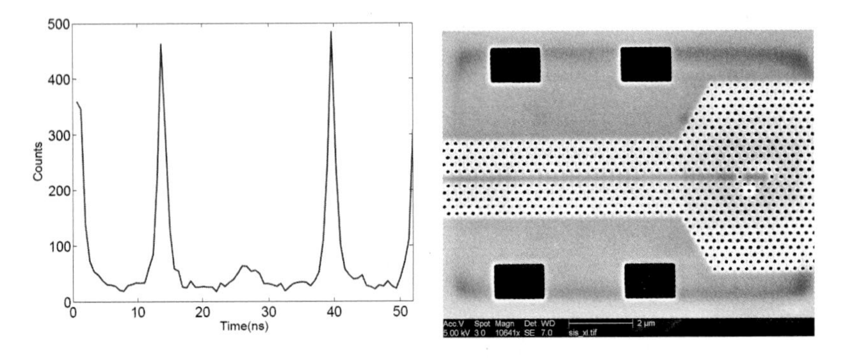

Fig. 17.10. Single photon source based on a quantum dot in photonic crystal. Left: measured second order photon correlation function for such a source. The suppression of the central peak indicates suppression of multiphoton pulses relative to an attenuated laser of the same intensity. Right: Quantum dot-photonic crystal cavity based single photon source coupled to a photonic crystal waveguide.

Triggered single photons can therefore be generated using quantum dots grown in a bulk semiconductor material, but the efficiency of such a system is poor, since the majority of emitted photons are lost in the semiconductor substrate. To increase the photon extraction efficiency and reduce the duration of single-photon pulses emitted from semiconductor systems, cavity quantum electrodynamics can be used [17]. We have demonstrated cavity-enhanced single photon sources based on both microposts [17, 50] and photonic crystal cavities [13], including photonic crystal cavity single photon sources integrated on a chip [16]. For the best of our sources, the multiphoton probability was suppressed to 2% (relative to an attenuated laser of the same intensity), photon pulse duration was as short as 100ps (implying possible 10GHz repetition rate), outcoupling efficiency into the collection lens was about 30%, and photon indistinguishability (overlap between consecutive photons emitted from such a source) around 80%. In Fig. 17.10 (Left) we show the experimental results for such a source based on a single QD in a photonic crystal. This figure shows the measurement of the second order photon correlation function, indicating the antibunching (suppression of the central peak), and thus the suppression of the multiphoton probability relative to an attenuated laser of the same intensity.

Another important component of quantum networks is a quantum channel for efficiently transferring information between spatially separated nodes [1]. For this reason, we have been pursuing single photon sources which are not only based on a single quantum dot in a PC cavity, but where the PC cavity output is redirected into the waveguide for efficient collection (as in the structure shown in Fig. 17.10 (Right)). In the rest of this section, we will describe single photon generation in such a structure, where emitted photons are outcoupled into the waveguide, and then transferred into another cavity on the chip, as in the structure shown in Fig. 17.11. This can be also viewed as a basic building block of a quantum network by the generation and transfer of single photons on a photonic crystal chip. A cavity-coupled QD single photon source is connected through a 25μm channel to an otherwise identical target cavity so that different cavities may be interrogated and manipulated independently (Fig. 17.11). Thanks to their near-minimum mode volume $V_{mode} \approx 0.74(\lambda/n)^3$, these end-cavities allow a large spontaneous emission rate enhancement. Among the 30 fabricated structures, we experimentally found the total Q values to fall in the range 1560 ± 550. This system provides a source of single photons with a high degree of indistinguishability (mean wavepacket overlap of $\sim 67\%$), 12-fold spontaneous emission rate enhancement, spontaneous emission coupling factor $\beta \sim 0.98$ into the cavity mode, and high-efficiency coupling into a waveguide. These photons are transferred into the target cavity with a target/source field intensity ratio of 0.12 ± 0.01 (up to 0.49 observed in structures without coupled QDs), showing the system's potential as a fundamental component of a scalable quantum network for building on-chip quantum information processing devices.

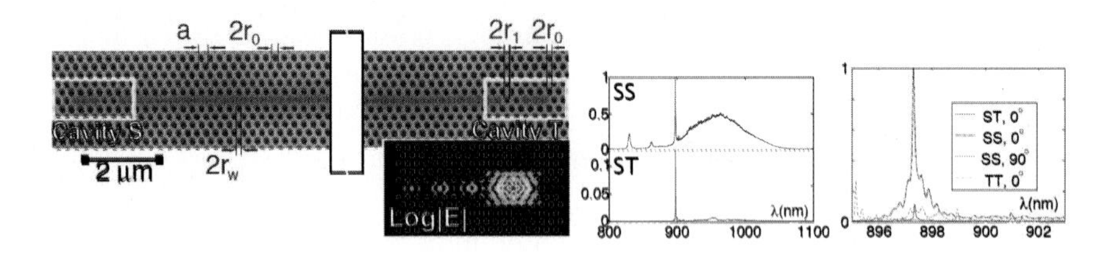

Fig. 17.11. Coupled cavities system. Left: The identical source (S) and target (T) cavities are connected via the 25μm waveguide. Design parameters for the PC circuit are: $a = 256$nm, $r_0 = 0.3a$ $r_1 = 0.25a$, $r_0 = 0.3a$, and $r_w = 0.25a$. Electric field pattern is shown in the inset. Middle: Broad emission in cavity S (plot 'SS') is filtered into the target cavity (plot 'ST'), indicating matched cavity resonances with Q=1000. Right: When the QD exciton at 897.3nm in cavity S is pumped (resonantly at 878nm, with 460μW and 1μm focal spot pump), the emission is observed from S ('SS') and T ('ST'). The cross-polarized spectrum from S shows nearly complete quenching of QD emission ('SS, 90°'). The line at 897.3nm is only observed when S is pumped.

A single-excition transition is coupled to cavity S at 897nm (see Fig. 17.11). The transition is driven resonantly through a higher-order excited QD state with a 878nm pump from a Ti-Sapph laser. The spontaneous emission rate enhancement is measured from the modified emitter lifetime, which is dominated by radiative recombination [13]. A direct streak camera measurement puts the modified lifetime at 116ps (Fig. 17.12(b)). Compared to the average lifetime of 1.4ns for QDs in the bulk semiconductor of this wafer, this corresponds to a Purcell enhancement of $F = 12$. We estimate that this value of F is about 13 times lower than for an ideally aligned dot, indicating spatial mismatch to 28% of the field maximum. The spontaneous emission coupling factor into the cavity mode is then $\beta = F/(F + f) \sim 0.98$, where $f \sim 0.2$ reflects the averaged spontaneous emission rate suppression into other modes due to the bandgap of the surrounding PC [13].

We characterized the exciton emission by measurements of the second-order coherence and indistinguishability of consecutive photons. The second-order coherence $g^{(2)}(t')$ is measured with a Hanbury-Brown and

Twiss interferometer [13]. When the QD in cavity S is pumped resonantly, then photons observed from S shows clear antibunching (Fig. 17.12(a)), with $g^2(0) = 0.35 \pm 0.01$. This value exceeds measurements for cavity-detuned QDs, where typically $g^{(2)}(0) < 0.05$, similar to previous reports [38]. The main contributor to the larger $g^{(2)}(0)$ for the coupled QD is enhanced background emission from nearby transitions and the wetting layer emission tail, which decays at $\sim$ 100ps time scales (see Fig. 17.12(b)) and is not completely filtered by our grating setup. The background emission is rather large in this study because of the high QD density of the sample (e.g., four times larger than that in the experiment [38]).

Because of the shortened lifetime of the cavity-coupled QD exciton, the coherence time of emitted photons becomes dominated by radiative effects and results in high photon indistinguishability [40]. We measured the indistinguishability using the Hong-Ou-Mandel (HOM) type setup [38]. The QD is excited twice every 13ns, with a 2.3ns separation. The emitted photons are directed through a Michelson interferometer with a 2.3ns time difference. The two outputs are collected with single photon counters to obtain the photon correlation histogram shown in the inset of Fig. 17.12(b). The five peaks around delay $\tau = 0$ correspond to the different possible coincidences on the beamsplitter of the leading and trailing photons after passing through the long or short arms $\mathcal{L}$ or $\mathcal{S}$ of the interferometer. If the two photons collide and are identical, then the bosonic symmetry of the state predicts that they must exit in the same port. This photon bunching manifests itself as antibunching in a correlation measurement on the two ports. This signature of photon indistinguishability is apparent in Fig. 17.12(b) in the reduced peaks near zero time delay. Following the analysis of [38], the data (inset Fig. 17.12(b)) indicate a mean wavefunction overlap of $I = 0.67 \pm 0.18$, where we adjusted for the imperfect visibility (88%) of our setup and subtracted dark counts in the calculation. Even with higher spontaneous emission rate enhancement, we expect that $I \lesssim 0.80$ for resonantly excited QDs [51] because of the finite relaxation time, measured here at 23ps by the streak camera.

We will now consider the transfer of single photons to the target cavity T. Experimentally, we verified photon transfer from S to T by spectral measurements (see Fig. 17.11): the exciton line is observed from T only if S is pumped. It is not visible if the waveguide or cavity T itself are pumped, indicating that this line originates from the QD coupled to cavity S and that a fraction of the emission is transferred to T. This emission has the same polarization and temperature-tuned wavelength dependence

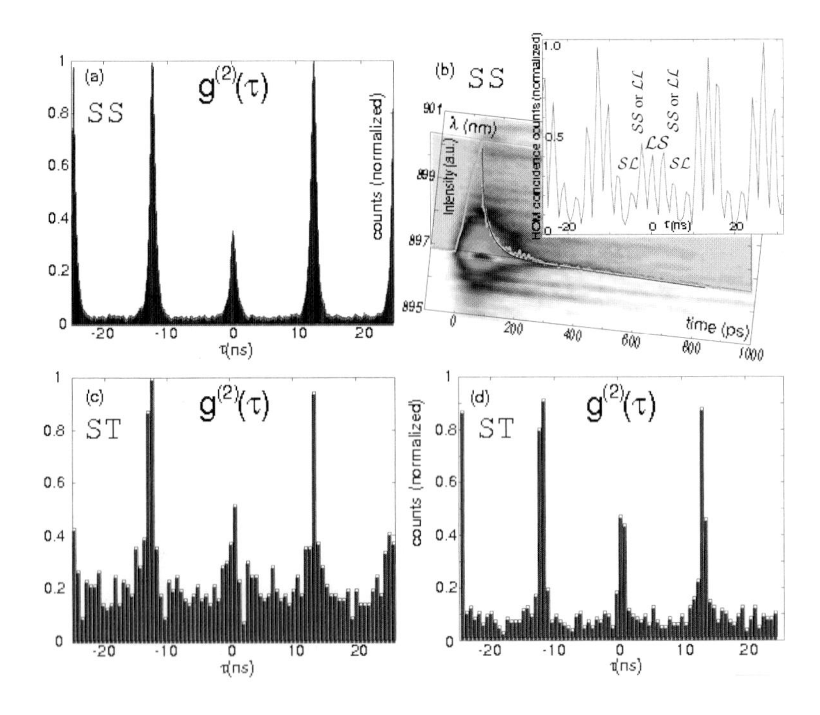

Fig. 17.12. Single photon source characterization. (a) Autocorrelation data when cavity S pumped and collected. (b) Streak camera data indicate exciton lifetime $\tau = 116$ps. The rise-time is measured at 23ps with a lower-density grating with higher time response (data not shown). *Inset:* Two-photon interference experiment. Colliding indistinguishable photons interfere, resulting in a decreased area of peak $\mathcal{LS}$. The area does not vanish largely because of non-zero $g^{(2)}(0)$ of the source. (c) Autocorrelation data when cavity S pumped and T is collected (with grating filter). (d) Cavity S pumped and T collected directly (no grating filter).

as emission from S. Photon autocorrelation measurements on the signal from T indicate the antibunching characteristic of a single emitter when S is pumped (Fig. 17.12(c)). The signal-to-noise ratio is rather low because autocorrelation count rates are ~ 0.014 times lower than for collection from S. Nevertheless, the observed antibunching does appear higher, in large part because the background emission from cavity S is additionally filtered in the transfer to T, as shown in Fig. 17.11(b). Indeed, this filtering through the waveguide/cavity system suffices to bypass the spectrometer in the HBT setup (a 10nm bandpass filter was used to eliminate room lights). The count rate is about three times higher while antibunching, $g^2(0) = 0.50 \pm 0.11$, is still clearly evident (Fig. 17.12(d)). The largest contribution to $g^2(0)$ comes from imperfectly filtered photoluminescence near the QD distribution peak seen in Fig. 17.11(c). This on-chip filtering will be essential in future quantum information processing applications and should also find uses in optical communications as a set of cascaded drop filters.

This system forms a basic building block of a quantum network consisting of a QD with large coupling to a high-Q/V cavity, which in turn is coupled to a target cavity via a waveguide. The coupled system functions as an efficient on-demand source of single photons with mean wavepacket overlap of $\sim 67\%$, spontaneous emission coupling factor $\beta \sim 0.98$ into the cavity mode, and high out-coupling efficiency into the waveguide. These single photons from cavity S are channelled to the target cavity, as confirmed by localized spectroscopic measurements. We measured a high photon transfer with a field intensity ratio of 0.12 ± 0.01 for this system. In other structures we measured field ratios up to 0.49 ± 0.04, though without coupled QDs. These efficiencies greatly exceed what is possible in off-chip transfer from a cryostat-mounted PC structure and demonstrate the great potential of this system as a building block of future on-chip quantum information processing systems, such as those described in the following section.

17.6. Quantum Dot-photonic Crystal Based Quantum Networks

One of the main limitations in quantum communication is the large losses induced by optical channels. These losses lead to a communication rate which is exponentially decaying with distance. In classical communication, these channel losses can be overcome by optical amplifiers placed along the path of the fiber. At each amplifier node, the signal is boosted to overcome the fiber losses. This solution is not possible in quantum communication because amplification injects quantum noise into the channel, which destroys the state of the qubit.

The exchange of quantum bits over long distances can be achieved in a different manner. The key is to create an entangled state between the two communication points. The method of quantum teleportation [52] can be used to exchange qubits between nodes. Entanglement can be generated over long distance by creating entanglement between a large number of intermediate nodes (or repeater stations), which are spaced by short distances along the communication path. Entanglement swapping [53] can then be used to create entanglement between the two end nodes. This procedure can be made fault tolerant by using entanglement purification protocols [54].

Quantum repeaters can be implemented in a variety of systems. All optical proposals [55], as well as proposals based on atomic systems [56] have been extensively studied. More recent proposals rely on optically controlled spins in semiconductor cavities interacting with coherent light [57, 58] (for the detailed description, please refer to the chapter by Ladd et al in this volume).

In this section, we describe a method for implementing entanglement, as well as a full Bell measurement on a quantum dot or an atomic system using only interaction with a coherent field [23, 24]. This leads to an extremely simple implementation of a quantum repeater that relies only on the building blocks we already demonstrated in the previous sections of this chapter. The scheme exploits the interaction of the cavity-dipole system, but does not require the strong coupling regime. As described in Section 4, our recent experimental demonstrations show that this regime is practically achievable with current technology.

To implement a full quantum repeater requires the capability to perform three important operations. First, one must be able to generate entanglement between dipole systems that are separated by large distances. Second, one must be able to measure the state of the individual dipole systems. Finally, an efficient quantum repeater requires the ability to perform a full Bell measurement between two dipole systems in order to implement an efficient entanglement swap and entanglement purification.

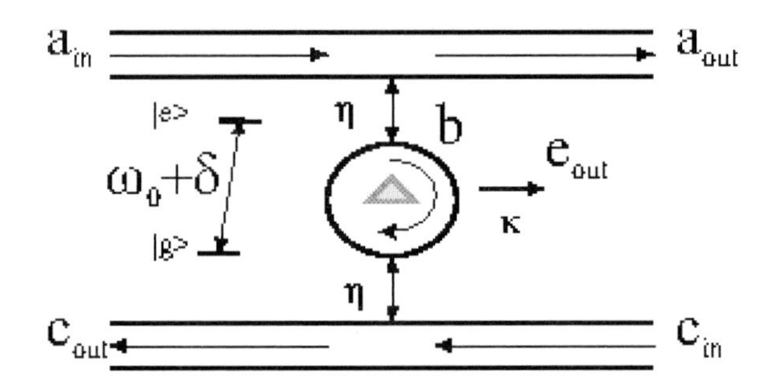

Fig. 17.13. Cavity- waveguide system for quantum repeaters.

Fig. 17.13 shows a schematic of the type of system we are considering. A cavity containing a single dipole emitter is evanescently coupled to two waveguides. The cavity is assumed to have a single relevant mode, which couples only to the forward-propagating fields (e.g., a whispering gallery mode). The dipole may be detuned by δ from cavity resonance, denoted ω_0, while g is coupling strength between the dipole and the cavity field. Both waveguides are assumed to have equal coupling rates into the cavity (η). The parameter η is the energy decay rate from the cavity into each waveguide. The operators $\hat{\mathbf{a}}_{in}$ and $\hat{\mathbf{c}}_{in}$ are the field operators for the flux of the two input ports of the waveguides, while $\hat{\mathbf{e}}_{in}$ is the operator for potential leaky modes due to all other losses such as out-of-plane scattering and material absorption. The bare cavity has a resonant frequency ω_0 and an energy decay rate κ in the absence of coupling to the waveguides (for simplicity, κ in this Section denotes the energy decay rate and not the cavity field decay rate, which is two times smaller). This decay rate is related to the cavity quality factor Q by $\kappa = \omega_0/Q$. For $\eta \gg \kappa$, the critical coupling is achieved and the input field from one waveguide is completely transmitted to the other [59]. The dipole operator σ_- has a decay rate γ, and $\hat{\mathbf{f}}$ is a noise operator which preserves the commutation relation.

We start with the Heisenberg operator equations for the cavity field operator $\hat{\mathbf{b}}$ and dipole operator σ_-, given by [60]. Assuming the cavity is excited by a weak monochromatic field with frequency ω, we calculate the response of $\hat{\mathbf{b}}$ and σ_- in frequency space. We assume that the cavity decay rate is much faster than the dipole decay rate, so that $\eta/\gamma \approx 0$. This is a realistic assumption for a quantum dot coupled to a photonic crystal cavity, but does not necessarily apply in atomic systems coupled to very high-Q optical resonators. We also consider the weak excitation limit, where the quantum dot is predominantly in the ground state ($\langle\sigma_z(t)\rangle \approx -1$ for all time). This condition basically states that the incoming photon flux $\langle\hat{\mathbf{a}}_{in}^\dagger\hat{\mathbf{a}}_{in}\rangle$ must be much smaller than the modified spontaneous emission decay rate of the emitter (in the limit that the cavity decay is dominated by the leakage into the waveguide described by η), and is well satisfied in the operating regime we are working in. In this limit the waveguide input-output relations are given by the expressions

$$\hat{\mathbf{a}}_{out} = \frac{-\eta\hat{\mathbf{c}}_{in} + \left(-i\Delta\omega + \frac{\kappa}{2} + \frac{g^2}{-i(\Delta\omega-\delta)+\gamma}\right)\hat{\mathbf{a}}_{in} - \sqrt{\kappa\eta}\hat{\mathbf{e}}_{in}}{-i\Delta\omega + \eta + \kappa/2 + \frac{g^2}{-i(\Delta\omega-\delta)+\gamma}} \tag{17.4}$$

$$\hat{\mathbf{c}}_{out} = \frac{-\eta\hat{\mathbf{a}}_{in} + \left(-i\Delta\omega + \frac{\kappa}{2} + \frac{g^2}{-i(\Delta\omega-\delta)+\gamma}\right)\hat{\mathbf{c}}_{in} - \sqrt{\kappa\eta}\hat{\mathbf{e}}_{in}}{-i\Delta\omega + \eta + \kappa/2 + \frac{g^2}{-i(\Delta\omega-\delta)+\gamma}} \tag{17.5}$$

where $\Delta\omega = \omega - \omega_0$.

Consider the case where the dipole is resonant with the cavity, so that $\delta = 0$. In the ideal case, the bare cavity decay rate κ is very small and can be set to zero. In this limit, when the field is resonant with the cavity and $g = 0$ we have $\hat{\mathbf{a}}_{in} = -\hat{\mathbf{c}}_{out}$, as one would expect from critical coupling. In the opposite regime, when $g^2 \gg (\eta + \kappa/2)\gamma$ we have $\hat{\mathbf{a}}_{in} = \hat{\mathbf{a}}_{out}$, so that the field remains in the original waveguide. This condition can be re-written as $F = g^2/(\eta + \kappa/2)\gamma \gg 1$, where F is the Purcell factor. Thus, in order to make the waveguide transparent (i.e., decouple the field from the cavity), we need to achieve large Purcell factors. However, we do not need the full normal mode splitting condition $g > \eta + \kappa/2$. When $\gamma \ll \eta + \kappa/2$ we can achieve transparency for much smaller values of g.

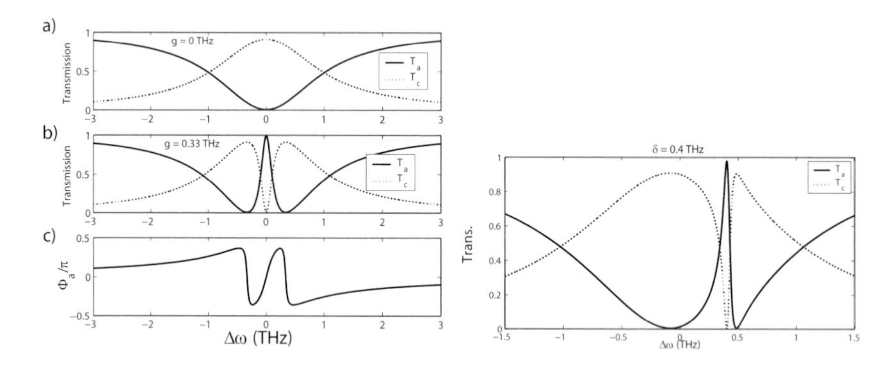

Fig. 17.14. Left: Probability for field in $\hat{\mathbf{a}}_{in}$ to transmit into $\hat{\mathbf{a}}_{out}$ and $\hat{\mathbf{c}}_{out}$ respectively. (a) Transmission with no dipole in cavity. (b) Transmission with a dipole in the cavity. (c) Phase imposed on transmitted field when a dipole is present in the cavity. Right: Transmission of waveguide when the dipole is detuned from the cavity by δ.

Fig. 17.14 plots the probability that $\hat{\mathbf{a}}_{in}$ transmits into $\hat{\mathbf{a}}_{out}$ and $\hat{\mathbf{c}}_{out}$. Assuming that the initial field begins in mode $\hat{\mathbf{a}}_{in}$, we define $\hat{\mathbf{a}}_{out}/\hat{\mathbf{a}}_{in} = \sqrt{T_a}e^{i\Phi_a}$ and $\hat{\mathbf{c}}_{out}/\hat{\mathbf{a}}_{in} = \sqrt{T_c}e^{i\Phi_c}$, and use cavity and dipole parameters that are appropriate for a photonic crystal cavity coupled to a quantum dot. We set $\eta = 1$THz which is about a factor of 10 faster than κ for a cavity with a quality factor of $Q = 10000$. We set $g = 0.33$THz, a number calculated from FDTD simulations of cavity mode volume for a single defect dipole cavity in a planar photonic crystal coupled to a quantum dot [20]. The dipole decay rate is set to $\gamma = 0.5$GHz, taken from experimental measurements [50].

Panel (a) of Fig. 17.14(Left) considers the case where the cavity does not contain a dipole. In this case $g = 0$, representing a system where the two waveguides are coupled by a drop filter cavity. The width of the transmission spectrum for the drop filter is determined by the lifetime of the cavity, which in our case is dominated by η. When a dipole is present in the cavity, the result is plotted in panel (b). In this case, a very sharp peak in the transmission spectrum appears at $\Delta\omega = 0$, with a width of approximately 0.1THz, although g is three times smaller than the cavity decay rate η. On resonance, the field does not couple to the cavity and is instead transmitted through the waveguide. We refer to this effect as the Dipole Induced Transparency (DIT), as the presence of a single dipole in the channel drop filter changes its transmission. In the high-Q regime where $g \gg \eta$ and in the limit of $\eta \gg \kappa/2$ (cavity losses are dominated by waveguides), the full width half maximum of the transmission dip is equal to $2g$ as expected by normal mode splitting. In the low-Q regime where $g \ll \eta$ and assuming that $\eta \gg \kappa/2$, the spectral width of the transmission peak is equal to g^2/η (the modified spontaneous emission rate of the dipole).

In panel (c) we plot Φ_a for the case where $g = 0.33$THz. The region near zero detuning exhibits very large dispersion, which results in a group delay given by $\tau_g = (\eta + \kappa/2)/g^2 = \tau_{mod}$, where τ_{mod} is the modified spontaneous emission lifetime of the dipole. One can show that the group velocity dispersion at zero detuning vanishes, ensuring that the pulse shape is preserved.

We now consider the effect of detuning the dipole. The transmission spectrum for a detuned dipole is plotted in Fig. 17.14(Right), where δ denotes the dipole detuning from the cavity resonance. Introducing a detuning in the dipole causes a shift in the location of the transmission peak, so that destructive interference occurs when the field frequency is equal to the dipole frequency. Thus, we do not have to hit the cavity resonance very accurately to observe DIT. We only need to overlap the dipole resonance within the cavity transmission spectrum.

The fact that we can switch the transmission of a waveguide by the state of a dipole in the low-Q regime can be extremely useful for quantum information processing. As one example, we now present a way in which DIT can be applied to engineering quantum repeaters for long distance quantum communication. In panel (a) of Fig. 17.15(Left) we show how DIT can be used to generate entanglement between two spatially separated dipoles. A weak coherent beam is split on a beamsplitter, and each port of the beamsplitter is then sent to two independent cavities containing dipoles. The waveguide fields are then mixed on a beamsplitter such that constructive interference is observed in ports $\hat{\mathbf{f}}$ and $\hat{\mathbf{h}}$. Each dipole is assumed to have

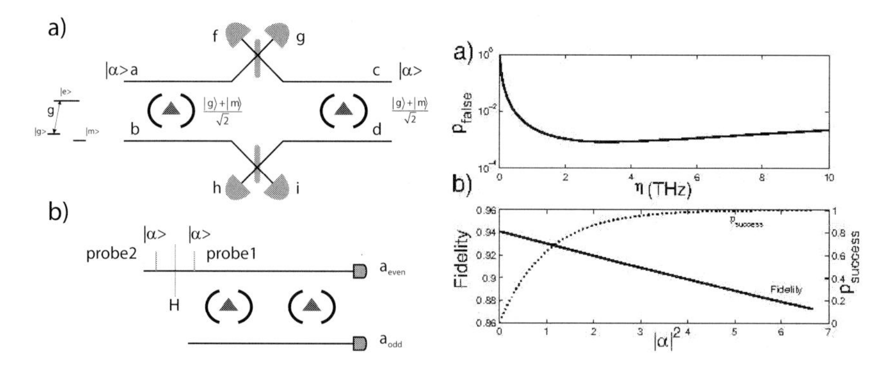

Fig. 17.15. Left: Application of DIT to quantum repeaters. (a) A method for generating entanglement between two dipoles using DIT. (b) A non-destructive Bell measurement. Right: (a) Probability of detecting even parity for an odd parity state as a function of η. (b) Solid line plots the fidelity of the state $(|gg\rangle \pm |mm\rangle)/\sqrt{2}$ after a parity measurement. Dotted line plots the probability that the measuring field contains at least one photon for detection.

three relevant states, a ground state, an excited state, and a long lived metastable state which we refer to as $|g\rangle$, $|e\rangle$, and $|m\rangle$ respectively. The transition from ground to excited state is assumed to be resonant with the cavity while the metastable to excited state transition is well off resonance from the cavity, and is thus assumed not to couple to state $|e\rangle$. The states $|g\rangle$ and $|m\rangle$ represent the two qubit states of the dipole.

When the dipole is in state $|m\rangle$, it does not couple to the cavity, which now behaves as a drop filter. Thus, we have a system that transforms $\hat{\mathbf{a}}_{in}^{\dagger}|g\rangle|0\rangle \rightarrow \hat{\mathbf{a}}_{out}^{\dagger}|g\rangle|0\rangle$ and $\hat{\mathbf{a}}_{in}^{\dagger}|m\rangle|0\rangle \rightarrow -\hat{\mathbf{c}}_{out}^{\dagger}|m\rangle|0\rangle$. This operation can be interpreted as a C-NOT gate between the state of the dipole and the incoming light. When the dipole is in a superposition of the two states, this interaction generates entanglement between the path of the field and the dipole state. After the beamsplitter, this entanglement will be transferred to the two dipoles. If the state of both dipoles is initialized to $(|g\rangle+|m\rangle)/\sqrt{2}$, it is straightforward to show that a detection event in ports $\hat{\mathbf{g}}$ or $\hat{\mathbf{i}}$ collapses the system to $(|g,m\rangle - |m,g\rangle)/\sqrt{2}$.

Another important operation for designing repeaters is a Bell measurement, which measures the system in the states $|\phi_{\pm}\rangle = (|gg\rangle \pm |mm\rangle)/\sqrt{2}$ and $|\psi_{\pm}\rangle = (|gm\rangle \pm |mg\rangle)/\sqrt{2}$. Panel (b) of Fig. 17.15(Left) shows how to implement a complete Bell measurement between two dipoles using only cavity waveguide interactions with coherent fields. The two cavities containing the dipoles are coupled to two waveguides. When a coherent field

$|\alpha\rangle$ is sent down waveguide 1, each dipole will flip the field to the other waveguide if it is in state $|m\rangle$, and will keep the field in the same waveguide if it is in state $|g\rangle$. Thus, a detection event at ports $\hat{\mathbf{a}}_{even}$ and $\hat{\mathbf{a}}_{odd}$ corresponds to a parity measurement. A Bell measurement can be made by simply performing a parity measurement on the two dipoles, then a Hadamard rotation on both dipoles, followed by a second parity measurement. This is because a Hadamard rotation flips the parity of $|\psi_+\rangle$ and $|\phi_-\rangle$, but does not affect the parity of the other two states.

The performance of the Bell apparatus is analyzed in Fig. 17.15(Right). Panel (a) plots the probability that an odd parity state will falsely create a detection event in port $\hat{\mathbf{a}}_{even}$, as a function of η. The probability becomes high at large η due to imperfect transparency. It also increases at small η because of imperfect drop filtering. The minimum value of about 10^{-3} is achieved at $\eta \approx 3$THz. In panel (b) of Fig. 17.15(Right) we plot both the fidelity and success probability of a parity measurement as a function of the number of photons in the probe field. The fidelity is calculated by applying the Bell measurement to the initial state $|\psi_i\rangle = (|g,g\rangle \pm |m,m\rangle)/\sqrt{2}$, and defining the fidelity of the measurement as $F = |\langle\psi_f|\,\psi_i\rangle|^2$, where $|\psi_f\rangle$ is the final state of the total system which includes the external reservoirs. The probability of success is defined as the probability that at least one photon is contained in the field. The fidelity is ultimately limited by cavity leakage, which results in "which path" information being leaked to the environment. This information leakage depends on the strength of the measurement which is determined by the number of photons in the probe fields. Using more probe photons results in a higher success probability, but a lower fidelity. To calculate this tradeoff, we use previously described values for cavity and reservoir losses, and set the coupling rate η to 4THz, which is where the probability of false detection is near its minimum. At an average of three photons, a fidelity of over 90% can be achieved with a success probability exceeding 95%. These numbers are already promising, and improved cavity and dipole lifetimes could lead to even better operation.

17.7. Conclusions and Future Prospects

In recent years, tremendous progress has been made in realizing many of the basic building blocks of practical quantum information processing systems. In this chapter, we have reviewed the quantum dot-photonic crystal based approach that our group has taken. The technique for selective tuning of quantum dots on chip that we have recently developed [22] enables tuning of individual dots by 1.8nm, and therefore addresses the problem of large quantum dot inhomogeneous broadening, which has been considered the main obstacle in practical implementation of such systems. We have successfully demonstrated cavity QED with a single quantum dot in photonic crystal both in the weak and strong coupling regimes, even in combination with the selective tuning technique [13, 22], which is crucial for implementing efficient nonclassical light sources and quantum gates. We have also demonstrated a rudimentary quantum network, by generating and transferring single photons on a photonic crystal chip, via a waveguide between two cavites [16]. Finally, the demonstrated basic building blocks are already sufficiently good to enable construction of more complex systems, including a quantum repeater based on the described platform [23, 24].

Acknowledgments

The authors would like to thank Prof. Pierre Petroff (UCSB) and Prof. Yoshi Yamamoto (Stanford) for providing quantum dot material used in the described experiments. Financial support for this work has been provided by the MURI Center for Photonic Quantum Information Systems (ARO/DTO program No. DAAD19-03-1-0199) and the ONR Young Investigator Award.

Edo Waks is currently with the Department of Electrical Engineering, University of Maryland, College Park, MD.

References

[1] J. I. Cirac, P. Zoller, H. J. Kimble, and H. Mabuchi, Quantum State Transfer and Entanglement Distribution among Distant Nodes in a Quantum Network, *Physical Review Letters.* **78**(16), 3221–24, (1997).

[2] L. Duan and H. Kimble, Scalable Photonic Quantum Computation through Cavity-Assisted Interactions, *Physical Review Letters.* **92**, 127902, (2004).

[3] C. Monroe, D. Meekhof, B. King, W. Itano, , and D. Wineland, Demonstration of a Fundamental Quantum Logic Gate, *Physical Review Letters.* **75**, 4714–4717, (1995).

[4] J. Chiaverini, D. Leibfried, T. Schaetz, M. Barrett, R. Blakestad, J. Britton, W. Itano, J. Jost, E. Knill, C. Langer, R. Ozeri, and D. Wineland, Realization of quantum error correction, *Nature.* **432**, 602–605, (2005).

[5] A. Imamoğlu, D. D. Awschalom, G. Burkard, D. P. DiVincenzo, D. Loss, M. Sherwin, and A. Small, Quantum Information Processing Using Quantum Dot Spins and Cavity QED, *Physical Review Letters.* **83**, 4204–4207, (1999).

[6] D. Fattal, *Single photons for quantum information processing.* (Ph.D. Thesis, Stanford University, 2005).
[7] W.Yao, R. Liu, and L. Sham, Theory of Control of the Spin-Photon Interface for Quantum Networks , *Physical Review Letters.* **95**, 030504, (2005).
[8] B. Blinov, D. Moehring, L. Duan, and C. Monroe, Observation of entanglement between a single trapped atom and a single photon, *Nature.* **428**, 153–157, (2004).
[9] Q. Turchette, C. Hood, W. Lange, H. Mabuchi, and H. J. Kimble, Measurement of conditional phase shifts for quantum logic, *Physical Review Letters.* **75**, 4710–4713, (1995).
[10] K. M. Birnbaum, A. Boca, R. Miller, A. D. Boozer, T. E. Northup, and H. J. Kimble, Photon blockade in an optical cavity with one trapped atom, *Nature.* **436**, 87–90, (2005).
[11] A. Rauschenbeutel, G. Nogues, S. Osnaghi, P. Bertet, M. Brune, J. M. Raimond, and S. Haroche, Coherent Operation of a Tunable Quantum Phase Gate in Cavity QED, *Physical Review Letters.* **83**, 5166–5169, (1999).
[12] G. Nogues, A. Rauschenbeutel, S. Osnaghi, M. Brune, J. M. Raimond, and S. Haroche, Seeing a single photon without destroying it, *Nature.* **400**, 239–242, (1999).
[13] D. Englund, D. Fattal, E. Waks, G. Solomon, B. Zhang, T. Nakaoka, Y. Arakawa, Y. Yamamoto, and J. Vučković, Controlling the Spontaneous Emission Rate of Single Quantum Dots in a Two-Dimensional Photonic Crystal, *Physical Review Letters.* **95**(013904), (2005).
[14] T. Yoshie, A. Scherer, J. Hendrickson, G. Khitrova, H. M. Gibbs, G. Rupper, C. Ell, O. B. Shchekin, and D. G. Deppe, Vacuum Rabi splitting with a single quantum dot in a photonic crystal nanocavity, *Nature.* **432**, 200–203, (2004).
[15] K. Hennessy, A. Badolato, M. Winger, D. Gerace, M. Atatüre, S. Gulde, S. Falt, E. Hu, and A. Imamoğlu, Quantum nature of a strongly coupled single quantum dot-cavity system, *Nature.* **445**, 896–899, (2007).
[16] D. Englund, A. Faraon, B. Zhang, Y. Yamamoto, and J. Vučković, Generation and Transfer of Single Photons on a Photonic Crystal Chip, *Optics Express.* **15**, 5550–5558, (2007).
[17] J. Vučković, C. Santori, D. Fattal, M. Pelton, G. Solomon, and Y. Yamamoto, *in Optical Microcavities, edited by K. Vahala.* (World Scientific, Singapore, 2004).
[18] D. Englund, I. Fushman, and J. Vučković, General Recipe for Designing Photonic Crystal Cavities, *Optics Express.* **12**(16), 5961–75, (2005).
[19] J. Vučković, M. Lončar, H. Mabuchi, and A. Scherer, Design of photonic crystal microcavities for cavity QED, *Physical Review E.* **65**, 016608, (2002).
[20] J. Vučković and Y. Yamamoto, Photonic crystal microcavities for cavity quantum electrodynamics with a single quantum dot, *Applied Physics Letters.* **82**(15), 2374–2376, (2003).
[21] P. Michler, *Single quantum dots: Fundamentals, Applications, and New Concepts.* (Topics in Applied Physics, Springer-Verlag, 2003).
[22] A. Faraon, D. Englund, I. Fushman, J. Vučković, N. Stoltz, and P. Petroff, Local Quantum Dot tuning on photonic crystal chips, *Applied Physics Let-*

ters. **90**, 213110, (2007).

[23] E. Waks and J. Vučković, Dipole induced transparency in drop-filter cavity-waveguide systems, *Physical Review Letters.* **96**(153601), (2006).

[24] E. Waks and J. Vučković, Dispersive properties and large Kerr non-linearities using Dipole Induced Transparency in a single-sided cavity, *Physical Review A.* **73**(041803(R)), (2006).

[25] A. Badolato, K. Hennessy, M. Atatüre, J. Dreiser, E. Hu, P. Petroff, and A. Imamoğlu, Deterministic Coupling of Single Quantum Dots to Single Nanocavity Modes, *Science.* **308**(5725), 1158 – 1161, (2005).

[26] A. Högele, S. Seidl, M. Kroner, K. Karrai, R. Warburton, B. Gerardot, and P. Petroff, Voltage-Controlled Optics of a Quantum Dot, *Physical Review Letters.* **93**(217401), (2004).

[27] D. Haft, C. Schulhauser, A. Govorov, R. Warburton, K. Karrai, J. Garcia, W. Schoedfeld, and P. Petroff, Magneto-optical properties of ring-shaped self-assembled InGaAs quantum dots, *Physica E.* **13**, 165–169, (2002).

[28] A. Kiraz, P. Michler, C. Becher, B. Gayral, A. Imamoğlu, L. Zhang, E. Hu, W. Schoenfeld, and P. Petroff, Cavity-quantum electrodynamics using a single InAs quantum dot in a microdisk structure, *Applied Physics Letters.* **78** (25), 3932–3934, (2001).

[29] S. Seidl, M. Kroner, A. Högele, K. Karrai, R. J. Warburton, A. Badolato, and P. M. Petroff, Effect of uniaxial stress on excitons in a self-assembled quantum dot, *Applied Physics Letters.* **88**(203113), (2006).

[30] W. Fon, K. Schwab, J. Worlock, and M. Roukes, Phonon scattering mechanisms in suspended nanostructures from 4 to 40 K, *Physical Review B.* **66** (045302), (2002).

[31] C. Thurmond, The standard thermodynamic functions for the formation of electrons and holes in Ge, Si, GaAs and GaP, *J. Electrochem. Soc.* **122** (1133), (1975).

[32] K. Hennessy, A. Badolato, A. Tamboli, P. M. Petroff, E. Hu, M. Atatüre, J. Dreiser, and A. Imamoğlu, Tuning photonic crystal nanocavity modes by wet chemical digital etching, *Applied Physics Letters.* **87**(021108), (2005).

[33] S. Strauf, M. T. Rakher, I. Carmeli, K. Hennessy, C. Meier, A. Badolato, M. J. A. DeDood, P. M. Petroff, E. L. Hu, E. G. Gwinn, and D. Bouwmeester, Frequency control of photonic crystal membrane resonators by monolayer deposition, *Applied Physics Letters.* **88**(043116), (2006).

[34] A. Rastelli, A. Ulhaq, S. Kiravittaya, L. Wang, A. Zrenner, and O. Schmidt, In situ laser microprocessing of single self-assembled quantum dots and optical microcavities, *Applied Physics Letters.* **90**(073120), (2007).

[35] H. J. Kimble, *in Cavity Quantum Electrodynamics, edited by P. Berman.* (Academic Press, San Diego, 1994).

[36] Y. Yamamoto and A. İmamoğlu, *Mesoscopic quantum optics.* (John Wiley & Sons, INC., New York, 1999).

[37] E. M. Purcell, Spontaneous emission probabilities at radio frequencies, *Physical Review.* **69**, 681, (1946).

[38] C. Santori, D. Fattal, J. Vučković, G. Solomon, and Y. Yamamoto, Indistinguishable photons from a single-photon device, *Nature.* **419**(6907), 594–597,

(2002).

[39] J. McKeever, A. Boca, A. D. Boozer, J. R. Buck, and H. J. Kimble, Experimental realization of a one-atom laser in the regime of strong coupling, *Nature.* **425**(6955), 268–271, (2003).

[40] A. Kiraz, M. Atatüre, and I. Imamoğlu, Quantum-dot single-photon sources: Prospects for applications quantum-information processing, *Phys. Rev. A.* **69**, 032305, (2004).

[41] D. Bouwmeester, J. Pan, K. Mattle, M. Eibl, H. Weinforuter, and A. Zeilinger, Experimental quantum teleportation, *Nature.* **390**, 575–9, (1997).

[42] E. Knill, R. Laflamme, and G. J. Milburn, A scheme for efficient quantum computation with linear optics, *Nature.* **409**, 4652, (2001).

[43] M. A. Nielsen and I. L. Chuang, *Quantum Computation and Quantum Information.* (Cambridge Univ. Press, Cambridge, 2000).

[44] P. Grangier, B. Sanders, and J. Vučković, *in Special issue of the New Journal of Physics on Single Photons on Demand.* vol. 6, 2004.

[45] P. Michler, A. Kiraz, C. Becher, W. V. Schoenfeld, P. M. Petroff, L. Zhang, E. Hu, and A. Imamoğlu, A quantum dot single-photon turnstile device, *Science.* **290**, 2282–2285, (2000).

[46] C. Santori, M. Pelton, G. Solomon, Y. Dale, and Y. Yamamoto, Triggered single photons from a quantum dot, *Physical Review Letters.* **86**(8), 1502–1505, (2001).

[47] V. Zwiller, H. Blom, P. Jonsson, N. Panev, S. Jeppesen, T. Tsegaye, E. Goobar, M. E. Pistol, L. Samuelson, and G. Bjork, Single quantum dots emit single photons at a time: antibunching experiments, *Applied Physics Letters.* **78**, 2476–2478, (2001).

[48] Z. Yuan, B. E. Kardynal, R. M. Stevenson, A. J. Shields, C. J. Lobo, K. Cooper, N. S. Beattie, D. A. Ritchie, and M. Pepper, Electrically driven single-photon source, *Science.* **295**, 102–105, (2002).

[49] E. Moreau, I. Robert, J. M. Gérard, I. Abram, L. Manin, and V. Thierry-Mieg, Single-mode solid-state single photon source based on isolated quantum dots in pillar microcavities, *Applied Physics Letters.* **79**(18), 2865–2867, (2001).

[50] J. Vučković, D. Fattal, C. Santori, G. Solomon, and Y. Yamamoto, Enhanced single photon emission from a quantum dot in a micropost microcavity, *Applied Physics Letters.* **82**, 3596–3598, (2003).

[51] J. Vučković, D. Englund, D. Fattal, E. Waks, and Y. Yamamoto, Generation and manipulation of nonclassical light using photonic crystals, *Physica E.* **31**(2), (2006).

[52] C. H. Bennett, G. Brassard, C. Crpeau, R. Jozsa, A. Peres, and W. K. Wootters, Teleporting an unknown quantum state via dual classical and Einstein-Podolsky-Rosen channels, *Physical Review Letters.* **70**, 1895–1899, (1993).

[53] M. Zukowski, A. Zeilinger, M. A. Horne, and A. K. Ekert, Event-ready-detectors: Bell experiment via entanglement swapping, *Physical Review Letters.* **71**, 4287–4290, (1993).

[54] C. H. Bennett, G. Brassard, S. Popescu, B. Schumacher, J. A. Smolin, and W. K. Wootters, Purification of Noisy Entanglement and Faithful Teleportation via Noisy Channels , *Physical Review Letters.* **76**, 722–725, (1996).

[55] H. Briegel, W. Dur, J. Cirac, and P. Zoller, Quantum Repeaters: The Role of Imperfect Local Operations in Quantum Communication, *Physical Review Letters.* **81**, 5932–5935, (1998).

[56] L. Duan, M. Lukin, J. Cirac, and P. Zoller, Long-distance quantum communication with atomic ensembles and linear optics, *Nature.* **414**, 413–418, (2001).

[57] L. Childress, J. M. Taylor, A. S. Srensen, and M. D. Lukin, Fault-tolerant quantum repeaters with minimal physical resources and implementations based on single-photon emitters, *Physical Review A.* **72**, 052330, (2005).

[58] T. D. Ladd, P. van Loock, K. Nemoto, W. J. Munro, and Y. Yamamoto, Hybrid quantum repeater based on dispersive CQED interactions between matter qubits and bright coherent light, *New Journal of Physics.* **8**, 184, (2006).

[59] C. Manolatou, M. Khan, S. Fan, P. Villeneuve, H. Haus, and J. Joannopoulos, Coupling of modes analysis of resonant channel add-drop filters, *IEEE Journal of Quantum Electronics.* **35**, 1322–1331, (1999).

[60] D. Walls and G. Milburn, *Quantum Optics.* (Springer, Berlin, 1994).

Chapter 18

High-Speed Quantum Repeaters and Quantum Computers with Optically Controlled Spins in Semiconductors

T. D. Ladd[1,2], K. Sanaka[1,2], K. M. C. Fu[1], S. Koseki[1], D. Press[1], S. M. Clark[1], K. De Greve[1], and Y. Yamamoto[1,2]

[1]*Edward L. Ginzton Laboratory, Stanford University, Stanford, CA 94305, USA*

[2]*National Institute of Informatics, 2-1-2 Hitotsubashi, Chiyoda-ku, Tokyo 101-8430, Japan*

Most designs for fault tolerant quantum communication networks and quantum computers are limited by the difficulty of transporting quantum information between distant qubits. In this chapter, we consider schemes to overcome this difficulty by coupling distant semiconductor spins in modest-Q microcavities via waveguided photons. We first discuss potential designs for fast quantum repeaters using practical physical resources, such as normal laser sources and optical homodyne measurement. We then indicate how similar resources may also form the basis of an architecture for quantum computers with promising elements for fault-tolerance, such as fast QND measurements, arbitrary non-local two-qubit gates, and a fast logical clock speed. The physical implementation of such architectures depends on the development of electron-spin and nuclear-spin qubits well coupled to optical cavities; we discuss preliminary experimental efforts to develop these qubits in several different semiconductors.

Contents

18.1. Introduction

Electron and nuclear spins in semiconductors provide promising candidates for qubits, but a large number of challenges must be overcome before they may be effectively used for fast quantum information processing on a large scale. Such qubits are always subject to some decoherence, and our methods for controlling them are always imperfect. The development of fault-tolerant architectures is critical for large-scale quantum communication networks and quantum computers. Most semiconductor-based schemes for quantum logic rely on nearest-neighbor spin-spin couplings, based on exchange or dipole-dipole couplings (for example, see Chapters 1 and 2). Fault tolerant architectures with such schemes are possible, but make high demands on error rates and result in low final computation speeds.

Improved error thresholds and faster rates are available in architectures that allow two distant qubits to couple directly and quickly. One way to allow such non-local couplings is to use waveguided light in a semiconductor chip, drawing from currently developing microphotonic technologies for classical high-speed networking. Optical pulses from controlled laser sources may interface with electron spins bound to impurities or embedded in quantum dots via a strong cavity-enhanced coupling to excitonic states. These pulses may move quantum information quickly around the chip, connecting distant spins. The controlled photonic interface in such architectures is called a quantum bus, or "qubus."

If the qubus is fed into a waveguide that extends over very long distances, such as an optical fiber, a network of qubus-based devices may be used to extend the distance of entangled quantum memories arbitrarily far, which may be applied to the distribution of absolutely secure secret keys for cryptography, or to the networking of distant quantum computers. Such a system of networked quantum memories is called a quantum repeater. We will begin our discussion with several schemes for quantum repeaters in Sec. 18.2.

The quantum logic required in a quantum repeater is relatively simple, and the number of qubits is small (10 to 100 per device). For quantum computation, arbitrarily complex fault-tolerant logic with a very large number of qubits is needed. The distant couplings and fast control times in qubus devices, which we discuss in Sec. 18.3, may allow such large-scale computation.

The implementation of such devices in real semiconductors still requires substantial research, but appears to be realistic in view of recent experimental developments. In Sec. 18.4, we will give some specific instances of candidate semiconductor systems with recent experimental progress.

18.2. Differential-phase-shift Quantum Repeater

A quantum repeater system must enable entanglement over a long distance, it must store this entanglement, and if the entanglement is noisy, it must allow the noise to be removed [1]. The first step toward such a system, which we discuss in Sec. 18.2.1, is an efficient interface between light and isolated semiconductor electron spins. There are several possible physical mechanisms for this interface, but those which couple the state of the electron spin to an optical phase seem particularly promising in terms of efficiency. We will discuss two ways this may be done in Sec. 18.2.2 and Sec. 18.2.3, and comment on the performance of the resulting quantum repeater system in Sec. 18.2.4.

18.2.1. *Excitonic interface to the qubus*

For high-fidelity, optically mediated interactions between stationary matter qubits, the qubit system should have at least three discrete states arranged into a Λ-type system, in which two metastable ground states are optically coupled to a single excited state (Fig. 18.1). In such a system, the excited state population can be kept low at all times to minimize decoherence due to spontaneous emission.

A natural Λ-system in a semiconductor is provided by a single trapped electron[a]. The two lower states are provided by the electron-spin Zeeman states in a magnetic field. The potential which traps the electron might be due to a quantum dot, or a single donor impurity in the semiconductor lattice. In either case, we will call this entity a dot. The excited state is provided by an exciton bound to the same potential; in quantum dots this

[a] A trapped hole is also possible but the resulting qubit would decohere more quickly.

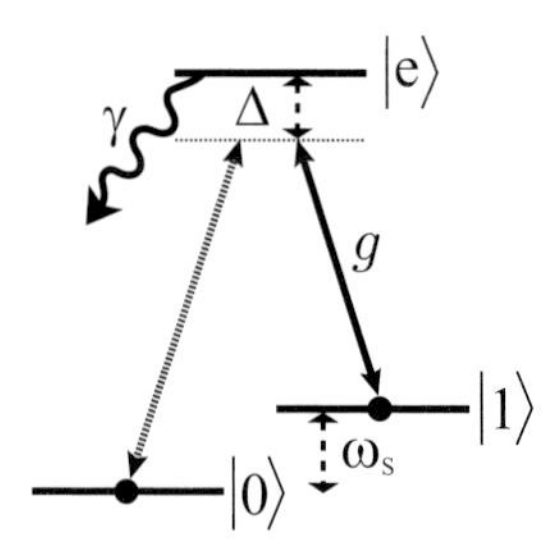

Fig. 18.1. A Λ-system of energy levels.

is referred to as a trion state, while in donor impurity systems it is called a bound-exciton state. The angular momentum of the lowest manifold of such states is determined by the $J = 3/2$ hole and the envelope wavefunction, since the two electron spins form a singlet state. In the Faraday geometry (incident light parallel to the magnetic field), the $m = \pm 3/2$ excited states, the lower of which we label $|e\rangle$, each couple to only one of the two electron spin states, which we label $|0\rangle$ and $|1\rangle$. In the Voigt geometry (incident light orthogonal to magnetic field), both spin states $|0\rangle$ and $|1\rangle$ may couple to the same excited state $|e\rangle$ with a single polarization of light.

A seminal proposal for using such a system for quantum networking was that of [2]. Here, a qubit stored in the two lower states of a Λ-system is transferred to a photon by introducing a carefully shaped control pulse of light which adiabatically changes the qubit state into a photon in a surrounding optical cavity. This photon leaks out of the cavity into a waveguide, and then may enter a second cavity. If the control pulse of light is time-reversed, a Λ-system in this second cavity may perfectly absorb the photon, ending up in the state of the first qubit.

To address the feasibility of this scheme in semiconductors, let us introduce a few physical parameters. The coupling g between the cavity photon and the optical transition in the dot is estimated by $g \approx k\sqrt{f/V}$. Here the constant k is $e/\sqrt{4\epsilon m_{\mathrm{e}}}$ where ϵ is the permittivity of the semiconductor, and e and m_{e} are the electron charge and mass. The unitless number f is the oscillator strength of the optical transition, which may be found from the ratio of the radiative decay rate γ normalized by that of a classical electron dipole emitter of frequency ω_0 and wavelength λ:

$$f = \frac{3}{(4\pi k)^2}\left(\frac{\lambda}{n}\right)^3 \omega_0 \gamma. \tag{18.1}$$

A critical factor for a high value of g is a small mode volume V of the cavity.

Nanofabrication techniques in semiconductors allow V to be made as small as $(\lambda/n)^3/2$. Another important parameter is the rate at which photons leak out of the cavity. This is given by $\kappa = \kappa_C + \kappa_L$, where κ_C is the rate at which photons couple to waveguide modes and κ_L is the rate at which photons are lost to leaky sidemodes or absorption in the cavity mirrors.

There are a number of figures of merit to be considered from these constants, which we now consider and discuss in the context of semiconductors. First is the quality factor of the cavity, $Q = \omega_c/\kappa$ for cavity frequency ω_c. The small size of semiconductor microcavites make them subject to surface defects, which prevents their Q from being as high as in larger cavities. More important for quantum networking than the absolute Q, though, is the ratio κ_C/κ. Keeping light in the cavity is less critical than making sure that all light that leaks out ends up in the waveguide.

The leakage rates κ and γ may be compared to the coupling rate g in the context of the emission spectrum of the dot/cavity system. When $\omega_0 = \omega_c$, the frequencies of the two normal modes of the coupled system are [3, 4]

$$\omega_{\pm} = \omega_0 - i(\kappa + \gamma)/4 \pm \sqrt{g^2 - (\kappa - \gamma)^2/16}. \tag{18.2}$$

When $g > |\kappa - \gamma|/4$, the system shows a splitting into two eigenmodes separated by a frequency called the vacuum Rabi splitting. This is the strong coupling regime.

Strong coupling is not needed for the coherent generation and capture of photons, but it makes such schemes far simpler. In the weak coupling regime, large detunings and very fast, critically shaped pulses must be used. Although strong coupling has been observed in semiconductor microcavity systems, it remains challenging to implement reliably. We will discuss progress in this direction in Sec. 18.4.1. Fortunately, there are other ways to network distant semiconductor qubits that make fewer demands on the microcavity. These rely principally on scattering the information-carrying photons off of cavities, rather than generating those photons in the cavities. For such schemes, as we will discuss in the following sections, the critical figure of merit is not $4g/|\kappa - \gamma|$, but rather the cooperativity parameter, $C = 4g^2/\gamma\kappa$. An immediate physical indication that this parameter may take on a high value in existing semiconductor microcavity/dot systems is the Purcell effect, in which the rate of radiation of an electric dipole is enhanced by the modified local density of photon states inside a cavity. This may be seen from Eq. (18.2) by taking $g, \gamma \ll \kappa$, in which case the $+$ solution is $\omega_0 - i(\gamma/2)(1 + C)$, indicating that the radiative damping of the dot is enhanced by the factor $1 + C$. C may be quickly estimated as

$C = 3/4\pi^2 \times (\lambda/n)^3 \times Q/V$. Semiconductor microcavity systems show a key advantage in achieving high values of C in comparison to larger cavities that might be used with trapped atoms or ions [5].

We now discuss how a semiconductor system in the weak coupling regime but with high values of Q/V may allow efficient generation of quantum entanglement.

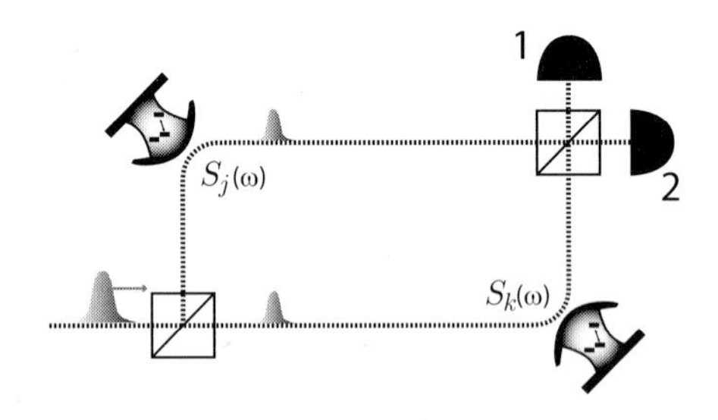

Fig. 18.2. Scattering off of distant cavities in an interferometer to generate entanglement.

18.2.2. *Single-photon-state probe and photon counting detection*

Let us consider a simple example of entanglement generation by a scattering experiment, referring to Fig. 18.2. Each dot/cavity system acts as an imperfect mirror in a Mach-Zehnder interferometer. The two qubits are each put into an equal superposition of their ground states $(|0\rangle + |1\rangle)/\sqrt{2}$ by some means (see Sec. 18.3.1 for one possibility). A pure single-photon wavepacket is introduced to the input. Its wavefunction is described by $\int d\omega \phi(\omega) a^\dagger_{\text{IN}}(\omega)\,|\Omega\rangle$, where $a^\dagger_{\text{IN}}(\omega)$ is the photon creation operator for waveguided photons of energy $\hbar\omega$ and $|\Omega\rangle$ is the vacuum state. Now suppose that each path of the interferometer transmits a photon of energy $\hbar\omega$ with complex transfer function $S_j(\omega)$, depending on the qubit state, so that after a single interaction with one qubit the resulting qubit/photon entangled state would be

$$|\psi\rangle = \frac{1}{\sqrt{2}} \int d\omega \phi(\omega) \Big[|0\rangle \otimes S_0(\omega) a^\dagger_{\text{IN}}(\omega)\,|\Omega\rangle + |1\rangle \otimes S_1(\omega) a^\dagger_{\text{IN}}(\omega)\,|\Omega\rangle \Big].$$

For example, if the photon is lost with probability p when the qubit is in state $|0\rangle$, then $S_0(\omega) \propto \sqrt{1-p}$. The above state is then no longer normalized indicating a reduced probability of measuring the photon. As another example, if the photon of any energy undergoes a phase shift by angle θ when the qubit is in state $|1\rangle$, then $S_1(\omega) = \exp(i\theta)$.

The resulting photonic state after mixing at the second beamsplitter but prior to detection may be written in terms of two-qubit states $|jk\rangle$ as

$$|\psi\rangle = \sum_{jk} |jk\rangle \int d\omega \phi(\omega) \left[\frac{S_j(\omega) + S_k(\omega)}{4} a_1^\dagger(\omega) + \frac{S_j(\omega) - S_k(\omega)}{4} a_2^\dagger(\omega) \right] |\Omega\rangle .$$

Therefore, a photon detected at detector 2 perfectly projects the qubits into the entangled Bell state $|\Psi^-\rangle = (|01\rangle - |10\rangle)/\sqrt{2}$. This occurs with probability

$$P_{\text{success}} = \int d\omega |\phi(\omega)|^2 \frac{|S_0(\omega) - S_1(\omega)|^2}{8}. \tag{18.3}$$

The scattering amplitude $S_j(\omega)$ includes the total transmission T through the waveguide. This transmission may be reduced by imperfect couplings from the waveguide to the device containing the cavity. $S_j(\omega)$ also includes the scattering amplitude of the dot/cavity system. This may be calculated using a Wigner-Weisskopf approach in conjunction with input-output theory. If we assume that only state $|1\rangle$ of the qubit is optically active (as in the Faraday geometry), and that the input is a single photon state or a weak coherent state which reflects off of the cavity, then

$$S_j(\omega) = \sqrt{T} \frac{(\kappa_{\text{L}} - \kappa_{\text{C}})/2 - i(\omega - \omega_{\text{c}}) + \delta_{j1} g^2/[\gamma_{\text{L}}/2 - i(\omega - \omega_{\text{c}} - \Delta)]}{(\kappa_{\text{L}} + \kappa_{\text{C}})/2 - i(\omega - \omega_{\text{c}}) + \delta_{j1} g^2/[\gamma_{\text{L}}/2 - i(\omega - \omega_{\text{c}} - \Delta)]},$$

where Δ is the detuning between the $|1\rangle \rightarrow |\text{e}\rangle$ transition frequency and the light frequency and γ_{L} is the rate of spontaneous emission into non-cavity modes. If we assume a long pulse centered at the cavity frequency, then

$$P_{\text{success}} \approx \frac{T}{2} \frac{\kappa_{\text{C}}}{\kappa} \frac{C^2}{(\gamma_{\text{L}}/\gamma + C)^2 + (2\Delta/\gamma)^2}. \tag{18.4}$$

The ideal regime is where $\kappa_{\text{C}} = \kappa$, meaning that cavity loss is completely dominated by the coupling to the waveguide. Further, if $\Delta = 0$, meaning that the $|1\rangle \rightarrow |\text{e}\rangle$ transition is resonant with the cavity mode, and if the co-operativity parameter C is much larger than γ_{L}/γ, then $S_0(\omega_{\text{c}}) = -S_1(\omega_{\text{c}})$, so that a qubit in state $(|0\rangle + |1\rangle)/\sqrt{2}$ effectively performs a controlled-phase gate on a photonic qubit in the number-state basis. The use of this remarkable π-phase shift with a two-level atom was indicated by Hofmann *et al.* [6] and proposed for a single-photon controlled-phase gate by Duan and Kimble [7]. If used via transmission through a cavity, this phase-shift changes the transparency of a waveguide, which has been proposed for the quantum repeater discussed in Chapter 17. In this ideal regime, photon detection at detector 1 also generates an entangled Bell state, in this case $|\Phi^-\rangle = (|00\rangle - |11\rangle)/\sqrt{2}$. The probability of success for each result is just the probability that the two qubits were found in the desired Bell state, which is 1/2, multiplied by the probability that the photon made it through the system, which is T.

If such a perfect differential phase shift of π can be obtained, then the resulting quantum entanglement scheme has the advantage that it is very robust to loss in the waveguides. If T becomes small due to a large distance between the two qubits, the rate of successful entanglement generation is decreased, but the fidelity of the resulting entanglement is not decreased. The tolerance of this system to optical loss at the price of speed holds even if large imperfections of the cavity system are in place. If C is somewhat low, a large detuning Δ is used to suppress qubit population changes, and cavity loss is included, this entanglement generation scheme still works. In this case the phase shift between $S_0(\omega_c)$ and $S_1(\omega_c)$ may be substantially less than π, and most scattered photons will be lost at the cavity. One may then choose to input the pulse into a non-cavity mode so that $S_0(\omega) = 0$. Such a scheme was proposed by Childress *et al.* [8]. In this case the probability of success may be extremely low, substantially slowing the resulting quantum repeater system.

A number of practical challenges are important to overcome in such a scheme. Foremost of these is the reduction of the fidelity of the measurement-induced entanglement due to fluctuating phase-shift differences between the two arms of the long-distance interferometer. The phase stabilization of the interferometer can be improved by reconfiguring it as in Fig. 18.3, which uses a fast optical switch to fold the spatial modes into two temporal modes. If short optical pulses are used, the two temporal modes can be nearly coincident in the long-distance fiber waveguide, so that the pulses from each arm of the interferometer see the same random phase shifts. However, this works well only if the pulses are short in comparison to the timescale of phase fluctuations in the channel. Other challenges include:

- The fidelity is degraded by false detector counts.
- An ideal or nearly ideal single-photon source may be an expensive component to such a system. An attenuated laser (coherent state) may be used instead, but then the probability of success is further reduced by a factor of the average photon number of the input state, which should

be less than one to suppress two-photon events.

- If imperfect optical switches are used, then fidelity will be degraded by their timing jitter.
- If short pulses are used, then the integral over the pulse shape of $|S_1(\omega) - S_0(\omega)|^2$ will be smaller than the $\omega = \omega_c$ approximation used above.
- The two dot/cavity systems may not be exactly identical, as we assumed in the equations above. This problem affects the fidelity when π phase shifts are degraded by imperfect cavities and when short pulses are used for phase stabilization.

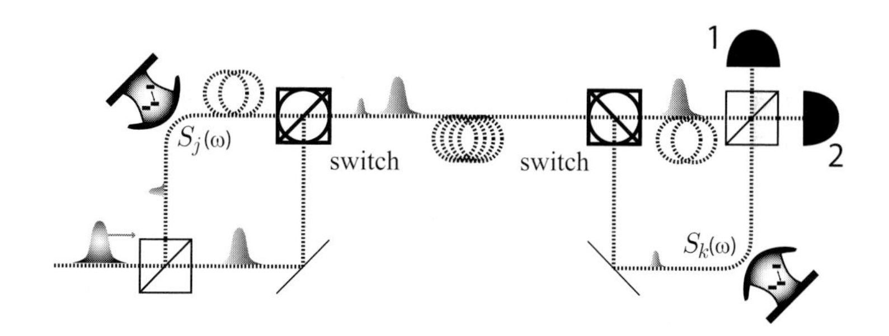

Fig. 18.3. Temporal interferometer for phase stabilized entanglement generation.

18.2.3. *Coherent state probe and homodyne detection*

Many of the practical problems of the single-photon repeater schemes are alleviated with another strategy which uses pulses of light that are much brighter than a single photon [9]. We will see that this strategy is more tolerant to differences between the two qubits, and that the requirements on the figures of merit of the cavity (κ_c/κ and C) need not be as high as those needed to achieve a π phase shift between $S_0(\omega)$ and $S_1(\omega)$. Critically, shorter pulses may be used, improving the ability to do phase stabilization in an interferometer of the form of Fig. 18.3. Since a large detuning Δ between the dot transition frequency and the light frequency is used, the shape of a fast optical pulse is not strongly perturbed by the state of the dot; only a small overall delay occurs. Finally, the input light is a coherent state from an ordinary laser, alleviating concerns of imperfect single photon sources, and the detection may be homodyne detection using ordinary linear detectors and a strong reference pulse, removing problems with detector efficiency and dark count rates.

The most important advantage of a strategy employing bright coherent light, however, is that the probability of success can be made quite high, and the resulting high rate of entanglement generation is nearly constant when the optical loss increases. The disadvantage of doing this, as we will discuss, is that the loss of so many photons in the long channel between qubits degrades the fidelity of that rapidly created entanglement. Such fidelity degradation must be compensated for by entanglement purification, which we discuss in the next section.

Like the single-photon-based scheme, the bright-coherent-state scheme works by scattering in an interferometer of the form of Fig. 18.3, but it relies on the very small phase shift that occurs between a dot/cavity system in state $|0\rangle$ versus one in state $|1\rangle$ when the pulse and cavity are far detuned from the dot transition ($\Delta \gg \gamma$). This small phase shift is resolved by homodyne detection, which uses a probe pulse (depicted by the smaller pulse in Fig. 18.3) and a strong local oscillator (LO) pulse (depicted by the larger pulse in Fig. 18.3) from the same laser source. When the LO pulse with a particular phase interferes with the probe pulse at the second beamsplitter, the difference signal between detectors 1 and 2 provides a continuous variable measurement of one of the quadrature components of the coherent state amplitude α. The amplitude α is proportional to the square root of the average number of photons in the pulse, so to resolve a very small phase shift, many photons are needed in the probe pulse. For such large photon numbers, Eq. (18.2.2) is no longer valid, because emission from the dot becomes a far more probable event. An improved analytic approximation of the ratio of scattering amplitudes may be derived from a master equation approach [10]. For slow pulses of duration σ_{P}, this approximation is

$$\frac{S_1(\omega_{\mathrm{c}})}{S_0(\omega_{\mathrm{c}})} \approx 1 + i\frac{4\kappa_{\mathrm{C}} g^2}{\kappa^2} \frac{\Omega + i\Gamma}{\Omega^2 + \Gamma^2 + 4\kappa_{\mathrm{C}} g^2 |\alpha|^2/\sigma_{\mathrm{P}}\kappa^2}, \tag{18.5}$$

where $\Gamma = (\gamma_{\mathrm{L}} + \kappa_{\mathrm{L}}')/2$, $\Omega = \Delta[1 + (\kappa_{\mathrm{L}}'/\kappa)]$, and κ_{L}' is the Purcell-enhanced emission into cavity-loss modes, given by $\kappa_{\mathrm{L}}' = (\kappa_{\mathrm{L}}/\kappa)\gamma C/(1 + 4\Delta^2/\kappa^2)$. If we set the detuning at the large value of $\Omega = \Phi\Gamma/\theta$, where $\Phi = (\gamma/\Gamma)(\kappa_{\mathrm{C}}/\kappa)C$, then

$$\frac{S_1(\omega_{\mathrm{c}})}{S_0(\omega_{\mathrm{c}})} \approx 1 + i\theta \frac{1 + i\Phi^{-1}}{1 + \Phi^{-1}|\alpha\theta|^2/\Gamma\sigma_{\mathrm{P}} + \Phi^{-2}}. \tag{18.6}$$

According to this approximate analysis, if $\Phi \gg 1$, then we see a phase shift of θ and negligible loss. The homodyne signal is proportional to $\alpha\theta$, which according to Eq. (18.6) may be as large as $\alpha\theta \sim \sqrt{\Gamma\sigma_{\mathrm{P}}\Phi}$. However, the master equation analysis also shows that the qubit undergoes dephasing by a factor of $\exp(-|\alpha\theta|^2/\Phi)$, indicating that fidelity is improved by using longer pulses and keeping $\alpha\theta \ll \sqrt{\Gamma\sigma_{\mathrm{P}}\Phi}$.

This analytic approximation assumed long pulses and simplifying selection rules. However, detailed numerical simulations indicate that values of $\alpha\theta$ as large as 2 may be obtained even if both $|0\rangle$ and $|1\rangle$ are optically coupled with equal oscillator strengths to $|\mathrm{e}\rangle$ and with pulses as short as a tenth of the vacuum spontaneous emission lifetime. In direct-bandgap semiconductor quantum dots or donor impurities, magnetic fields of several tesla are required to split the ground states enough to allow such large controlled phase shifts. Typical values of θ for these situations are on the order of 0.01.

If optical loss is not too high, such a small phase shift θ is sufficient for generating entanglement. A detailed analysis appears in Ref. [10]. To summarize, suppose as before that two distant qubits are prepared in coherent superposition states $(|0\rangle + |1\rangle)/\sqrt{2}$, and a probe pulse in coherent state $|\alpha\rangle$ (with α real) interacts with both qubits sequentially, as indicated by Fig. 18.3. Defining $S_0(\omega) \approx 1 - i\theta/2$, an interaction of the probe pulse with the first qubit yields the entangled state $(|0\rangle|\alpha(1 - i\theta/2)\rangle + |1\rangle|\alpha(1 + i\theta/2)\rangle)/\sqrt{2}$. Now suppose the second qubit is slightly different than the first, and causes a slightly different phase shift θ'. Then we may interfere the probe pulse with a reference pulse at a beamsplitter to achieve an overall displacement of the coherent state amplitude by $\alpha(\theta/\theta' - 1)$. After the displaced probe pulse interacts with the second qubit, the combined qubit/qubus wave function before detection may be described as

$$|\psi\rangle = \frac{1}{2}\Big[|00\rangle|\alpha(\theta/\theta' - i\bar{\theta})\rangle + |11\rangle|\alpha(\theta/\theta' + i\bar{\theta})\rangle + \Big(|01\rangle + |10\rangle\Big)|\alpha\theta/\theta'\rangle\Big],$$

where $\bar{\theta} = (\theta + \theta')/2$ and contributions of order θ^2 and higher have been ignored for simplicity. The homodyne detection attempts to measure the imaginary part of the coherent state amplitude, but only does so up to a Gaussian fluctuation with unity standard deviation. We therefore decide that projection into the Bell state $|\Psi^+\rangle = (|01\rangle + |10\rangle)/\sqrt{2}$ is successful when the result of the homodyne measurement is less than some cut-off value p_{c}, which results in perfect projection only in the limit $p_{\mathrm{c}} \to 0$. In an ideal system, however, we could increase the fidelity of this projection by increasing the average photon number, thus increasing $\alpha\bar{\theta}$, and allowing a high value of p_c. In this case the probability of success is approximately $\mathrm{erf}(\sqrt{2}p_c)/2 \to 1/2$. However, optical loss will slightly reduce this proba-

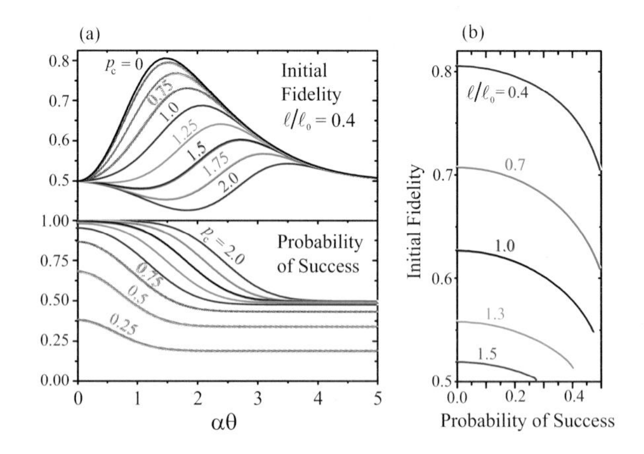

Fig. 18.4. (a) Fidelity of generated quantum entanglement and probability of success as a function of $\alpha\theta$ for several values of $T = \exp(-\ell/\ell_0)$. (b) Fidelity at optimum value of $\alpha\theta$ for each value of the probability of success.

bility of success. Photons lost in the waveguide or in the cavity degrade the fidelity by a factor proportional to $\exp[-|\alpha\theta|^2(1-T)/2]$ for total transmission T. This consideration of loss suggests that $\alpha\bar{\theta}$ should be made high enough to allow us to effectively project the desired Bell state, but not too high to completely decohere it. The fidelity as a function of $\alpha\theta$ is shown in Fig. 18.4, as well as the optimum fidelity for each choice of p_c. For $T = 0.67$, which corresponds to the transmission of 10 km of telecom fiber, a Bell state with fidelity 0.77 can be projected with a 0.35 probability of success. This probability of success is orders of magnitude greater than single-photon based schemes, where every form of loss (including the source, the channel, the cavity, and the detector) reduces such probability. A high probability of success is important for the speed of a long-distance quantum repeater system, because the rate at which one may repeat attempts to generate entanglement is limited to the time for the optical pulse to travel the long distance between repeaters.

18.2.4. *System performance*

When the entire quantum repeater system is considered, is it better to generate lower fidelity entangled pairs with a higher probability of success, as in the scheme discussed in Sec. 18.2.3? Or, is it better to generate higher fidelity entanglement at a slower rate, as in the schemes based on single photons discussed in Sec. 18.2.2? The answer is not obvious. In either system, the imperfect fidelity of entangled pairs requires a form of error correction called entanglement purification, which uses many imperfect entangled pairs to distill a few nearly perfect entangled pairs [1, 11, 12]. A system with higher success probability but lower fidelity may or may not be faster, depending on how long it takes to implement sufficient entanglement purification.

Monte-Carlo simulations of quantum repeater networks, including probabilistic entanglement generation, entanglement purification, and entanglement swapping, give some indication of the trade-offs between probability of success and fidelity. Entanglement generation, purification, and swapping all require classical communication of measurement results, and the time to complete this communication provides the limiting timestep in the simulation. The final rate of long-distance entanglement generation will depend on the entanglement purification protocol; one simple example is a "greedy" protocol, in which entangled pairs are purified according to the best resources available at any particular timestep. Simulations using this protocol indicate that an entanglement scheme which creates an initial entanglement fidelity of 77% with probability of success 35%, as described in Sec. 18.2.3, performs at a comparable speed to a single-photon based scheme which creates entanglement with 96% fidelity and probability of success of 0.5% [10]. This simulation used 129 repeater stations with station-to-station optical loss corresponding to 10 km of telecom fiber. Each station has 16 qubits multiplexed with the same number of communication channels. The final long-distance entanglement fidelity is 95% for this simulation and the entanglement rates are between 1 to 10 bits per second. The dependence of this final rate has a roughly linear dependence on both the initial probability of success and the number of multiplexed qubits in each station.

However, this result is only one particular example. The achievable rates depend critically on many factors besides probability of successful entanglement generation and initial entanglement fidelity. How much error is present in the quantum logic for entanglement purification and swapping? How does the qubit decoherence time compare to the classical communication speed? What final fidelity is needed for the long-distance entanglement? For each answer to these assorted questions, the ideal choice of purification protocol will vary. Both the specific quantum operations chosen and the scheduling of purification need to be optimized for each entanglement generation scheme. A complete analysis of optimized quantum repeater protocols for broad ranges of parameters of initial entanglement generation, gate error, memory error, and so on remains an open research avenue of great importance to the future of quantum communication.

18.3. Qubus Quantum Computer

The generation of entanglement of distant electron spins discussed above is ultimately probabilitistic. Although such probabilistic gates may be used for quantum *computation* at the expense of time and memory resources, deterministic gates would ultimately be faster and easier to bring to a fault-tolerant regime. In this section, we discuss concepts for building fast, deterministic logic gates, beginning with arbitrary single-qubit rotations in Sec. 18.3.1 and then two-qubit controlled phase gates in Sec. 18.3.2; these two resources together are enough for universal logic. Prospects for scalability are discussed in Sec. 18.3.3.

18.3.1. *Single qubit gate by single broadband optical pulses*

If a qubit is defined by an electron spin in a magnetic field, single qubit rotations may be performed using strong microwave pulses resonant with the magnetic dipole transition, as in electron spin resonance (ESR). However, such a method has several problems. First, long-wavelength microwaves cannot be easily confined to a single qubit, but rather will require extra technology such as magnetic field gradients to localize single-qubit rotations. Second, it is difficult to obtain broadly tunable, high-power coherent sources of microwaves, especially at the very high frequencies needed for high-magnetic-field operation where the controlled optical phase shifts discussed in Sec. 18.2.3 work well. Even if a source is available, such a direct rotation should be much slower than the Larmor frequency ω_{s}.

Using two focussed laser beams to cause a stimulated Raman transition between the two ground states alleviates the problem of localizing the rotation, but does not improve the speed, since the optical pulses should still be much slower than the Larmor period. Additionally, precise phase control of the two optical pulses is required [13].

A simpler and faster way to achieve arbitrary single-spin rotations is available with high-oscillator-strength semiconductor quantum dot systems [14]. This scheme allows a general qubit rotation using a single, timed, ultra-fast broadband pulse. For simplicity, we suppose equal optical matrix elements for both Zeeman ground states to state $|e\rangle$ (as in the Voigt geometry). The interaction-picture optical dipole Hamiltonian is then

$$\mathcal{H}(t) = \frac{\Omega(t)}{2}\left[e^{i\Delta t}\,|e\rangle\langle 1| + e^{i(\Delta+\omega_s)t}\,|e\rangle\langle 0| + \text{h.c.}\right], \tag{18.7}$$

where h.c. refers to Hermitian conjugate, and $\Omega(t)$ is the normalized optical pulse shape, including the oscillator strength and the amplitude of the optical pulse. If Δ is large, adiabatic elimination of the excited state from the master equation yields the approximate two-state Hamiltonian

$$\mathcal{H}_{\text{eff}}(t) = G(t) - \mathbf{B}_{\text{eff}} \cdot \mathbf{S}, \tag{18.8}$$

where $\mathbf{S}$ is the electron spin operator and $G(t)$ is an irrelevant overall phase factor. The effective field $\mathbf{B}_{\text{eff}}$ has a small z-component $\delta|\Omega(t)|^2/(4\Delta^2+\gamma^2)$, which is negligible in comparison to the transverse components

$$B^x_{\text{eff}} - iB^y_{\text{eff}} = -2\frac{\Delta}{4\Delta^2+\gamma^2}|\Omega(t)|^2 e^{i\omega_s t}. \tag{18.9}$$

The pulse itself is very short; $|\Omega(t)|^2$ may be nonzero for a time much shorter than the Larmor period $2\pi/\omega_s$. Despite such a short pulse, the effective angle of rotation caused by the pulse, given approximately by the time-integral of $|\Omega(t)|^2/2\Delta$, can be very large, exceeding π. These equations indicate that if this ultra-fast pulse is peaked at time t_0, the phase of the rotation will be $\exp(i\omega_s t_0)$, allowing one to control the rotation axis in this rotating frame by choosing when the pulse arrives.

The underlying concept here is that a fast pulse contains many instances of the two frequencies needed to form a Raman resonance, and the phase of the resulting rotation depends on the phase difference between these components. If a short, Fourier-transform-limited pulse arrives at a particular time with a Fourier transform $\Omega(\omega)$ that can be considered entirely real and positive, then there is no phase shift between its Fourier components, and we define that the rotation it achieves is about the x-axis. If the pulse is delayed by a quarter of a Larmor period $\pi/2\omega_s$, then the same pulse in the Fourier domain acquires a phase factor $\Omega(\omega) \to \Omega(\omega)\exp(i\omega\pi/2\omega_s)$, so that frequency components separated by ω_s receive a phase shift of $\exp(i\pi/2)$, altering the axis of rotation by $\pi/2$. Rotations of arbitrary angle about x, y, and $-x$ are sufficient to generate any desired SU(2) operation of a single qubit.

In practice, such fast rotations might be implemented by a high-power semiconductor mode-locked laser system whose pulse period is tuned to the Larmor frequency of the electron spins, and rotations about x, y, and $-x$

axes are independently achieved by three fixed optical delay lines. Under such conditions, it is reasonable for the physical clock cycle for single qubit control to be on the order of 100 GHz.

This concept has been tested by direct numerical solution of time-dependent master equations [14]. For example, if we use the realistic quantum dot parameters $\gamma = (200 \text{ ps})^{-1}$ (see Sec. 18.4.1), electron dephasing time $T_2 = 10$ μs (see Chapters 5, 9–12 and Refs. [15–17]), and $\omega_S/2\pi = 100$ GHz, and if we assume a Fourier-transform-limited Gaussian pulse detuned by 10 THz with 100 fs full-width half-maximum, then both π- and $\pi/2$-pulses have fidelity greater than 0.999 for applied pulse energy densities of 14 μJ/cm^2 and 5 μJ/cm^2 respectively. These are reasonable values for the energy output of mode-locked semiconductor lasers followed by optical amplifiers. If we allow 5% intensity fluctuations in the pulses, the fidelity decreases to 0.995 for π-pulses and 0.998 for $\pi/2$-pulses. The high fidelity of the single-pulse Raman rotation is a direct consequence of the speed of the pulse; all relaxation and decoherence processes occur at a time scale much slower than the pulse time, including dephasing effects introduced by the pulses themselves.

18.3.2. *Two-qubit gates by controlled phase shifts*

Two-qubit gates may be accomplished with a speed comparable to the single qubit rotations. As discussed in Sec. 18.2.3, a pulse of width on the order of $0.1\gamma^{-1}$ can cause a state-dependent optical phase shift θ on a laser pulse consisting of about 10^5 photons. We may also mix that optical pulse with a reference pulse to achieve an arbitrary optical displacement. These are sufficient resources to achieve deterministic quantum logic. There are several ways to do this; for details, see [18]. Briefly, a single optical pulse alternatingly interacts twice with each of two qubits, developing entanglement between the state of the qubit and the optical phase after each interaction. The pulse also receives appropriate optical displacements between interactions. These displacements serve to move the coherent state amplitude through a closed trajectory in phase space which encloses a finite area. This introduces a geometric phase proportional to that area, which in turn depends on the collective states of the two qubits. Ignoring single qubit phase shifts, a geometric phase of π may be developed if and only if both qubits are in state $|1\rangle$. This is a controlled phase gate.

The time-duration of such a gate is limited by the length of the optical pulse and the time for light to propagate between the two qubits several times. As discussed in Sec. 18.2.3, sufficiently large optical phase shifts are attainable with pulses as short as $0.1\gamma^{-1}$. For high-oscillator-strength quantum dots, which we will discuss in Sec. 18.4.1, this enables pulses on the order of 10 ps. This is comparable to the light propagation time for reasonably sized on-chip photonic devices. These considerations suggest the possibility of a two-qubit logic-gate cycle of 1 to 10 GHz.

18.3.3. *System performance*

Controlled-phase gates in combination with the single qubit rotations discussed in Sec. 18.3.1 provide the elements for universal logic that could be incorporated into a design for a fast quantum computer. However, the next important question to ask is whether such a device can be made scalable and fault tolerant.

One critical element for fault tolerance is the ability to rapidly measure qubits. Fortunately, the controlled phase shifts for which these qubits will be optimized provides a fast, optical probe of the qubit state (also see Chapter 6). Combined with homodyne detection, this allows a rapid, robust means for quantum nondemolition measurement of the qubit, useful for both syndrome measurements in quantum error correction and qubit initialization.

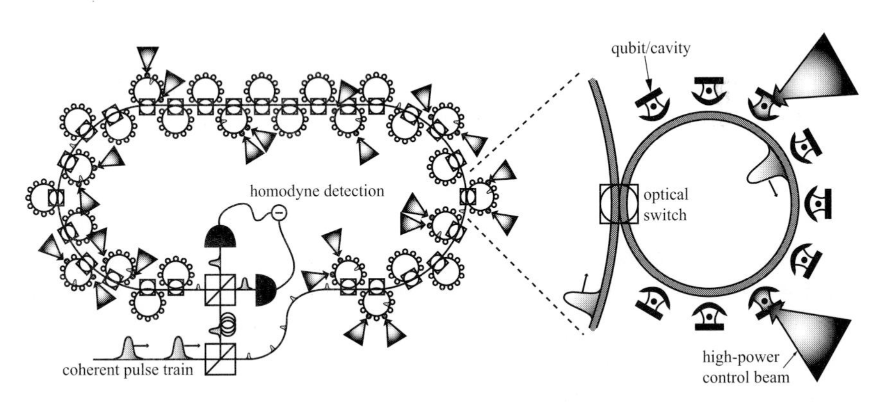

Fig. 18.5. Conceptual schematic for a photonic semiconductor quantum computer based on qubus concepts.

The next critical element is a large number of logically coupled ancilla qubits. Since each qubit has its own microcavity, the area taken by a single qubit may be several micrometers, and so incorporating thousands of qubits on a single semiconductor chip results in optical waveguides which may be centimeters long. Such a long semiconductor waveguide may be quite lossy. Unfortunately, a critical limitation to deterministic qubus gates such as those discussed in Sec. 18.3.2 is their sensitivity to optical loss [10]. Although the efficiency of on-chip photonic waveguides and switches is continually improving, it is likely that light traversing a chip containing thousands of qubits of this kind will suffer too much loss for deterministic qubus gates between arbitrary qubits. Nonetheless, one might imagine that deterministic gates can be used for small clusters of a few qubits arranged around a low-loss ring waveguide, as indicated in Fig. 18.5. Coherent pulses are coupled into the waveguides via beam-splitting switches to achieve qubus gates. The small rings are coupled together by a longer waveguide, where probabilistic entanglement generation as described in Sec. 18.2.3 allows the quantum teleportation of logical gates between distant qubits in a robust, verifiable way [19]. If the input pulse train and optical switch network is fast enough, different rings may operate and couple concurrently.

Specific qubits on a ring might be selected for quantum logic using an optical Kerr switch, which uses the optical nonlinearities of the semiconducting substrate. In this scheme, cavities are initially far off resonance from the qubus pulses, but bright, focussed, mid-gap light applied from above rapidly shift the cavities into resonance, activating the dots/cavities one at a time for quantum logic. Focussed laser light from above also achieves single-qubit rotations.

Detailed analyses of the feasibility of such designs are of course needed, but it is clear that the ultimate logical clock speed of such a computer will be several orders of magnitude slower than the physical clock speed. Although quantum computers offer exponential speed-up over classical computers, their ability to surpass fast networks of parallel classical computers will depend on this logical speed being fast, at least kHz to MHz in order to surpass current classical computers in factoring problems lasting less than a year [20]. Therefore, the ability to do rapid quantum logic as discussed above is important for the ultimate goal of a useful, scalable quantum computer.

18.4. Hardware

So far, we have discussed ways to implement quantum repeaters and quantum computers using optically bright Λ-systems in semiconductors with high cooperativity factors, C. To date, however, such schemes have not been implemented, but much progress has been made in recent years in achieving the needed qubit interfaces. Many of the chapters of this book have indicated such progress. Here, we will indicate just a few of the research directions actively pursued for achieving these proposals: the increase of the oscillator strength of the quantum dots, the incorporation of those dots in cavities with high cooperativity factors, and the use of donor-bound impurities to improve the homogeneity of the dots.

18.4.1. *Quantum dots in microcavities*

High-oscillator-strength emitters are important for achieving strong coupling to a cavity, and also to enable the rapid optically-controlled single-qubit rotations discussed in Sec. 18.3.1. Currently, the most advanced candidate for high oscillator strengths in semiconductors is the large area quantum dot. Oscillator strength is generally proportional to the quantum dot size. Smaller dots, such as Stranski-Krastanow-grown self-assembled InAs quantum dots, have oscillator strengths of about 6. Larger dots may be defined by the fluctuating interface between GaAs quantum wells and AlGaAs layers; these have provided experimental confirmation that larger dot size ($\sim$100 nm) leads to order-of-magnitude larger oscillator strength ($\sim$75) [21]. However, there is a concern that such interface-fluctuation quantum dots, if incorporated into a microcavity, would unavoidably limit the cavity Q because of the background absorption from the quantum well region [22].

For this reason, large quantum dots with an energetically separated wetting layer would be a better choice for cavity QED applications. One approach to achieve this is the GaAs quantum dot grown by droplet epitaxy. The droplet-epitaxy method [23] is a molecular-beam epitaxy (MBE) growth method for III-V quantum dots based on the group-III droplet formation on a group-III stabilized surface and then subsequent direct incorporation of group-V atoms into the droplet. Since this method does not depend on the lattice mismatch, the quantum dot properties such as size, density, and thickness of the wetting layer can be adjusted by changing the growth conditions.

Single, round-shaped quantum dots with average diameter 100 nm were grown with this technique and isolated by the fabrication of 400 nm mesa structures. The radiative lifetime, measured by a streak camera, is used to calculate the oscillator strength using Eq. (18.1). This lifetime is found to vary between 100 and 500 ps from dot to dot. The streak camera data

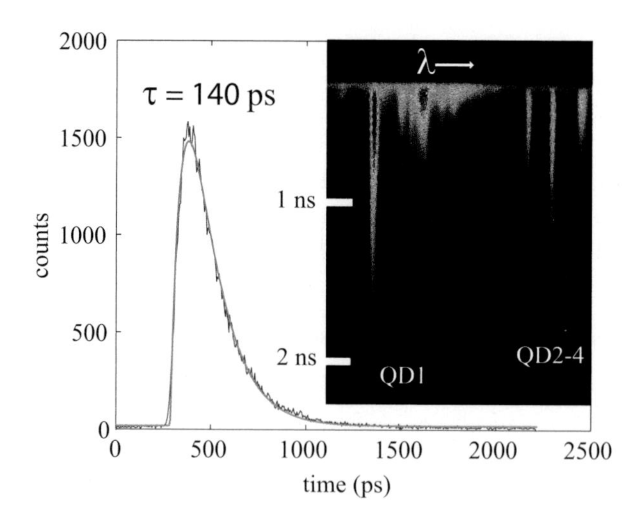

Fig. 18.6. Raw streak image from 668 nm to 682 nm, and the time vs. intensity plot for quantum dot 4 (QD4,λ=676.4 nm) obtained from the streak image.

shown in Fig. 18.6 shows the radiative decay for one particular dot emitting at $\lambda = 676.4$ nm with $\tau = \gamma^{-1} = 140$ ps. This corresponds to an oscillator strength of $f = 65$. These measurements used an excitation wavelength of 405 nm, which pumps free carriers into the $Al_{0.31}Ga_{0.69}As$ barrier layer. When this sample was illuminated by another laser diode with an excitation wavelength of 650 nm, which is unable to pump free carriers into the AlGaAs barrier, no photoluminescence was observed from the dot. This means that the energy level of the wetting layer is either higher than the photon energy at 650 nm or is not present at all. This result indicates that GaAs quantum dots grown by the droplet epitaxy method are promising candidates to yield strong coupling inside optical microcavities.

So far, such dots have not been introduced to an optical microcavity. There have, however, been several recent demonstrations of strong coupling between (In,Ga)As quantum dots and either micropillar [24], photonic crystal [25, 26], or microdisk [27] cavities. As an example, the parameters for one observed InGaAs quantum dot in a micropillar cavity, discussed in Chapter 13, are $g/2\pi = 8.5$ GHz and $\kappa^{-1} = 8$ ps. Using an independent measurement of the quantum dot lifetime to find $\gamma^{-1} = 690$ ps, the cooperativity parameter is calculated to be $C = 61$, indicating great potential for the proposals above. Comparable parameters are available from cavities made from defects in photonic crystals, as discussed in Chapter 17. Photonic crystal systems also show great promise for monolithic integration on a single semiconductor chip, efficient cavity/waveguide coupling, and efficient optical switching.

18.4.2. *Donor-bound excitons*

Quantum dots in microcavities have shown that high cooperativity factors are attainable in semiconductors, but unfortunately quantum dots have a fundamental limitation that might never be fully resolved by engineering advances. This limitation is that, unlike atoms, every quantum dot is different. Their energy levels, oscillator strengths, electron g-factors, and selection rules all vary with the size and shape of each dot. Although the proposals outlined above are somewhat tolerant to inhomogeneity, the extremely broad inhomogeneous distribution of most quantum dot systems may prohibit scalability.

The needed replacement for the quantum dot may be a feature already present in most semiconductors. The tremendous progress in the purification of semiconductors over the last fifty years has given rise to the discovery of a number of optically active complexes previously obscured by the presence of large-scale inhomogeneous broadening. One such feature, which we briefly mentioned in Sec. 18.2.1, is the emission of radiation due to recombination of excitons bound to impurities. The initial prediction of the existence of exciton-impurity complexes by Lampert in 1958 [28] was soon experimentally confirmed by the observation of sharp emission lines in the photoluminescence spectra of Si [29] and CdS [30]. Many reports have since been published on bound-exciton complexes in a whole variety of different semiconductor systems, both in bulk and in heterostructures (e.g. quantum wells [31]). For an excellent review on the earlier work on bound-exciton complexes, we refer to Ref. [32].

18.4.2.1. D^0:GaAs

The homogeneity of the GaAs donor-bound exciton system is evident in the magneto-photoluminescence spectrum of a bulk, MBE-grown sample of GaAs shown in Fig. 18.7. The brightest transitions, shown in red, are transitions from donor-bound exciton states (D^0X) to neutral donor ground states (D^0). The diamagnetic shift and paramagnetic splitting of these transitions are clearly observed as the magnetic field is increased, indicative of the energy separation of the qubit D^0 states $|0\rangle$ and $|1\rangle$. The excited D^0X

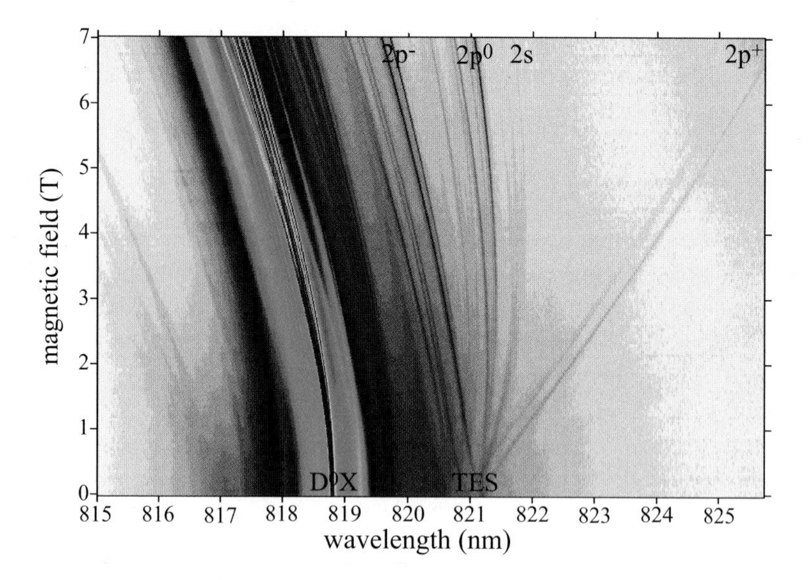

Fig. 18.7. Bulk magneto-photoluminescence spectra of the GaAs D^0-D^0X system.

state splits into many levels with energies determined by the orbital angular momentum and spin of the unpaired hole [33], but the lowest of these may be used as state $|e\rangle$ of the Λ-system discussed in Sec. 18.2.1. Relaxation of the D^0X states into excited hydrogenic 2p and 2s D^0 states, called two-electron satellite (TES) transitions, are also observed in Fig. 18.7. These transitions provide a convenient way to monitor the population of state $|e\rangle$ while resonantly exciting the Λ-system transitions.

One experiment demonstrating the coherent manipulation of ensembles of such Λ-systems is the coherent population trapping (CPT) into a qubit-superposition state [34]. To observe CPT, a single frequency Ti:Sapphire laser is scanned over the $|0\rangle \leftrightarrow |e\rangle$ transition while a single frequency diode laser is fixed on the $|1\rangle \leftrightarrow |e\rangle$ transition. As shown by Fig. 18.8, on two-photon resonance there is a pronounced dip in the photoluminescence of the $D^0X \rightarrow 2p^- D^0$ TES line, and thus the population of the excited state $|e\rangle$. This is a signature of the CPT state. The width and depth of the dip are determined by the electron spin coherence time and the diode laser power. A fit to the full three-level master equation indicates 1–2 ns electron dephasing times in this system. This result is similar to results obtained in charged III-V quantum dots (see Chapters 9–12) and the dephasing in both systems is believed to be due to the electron hyperfine interaction with the

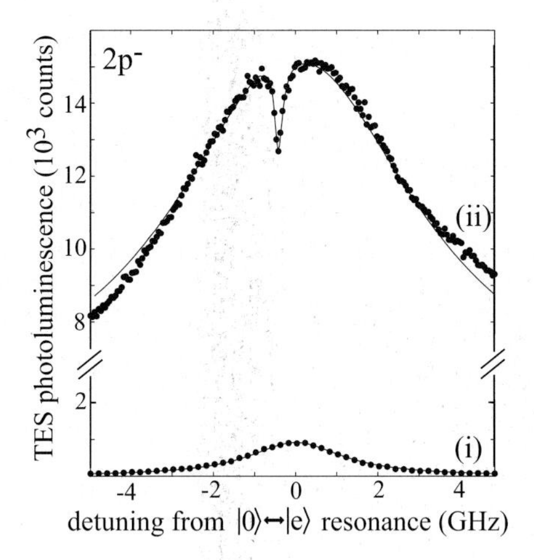

Fig. 18.8. TES photoluminescence intensity from the D^0X state to the $2p^-D^0$ state. (i) With no laser exciting $|1\rangle \leftrightarrow |e\rangle$, the 2 GHz ensemble linewidth is seen. (ii) With a second laser resonant on the $|1\rangle \leftrightarrow |e\rangle$ transition, coherent population trapping is observed. The solid line is a fit to a three level density matrix model.

inhomogeneous nuclear-spin environment.

Decoherence due to hyperfine couplings to nuclei is a fundamental problem with quantum dots and semiconductor impurities in III-V semiconductors. Inhomogeneous dephasing due to nuclei could be eliminated with a spin-echo technique using the single-qubit rotations discussed in Sec. 18.3.1, but decoherence due to nuclear spin diffusion would persist. If the nuclei could be fully polarized, then not only would this decoherence source vanish, but the substrate nuclei could be used as a long-lived quantum memory (see Chapters 3–4, 8–12). Unfortunately, while nuclear polarizations much higher than thermal polarizations are routinely observed, the high polarizations needed to significantly reduce decoherence are currently unattainable.

18.4.2.2. ^{31}P:Si

A promising solution to the problem of nuclear decoherence is to use a substrate consisting of spin-0 nuclei. This is available in isotopically purified ^{28}Si, where the isotopic purification increases the coherence time of electron spins bound to ^{31}P impurities to an astounding value of 60 ms [35]. Even longer coherence times of many hours may be expected for the ^{31}P nuclear spin itself. This time has yet to be measured, although NMR measurements of ^{29}Si nuclei isolated from dipolar noise via radio-frequency pulse sequences puts an experimental lower bound of 25 seconds on this coherence time [36].

Such long coherence times may provide a boost to fault-tolerant quantum computers, as the fidelity of transferring an electron spin coherence to a nuclear spin coherence and back using pulsed electron-nuclear double resonance (ENDOR) techniques may be higher than the fidelity of correcting electron-spin decoherence for the equivalent number of error correcting cycles. For quantum repeaters, in which error correction may not be implemented at all and very small errors are tolerated rather than corrected, such long decoherence times are critical, as qubit entanglement must remain coherent over the time-scale of generating long-distance entanglement, which may easily be many seconds.

Experimental progress toward demonstrating such nuclear memory is hindered by the ability to measure the state of single nuclei. Here, ^{31}P:Si may prove a critical testing ground for such ideas. The homogeneous linewidth of the donor bound exciton transitions in samples of nearly perfect ^{28}Si are extremely sharp. They are so sharp that the hyperfine splitting due to the ^{31}P may be resolved in photoluminescence excitation and photocurrent experiments [37]. These experiments still measure large ensembles of ^{31}P, and unfortunately the prospects of an optical measurement of a single ^{31}P nucleus in a bulk crystal are not good due to the optically dim emission from this indirect bandgap system. However, silicon has another advantage: extremely advanced microprocessing technology, which has enabled record-high Q values for photonic crystal silicon microcavities [38, 39]. These Q values of over 10^6 reported in silicon are for microcavities with mode volumes on the order of a cubic wavelength in the material. In principle, they could enhance the sharp zero-phonon line with Purcell/cooperativity factors C on the order of 10^5, bringing the kHz radiative rate into the MHz range. Especially if ultra-sensitive superconducting transition-edge sensors are used, the state of a single phosphorous nuclear spin could be detected in a timescale shorter than its relaxation time under measurement conditions [40]. High-Q microcavities may also enable entanglement distribution as described in Sec. 18.2.3 for this system.

18.4.2.3. ^{19}F:ZnSe

The small oscillator strength of the silicon system, however, prevents the use of ^{31}P:Si for ultra-high speed single qubit rotations as in Sec. 18.3.1 or comparably fast two-qubit logic gates. An ideal neutral donor system would have a nuclear-spin-free substrate *and* a large oscillator strength. Such a system may be found in the group-II-VI system of ZnSe. The isotopes that have non-zero nuclear spins are less than 10% abundant in naturally existing Zn and Se compositions, and isotopic purification could generate a spin-free substrate.

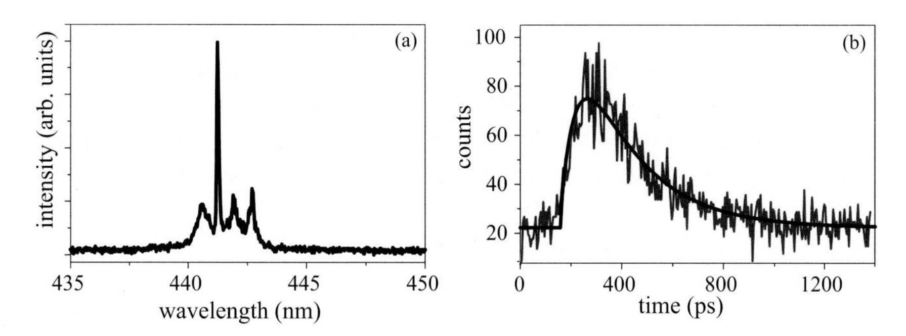

Fig. 18.9. Time-resolved spectroscopy of bound-exciton transitions for isolated ^{19}F impurities in ZnSe. (a) Individual peaks in photoluminescence spectrum correspond to a few resolvable impurities in a 100 nm mesa structure. (b) Streak-camera data indicates a 200 ps lifetime.

Experiments have already been performed to isolate single acceptors in ZnSe [41], but acceptor ground states decohere rapidly, motivating a search for appropriate, stable, isolated neutral donors. The ^{19}F donor is a good choice, because this nucleus is 100% spin-1/2 and possesses a high magnetic moment, second only to protons, allowing a strong hyperfine coupling comparable to the ^{31}P:Si system. Measurements to verify the growth and luminescence of this donor system are described in [42]. These ZnSe samples were grown by molecular beam epitaxy on GaAs substrates. To isolate single impurities, ^{19}F-doped ZnSe layers of 5 nm width are sandwiched between two 50 nm ZnMgSe barriers, which have a larger bandgap. Multiple mesa structures are fabricated by e-beam lithography and wet-chemical etching. Figure Fig. 18.9a shows the photoluminescence spectra from a small mesa structure with a diameter of about 100 nm. It shows emission peaks corresponding to individual ^{19}F-donor bound excitons. Figure 18.9b shows the exciton-decay lifetime of such an isolated ^{19}F-donor bound exciton. The measured lifetime of about 200 ps implies that the ^{19}F:ZnSe has a comparable oscillator strength to quantum dots.

While such emitters show great promise for homogeneous, coherent, bright emitters, they have yet to be incorporated into a high-Q microcavity, as materials processing and fabrication techniques are not as advanced in ZnSe as in silicon or GaAs. The development of this and other materials well suited for quantum computers remains an ongoing research effort.

18.5. Conclusion

The development of quantum dots with high oscillator strength, microcavities with high cooperativity factors, and isolated neutral donor impurities with high homogeneity represents a first stage in the long-term development of fast quantum computers and quantum repeaters. The implementation of local and nonlocal quantum circuits with many qubits connected by a qubus will also require substantial development in photonic waveguiding and switching. Fortunately, the use of such technologies extends far beyond a single application such as quantum key distribution, and the technologies available in semiconductor nanophotonics improve every day. It is our belief that the convergence of directed research in individual qubit/cavity systems, integrated photonic circuits, and improved ideas for fault-tolerant quantum computing architectures will converge to allow fast, scalable, and useful devices in the near future.

Acknowledgments

This work was supported by NICT, MEXT, JST/SORST, and MURI (ARMY, DAAD 19-03-1-0199)

References

[1] W. Dür, H. J. Briegel, J. I. Cirac, and P. Zoller, *Phys. Rev. A.* **59**, 169, (1999).
[2] J. I. Cirac, P. Zoller, H. J. Kimble, and H. Mabuchi, *Phys. Rev. Lett.* **78**, 3221, (1997).
[3] L. C. Andreani, G. Panzarini, and J. M. Gerard, *Phys. Rev. B.* **60**, 13276–13279, (1999).
[4] S. Rudin and T. L. Reinecke, *Phys. Rev. B.* **59**, 10227–10233, (1999).
[5] D. Englund *et al.*, *Phys. Rev. Lett.* **95**, 013904, (2005).
[6] H. F. Hofmann, K. Kojima, S. Takeuchi, and K. Sasaki, *J. Opt. B: Quantum Semiclass. Opt.* **5**, 218, (2003).
[7] L. M. Duan and H. J. Kimble, *Phys. Rev. Lett.* **92**, 127902, (2004).
[8] L. Childress, J. M. Taylor, A. S. Sørensen, and M. D. Lukin, *Phys. Rev. A.* **72**, 52330, (2005).
[9] P. van Loock, T. D. Ladd, K. Sanaka, F. Yamaguchi, K. Nemoto, W. J. Munro, and Y. Yamamoto, *Phys. Rev. Lett.* **96**, 240501, (2006).
[10] T. D. Ladd, P. van Loock, K. Nemoto, W. J. Munro, and Y. Yamamoto, *New J. Phys.* **8**, 184, (2006).

[11] C. H. Bennett, G. Brassard, S. Popescu, B. Schumacher, J. A. Smolin, and W. K. Wootters, *Phys. Rev. Lett.* **76**, 722, (1996).
[12] D. Deutsch, A. Ekert, C. Macchiavello, S. Popescu, and A. Sanpera, *Phys. Rev. Lett.* **77**, 2818, (1996).
[13] W. Yao, R.-B. Liu, and L. J. Sham, *Phys. Rev. Lett.* **95**, 030504, (2005).
[14] S. M. Clark and K-M. C. Fu and T. D. Ladd and Y. Yamamoto, *Phys. Rev. Lett.* **99**, 040501, (2007).
[15] J. R. Petta *et al.*, *Science.* **309**, 2180, (2005).
[16] M. Bayer *et al.*, *Phys. Rev. B.* **65**, 195315, (2002).
[17] A. Greilich *et al.*, *Science.* **313**, 341, (2006).
[18] T. P. Spiller, K. Nemoto, S. L. Braunstein, W. J. Munro, P. van Loock, and G. J. Milburn, *New J. Phys.* **8**, 30, (2006).
[19] D. Gottesman and I. L. Chuang, *Nature.* **402**, 390, (1999).
[20] R. Van Meter, K. M. Itoh, and T. D. Ladd, *Proc. Mesoscopic Superconductivity and Spintronics (MS+S2006).* (2006). quant-ph/0507023.
[21] J. Hours, P. Senellart, E. Peter, A. Cavanna, and J. Bloch, *Phys. Rev. B.* **71**, 161306, (2005).
[22] B. Gayral, J. Gerard, A. Lemaitre, C. Dupuis, L. Manin, and J. Pelouard, *Appl. Phys. Lett.* **75**, 1908, (1999).
[23] K. Watanabe, S. Tsukamoto, Y. Gotoh, and N. Koguchi, *J. Cryst. Growth.* **227-228**, 1073, (2001).
[24] J.P. Reithmaier *et al.*, *Nature.* **432**, 197, (2004).
[25] T. Yoshie *et al.*, *Nature.* **432**, 200, (2004).
[26] K. Hennessy *et al.*, *Nature.* **445**, 896, (2007).
[27] E. Peter, P. Senellart, D. Martrou, A. Lemaitre, J. Hours, J. M. Gérard, and J. Bloch, *Phys. Rev. Lett.* **95**, 067401, (2005).
[28] M. A. Lampert, *Physical Review Letters.* **1**, 450, (1958).
[29] J. Haynes, *Physical Review Letters.* **4**, 361, (1960).
[30] D. Thomas and J. Hopfield, *Physical Review Letters.* **7**, 316, (1961).
[31] R. Miller and D. Kleinman, *Journal of Luminiscence.* **30**, 520, (1985).
[32] D. Herbert, *Journal of Physics C: Solid State Physics.* **10**, 3327, (1977).
[33] V. A. Karasyuk, D. G. S. Beckett, M. K. Nissen, A. Villemaire, T. W. Steiner, and M. L. W. Thewalt, *Phys. Rev. B.* **49**, 16381, (1994).
[34] K. M. C. Fu, C. Santori, C. Stanley, M. C. Holland, and Y. Yamamoto, *Phys. Rev. Lett.* **95**, 187405, (2005).
[35] A. M. Tyryshkin, S. A. Lyon, A. V. Astashkin, and A. M. Raitsimring, *Phys. Rev. B.* **68**, 193207, (2003).
[36] T. D. Ladd, D. Maryenko, Y. Yamamoto, E. Abe, and K. M. Itoh, *Phys. Rev. B.* **71**, 14401, (2005).
[37] A. Yang *et al.* **97**, 227401, (2006).
[38] B.-S. Song, S. Noda, T. Asano, and Y. Akahane, *Nature Materials.* **4**, 207, (2005).
[39] T. Tanabe, M. Notomi, E. Kuramochi, A. Shinya, and H. Taniyama, *Nature Photonics.* **1**, 49, (2007).
[40] K.-M. C. Fu, T. D. Ladd, C. Santori, and Y. Yamamoto, *Phys. Rev. B.* **69**, 125306, (2004).

[41] S. Strauf, P. Michler, M. Klude, D. Hommel, G. Bacher, and A. Forchel, *Phys. Rev. Lett.* **89**, 177403, (2002).
[42] A. Pawlis, K. Sanaka, S. Götzinger, Y. Yamamoto, and K. Lischka, *Semicond. Sci. Technol.* **21**, 1412, (2006).

Chapter 19

Quantum Dots: Single-Photon Sources for Quantum Information

Matthias Scholz, Thomas Aichele, and Oliver Benson

Nanooptics, Department of Physics, Humboldt University Berlin, Hausvogteiplatz 5-7, 10117 Berlin, Germany

In this chapter we report on applications of single photon sources based on semiconductor quantum dots. After a brief introduction we introduce quantum cryptography and multiplexing as a first step towards *single photonics*. We also describe the implementation of a single photon source in a linear optics quantum computing experiment.

Contents

19.1. Introduction

Photons are ideal carriers to transmit quantum information over large distances due to their low interaction with the environment. In 1984, Bennett and Brassard proposed a secret key-distribution protocol [1] that uses the single-particle character of a photon to avoid any possibility of eavesdropping on an encoded message (for a review see [2]). Also, the implementation of efficient quantum gates based on photons and linear optics was proposed [3, 4] and demonstrated [5, 6]. Besides, photons have been proposed as information carriers in larger networks [7] between processing nodes of stationary qubits, like ions [8–11], atoms [12, 13], quantum dots (QDs) [14, 15], and Josephson qubits [16, 17].

Applications of linear optics in quantum information processing require the reliable generation of single- or few-photon states on demand. However, due to their bosonic character, photons tend to appear in bunches. This characteristics hinders the implementation of classical sources particularly in quantum cryptographic systems since an eavesdropper may gain partial information by a beam splitter attack. Similar obstacles occur for linear optics quantum computation where photonic quantum gates [3], quantum repeaters [18], and quantum teleportation [19] require the preparation of single- or few-photon states *on demand* in order to obtain reliability and high efficiency.

A promising process for single-photon generation is the spontaneous emission from a single quantum emitter. Numerous emitters have been used to demonstrate single-photon emission [21]. Single atoms or ions are the most fundamental systems [22, 23]. Other systems capable of single-photon generation are single molecules and single nanocrystals [24–26]. However, their drawback is a susceptibility for photo-bleaching and blinking [27]. Stable alternatives are nitrogen-vacancy defect centers in diamond [28, 29], but they show broad optical spectra together with comparably long lifetimes (≈ 12 ns).

In this chapter, we focus on single-photon generation from self-assembled single QDs. QDs are few-nanometer sized semiconductor structures which resemble features known from atoms, like discrete emission spectrum and electronic structure, and which are therefore often cited as *artificial atoms*. Most experiments with QDs have to be conducted at cryogenic temperature in order to reduce electron-phonon interaction and thermal ionization. High count rates can be obtained due to their short transition lifetimes, and their spectral lines are nearly lifetime-limited with material systems covering the ultraviolet, visible, and infrared spectrum. QDs also gain attractiveness by the possibility for electric excitation [30] and the implementation in integrated photonic structures [31].

This chapter is organized as follows. Section 19.2 provides an introduction to electronic properties and the radiative decay of single QDs. Section 19.3 covers the characterization and measurement of single-photon states. In section 19.4, we show decay cascades in single QDs. Their application in a multiplexed quantum cryptography setup is demonstrated in section 19.5. Section 19.6 describes the realization of a two-qubit Deutsch-Jozsa algorithm [32] that makes use of different degrees of freedom of one single photon. The setup was constructed in a way that it becomes resistant to noise induced by phase fluctuations. Section 19.7 concludes with a short summary.

19.2. Single Quantum Dots

All experiments in this chapter were performed with self-assembled QDs fabricated by Stranski-Krastanov growth [33]. Overgrowth of QDs is required to obtain high-efficient and optically stable emission. As it is impossible to characterize the exact shape or material composition of an overgrown QD a priori, certain simplified models for the electronic structure are often used. Figure 19.1 shows a spherical potential which traps a single electron-hole pair (exciton, figure 19.1a) or two electron-hole pairs (biexciton, figure 19.1b). In figure 19.1c, the possible decay paths of a biexciton via the so-called bright excitons are depicted. A level splitting is shown for clarity. The recombination of a carrier pair leads to the emission of a single photon, generally at different wavelengths for exciton and biexciton due to Coulomb interaction.
The image of an ensemble of InP QDs is displayed in figure 19.2a. Figure 19.2b shows the spectrum of a single InP QD in a GaInP matrix with two dominant spectral lines originating from exciton and biexciton decay. This material system can generate single photons in the 620–750 nm region

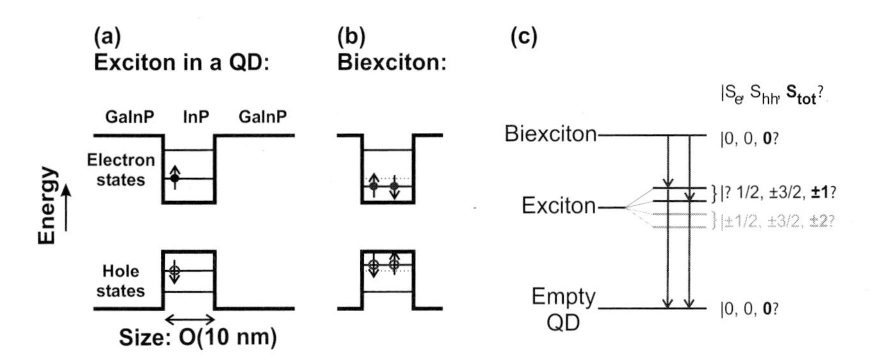

Fig. 19.1. Excitations in a QD: (a) An electron-hole pair forms an exciton. (b) Two electron-hole pairs form a biexciton, generally at an energy different from the exciton. (c) Schematic term scheme for the exciton and biexciton decay cascade.

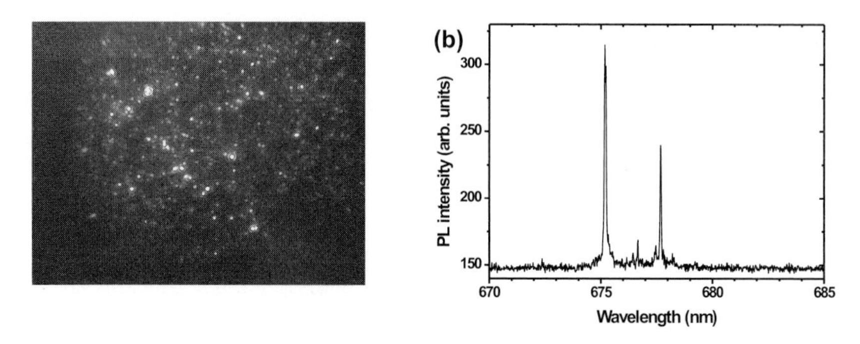

Fig. 19.2. (a) Micro-photoluminescence image of InP QDs in GaInP. (b) Spectrum of a single InP/GaInP QD with spectral lines of exciton and biexciton decay.

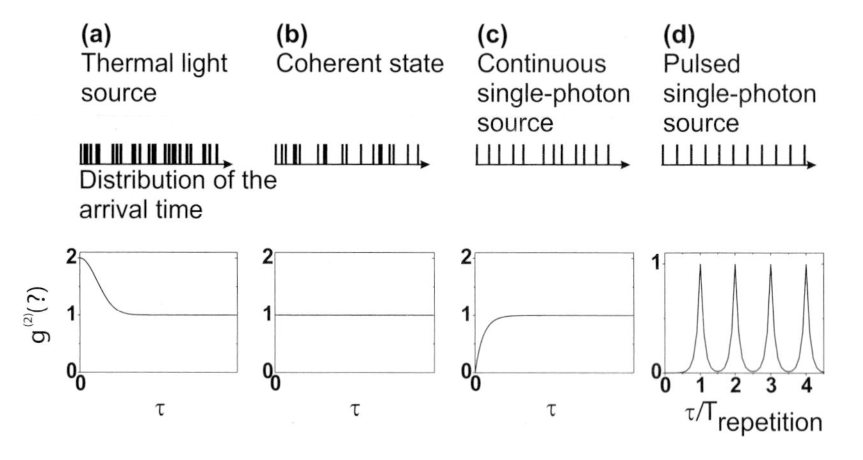

Fig. 19.3. $g^{(2)}$ function of (a) a thermal light source, (b) coherent light, (c) a continuously driven single-photon source, and (d) a pulsed single-photon source.

which perfectly fits the maximum detection efficiency of commercial silicon avalanche photo diodes (APD) with over 70% at wavelengths around 700 nm.

19.3. Single-Photon Generation

19.3.1. *Correlation measurements*

The single-photon character of the photoluminescence from a single QD can be tested by measuring the normalized second-order correlation function $g^{(2)}(t_1, t_2)$ via detecting the light intensity $\langle \hat{I}(t) \rangle$ at two points in time. For stationary fields, it reads

$$g^{(2)}(\tau = t_1 - t_2) = \frac{\langle : \hat{I}(0)\hat{I}(\tau) : \rangle}{\langle \hat{I}(0) \rangle^2}$$

where :: denotes normal operator ordering. For classical fields, $g^{(2)}(0) \geq 1$ and $g^{(2)}(0) \geq g^{(2)}(\tau)$ hold [34] which prohibits values smaller than unity. For thermal light sources, there is an increased probability to detect a photon shortly after another. This bunching phenomenon leads to $g^{(2)}(0) \geq 1$ (figure 19.3a). Coherent states show $g^{(2)}(\tau) = 1$ for all τ according to a Poisson photon number distribution (figure 19.3b). A single-photon state shows the anti-bunching effect of a sub-Poisson distribution with $g^{(2)}(0) \leq 1$ (figure 19.3c). For a Fock state $|n\rangle$, one has $g^{(2)}(0) = 1 - 1/n$ which yields $g^{(2)}(0) = 0$ for the special case of $|n = 1\rangle$. The pulsed excitation of a single-photon source leads to a peaked structure in $g^{(2)}(\tau)$ with a missing peak at zero delay indicating single-photon emission (figure 19.3d).

In order to circumvent the detector dead time ($\approx$ 50 ns for APDs [35]), the second-order correlation function is measured in a Hanbury Brown-Twiss arrangement [36] as depicted in figure 19.4, consisting of two photo detectors and a 50:50 beam splitter. A large number of time intervals between detection events is measured and binned together in a histogram.

19.3.2. *Micro-photoluminescence*

Optical investigation of QDs is usually performed in a micro-photoluminescence (PL) setup which combines excellent spatial resolution with high detection efficiency. Samples with low densities of $10^8 - 10^{11}$

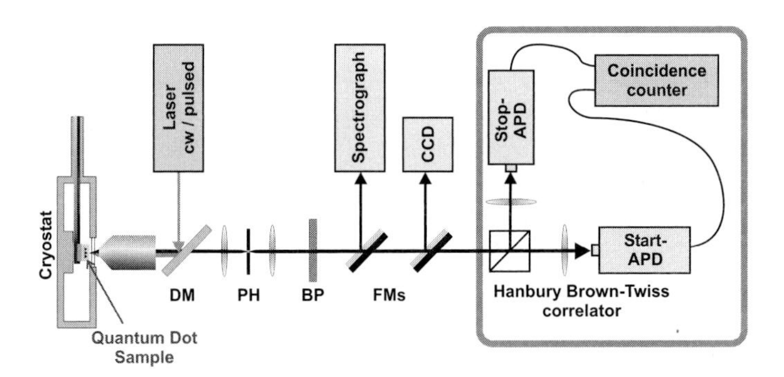

Fig. 19.4. Micro-PL setup (FM: mirrors on flip mounts, DM: dichroic mirror, PH: pinhole, BP: narrow bandpass filter, APD: avalanche photo diode)

dots/cm^2 are required to isolate a single dot. In our setup, the sample is stored at 4 K inside a continuous-flow liquid Helium cryostat (figure 19.4). The dots are excited either pulsed (Ti:Sa, pulse width 400 fs, repetition rate 76 MHz, frequency-doubled to 400 nm) or continuously (Nd:YVO_4, 532 nm). The microscope system (NA = 0.75) reaches a lateral resolution of 500 nm, and further spatial and spectral filtering selects a single transition of a single QD. The PL can be imaged on a CCD camera (see also figure 19.2a), on a spectrograph (figure 19.2b), or can be sent to a Hanbury Brown-Twiss setup with a time resolution of about 800 ps to prove single-photon emission.

19.3.3. *InP quantum dots*

The sample used in our experiment was grown by metal-organic vapor phase epitaxy. An aluminum mirror was evaporated on the sample which was then thinned down to 400 nm and glued on a Si-substrate (see figure 19.5a). Figure 19.5b depicts the spectrum of a single InP QD on the sample under weak excitation showing only one dominant spectral line. Its linear dependance on the excitation power identifies it as an exciton emission. A 1-nm bandpass filter was used to further reduce the background. Figure 19.6 shows the corresponding cw correlation measurement at a count rate of 1.1×10^5 and 10 K. The function is modeled as a convolution of the expected shape of the ideal correlation function $g^{(2)}(\tau) = 1 - \exp(-\gamma\,\tau)$ and a Gaussian distribution according to a 800 ps time resolution of the detectors. The width of the anti-bunching dip (2.3 ns) depends on both the transition lifetime and the excitation timescale. Equivalent measurements at pulsed excita-

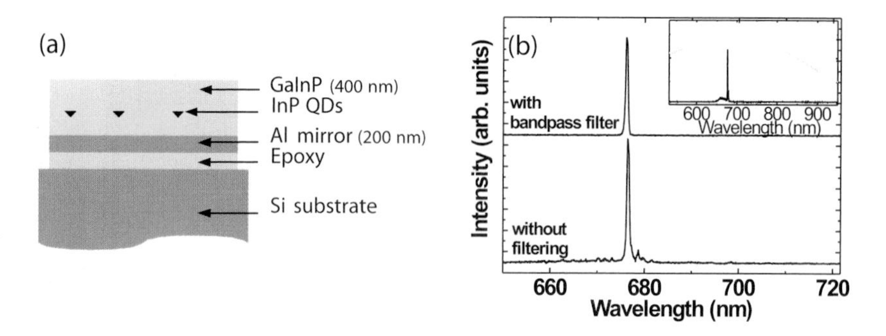

Fig. 19.5. (a) Structure of the InP/GaInP sample. (b) PL spectra from a single InP QD. An offset was added to separate the graphs. The inset shows a wider spectrum together with the APD detection efficiency.

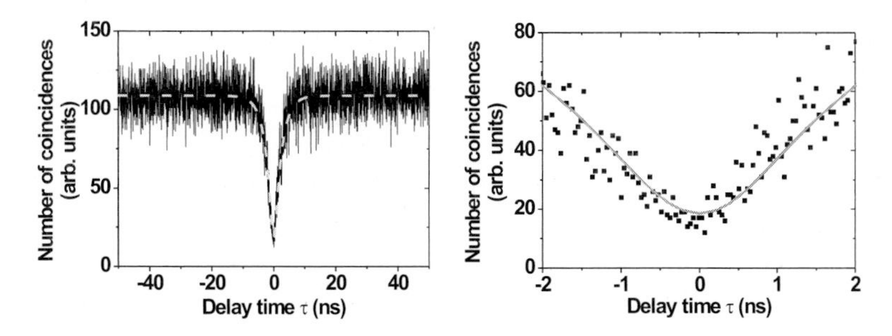

Fig. 19.6. $g^{(2)}$ function at continuous excitation. The fit function corresponds to an ideal single-photon source with limited time resolution. The right image shows a zoom into the central dip region.

tion show a vanishing peak at zero delay (figure 19.7a) which again proves single-photon generation. At rising temperature, phonon interactions lead to an increased incoherent background with other spectral lines overlapping the filter transmission window (figure 19.7b). However, the characteristic anti-bunching was observed up to 50 K [37].

19.4. A Multi-color Photon Source

A great advantage of QDs as photon emitters is their potential to create more complex states of light consisting of few photons. For example, by trapping two or more electron-hole pairs (compare figure 19.1), photon cascades can be produced. These cascades can be used to enhance the transmission rate of cryptographic systems usually limited by spontaneous lifetime. Previous experiments have studied cascaded decays in InAs [20, 38, 39] and II-VI QDs [40, 41]. Very recently, also entangled photon pair

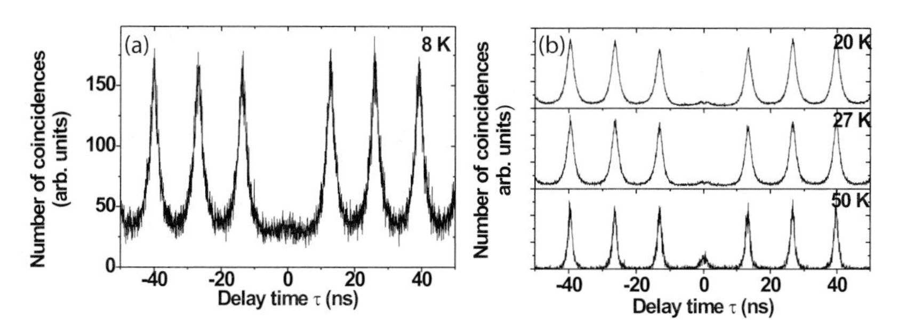

Fig. 19.7. $g^{(2)}$ function of a single InP QD at pulsed excitation: (a) at 8 K and (b) between 20 and 50 K.

generation has been demonstrated [42, 43] following earlier proposals [44]. In order to study correlations between photons emitted in a cascade, the *cross-correlation function*

$$g^{(2)}_{\alpha\beta}(\tau) = \frac{\langle : \hat{I}_\alpha(t)\hat{I}_\beta(t+\tau) : \rangle}{\langle \hat{I}_\alpha(t)\rangle\langle \hat{I}_\beta(t)\rangle}$$

has to be measured with a Hanbury Brown-Twiss configuration modified by an interference filter in each arm. The filters can be tuned, e.g., to the exciton and biexciton transition lines in a two-photon cascade indicated by the index α and β in $g_{\alpha\beta}$, respectively. Cross-correlation measurements confirm assignments of spectral lines to exciton, biexciton, and triexciton transitions by the strong asymmetric shape of the correlation function (figure 19.8). If a biexciton decay starts the measurement and an exciton stops it, photon bunching occurs since an exciton photon is predominantly emitted shortly after a biexciton photon in a cascade. The opposite holds for switched start and stop channels. In this case, the detection of the exciton photon projects the QD to its empty (ground) state. The required re-excitation process becomes manifest in an anti-bunching dip. Biexciton-triexciton correlations can be explained equivalently, but on different timescales. Since cross-correlation functions of two independent transitions show no (anti-)bunching, the results in figure 19.8 confirm a three-photon cascade from the triexciton to the ground state of a single QD. The dashed lines in figure 19.8 are a fit to the rate equation model described in [45, 46].

19.5. Multiplexed Quantum Cryptography on the Single-Photon Level

19.5.1. *A single-photon add/drop filter*

For quantum communication protocols, a single-photon source with high repetition rate and collection efficiency is desired. Beside passive elements like mirrors and solid immersion lenses, resonant techniques exploiting the Purcell effect can greatly enhance the emission rate [47, 48] if the source relies on the decay of an excited state. A well-established technique from classical communication is *multiplexing*. Each signal is marked with a physical label, like its wavelength, and identified at the receiver using filters tuned to the carrier frequencies. For single photons, their wavelength can be used as their distinguishing label and polarization to encode quantum information [49]. If the two photons from a biexciton-exciton cascade pass a Michelson interferometer, constructive and destructive interference will

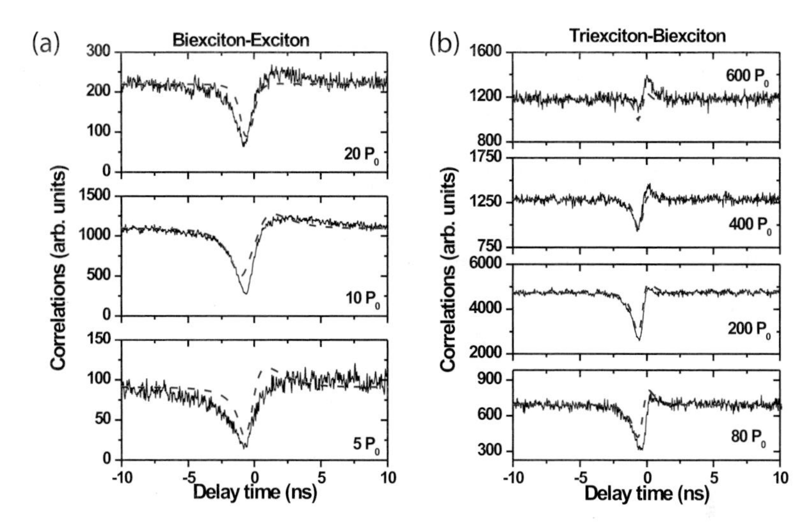

Fig. 19.8. Cross-correlation functions between the (a) exciton and biexciton line and (b) biexciton and triexciton line.

be observed at the two output ports by proper choice of the relative path difference (figure 19.9). These photons are fed into an optical fiber each, and one of them is delayed by half the repetition rate of the excitation laser. Recombination at a beam splitter then leads to a photon stream with a doubled count rate. This is also reflected by auto-correlation measurements with a spacing of only 6.6 ns in the peak structure (see figure 19.9d) [49].

19.5.2. *Application to quantum key distribution*

Multiplexing can enhance the bandwidth of quantum communication systems which use single-photon sources based on spontaneous emission. In our experiment, we implemented the BB84 protocol [1, 2] where the exact photon wavelength is unimportant. Quantum cryptography has been realized with single-photon states from diamond defect centers [50] and single QDs [51]. Figure 19.10 shows an implementation of interferometric multiplexing in the BB84 protocol using a cascaded photon source. Behind an exciton-biexciton add/drop filter, Alice randomly prepares the photons' polarization. Bob's detection consists of a second EOM, an analyzing polarizer, and an APD. In our experiment, all EOM and APD operation is performed automatically by a LabVIEW program. A visualization of a successful secret transmission of data is depicted in figure 19.11. After exchange of the key, Alice encrypts her data by applying an XOR operation between every image bit and the sequence of random bits (the secret key) which she has previously shared with Bob by performing the BB84 protocol. The encoded image (shown in figure 19.11b) is transmitted to Bob over a distance of 1 m with 30 bits/s. Another XOR operation by Bob reveals the original image with an error rate of 5.5%.

19.6. Deutsch-Jozsa Algorithm with Single Photons from a Single Quantum Dot

In reference [3], Knill *et al.* propose the realization of a universal set of quantum gates for quantum computation, using linear optical elements and single-photon detectors. This scheme requires triggered, indistinguishable single photons. Two-photon quantum gates (c-NOT gates) and basic quantum algorithms have been realized using photon pairs created by parametric down conversion [52, 53]. Here, we report the on demand operation of a two-qubit Deutsch-Jozsa algorithm [32] using our deterministic single-

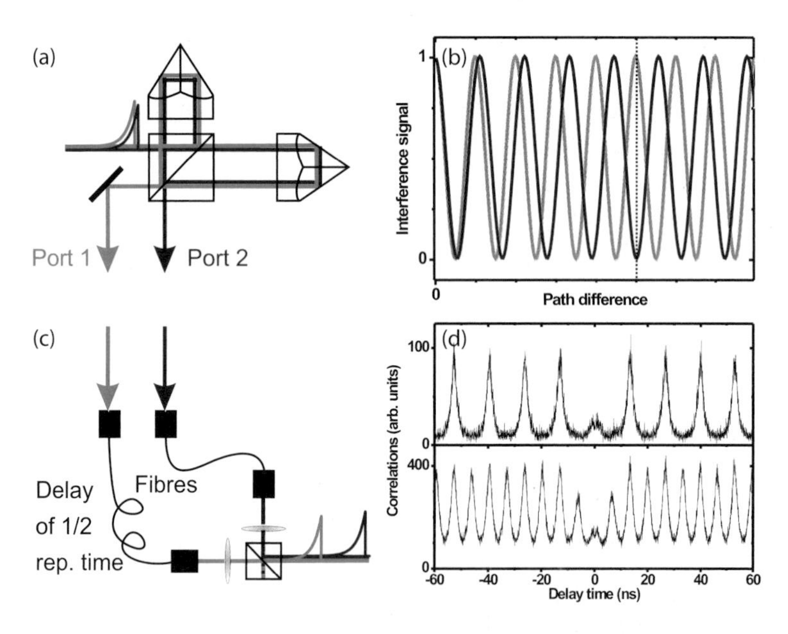

Fig. 19.9. (a) Michelson interferometer with two output ports as a single-photon add/drop filter. (b) Interference pattern for two distinct wavelengths versus the relative arm length. (c) Merging the separated photons with one path delayed by half the excitation repetition rate. (d) Intensity correlation of the exciton spectral line and the multiplexed signal.

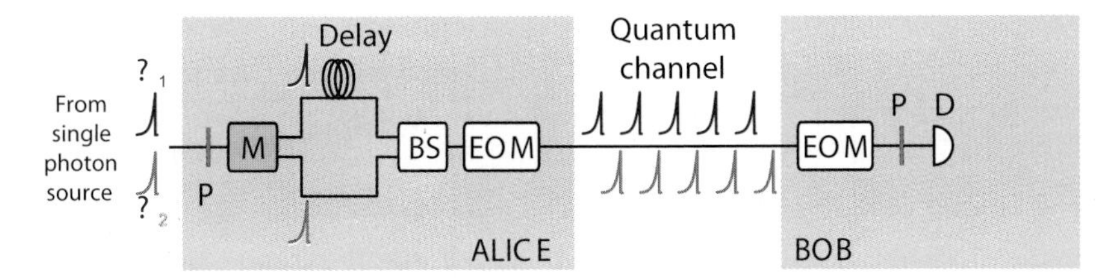

Fig. 19.10. Possible implementation of the multiplexer using the BB84 protocol. For a detailed description of the setup see text.

photon source. Previous all-optical experiments were restricted to emulate the Deutsch-Jozsa algorithm with attenuated classical laser pulses [54].
An intuitive and often cited interpretation compares this algorithm with a "coin tossing" game: While a classical observer has to check both sides of a coin in order to tell if it is fair, i.e. if its two sides are different (head and tail), or unfair with two equal sides (head–head or tail–tail), quantum computation can perform this task with a single shot. In the simplest case, two qubits are needed to store query and answer, respectively.
Here, we apply the proposal of [55] in order to implement two qubits using different degrees of freedom of a single photon (see also reference [56]). The first qubit corresponds to the which-way information about the photon path inside an Mach-Zehnder interferometer, namely the spatial modes a and b (figure 19.12a). The second qubit is implemented via the photon's polarization state. With these definitions of our two qubits, we are able to construct the proper states for a Deutsch-Jozsa algorithm. The initial state is realized by sending a photon into a linear superposition of the two spatial modes using a non-polarizing 50:50 beam splitter after preparing it in a superposition of horizontal and vertical polarization.
In a next step, a unitary transformation is realized by selectively adding a

Fig. 19.11. Visualization of the quantum key distribution. Alice's original data (a), a photography taken out of our lab window in Berlin, is encrypted and sent to Bob (b). After decryption with his key, Bob obtains image (c).

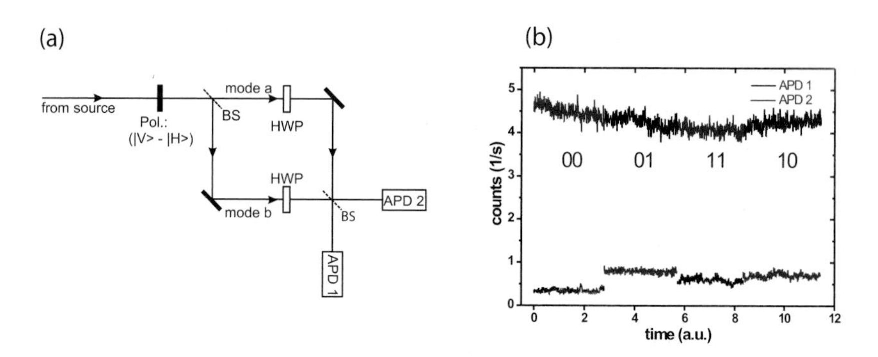

Fig. 19.12. (a) Setup of the Deutsch-Jozsa experiment. Pol: Polarizer, BS: beam splitter, HWP: half-wave plate (b) Detector count rates of APD 1 and 2 for the four different combinations of the HWPs inside the interferometer: 00 no HWP, 10: HWP in mode a, 01: HWP in b, 11: HWP in a and b.

half-wave plate (HWP) to each mode. A HWP in only one of the two paths corresponds to a balanced situation representing fair coins in the example above. In contrast, no HWP or a HWP in either path correspond to the constant situations, illustrated by the two possible false coins. Physically, the difference between these two situations is an additional local phase difference between the two paths, 0 for the constant and π for the balanced case. Recombining the two interferometer arms on a second 50:50 beam splitter implements a final Hadamard gate on the first qubit. The two interferometer outputs are monitored with one APD each. Constructive interference at the top (bottom) output indicates the balanced (constant) situation, which can thus be distinguished by only one detection event of one single photon. Figure 19.12b shows the outcome of many detection events (albeit only one would be enough) for the four possible combinations. In these measurements an overall success probability of 79% was achieved. This was mainly limited by a slight temporal and spatial mode mismatch in the interferometer.
A fundamental problem of quantum computation is dephasing, i.e. the tendency of a qubit to loose or change its quantum information due to coupling to the environment. A way to solve this might be active quantum error correction [57] or the use of decoherence-free subspaces of the Hilbert space spanned by several qubits [58]. In our experiment, a modified setup insensitive to phase noise has been achieved by extending the setup to the scheme in figure 19.13a. Horizontal and vertical polarization are separated and merged at polarizing beam splitters (PBS) so that the modes, that in-

terfere at the final beam splitter, travel along the same path. Thus, phase noise in the central interferometer does not cause perturbations of the algorithm but only lead to a global phase change [56].
In our experiment, we simulate phase noise by modulating piezo-electric mirror mounts. The left graph in figure 19.13b displays the detection signals for two combinations (one constant and one balanced) of the HWPs in the first interferometer, without any *artificial* noise. Modulating piezo 2 in the central interferometer does also not affect the count rates, as expected (middle graph). For comparison, phase noise induced by piezo 3 in the final interferometer changes the APD signals substantially (right graph). The same happens when modulating piezo 1 (not shown). Thus, using this extended experimental setup, we showed the robustness of the algorithm against phase noise when information is properly encoded in an unaffected superposition of polarization states.

19.7. Summary

In this chapter, we have reviewed single-photon generation from single InP QDs and its application to quantum information processing. Single-photon statistics and cross-correlations of various transitions from multi-excitonic states in the visible spectrum around 690 nm have been demonstrated. We described a proof-of-principle experiment that shows how multi-photon generation from single QDs may find applications in quantum cryptography devices to enhance their maximum bandwidth. In this experiment, multiplex-

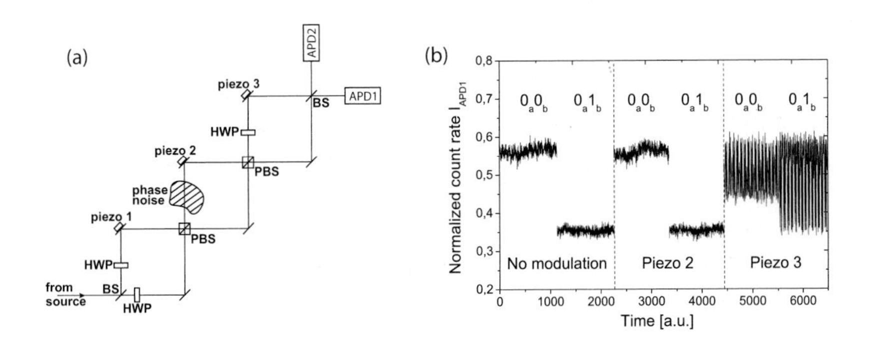

Fig. 19.13. (a) Extended setup. (P)BS: (polarizing) beam splitter. (b) Detector count rates for a constant (00) and a balanced (01) HWP combination. Left graph: No modulation of the piezos, middle graph: modulation on piezo 2, right graph: modulation on piezo 3.

ing on the single-photon level and its implementation in the BB84 quantum key distribution protocol has been accomplished for the first time. Another important application for single photon sources is quantum computation. We have demonstrated an experimental realization of the Deutsch-Jozsa algorithm with linear optics using our quantum dot single-photon source. By encoding on adequate qubit bases in an extended experimental setup, the robustness of the setup against phase noise was proven.
Our experiments show that today's single-photon sources based on QDs have reached the level of sophistication to be ready-to-use non-classical light sources. The *on demand* character of the emission is extremely useful in all-solid state implementations of quantum information devices. Single QDs will therefore play an important role in the field of quantum cryptography and quantum computation, but also for realizing interfaces in larger quantum networks.

Acknowledgments

The authors want to thank W. Seifert for providing the QD sample and V. Zwiller, G. Reinaudi, and J. Persson for valuable assistance. This work was supported by Deutsche Forschungsgemeinschaft (SFB 296) and the European Union (EFRE). M. Scholz acknowledges financial support from Ev. Studienwerk Villigst and Deutsche Telekom Stiftung and T. Aichele by DAAD.

References

[1] C.M. Bennett and G. Brassard, in *Proc. IEEE Conference on Computers, Systems, and Signal Processing in Bangalore, India* (IEEE, New York, 1984), p. 175
[2] N. Gisin, G. Ribordy, W. Tittel, and H. Zbinden, Rev. Mod. Phys. **74**, 145 (2002)
[3] E. Knill, R. Laflamme, and G.J. Milburn, Nature **409**, 46 (2001)
[4] D. Gottesman and I.L. Chuang, Nature **402**,390 (1999)
[5] S. Gasparoni, J.-W. Pan, P. Walther, T. Rudolph, and A. Zeilinger, Phys. Rev. Lett. **93**, 020504 (2004)
[6] N. Kiesel, C. Schmid, U. Weber, R. Ursin, and H. Weinfurter, Phys. Rev. Lett. **95**, 210505 (2005)
[7] L. Tian, P. Rabl, R. Blatt, and P. Zoller, Phys. Rev. Lett. **92**, 247902 (2004)
[8] J.I. Cirac and P. Zoller, Phys. Rev. Lett. **74**, 4091 (1995)
[9] J.I. Cirac and P. Zoller, Physics Today **57**, 38 (2004)

[10] F. Schmidt-Kaler, H. Häffner, M. Riebe, S. Gulde, G.P.T. Lancaster, T. Deuschle, C. Becher, C.F. Roos, J. Eschner, and R. Blatt, Nature **422**, 408 (2003)

[11] D. Leibfried, B. Demarco, V. Meyer, D. Lucas, M. Barrett, J. Britton, W.M. Itano, B. Jelenkovic, C. Langer, T. Rosenband, et al., Science **422**, 412 (2003)

[12] D. Kielpinski, C. Monroe, and D.J. Wineland, Nature **417**, 709 (2002)

[13] J.I. Cirac and P. Zoller, Nature **404**, 579 (2000)

[14] D. Loss and D.P. DiVicenzo, Phys. Rev. A **57**, 120 (1998)

[15] W.G. van der Wiel, S.D. Franceschi, J.M. Elzerman, T. Fujisawa, S. Tarucha, and L.P. Kouwenhoven, Rev. Mod. Phys. **75**, 1 (2003)

[16] J.E. Mooij, T.P. Orlando, L. Levitov, L. Tian, C.H. van der Wal, and S. Lloyd, Science **285**, 1036 (1999)

[17] Y.A. Pashkin, T. Yamamoto, O. Astafiev, Y. Nakamura, D.V. Averin, and J.S. Tsai, Nature **421**, 823 (2003)

[18] H.-J. Briegel, W. Dür, J.I. Cirac, and P. Zoller, Phys. Rev. Lett **81**, 5932 (1998)

[19] D. Bouwmeester, J.-W. Pan, K. Mattle, M. Eibl, H. Weinfurter, and A. Zeilinger, Nature **390**, 575 (1997)

[20] E. Moreau, I. Robert, L. Manin, V. Thierry-Mieg, J.M. Gérard, and I. Abram, Phys. Rev. Lett. **87**, 183601 (2001)

[21] *New Journal of Physics* (special issue on single-photon sources) **6** (2004)

[22] A. Kuhn, M. Hennrich, and G. Rempe, Phys. Rev. Lett. **89**, 067901 (2002)

[23] M. Keller, B. Lange, K. Hayasaka, W. Lange, and H. Walther, Nature **431**, 1075 (2004)

[24] C. Brunel, B. Lounis, P. Tamarat, and M. Orrit, Phys. Rev. Lett **83**, 2722 (1999)

[25] B. Lounis and W.E. Moerner, Nature **407**, 491 (2000)

[26] P. Michler, A. Imamağlu, M.D. Mason, P.J. Carson, G.F. Strouse, and S.K. Buratto, Nature **406**, 968 (2000)

[27] M. Nirmal, B.O. Dabbousi, M.G. Bawendi, J.J. Macklin, J.K. Trautman, T.D. Harris, and L.E. Brus, Nature **383**, 802 (1996)

[28] C. Kurtsiefer, S. Mayer, P. Zarda, and H. Weinfurter, Phys. Rev. Lett. **85**, 290 (2000)

[29] A. Beveratos, S. Kühn, R. Brouri, T. Gacoin, J.-P. Poizat, and P. Grangier, Eur. Phys. J. D **18**, 191 (2002)

[30] Z.Yuan, B.E. Kardynal, R.M. Stevenson, A.J. Shields, C.J. Lobo, K. Copper, N.S. Beattie, D.A. Ritchie, and M. Pepper, Science **295**, 102 (2002)

[31] A. Badolato, K. Hennessy, M. Atatre, J. Dreiser, E. Hu, P.M. Petroff, and A. Imamoğlu, Science **308**, 1158 (2005)

[32] D. Deutsch and R. Jozsa, Proc. Roy. Soc. Lond. A **439**, 553 (1992)

[33] D. Bimberg, M. Grundmann, and N. Ledentsov, *Quantum dot heterostructures* (Wiley, Chichester, UK, 1988)

[34] L. Mandel and E. Wolf, *Optical coherence and quantum optics* (Cambridge University Press, 1995)

[35] *Single-photon counting module - SPCM-AQR series specifications*, Laser

Components GmbH, Germany (2004)
[36] R. Hanbury Brown and R.Q. Twiss, Nature **178**, 1046 (1956)
[37] V. Zwiller, T. Aichele, W. Seifert, J. Persson, and O. Benson, Appl. Phys. Lett. **82**, 1509 (2003)
[38] D.V. Regelman, U. Mizrahi, D. Gershoni, E. Ehrenfreund, W.V. Schoenfeld, and P.M. Petroff, Phys. Rev. Lett. **87**, 257401 (2001)
[39] C. Santori, D. Fattal, M. Pelton, G.S. Salomon, and Y. Yamamoto, Phys. Rev. B **66**, 045308 (2002)
[40] S.M. Ulrich, S. Strauf, P. Michler, G. Bacher, and A. Forchel, Appl. Phys. Lett. **83**, 1848 (2003)
[41] C. Couteau, S. Moehl, F. Tinjod, J.M. Gérard, K. Kheng, H. Mariette, J.A. Gaj, R. Romestain, and J.P. Poizat, Appl. Phys. Lett. **85**, 6251 (2004)
[42] R.M. Stevenson, R.J. Young, P. Atkinson, K. Cooper, D.A. Ritchie, and A.J. Shields, Nature **439**, 179 (2006)
[43] N. Akopian, N.H. Lindner, E. Poem, Y. Berlatzky, J. Avron, D. Gershoni, B.D. Gerardot, and P.M. Petroff, Phys. Rev. Lett **96**, 130501 (2006)
[44] O. Benson, C. Santori, M. Pelton, and Y. Yamamoto, Phys. Rev. Lett. **84**, 2513 (2000)
[45] J. Persson, T. Aichele, V. Zwiller, L. Samuelson, and O. Benson, Phys. Rev. B **69**, 233314 (2004)
[46] T. Aichele, V. Zwiller, M. Scholz, G. Renaudi, J. Persson, and O. Benson, Proc. SPIE **5722**, 30 (2005)
[47] G. Solomon, M. Pelton, and Y. Yamamoto, Phys. Rev. Lett. **86**, 3903 (2001)
[48] J.M. Gérard, B. Sermage, B. Gayral, B. Legrand, E. Costard, and V. Thierry-Mieg, Phys. Rev. Lett. **81**, 1110 (1998)
[49] T. Aichele, G. Reinaudi, and O. Benson, Phys. Rev. B **70**, 235329 (2004)
[50] A. Beveratos, R. Brouri, T. Gacoin, A. Villing, J.-P. Poizat, and P. Grangier, Phys. Rev. Lett. **89**, 187901 (2002)
[51] E. Waks, K. Inoue, C. Santori, D. Fattal, J. Vučković, G.S. Solomon, and Y. Yamamoto, Nature **420**, 762 (2002)
[52] J.L. O'Brien, G.J. Pryde, A.G. White, T.C. Ralph, and D. Branning, Nature **426**, 264 (2003)
[53] P. Walther, K.J. Resch, T. Rudolph, E. Schenck, H. Weinfurter, V. Vedral, M. Aspelmeyer, and A. Zeilinger, Nature **434**, 264 (2005)
[54] S. Takeuchi, Phys. Rev. A **62**, 032301 (2000)
[55] J. Cerf, C. Adami, and P.G. Kwiat, Phys. Rev. A **57**, 1477 (1998)
[56] M. Scholz, T. Aichele, S. Ramelow, and O. Benson, Phys. Rev. Lett. **96**, 180501 (2006)
[57] J. Chiaverini, D. Leibfried, T. Schaetz, M.D. Barrett, R.B. Blakestad, J. Britton, W. Itano, J.D. Jost, E. Knill, C. Langer, *et al.*, Nature **432**, 602 (2004)
[58] D.A. Lidar, I.L. Chuang, and K.B. Whaley, Phys. Rev. Lett. **81**, 2594 (1998)

Author Index

Subject Index